Second Edition

International Labeling Requirements
for
Medical Devices,
Medical Equipment,
and **Diagnostic Products**

Second Edition

International Labeling Requirements
for
Medical Devices,
Medical Equipment,
and **Diagnostic Products**

Charles Sidebottom

CRC Press
Taylor & Francis Group
Boca Raton London New York

CRC Press is an imprint of the
Taylor & Francis Group, an **informa** business

CRC Press
Taylor & Francis Group
6000 Broken Sound Parkway NW, Suite 300
Boca Raton, FL 33487-2742

First issued in paperback 2019

© 2003 by Chapman & Hall/CRC
CRC Press is an imprint of Taylor & Francis Group, an Informa business

No claim to original U.S. Government works

ISBN-13: 978-0-8493-1850-4 (hbk)
ISBN-13: 978-0-367-39516-2 (pbk)

This book contains information obtained from authentic and highly regarded sources. Reasonable efforts have been made to publish reliable data and information, but the author and publisher cannot assume responsibility for the validity of all materials or the consequences of their use. The authors and publishers have attempted to trace the copyright holders of all material reproduced in this publication and apologize to copyright holders if permission to publish in this form has not been obtained. If any copyright material has not been acknowledged please write and let us know so we may rectify in any future reprint.

Except as permitted under U.S. Copyright Law, no part of this book may be reprinted, reproduced, transmitted, or utilized in any form by any electronic, mechanical, or other means, now known or hereafter invented, including photocopying, microfilming, and recording, or in any information storage or retrieval system, without written permission from the publishers.

For permission to photocopy or use material electronically from this work, please access www.copyright.com (http://www.copyright.com/) or contact the Copyright Clearance Center, Inc. (CCC), 222 Rosewood Drive, Danvers, MA 01923, 978-750-8400. CCC is a not-for-profit organization that provides licenses and registration for a variety of users. For organizations that have been granted a photocopy license by the CCC, a separate system of payment has been arranged.

Trademark Notice: Product or corporate names may be trademarks or registered trademarks, and are used only for identification and explanation without intent to infringe.

Library of Congress Card Number 2003046076

Important Disclaimer

The material in this publication is of the nature of general comment and opinion only. It is not offered as advice in regard to the subject matter covered and should not be taken as such. The author and the publisher disclaim all liability to any person in respect to anything and the consequences of anything done or omitted to be done wholly or partly in reliance upon the whole or any part of the contents of this publication. The reader should not act or refrain from acting on the basis of any matter contained in this publication. If legal advice or other expert assistance is required, the services of a competent professional should be sought.

Library of Congress Cataloging-in-Publication Data

Sidebottom, Charles B.
    International labeling requirements for medical devices, medical equipment, and
diagnostic products/ Charles B. Sidebottom. -- 2nd ed.
        p. ; cm.
    Includes bibliographical references and index.
    ISBN 0-8493-1850-5
        1. Medical instruments and apparatus--Packaging. 2. Medical instruments and
apparatus--Labeling. I. Title.
    [DNLM: 1. Product Labeling--legislation & jurisprudence. 2. Product
Labeling--standards. 3. International Cooperation. 4. Reference Standards. WA 33.1
S568i 2003]
    R857.P33S53 2003
    343′.0855681761--dc21   2003046076

**Visit the Taylor & Francis Web site at
http://www.taylorandfrancis.com**

**and the CRC Press Web site at
http://www.crcpress.com**

Foreword

The publication of the *International Labeling Requirements* in 1994 has been an asset to the medical device industry. It is widely recognized that labeling instructions on the appropriate use of a device is as important as the device itself in ensuring the device's safe and effective use. Labeling requirements vary across the spectrum of medical devices and regulatory systems as well. The first edition presented those requirements in the environment of the early 1990s.

But medical devices evolve. New diagnostic and therapeutic devices have been introduced and existing devices continue to expand and refine their capabilities. Similarly, our understanding of what information to provide and how to provide it continues to evolve. Finally, the regulatory environment has evolved in existing regions and expanded significantly in Eastern Europe, Asia, and Latin America. Thus the publication of a second edition of *International Labeling Requirements* is timely.

I have spent the bulk of my career as a regulatory professional in the medical device industry. I have come to understand the differing expectations of users (health care professionals and lay persons, regulators, and the industry) for medical device labeling. As Mr. Sidebottom noted in the Preface to the First Edition, this book is not and cannot be a "cookbook". Nevertheless, the Second Edition continues to provide practical guidance for industry to facilitate international regulatory approvals and appropriately educate users in today's environment.

Charles H. Swanson, Ph.D.
Vice President and Chief Regulatory Officer
Medtronic, Inc.
August 2002

Foreword (To the First Edition)

I can still remember the feeling I had when I saw a young child being kept alive by the external, wearable, battery-powered transistor pacemaker I had developed. It was an unbelievably moving experience. Besides being immensely satisfied and fulfilled, when I saw my device lying next to that child I was nearly overwhelmed with a feeling of responsibility. We had demonstrated the safety and effectiveness of this device in the animal physiology lab; and yet, seeing it sustain human life was an extremely sobering experience.

The first prototype was eventually used on more than 100 children. Initially, the other engineers who worked on this device and I were on hand to advise the physicians about its use, so labeling needs were minimal. We soon realized, however, that to ensure the safety and effectiveness of this device, we needed to label indications and contraindications for use.

That was in 1957, long before the Cooper Committee and the Food and Drug Administration had tackled the issue of medical device labeling. Even in those infant years of the medical device industry, most of us realized that device manufacturers had a duty to both the patient and the physician to make certain our products were safe and were used properly.

This concern continues today. The primary purpose of medical device labeling is to protect patients, both from unproven devices that could be dangerous and from unsafe applications of approved devices. A secondary purpose is to ensure that patients derive full value from medical devices. The more patients understand about how devices work in context of their physiological needs, the more effective devices are. Labeling not only helps the doctor or health professional achieve the full utility of the device, but also assists him or her in educating the patients to maximize the device's therapeutic benefits.

Yet there is another emerging purpose for medical device labeling: to enable device users to apply these new therapies in the most cost-effective manner. Given the current global trend of scrutinizing healthcare expenditures, this growing cost consciousness is extremely important in device design and labeling. Far from being annoying hurdles to medical device approval, labeling requirements serve a much-needed societal purpose: ensuring safe, cost-effective therapies and maintaining the quality of patient life.

While device approval times in some countries continue to be inordinately long, a solid understanding of national labeling requirements will go a long way in helping device manufacturers secure faster approvals and, at the same time, assure that new therapies are safe and effective. Charles Sidebottom's book will be a valuable tool in building this essential understanding.

Earl E. Bakken
Founder, Medtronic Inc.
Waikoloa, Hawaii
August 1994

Preface

To begin, it may be useful to indicate what kind of book this is—its scope, its overall intent, and its limitations—and what kind of book it is not. It is not a cookbook. It contains no easy-to-follow recipes that guarantee success. After all, the term "medical device" spans an extremely broad range of diagnostic, therapeutic, and prosthetic devices—from tongue depressors to magnetic resonance imaging (MRI) systems, implantable cardiac pacemakers to tissue heart valves. These devices employ diverse technologies ranging from sophisticated computer systems to genetically engineered tissue. These devices are intended to be used by people with a wide variety of backgrounds and training. The potential users extend from laypersons to highly trained healthcare professionals. Combine this diversity with the mixture of regulatory approaches practiced by the responsible authorities in the major markets, and you have a sure-fire prescription for confusion. Delineating all of the permutations and combinations of requirements and acceptable solutions is certainly beyond the scope of any single book.

This book is about the regulation of labeling for medical devices in the major medical device markets. Like the law or diplomacy, there is no last word in regulation. Regulation is a process that continues to evolve. It relies on precedent, but each case is different. While regulatory systems strive to be predictable, often they are not. It would be presumptuous to suggest that I have the answers to all the confounding labeling problems that the reader will encounter.

Furthermore, I work on a very broad canvas in writing this book. Some regulatory requirements are very prescriptive and leave little room for interpretation—"The label must bear the word STERILE." Other regulatory requirements—adequate directions for use is an example—require a detailed understanding of the design of the device, its potential users, and the clinical situation in which it will be used. These topics could be the subject of a multivolume treatise that would need to be supplemented periodically with new interpretations and recent developments. Thus, I do not presume to tell all that there is to know about any of the subjects I discuss.

Ultimately, the fundamental purpose of labeling is to enable the user of a medical device, be that user a healthcare professional or a layperson, to apply the device safely for the purposes intended by the manufacturer. This book presupposes that the reader understands the product in question, the potential users, and what "safely for the purposes intended" means within the clinical environment.

The first edition of this book was completed in late 1994. That edition covered regulatory requirements for labeling of medical devices in the highly regulated markets of Australia, Japan, North America, and the European Union (EU). At the time, this group accounted for a little over 92 percent of the world market for medical technology. In the ensuing seven years, the world market has grown by nearly 30 percent. This group still accounts for 88 percent of the consumption and nearly 100 percent of the production of medical technology. However, the importance of other markets, particularly East Asia and Eastern Europe, continues to grow. In addition, the number of countries adopting regulatory schemes that go beyond simple product registration is also growing. This trend is particularly strong in East Asia and South America. In Eastern Europe, regulation is also increasing, and it seems to be following the EU model, largely because many of these countries hope to join that economic union.

In this edition, I have updated the regulatory requirements covered in the first edition. There have been several major changes, such as a major overhaul of the Pharmaceutical Affairs Law (PAL) in Japan, the revision of the Australian Therapeutic Goods Act to implement mutual recognition with the European Union, and a significant update to the Canadian Medical Device Regulations (CMDR) to increase harmonization with both US and European regulations. In Europe, the In Vitro Diagnostic Device Directive (IVDD) has come into force and a new directive with a direct bearing on the regulation of medical devices was approved—the Radio and Telecommunications Equipment Directive. In the United States, the regulations applicable to labeling have not changed significantly. However, the Food and Drug Administration (FDA) continues to publish guidance documents that outline their current thinking on what is required for particular devices to provide "adequate directions for use." There have also been significant developments relating to "off-label use" and to the Internet as a vehicle for distributing information about medical devices. Under the provision of the Food and Drug Administration Modernization Act of 1997 (FDAMA), the FDA has implemented many process changes including a program to recognize consensus standards and accept a declaration of conformity to those standards including the labeling requirements that they may specify.

More change should be expected in the near future. Australia is preparing a new handbook on requirements to implement the 2002 changes to the Therapeutic Goods Act. In Japan, the Ministry of Health, Labor, and Welfare (MHLW) is working on yet another revision of the PAL.

Since the first edition was published, several new regulatory systems have been developed or are now being seriously enforced. In this edition, I have added information on the most mature of these systems. These include Argentina, Brazil, the People's Republic of China (PRC), the Republic of Korea, and Thailand.

Hopefully, you will find this book a good starting point and a useful guide to assist you in your professional work. I have drawn upon my experiences working as a product development manager and, more recently, working in regulatory affairs. However, I do not purport to advise readers about problems I have not had the opportunity to analyze. So, if you think you have a problem—or if you know you have one—this book should be a helpful beginning to enlighten you, but it is not a replacement for a good lawyer or other professional adviser. To help you make this decision, and to implement your decision in your own work environment, I hope this book will become a useful aid that you will reach for often and find a reliable guide in the business of medical device labeling.

Charles Sidebottom
Plymouth, Minnesota

Acknowledgments

My debt to the various authors, committees, and literally hundreds of dedicated professionals who have labored diligently to produce the reference material will be apparent, and I extend my thanks to all of them.

I appreciate the special efforts of my colleagues from around the world who provided background material, asked critical questions, and critiqued the drafts of the first or second editions of this work: Adriana Gonzales (Argentina), Lesley C. Pink (Australia); Dr. John Gams (Canada); Anita Li (PRC), Dr. Hwal Suh (Republic of Korea), Julie Reyes-Sandoval and Maria Alvarado (Mexico); Kenji Chikuni and Kazuo Murata (Japan); and Maarten Roelofs Heyrmans, Auke Poutsma, and Jacques Thielen (European Union). I would also like to thank Dr. Charles Swanson and Mr. Doyle (Chip) Whitacre for their support and encouragement. A special note of appreciation to Robert C. Flink for providing a steady stream of good ideas and for sharing his insights into medical device regulation.

Most of all, I would like to thank Carolyn Sidebottom for her unfailing support in this project. She was editor, proofreader, and cheerleader who understood when night after night I would disappear for hours on end to work on this project. She may also be the only person in the world besides myself who has read every word of this work not once, not twice, but at least three times. Without her, this book could not have been completed. It is to her that this work is dedicated.

Full source details of material cited in this work are given in the reference section. The reader's attention is directed to these references by the abbreviated citations within the text.

The reference section is arranged alphabetically according to the abbreviation used in the text citations. Readers will find that many references are cited several times.

The system of citation and reference used in this work has been devised to increase text readability and to assure full recoverability of the works cited.

The reader will note three variations in references. Two different citations of page numbers are used in the text, as follows: "(Lowery, Puleo p. 1–6)" indicates citation to a section and single page; "(Backinger pp. 18–30)" indicates reference to a sequence of pages. A third type of reference is frequently used when referencing regulations or works published on the Internet, as follows: "(CC&CR, §15(4))" indicates a reference to a chapter, section, clause or numbered paragraph within the work.

Credits

Certain material in this book is subject to international copyright protection. That material is reproduced with the permission of the copyright owner.

Material from *Australian Medical Device Requirements version 4 (DR4) under the Therapeutic Goods Act 1989* is reproduced by permission of the Therapeutic Foods Administration of the Department of Health and Aging. Material from the Therapeutic Goods Regulations is used by permission of the Commonwealth Copyright Administration. All legislation herein is reproduced by permission but does not purport to be the official or authorized version. It is subject to Commonwealth of Australia copyright.

Extracts from British Standards are reproduced with the permission of BSI under license number 2003/SK013. British Standards can be obtained from BSI Customer Services, 389 Chiswick High Road, London W4 4AL, United Kingdom. (Tel +44 20 8996 9001).

The author thanks the International Electrotechnical Commission (IEC) for permission to reproduce information from its International Standards IEC 60417-1 and IEC 61010-1. All such extracts are copyright of IEC, Geneva, Switzerland. All rights reserved. Further information on the IEC is available from www.iec.ch. IEC has no responsibility for the placement and context in which the extracts and contents are reproduced by the author; nor is IEC in any way responsible for the other content or accuracy therein.

The symbols taken from ISO 7000, Graphical symbols for use on equipment — Index and synopsis, are reproduced with the permission of the International Organization for Standardization, ISO. This standard can be obtained from any ISO member and from the Web site of the ISO Central Secretariat at the following address: www.iso.org. Copyright remains with ISO.

The author thanks Yakuji Nippo, Ltd. for permission to use Table 19.1, Table 19.2, Table 19.3, and the tables in Appendix E from *Guide to Medical Device Registration in Japan, 6th Edition* and the Japanese Standards Association for permission to use the material in Table 19.4 and Table 19.5 from JIS T 1001.

Table 28.1 on paper sizes is reproduced with the permission of the author, Mr. Markus Kuhn.

Table 27.1 includes material from *Effective Writing for Engineers, Managers and Scientists, 2nd Edition* by H.J. Tichy and is used with the permission of the publisher John Wiley & Sons, Inc.

List of Figures

FIGURE 1.1 Global medical device market 2000...5

FIGURE 8.1 Example of a hazard symbol on a large label 120

FIGURE 8.2 Example of a hazard symbol on a small label..121

FIGURE 8.3 Explosive symbol..124

FIGURE 8.4 Flammable symbol...125

FIGURE 8.5 Corrosive symbol...127

FIGURE 8.6 Toxic symbol..131

FIGURE 9.1 X-ray warning symbol..137

FIGURE 9.2 Alternate x-ray warning symbol..139

FIGURE 9.3 Ultrasonic radiation warning symbol 142

FIGURE 11.1 CCIB safety certification mark...157

FIGURE 11.2 Basic CCC mark 165

FIGURE 11.3 Basic CCC mark indicating type of certificate 165

FIGURE 11.4 Registered-product standard code number layout..167

FIGURE 11.5 Form of the registration number for medical products produced in China 169

FIGURE 11.6 Form of the registration number for imported products..170

FIGURE 11.7 China certification commission for medical devices (CMD) mark................................170

FIGURE 11.8 Examples of Chinese dates...171

FIGURE 14.1 CE conformity marking...206

FIGURE 14.2 Alert sign..209

FIGURE 18.1 Example of a trade name containing Japanese and alphanumeric characters..................297

FIGURE 18.2 Table of reagents...306

FIGURE 23.1 Warning logotype A..395

FIGURE 23.2 Warning logotype B..396

FIGURE 23.3 Declaration of conformity logo. 410

FIGURE 27.1 Example of a warning statement...463

List of Tables

TABLE 2.1 Products Whose Labels Identify Them as Atoxic, Sterile, and Free of Pyrogens...............................10

TABLE 2.2 Products Included in Article 1 of the Degree No. 2,505/85...10

TABLE 2.3 List of Reference Countries...11

TABLE 4.1 Diseases, Conditions, Ailments, and Defects for Which Advertising of Serious Forms
is Restricted...31

TABLE 4.2 Registrable Therapeutic Devices 34

TABLE 4.3 List of Materials Designated Biological Materials by the Australian Quarantine
and Inspection Service (AQIS)..41

TABLE 7.1 Diseases, Disorders, or Abnormal Physical States 75

TABLE 7.2 Medical Devices Other Than IVDs Classified According to Rule 16 of Schedule 1, Part 1 76

TABLE 7.3 IVDs Classified According to Rule 9 of Schedule 1, Part 2 77

TABLE 8.1 Medical Devices Classified as Implants..105

TABLE 8.2 Menstrual Tampon Absorbency Identification 108

TABLE 8.3 Marking on the Outside of Medical Electrical Equipment...114

TABLE 8.4 Colors for Indicator Lights and Push Buttons and their Recommended Meanings 115

TABLE 8.5 Minimum Hazard Symbol and Signal Word Size...122

TABLE 8.6 Minimum Height and Body Size of Type...122

TABLE 8.7 Required Information for Devices with Substances Classified as Corrosive....................................128

TABLE 8.8 Required Information for Devices with Substances Classified as Very Corrosive 130

TABLE 8.9 Required Information for Devices with Substances Classified as an Irritant 131

TABLE 8.10 Required Information for Devices with Substances Classified as Toxic..132

TABLE 8.11 Required Information for Devices with Substances Classified as Harmful 133

TABLE 11.1 Specification for Standard-Size CCC Mark 165

TABLE 12.1 KFDA Authorized Testing Agencies 176

TABLE 12.2 KFDA Clinical Trial Agencies 177

TABLE 14.1 Official Languages of the European Union 197

TABLE 14.2 Key Technical Harmonization Standards for Medical Device Labeling...199

TABLE 14.3 Markings to Facilitate Reuse and Recovery of Packaging and Packaging Waste 213

TABLE 15.1 Examples of Protective-Equipment Classification 216

TABLE 15.2 Examples of Medical Devices/Medicinal Products 217

TABLE 15.3 Graphical Symbols for Use in Labeling Medical Devices 219

TABLE 15.4 Units Outside the International System that Can be Used on Equipment..221

TABLE 15.5 Marking on the Outside of Medical Electrical Equipment...230

TABLE 15.6 Colors for Indicator Lights and Push Buttons and their Recommended Meanings 231

TABLE 16.1 Examples of Active Implantable Medical Devices 238

TABLE 16.1 Examples of Recommended Storage-Temperature Intervals..253

TABLE 16.2 Symbols for Marking *In Vitro* Diagnostic Instruments...257

TABLE 16.3 Hazard Symbols and Indications of Danger 269

TABLE 16.4 Minimum Hazard-Label Sizes..271

TABLE 18.1   Summary of the Regulatory Requirements for the Four Classes of Medical Devices in Japan 277

TABLE 18.2   Designated Medical Devices in Japan 278

TABLE 18.1   Medical Devices Subject to Japanese Mandatory Performance Standards 293

TABLE 18.2   Medical Devices Subject to Japanese Approval Standards 293

TABLE 18.3   Warnings and Directions by Device Category 296

TABLE 18.4   Markings on the Outside of Medical Electrical Equipment 301

TABLE 18.5   Colors for Indicator Lights and Push Buttons and their Recommended Meanings 302

TABLE 21.1   Type Size as a Function of Label Area for OTC Devices 343

TABLE 21.2   U.S. Abbreviations for Weights and Measures 344

TABLE 21.3   OTC Device Labeling in Terms of Weight or Liquid Measure 344

TABLE 21.4   OTC Device Labeling in Terms of Linear or Area Measure 345

TABLE 21.5   U.S. Symbols And Abbreviations for Weights and Measures 346

TABLE 21.6   SI and Customary Inch-Pound Conversion Factors 347

TABLE 22.1   Minimum Type Size Requirements 379

TABLE 25.1   Menstrual Tampon Terms of Absorbency 448

TABLE 25.2   Minimum Type Size Requirements 449

TABLE 27.1   Examples of Cluttered Language 466

TABLE 28.1   International (ISO) Paper Sizes 473

TABLE 28.2   Safety Colors and Contrast Colors 476

TABLE A.1   U.S. Department of Commerce NAICS Classifications 480

TABLE B.1   Goods Declared to Not be Therapeutic Goods 485

TABLE B.2   Goods Declared to be Drugs 486

TABLE B.3   Therapeutic Goods Exempt From Registration or Listing on the Australian Registry of Therapeutic Goods 486

TABLE B.4   Therapeutic Goods Exempt from the Operation of Part 3 of the Act Subject to Conditions 487

TABLE B.5   Specific Listable Device Policies 489

TABLE B.6   Goods Exempt From Licensing to Manufacturer 489

TABLE B.7   Persons Exempt from Licensing as Manufacturers 490

TABLE C.1   International Symbols for Medical Device Marking 492

TABLE D.1   KFDA Categories of Medical Devices 497

TABLE E.1   Japanese MHWL Categories of Medical Devices 518

TABLE E.2   Japanese MHLW Medical Devices Exempted From Approval on a Product-by-Product Basis 524

TABLE E.3   Japanese MHLW Japanese Industrial Standards (JIS) for Medical Devices 527

TABLE E.4   Japanese MHLW Medical Devices Subject to the Partial License (Kubun-kyoka) System 534

TABLE E.5   Japanese MHLW Medical Devices Exempted from the Requirements of Medical Device Good Manufacturing Practice (GMP) 535

TABLE E.6   Japanese MHLW Medical Devices to be Approved by Prefectural Governor 537

TABLE E.7   Japanese MHLW Devices for Which Labeling Exceptions are Permitted 538

TABLE F.1   U.S. DHHS Medical Device Classifications 540

Contents

PART I Negotiating a Common Understanding

Chapter 1 The Global Market for Healthcare Technology 3

Technology and the Healthcare Delivery System..3
Industry and the Global Market 4
Industry Composition 4
Industry and Regulation 5

PART II Argentina and Brazil

Chapter 2 Argentina..9

Background and General Intent of the Law 9
Scope of the Regulations...9
Bringing Devices to Market in Argentina...10
Importing Products...11
Technical Director...12
Adulteration and Misbranding..12
General Requirements for Labeling of Medical Devices 13
Packaging and Labeling Materials 13
Considerations for Sterile Packing 14
Arrangement of Labels 14
Package Label Contents 14
Language Requirements...14
Device Reuse 15
Device Reprocessing 15
Things to Remember 15

Chapter 3 Brazil...17

Background and General Intent of the Law 17
Scope of the Regulations...17
Bringing Devices to Market in Brazil...18
Product Registration..18
Import of Medical Products...19
Misbranding...20
General Labeling Provisions 20
Language Requirements...20
Minimum Labeling Requirements 20
Things to Remember 20

PART III Australia

Chapter 4 The Therapeutic Goods Act of Australia 25

Background and General Intent of the Law 26

The Therapeutic Goods Amendment of 2002..26
Scope of the TGA Regulations 27
The Regulations 27
Labels and Labeling 27
Labeling and Advertising 28
Adulteration and Misbranding..28
False or Misleading Labeling 29
Advertising and Promotion...29
Bringing Devices to Market in Australia 32
    Registration of Therapeutic Devices...33
    Claiming Equivalence to a Registered Device 35
    Listing of Therapeutic Devices..37
    Claiming Equivalence to a Listed Device 38
    Licensing of Therapeutic Goods Manufacturers...39
    Commencing the Supply of a Registered or Listed Device 40
Export of Therapeutic Devices...40
Import of Therapeutic Devices...40
Standards For Therapeutic Devices..41
Exemptions For Special and Experimental Uses..42
    Investigational Use 42
        Clinical Trial Notification (CTN) Scheme...42
        Clinical Trial Exemption (CTE) Scheme...43
    Custom Devices 43
    Goods Exempted for Regulatory Purposes 44
Things to Remember 44

Chapter 5   General Therapeutic Device Labeling in Australia...45

Misbranding..45
Adequate Directions for Use...45
General Requirements for Labels for Therapeutic Devices 46
    Presentation Requirements (§4(2)(c))...46
    General Labeling Requirements 47
    Name of the Therapeutic Device (§7(a))...47
    Name and Place of Business (§ 7(b))..47
    Batch or Serial Number (§7(c))..48
    Multiple Devices in a Package (§7(d))...48
    Labeling of Sterile Devices and Nonsterile Implantable Therapeutic Devices........................48
    Unit Packages (§10)..49
    Outer Packages (§13)..50
    Small Packages (§11)..50
    Individually Wrapped Goods (§12)...50
    Transparent Packages (§§14 and 15)..51
In Vitro Diagnostic (IVD) Goods 51
    Devices Supplied for Home Use and as a Commonwealth Pharmaceutical Benefit 52
    Devices Incorporating Material of Human Origin...52
    Devices Used for the Diagnosis of Infection with Human Immunodeficiency Virus (HIV)
    and Hepatitis C Virus (HCV)..52
Application of the Registration Number...52
Labeling of Components and Kits 52
Information to be Supplied For Specific Registrable Devices 53
    Heart-Valve Prostheses..53
    Active Implantable Medical Devices (AIMDs)...53
    Drug-Infusion System (Powered, nonimplantable)...54
    Breast Prostheses (Not Saline or Water) 54

Therapeutic Devices of Animal Origin 54
Therapeutic Devices of Human Origin 54
Intraocular Lenses (IOLs)...55
Intrauterine Contraceptive Devices (IUCDs) 55
Barrier Contraceptive Devices..56
Saline Breast Prostheses 56
HIV/HVC IVD Kits 56
Things to Remember 56

Chapter 6   The Therapeutic Goods Amendment of 2002...59

Scope of the New Regulation...60
The Approach..60
Device Classification...60
Essential Principles 61
Standards 61
Australian Register of Therapeutic Goods (ARTG)..61
Export of Medical Devices 62
Import of Medical Devices 62
Postmarket Requirements 62
Essential Principles for Labeling 62
General Requirements..63
Language Requirement 63
Labeling Format 63
Label Requirements 63
Instructions-for-Use Requirements 64
Medical Devices Used for a Special Purpose 66
Custom-Made Medical Devices..66
Devices Intended for Clinical Investigations or Experimental Purposes.................................66
System or Procedure Pack 67
Things to Remember 67

PART IV   Canada and Mexico

Chapter 7   The Food and Drugs Act of Canada 71

Background and General Intent of the Law 71
Scope of the Therapeutic Product Directorate Regulations...73
The Regulations 73
Labels and Labeling 73
Labeling and Advertising 74
Adulteration and Misbranding..74
False or Misleading Labeling..74
Advertising and Promotion...75
Bringing Devices to Market in Canada...76
Device Classification..76
Establishment License..77
Distribution Records 78
Complaint Handling 78
Mandatory Problem Reporting 78
Implant Registration...79
Safety and Effectiveness Requirements...79
Class I Medical Devices 81

Class II, III and IV Medical Devices 81
    Device License 81
    General Requirements for a Medical Device License Application.......................................81
    Class II Medical Device License Application 82
    Class III Medical Device License Application..82
    Class IV Medical Device License Application..83
    Amended Medical Device License Application 84
    Issuance of License 85
    Foreign Manufacturers 85
Standards for Medical Devices 85
    Recognition of Standards...86
    Use of Recognized Standards..87
Investigational Device Exemption (IDE) 88
Sale of Custom-made Devices and Medical Devices For Special Access..................................89
Export of Medical Devices...90
Things to Remember 90

Chapter 8   General Medical Device Labeling in Canada 93

Misbranding...93
General Labeling Requirements..93
    Device Identification 94
    Instructions for Use (IFU) 95
    Sterile Devices or Devices with a Limited Life...96
    Class III or IV Devices 97
Prominence of Required Information...97
Requirements for Medical Devices Intended to Be Sold to the General Public at a Self-Service Display 97
Language Requirements 97
In Vitro Diagnostic (IVD) Devices..98
    Package Insert 98
        Device Identification 98
        Instructions for Use (IFU)...98
        Sterile Devices or Devices with a Limited Life 101
        Class III or IV Devices 102
        Other Material 102
    Immediate Container Label 102
        Device Identification 102
        Instructions for Use (IFU)...102
        Sterile Devices or Devices with a Limited Life 103
        Class III or IV Devices 103
    Reagent Label 103
        Device Identification 103
        Instructions for Use (IFU)...104
        Sterile Devices or Devices with a Limited Life 104
        Class III or IV Devices 104
    Labeling for IVD Devices Containing Explosive Materials or Components.......................104
Implant-Registration Card 105
Labeling for Investigational Devices...106
Labeling of Custom-made Devices and Medical Devices for Special Access 106
Export of Medical Devices..107
Special Labeling Requirements for Specific Devices...107
    Labeling for Soft Contact Lenses...107
    Labeling for Menstrual Tampons..108
    Labeling for Contraceptive Devices 109
        Individual Containers for Contraceptives Devices 109

Contraceptive Effectiveness 109
Prophylactic Effectiveness..110
Individual Containers for Synthetic Condoms...110
Instructions for Use (IFU) of a Synthetic Condom 110
Labeling for Medical Gloves..110
Examination Gloves Only (Sterile And Non-Sterile)...110
Material of Manufacture...111
Powdered Natural Rubber Latex..111
Powderless Natural Rubber Latex..111
Natural-Rubber Latex with a Thin Inner Polyurethane Coating..112
Bilayer Neoprene-Latex Rubber 112
Neoprene...112
Polyolefin or Other Thin Film Copolymer 112
Vinyl 112
Hydrocarbon Polymer 112
Labeling for Medical Electrical Equipment 112
Markings on Electromedical Equipment 113
Documents Accompanying Electromedical Equipment 116
Instructions for Use (IFU)..116
Technical Description..117
Guidance on Labeling Devices in Pressurized Containers 118
Principal Display Panel..118
Manner of Disclosing Required Information 119
Placement of Information on Display Panels 119
Hazard Symbols 121
Signal Words...121
Primary Hazard Statements 122
Additional Hazard Statements, Negative and Positive Instructions, and First Aid Statements........122
Small Packages 122
Labeling for Home-Use Devices 123
Devices with Contents under Pressure 123
Labeling for Flammable Products 124
Devices with Combustible or Spontaneously Combustible Contents....................................125
Devices with Flammable Contents 126
Devices with Very Flammable Contents...126
Labeling for Corrosive Products...127
Devices with Corrosive Contents...127
Devices with Very Corrosive Contents 129
Devices with Contents Classified as an Irritant...129
Labeling for Toxic Products 129
Devices with Contents Classified as Toxic 129
Devices with Contents Classified as Harmful 132
Things to Remember 134

Chapter 9   Radiation-Emitting Device Labeling in Canada 135

Background and General Intent of the Law 135
The Regulations...135
Labels and Labeling 136
Labeling Requirements for Radiation-Emitting Products...136
Dental X-ray Equipment with an Extra-Oral Source..136
Labeling..137
Instructions for Use (IFU)...138
Photofluorographic X-Ray Equipment 138
Labeling..138

Diagnostic X-Ray Equipment...139
    Labeling 139
    Instructions for Use (IFU)..140
  Ultrasound-Therapy Devices 141
    Labeling 141
Things to Remember 142

Chapter 10 Mexico...143

Background and General Intent of the Law 144
Scope of the Regulations...144
Bringing Devices to Market in Mexico 145
  Product Registration...145
  Qualifying for Sale to the Mexican Government....................................147
Misbranding 149
General Labeling Provisions 149
  Types of Packaging...149
  Minimum Labeling Requirements...150
Things to Remember 151

PART V China, Korea, and Thailand

Chapter 11 People's Republic of China..155

Background and General Intent of the Law 155
Scope of the Regulations...156
Misbranding...157
Bringing Devices to Market in China 158
  Medical Devices Manufactured in China...158
    Medical Device Registration...158
    Class I Devices 158
    Class II and III Devices 159
    Drug/Device Combinations...160
  Medical Devices Imported into China..160
  Business Licensing and Quality-System Approval 162
    Production License 162
    Operation License...162
    Quality System 163
  Registration Certificate Alteration 163
  Compulsory Product Certification System (CPCS) 164
  Medical Product Standards 166
  Postmarket Surveillance..167
  Medical Devices Exported from China 167
Investigational Use 168
Advertising and Promotion..168
General Labeling Provisions 168
  Product Manual 168
  The Acknowledgment Symbol for Medical Equipment...........................169
The Chinese Language 169
  Chinese Writing 170
  The Chinese Calendar..171
Things to Remember 171

Chapter 12 Republic of Korea..173

Background and General Intent of the Law 173

Scope of the Regulations..174
Labeling and Advertising 175
Bringing Devices to Market in Korea...175
    Premarket Notification 175
    Premarket Approval..175
    Clinical Trials...176
    Importing Products...176
    Grandfathered Products..177
General Requirements for Labeling of Medical Devices.........................178
    Adequate Directions for Use 178
    Package Label 178
Things to Remember 178

Chapter 13 Thailand 181

Background and General Intent of the Law 181
Scope of the Regulations...181
Adulteration and Misbranding...182
Bringing Devices to Market in Thailand 183
    Device Registration..183
    Device Notification 184
    General Medical Devices 184
    Required Postmarket Reporting 184
Thai Labeling Requirements..185
    Labels 185
    Accompanying Documents 185
    Language Requirements..186
    Expiration Dates 186
Advertising...186
In Vitro Diagnostic (IVD) Devices...186
Things to Remember 187

PART VI European Union

Chapter 14 The European Medical Device Directives 191

Background and General Intent of the Law 192
Mutual Recognition Agreements..193
    Australia 194
    Switzerland...194
Scope of the European Union Regulation 194
    The Medical Device Directive (MDD)...195
    Active Implantable Medical Device Directive (AIMDD) 195
    In Vitro Diagnostic Device Directive (IVDD)................................195
    Radio Equipment and Telecommunications Terminal Equipment Directive (R&TTED)........................195
Labels and Labeling 196
Language Requirements 197
Standards..197
Bringing a Device to Market in the European Union 198
    Devices Conforming to the Essential Requirements.........................200
        Conformity Assessment 200
        Notified Bodies 200
        Competent Authorities...201
    Devices for Clinical Evaluations 202

       Devices for Performance Evaluation or Reevaluation 202
       Custom-Made Devices 202
       Devices for Trade Fairs, Exhibitions, and Demonstrations 203
   Postmarket Surveillance and Vigilance 203
       Postmarket Surveillance 203
       Vigilance 204
       The Safeguard Clause 205
   CE Marking of Conformity 206
       CE Conformity Marking on Devices Covered by the MDD 207
       CE Conformity Marking on Devices Covered by the AIMDD 207
       CE Conformity Marking on Devices Covered by the IVDD 208
       CE Conformity Marking on Devices Covered by the R&TTED 208
       CE Conformity Marking on Devices Covered by Overlapping Directives 209
   Investigational Device Marking 210
   Performance-Evaluation Device Marking 210
   Custom-Made Device Marking 210
   Other Directives of Interest 211
       Proprietary Medicinal Products 211
       Low-Voltage Equipment 211
       Machinery 211
       Dangerous Substances 211
       Packaging and Packaging Waste 212
   Local Requirements 213
   Things to Remember 213

Chapter 15 The Medical Device Directive (MDD) 215

   Device Classification 217
   Misbranding 218
   General Labeling Provisions 218
       Controls and Displays (§12.9) 220
       Intended Purpose (§13.4) 220
   Devices with a Measuring Function 221
       Measurement, Monitoring, and Display Scales (§10.2) 221
       Units of Measure (§10.3) 221
   Particulars on the Label 221
       Manufacturer Identification (§13.3(a)) 222
       Identity of the Device (§13.3(b)) 222
       Sterile Device Marking (§§13.3(c), 13.3(m), and 8.7) 222
       Product Identification (§§13.3(d) and 13.5) 223
       Expiration Dating (§13.3(e)) 223
       Year of Manufacture (§13.3(l)) 223
       Single-Use Devices (§13.3(f)) 223
       Storage and Handling Conditions (§13.3(i)) 223
       Special Operating Instructions (§13.3(j)) 224
       Warnings and Precautions (§13.3(k)) 224
       Markings for Special-Purpose Devices (§§13.3(g) and 13.3(h)) 224
       Devices Incorporating Stable Human Blood Derivatives (§13.3(n)) 224
   Instructions for Use (IFU) 224
       Particulars from the Label (§13.6(a)) 225
       Performance Intended by the Manufacturer (§13.6(b)) 225
       Connection to Other Medical Devices (§§9.1 and 13.6(c)) 225
       Installation and Maintenance (§13.6(d)) 225
       Reciprocal Interference (§13.6(f)) 226
       Sterile Packaging (§13.6(g)) 226

Reusable Devices (§13.6(h)) 226
Device Preparation (§13.6(i)) 226
Radiation-Emitting Devices (§§11.4.1 and 13.6(j)) 226
Implantable Devices (§13.6(e)) 226
Patient Information 227
Changes in Performance (§13.6(k)) 227
Exposure to Environmental Conditions (§13.6(l)) 227
Administration of Medicinal Products (§13.6(m)) 227
Disposal of the Device (§13.6(n)) 227
Medicinal Substances Incorporated into the Device (§13.6(o)) 227
Measuring Accuracy (§13.6(p)) 227
Special Labeling Requirements for Specific Devices 228
Markings on Medical Electrical Equipment 228
Documents Accompanying Medical Electrical Equipment 232
Instructions for Use (IFU) 232
Technical Description 233
Medical Electrical Systems 234
Things to Remember 235

Chapter 16 The Active Implantable Medical Device Directive (AIMDD) 237
Misbranding 238
General Labeling Provisions 238
Product Identification (§11) 239
Noninvasive Identification (§12) 239
Controls and Displays (§13) 239
Particulars on the Sterile Package Label 240
Manufacturer Identification (§14.1(iii)) 240
Identity of the Device (§14.1(iv)) 240
Sterile Device Marking (§§14.1(i), 14.1(ii), and 14.1(vii)) 240
Expiration Dating (§14.1(ix)) 241
Date of Manufacture (§14.1(viii)) 241
Connection to Other Devices (§9.(iv)) 241
Markings for Special-Purpose Devices (§§14.1(v) and 14.1(vi)) 241
Particulars on the Sales Package Label 242
Manufacturer Identification (§14.2(i)) 242
Identity of the Device (§§14.2(ii), 14.2(iii), and 14.2(iv)) 242
Sterile Device Marking (§14.2(vii)) 243
Expiration Dating (§14.2(ix)) 243
Date of Manufacture (§14.2(viii)) 243
Storage and Handling Conditions (§14.2(x)) 243
Connection to Other Devices (§9(iv)) 243
Device Containing Radioactive Substances (§8(v)) 243
Markings for Special-Purpose Devices (§§14.2(v) and 14.2(vi)) 243
Instructions for Use (IFU) 244
Particulars from the Labels (§15(ii)) 244
Performances Intended by the Manufacturer (§15(iii)) 244
Selecting a Suitable Device (§15(iv)) 245
Device Operation (§15(v)) 245
Reciprocal Interference (§§15(vii) and 8(iv)) 245
Sterile Packaging (§15(viii)) 245
Reusable Devices (§15(ix)) 246
Implantable Devices (§15(vi)) 246
Year of Authorization to Affix the CE Mark (§15(i)) 246
Device Containing Radioactive Substances (§8(v)) 246

Medicinal Substances Incorporated into the Device (§10.) 246
Noninvasive Identification (§12.) 246
Patient Information 246
Lifetime of the Energy Source (§15(x)) 247
Changes in Performance (§15(xi)) 247
Exposure to Environmental Conditions (§15(xii)) 247
Administration of Medicinal Products (§15(xiii)) 247
Things to Remember 247

Chapter 17 The *In Vitro* Diagnostic Device Directive (IVDD) 249
Improper Labeling 249
General Labeling Provisions 250
Intended Purpose (§8.5) 250
IVDs Intended for "Self-Testing" (§7) 250
Product Identification (§8.6) 250
Devices with a Measuring Function 250
Measurement, Monitoring, and Display Scales (§3.6) 251
Units of Measure (§9.2) 251
Particulars on the Label of Reagents 251
Immediate Container 251
Product Name (§13.4(b)) 252
Manufacturer (Supplier) (§8.4(a)) 252
Lot Number (§8.4(d)) 252
Expiration Date (§8.4(e)) 252
Contents (§8.4(b)) 252
Intended Use (§§8.4(b), 8.4(g), and 8.4(k)) 252
Cautionary Statements (§8.4(j)) 252
Storage Information (§8.4(h)) 253
Sterile Device Marking (§8.4(c)) 253
Markings for Investigational Use (§8.4(f)) 253
Outer Container 253
Product Name (8.4(b)) 253
Manufacturer (Supplier) (§8.4(a)) 253
Lot Number (§8.4(d)) 254
Expiration date (§8.4(e)) 254
Contents (§8.4(b)) 254
Identity of the Device and Intended Use (§§8.4(b), 8.4(g), and 8.4(k)) 254
Cautionary Statements (§8.4(j)) 254
Storage Information (§8.4(h)) 254
Special Operating Instructions (§ 8.4(i)) 255
Sterile Device Marking (§8.4(c)) 255
Markings for Investigational Use (§8.3(f)) 255
Particulars on the Label of IVD Instruments and Equipment 255
Product Name (§13.4(b)) 256
Manufacturer (Supplier) (§8.4(a)) 256
Lot Number (§8.4(d)) 256
Expiration Date (§8.4(e)) 256
Intended Use (§§8.4(b), 8.4(g), and 8.4(k)) 257
Warning Markings (§8.4(j)) 257
Other Markings (§8.4(i)) 258
Markings for Investigational Use (§8.4(f)) 260
Instructions for Use (IFU) 260
Instructions for Use for Reagents 261
Particulars from the Labels (§8.7(a)) 261

Application and Intended Use (§8.7(d)) 261
Composition of Reagents (§8.7(b)) 261
Additional Materials and Devices (§8.7(e)) 261
Methodology (§8.7(h)(1)) 261
Performance Criteria, Limitations, and Possible Errors (§§8.7(d), 8.7(h)(2), and 13.7(h)(4)) 262
Reagent Preparation (§8.7(h)(3)) 262
Storage and Shelf Life after Opening (§§8.7(c) and 8.7(j)) 262
Specimens (§8.7(f)) 262
Test Procedure (§8.7(g)) 262
Reading and Explanation of Results (§§8.7(i), 8.7(k)(1), 8.7(k)(2), and 8.7 (l)) 262
Follow-up Action (§§8.7(t)(1), 8.7(t)(3), and 8.7(t)(4)) 263
Precautions and Warnings (§8.7(s)) 263
Sterile Package (§§8.7(o) and 8.7(p)) 263
Radiation-Emitting Products (§5.3) 263
Literature References (§§8.7(h), 8.7(i), 8.7(k), and 8.7(l)) 263
Particular Information that May Be Omitted (§8.7(t)(2)) 263
Date of Issue for the Instructions for Use (IFU) (§8.7(u)) 264
Instructions for Use (IFU) for Instruments and Equipment 264
Particulars from the Labels (§8.7(a)) 264
Application and Intended Use (§8.7(d)) 264
Additional Materials (§8.7(e)) 265
Methodology (§8.7(h)(1)) 265
Performance Criteria, Limitations, and Possible Errors (§§8.7(h)(2) and 8.7(h)(4)) 265
Specimens (§8.7(f)) 265
Instrument Operation and Test Procedure (§8.7(g)) 265
Reading and Explaning of Results (§§8.7(i) and 8.7(l)) 266
Follow-Up Action (§§8.7(t)(1), 8.7(t)(3) and 8.7(t)(4)) 266
Internal Quality Control (§8.7(k)) 266
Literature References (§§8.7(h), 8.7(i), 8.7(k), and 8.7(l)) 266
Installation, Calibration, and Changes in Performance (§§8.7(n), 8.7(q), and 8.7(j)) 266
Technical Specification (§§8.7(h)(2) and 8.7(r)) 268
Particular Information that May Be Omitted (§8.7(t)(2)) 268
Date of Issue for the Instructions for Use (IFU) (§8.7(u)) 268
Devices Incorporating Dangerous Substances 269
Hazardous Substance Labeling (Article 23) 269
Implementation of Labeling Requirements (Article 24) 270
Exemptions from Labeling and Packaging Requirements (Article 25) 271
Things to Remember 271

PART VII Japan

Chapter 18 The Pharmaceutical Affairs Law of Japan 275

1994 Revision to the PAL 275
The Regulations 276
Enforcement Ordinance of the Pharmaceutical Affairs Law (EOPAL) 278
Enforcement Regulations of the Pharmaceutical Affairs Law (ERPALs) 278
Adulteration and Misbranding 279
False or Misleading Labeling 279
Advertising and Promotion 279
Exhibitions for Specialists Promoting Academic Research 280
Exhibitions for the General Public Promotion of Scientific/Technical Issues and/or Related Industry 281
Exhibitions for the General Public Providing General Information 281

In Vitro Diagnostic (IVD) Products 282
Bringing Devices to Market In Japan 282
    Medical Device Manufacturer's (Importer's) License 282
    Medical Device Approval 283
    Determine Substantial Equivalence and Reexamination...284
    Changes in Approved Devices..285
    Medical Device Approval for Imported Products 286
    Role of the Prefecture Governments 286
    Medical Device Approval for Products Manufactured by Foreigners 287
Designated Medical Devices 287
Medical Device Vigilance...287
    Adverse Event Reporting..288
    Device Tracking 288
Things to Remember 288

Chapter 19 General Medical Device Labeling in Japan 291

Misbranding 291
General Labeling Provisions 291
    Immediate Container (PAL Article 63) 292
        Name and Address of the Manufacturer or Importer (ERPALs Article 61)..............292
        Performance Standards Established by Ordinance (ERPALs Article 60–2, Item 1).......292
        Approval Standards Established by Ordinance (ERPALs Article 60–2, Item 1)............293
        Medical Devices Manufactured in Foreign Countries (ERPALs Article 60–2, Item 2)...293
    Package Insert (PAL Article 52)..294
        Warnings and Directions for Use (PAL Article 52, Item 1) 294
        Warnings and Directions for Use by Device Category (PAL Article 52, Item 3)............296
        Warnings and Directions for Use Specified by Ordinance (PAL Article 52, Item 4)......296
        Prominence of Required Statements (PAL Article 53) 296
    Language Requirement (ERPALs Articles 58 and 62) 296
Trade Names..297
Export of Medical Devices (ERPALS Article 66)...297
Import of Medical Devices (ERPALS Articles 53–2, Item 3, and 60–2, Item 2) 298
Testing of Medical Devices (ERPALS Article 43)..298
Clinical Trials (Erpals Articles 67, 68, 69–2, and 70)..299
Warnings and Directions for Use by Device Category (PAL Article 52, Item 3) 299
    Electrical Medical Equipment 300
    Ultrasonic Diagnostic Equipment...303
    Radiation-Related Apparatus 303
    Surgical Laser Apparatus...303
    Dental Materials..304
    Disposable Products 304
    Small Steel Devices 305
    Contact Lenses 305
    Intraocular Lenses (IOLs)..305
    Electrical Therapy Apparatus for Household Use...306
Labeling Requirements For *In Vitro* Diagnostic (IVD) Products...306
    Inner Label or Wrapper of an *In Vitro* Diagnostic Reagent (ERPALs Article 56–2) 307
    Package Insert for *In Vitro* Diagnostic (IVD) Reagents (ERPALs Article 57) 308
    Outer Container Label or Wrapper of *In Vitro* Diagnostic (IVD) Reagents (ERPALs Article 56–2)........309
Labeling Requirements For Metals for Dental Use (ERPALS Article 60–3)...............................309
Labeling Requirements For Household-Use Devices...310
Things to Remember 310

PART VIII United States

Chapter 20 The Federal Food, Drug, and Cosmetic Act 315

Background and General Intent of the Law 316
Related Laws 321
Scope of the FDA Regulations 322
The Regulations 322
Labels and Labeling 323
Labeling and Advertising 324
Adulteration and Misbranding 324
False or Misleading Labeling 325
Advertising and Promotion 326
Promotion of Off-Label Uses 329
Reprints of Scientific Articles 330
Continuing Medical Education 333
Advertising on the Internet 334
Things to Remember 336

Chapter 21 General Device Labeling in the United States 337

Misbranding 338
General Labeling Provisions 338
Name and Place of Business (§801.1) 338
Intended Uses (§801.4) 339
Adequate Directions for Use (§801.5) 340
Misleading Statements (§801.6) 340
Prominence of Required Statements (§801.15) 340
Spanish Language (§801.16) 341
Labeling Requirements for Over-the-Counter (Nonprescription) Devices 341
Principal Display Panel (§801.60) 341
Statement of Identity (§801.61) 342
Declaration of Net Quantity of Contents (§801.62) 342
Customary Inch-Pound Declarations 344
SI Declarations 345
Conversion between Systems 345
Use of the Term "Net" 346
Use of the Term "Weight" 347
Exemptions from Adequate Directions For Use 347
Prescription Devices (§801.109) 348
Indications for Use 348
Contraindications 349
Warnings 349
Precautions 350
Adverse Reactions 350
Retail Exemption for Prescription Devices (§801.110) 351
Medical Devices Having Commonly Known Directions (§801.116) 351
In Vitro Diagnostic Products (§801.119) 351
Medical Devices for Processing, Repacking, or Manufacturing (§801.122) 351
Medical Devices for Use in Teaching, Law Enforcement, Research, and Analysis (§ 801.125) 351
Medical Devices: Expiration of Exemptions (§801.127) 352
Other Exemptions 352
Medical Devices: Processing, Labeling, or Repacking (§801.150) 352
Sterile Devices 353

Reprocessed Single-Use Devices 355

Electromagnetic Compatibility Labeling 355

Electronic Labeling...356

Special Requirements for Specific Devices 356

Export of Medical Devices...357

    Export of Approved Devices 357

    Export of Unapproved Devices 357

    Other Provisions of the ER&EA 359

        Exporting for Investigational Use 359

        Exporting for Marketing or in Anticipation of Foreign Marketing Approval 360

        Devices Intended for Treatment of Non-US Diseases 360

Import of Medical Devices..360

Banned Devices 363

    Labeling (§895.25)..363

    Veterinary Use (§895.1(d)) 364

Things to Remember 364

Chapter 22 *In Vitro* Diagnostic Product Labeling...365

    General IVD Labeling Provisions 366

    Labeling on the Immediate Container (§809.10(a))...366

    Labeling Requirements for Package Inserts (§809.10(b)) 367

Special Cases 370

    Exemptions from IVD Product Labeling Requirements (§809.10(c)) 371

    Labeling for General-Purpose Laboratory Reagents and Equipment (§809.10(d)) 371

Product Class Models for IVD Products 372

    Clinical Chemistry/Toxicology 372

    Clinical Microbiology/Immunology 373

    Clinical Hematology/Pathology...373

Home-Use IVD Products...374

Hazardous Substances 375

    Labeling Requirements 376

    Prominence, Placement, and Conspicuousness of Labeling...376

        Prominent Label Placement 377

    Area of Principal Display Panel 378

    Type-Size Requirements 379

    Accompanying Documents 380

    Outer Container or Wrappings..380

    Caustic Poisons 380

Storage Instructions and Expiration Dates...381

Things to Remember 381

Chapter 23 Radiation-Emitting Device Labeling..383

General Labeling Requirements for Electronic Products..383

    Product Certification (§1010.2) 384

    Product Identification (§1010.3)...384

    Variances (§1010.4) 384

    Exemptions for Products Intended for US Government Use (§1010.5)...385

    Export of Electronic Products (§1010.20)..385

Ionizing Radiation-Emitting Products..385

    Television Receivers (§1020.10)...385

    Cold-Cathode Gas Discharge Tubes (§1020.20) 386

    Diagnostic X-Ray Systems and their Major Components (§1020.30) 386

        Identification of Components (§1020.30(e))..386

Information Provided to the Assembler (§1020.30(g)) 387
Information Provided to the User (§1020.30(h)) 387
Warning Label (§1020.30(j)) 389
Repair of Components (§1020.30(d)(2)(iv)) 389
Radiographic Equipment (§1020.31) 389
Fluoroscopic Equipment (§1020.32) 390
Computed Tomography (CT) Equipment (§1020.33) 390
Cabinet X-Ray Systems (§1020.40) 393
Microwave and Radio-Frequency-Emitting Products 394
Light-Emitting Products 394
Laser Products (§1040.10) 394
Labeling Requirements (§1040.10(g)) 395
User Information (§1040.10(h)(1)) 399
Purchasing and Servicing Information (§1040.10(h)(2)) 400
Specific-Purpose Laser Products (§1040.11) 400
Medical Laser Products (§1040.11) 401
Surveying, Leveling, or Alignment Laser Products 403
Demonstration Laser Products 403
Sunlamp Products and Ultraviolet Lamps Intended for Use in Sunlamp Products (§1040.20) 403
Sunlamp and Ultraviolet Lamp Labels (§1040.20(d)) 403
Information Provided to the User (§ 1040.20(e)) 404
High-Intensity Mercury Vapor Discharge Lamps (§1040.30) 405
Sonic, Infrasonic, and Ultrasonic Radiation-Emitting Products 405
Ultrasound Therapy Products (§1050.10) 405
Discovery of a Product Defect or Failure to Comply 406
Radio Frequency Emitting Devices 407
Radio-Frequency Radiators 407
Intentional Radiators (§15.201) 408
Unintentional Radiators (§15.101) 409
Incidental Radiators (§15.13) 409
Industrial, Scientific, and Medical Equipment 409
Things to Remember 410

Chapter 24 Bringing Devices to Market in the United States 413

Premarket Notification (510(K)) 414
Premarket Notification Review Program 414
Submitting a Premarket Notification (510(k)) for a Change to an Existing Device 415
The Meaning of Intended Use 416
PMA Approval 417
PMA Supplements (§814.39) 418
Special PMA Supplement (§814.39(d)) 418
Licensing of a PMA Approval 418
Final PMA Labeling Review 419
Investigational Device Exemptions (IDE) 419
Labeling of Investigational Devices (§812.5) 420
Prohibited Practices (§§812.5(b) and 812.7) 420
Advertising of an Investigational Device (§812.7) 421
Advertising for Investigators 421
Recruiting Study Subjects 422
Manufacturing Practices for Investigational Devices (§812.20(b)(3)) 423
Exempted IDE Investigations (§812.2(c)) 423
Investigational Use of Marketed Products 424
Investigational Exemption for Intraocular Lenses (IOLS) 424
Custom Devices (§812.3(b)) 425

Humanitarian Device Exemption (HDE) 425
Things to Remember 426

Chapter 25 Good Manufacturing Practice in Labeling..429

Gmp Applications..430
GMP Exemptions..430
Custom-Device Manufacturers...431
Contract Manufacturers...431
Contract Testing Laboratories...431
Component Manufacturers...431
Remanufacturers..432
Repackers and Relabelers 432
Specification Developer 433
QSR Labeling Controls 433
Content Development and Approval (§820.30)..434
Document Controls (§820.40) 434
Document Approval and Distribution (§820.40(a))...434
Document Changes (§820.40(b)) 434
Label Integrity (§820.120(a)) 435
Purchasing (§820.50) 435
Receipt and Inspection (§§820.80 and 820.120(b))..435
Handling (§§820.60 and 820.140)...436
Storage (§§820.120(c) and 820.150) 436
Labeling and Packaging Operations (§§820.70 and 820.120(d)) 437
Nonconforming product (§820.90) 437
Distribution (§§820.80(d) and 820.160) 437
Traceability (§§820.65 and 820.120(e)) 438
Complaint Handling (§820.198) 438
Overlabeling..439
Shipping for Processing...439
Things to Remember 439

Chapter 26 Special Labeling Requirements for Specific Devices in the United States 441

Labeling For Contact Lens Solutions and Tablets (§800.12(C))..441
Use-Related Statements...441
Labeling of Articles Intended for Lay Use in Repairing and/or Refitting of Dentures (§801.405) 441
Maximum Acceptable Level of Ozone (§801.415)...443
Hearing-Aid Devices: Professional and Patient Labeling (§801.420)....................................443
Hearing Aids: Conditions for Sale (§801.421)...446
User Labeling for Menstrual Tampons (§801.430)...447
Prescription and Restricted Devices Containing or Manufactured with Ozone-Depleting Substances
(§§801.417 and 801.433) 448
User Labeling for Latex Condoms (§801.435) 450
User Labeling for Devices that Contain Natural Rubber (§801.437)......................................450
Contraceptives and Sexually Transmitted Diseases (STDs)..451
Protection from Pregnancy 452
Protection from Sexually Transmitted Diseases (STDs)...452
Electromagnetic Interference...453
Electrically Powered Wheelchairs 453
Label on the Product...454
Accompanying Product Literature 454
Implantable Pacemakers/Defibrillators 457
Cellular Telephones 457

Electronic Article Surveillance (Theft-Prevention) Systems 458
Things to Remember 458

PART IX   Development of Device Labels

Chapter 27 Reducing Labeling Problems 461

Consider the Audience.. 461
    Lay Users 461
    Professional Users 462
Organize the Information 462
Write to the Reader 464
Choose Words Carefully... 465
Avoid Clutter... 465
Writing Instructions.. 465
    Text.. 466
    Flowchart.. 467
    List... 467
Warnings and Cautions.. 468
Labeling Evaluation... 469
Limiting Liability 469
Things to Remember 470

Chapter 28 Designing Good Labeling 471

Conditions for Use... 471
    Device Labels.. 471
    Accompanying Documentation.. 471
Layout.. 472
Physical Attributes... 472
Type Fonts and Size 474
Highlighting... 474
White Space.. 474
Illustrations and Graphics... 475
Color 475
Things to Remember 476

Appendix A: U.S. Department of Commerce Medical Device Product Categories/Classification............................ 479

Appendix B: Australian Department of Health and Aging Classification of Devices 483

Appendix C: International Symbols for Medical Device Marking 491

Appendix D: Korean Medical Device Classification 497

Appendix E: Japanese Medical Device Classification 517

Appendix F: U.S. Department of Health and Human Services Medical Device Classification 539

Appendix G: Useful Web Sites.. 545

Glossary of Acronyms...547

References 553

Index 567

Part I

Negotiating a Common Understanding

1 The Global Market for Healthcare Technology

It is important to establish a clear definition of healthcare technology. Some organizations employ a broad definition, embracing most aspects of health treatment, including pharmaceuticals, devices, and medical procedures. Such a broad treatment is appropriate for many purposes; however, a more focused definition of healthcare technology needs to be adopted for this book. The Health Care Technology Institute defines healthcare technology as those products that fall into three specific categories: medical devices, diagnostic products, and healthcare information systems (Briones p. 3).

The medical device category ranges from simple products like tongue depressors to complex, highly sophisticated devices like implanted defibrillators.

Diagnostic products are those products used to detect or diagnose diseases. This category includes X-ray machines, Magnetic Resonance Imaging (MRI) systems , electrocardiographs, and automated laboratory test systems. The diagnostic products category also includes some *in vitro* diagnostic (IVD) products. IVDs are included because they are typically regulated as diagnostic devices in a number of the major markets.

Healthcare information systems are computerized systems used to keep track of patient and financial information in healthcare facilities. These information systems maintain patient records, provide data useful in treating patients, and maintain laboratory test results. In this book, all three categories will be referred to simply as "medical" or "therapeutic" devices since that is how the agencies responsible for protecting public health in the major markets classify them for regulatory purposes.

TECHNOLOGY AND THE HEALTHCARE DELIVERY SYSTEM

Traditionally, the vast majority of healthcare in industrialized nations has been delivered in acute care hospitals or physicians' offices. While a substantial portion of care will continue to be delivered in these settings, there has been a shift in the site of care for many patients. This shift is the result of many factors, ranging from the incentives embodied in the reimbursement systems and changes in physicians' practice patterns to advances in healthcare technology. New and emerging healthcare technologies have helped to make a wide range of diagnostic and treatment options possible. These include freestanding outpatient surgery and ambulatory centers, rehabilitation facilities, and even the home.

Today's patient benefits from a broad array of technologies that was not present a few years ago. Many hospitals offer interventional radiology that allows physicians to examine organs without the need for exploratory surgery. Physicians can plan and execute procedures with less invasive techniques. For example, fiberoptic technology allows surgeons to perform gallbladder surgery with only small incisions. Recovery is much quicker, reducing both the time spent in the hospital and the time spent recovering at home. Laboratory tests that once took days can now be completed by automated laboratory systems while the patient waits in the physician's office.

The care that patients receive in hospitals and physicians' offices has been transformed by healthcare technology. These same technologies have helped to facilitate treatment in nontraditional settings. Examples of technologies that have made care in such locations possible are MRI and

Computerized Axial Tomography (CAT) scans, extracorporeal shock-wave lithotripsy, outpatient intravenous therapy, cardiac catheterization, and improvements in anesthesia and surgical techniques. Clearly, technology-intensive diagnostic and therapeutic intervention has become increasingly integral to the treatment of patients.

INDUSTRY AND THE GLOBAL MARKET

The demand by patients and physicians for sophisticated treatment options has created a global market for healthcare technology that exceeded $120 billion in 2000 (Figure 1.1). Despite the international economic slowdown affecting most of the developed countries that are major consumers of healthcare technology, the demand for medical devices continues to rise. The aging population, the upgrading of healthcare systems, and the creation of a single market in Europe are some factors contributing to the continued growth of the global market.

Through innovation, intensive research and development, and aggressive marketing of its products, the medical device industry has generated strong economic growth. The United States Department of Commerce tracks the performance of the industrial sectors of the economy using the North American Industry Classification System (NAICS). There are six NAICS categories that encompass the majority of healthcare technology products. Appendix A provides a listing of the products that make up these categories.

Innovation has enabled the industry to create a world market that grew to over $120 billion in 2000. The United States (US) is still the largest market, accounting for approximately 51 percent of the total world sales of medical devices in 2000 (see Figure 1.1). Western Europe, consisting of the countries of the European Union (EU) and the European Free Trade Association (EFTA), consumed 22 percent of the world production of medical devices. Japan accounted for 12 percent of the total consumption, and the rest of the world, including the fast-growing markets on the Pacific Rim, consumed 15 percent of the total production in 2000. The United States, the EU, and Japan produce more than they consume and have trade surpluses. The rest of the markets consume more than they produce and are net importers of medical devices.

If the industry is going to continue its pattern of growth in the next decade, much of the growth will have to take place outside of the already established markets. Emerging markets in Eastern Europe and the newly industrialized countries of the Pacific Rim will account for much of the expansion of the healthcare technology market, as they will for other product sectors. As these countries attend to the growing needs of their populations, the demand for products should continue to increase.

However, these markets are changing, too. Technology has facilitated instant communication, transport, and travel, making goods and services easily accessible and affordable to the world's most isolated places. Suddenly, no place and no person is isolated. All parts of the world want the advanced products that the most developed markets already have. These products are wanted in their most advanced states of functionality, quality, reliability, service levels, and price competitiveness. Gone are the days when last year's models or used equipment taken in trade could be sold in emerging markets. Gone are the days when prices, margins, and profits abroad were generally better than those at home.

INDUSTRY COMPOSITION

Given the size of the world market in medical devices, one might conclude that large companies dominate the industry. While the industry includes large multinational corporations that employ 80,000 or more workers, it is estimated that more than 70 percent of the 11,000 plus companies

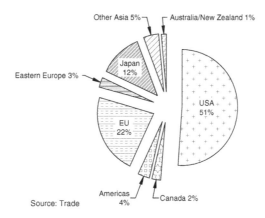

FIGURE 1.1 Global medical device market 2000. (Billion US Dollars.)

registered with the US Food and Drug Administration (FDA) employ less than 50 workers. Another 25 percent employ between 50 and 500 individuals. Only about 5 percent of FDA-registered medical device manufacturers have more than 500 employees.

Clearly this industry is dominated, not by the multinationals, but by small entrepreneurial companies that have succeeded by bringing a steady stream of innovative products to market. Increasingly, however, these companies are being required to think beyond their own domestic markets in order to survive. Reimbursement for new technology, particularly in the absence of data on long-term effectiveness, is one issue. Product regulatory cycles are another.

INDUSTRY AND REGULATION

Regulation of medical devices is important in virtually all the major markets. Europe, long encumbered by a patchwork of regulatory systems, has harmonized much of the product approval process through the implementation of three directives that align the essential requirements for safety across all of the member states. The Active Implantable Medical Device Directive (AIMD) came into force on January 1, 1993. The Medical Device Directive (MDD) came into force on January 1, 1995. The third directive covering IVD products, the In Vitro Diagnostic Device Directive (IVDD), came into force on December 7, 1998.

The European system focuses on the essential safety of the device and on proof that the device achieves the performance claimed by the manufacturer. Evaluation of devices is delegated to independent review bodies that perform their tasks on a fee-for-service basis. Decisions about the efficacy of the device are largely left to the marketplace.

Most of the power to regulate medical devices in the United States is assigned to the FDA and is designed to assure that all medical devices are both safe and effective. The type of review a medical device undergoes depends, for the most part, upon the potential risk that the device poses to the patient. For those products that pose the most significant potential risk, the FDA requires an extensive and often lengthy premarket evaluation to determine if they are safe and effective. Japanese, Australian, and Canadian regulation of healthcare technology falls some place between the EU and the FDA.

Most other countries simply require that medical devices be registered with the public health ministry prior to sale. For imported products, these countries rely heavily on product approvals from the regulatory entities in the country where the device was manufactured. Mexico is a prime example of a country that maintains a registry of medical devices.

Regardless of their other similarities or differences, one element that all of the major regulatory systems have in common is relatively specific requirements for labeling of the products. These requirements vary from specific messages that must be prominently displayed to broad general requirements such as "adequate directions for use." Whether requirements are specific or general, labeling is one area where even a minor error can lead to serious problems.

A failure on the part of a manufacturer, importer, or distributor to label a product properly can lead, and often has led, to delays in commercializing the device when prior regulatory approval is required to enter the market. At the other end of the spectrum, falsely labeling or branding (i.e., misbranding) a device, omitting required information, including false or misleading information, leaving out pertinent or material facts, or failing to exercise proper control over labeling operations in manufacturing is a violation of the law in the major markets. The regulatory authorities have a range of remedies available, including mandatory field actions to correct errors, product recalls, and civil and criminal penalties against the guilty party.

When developing labeling, it is important to remember that the fundamental function of labeling is to enable the user of a medical device to apply the device safely for the purposes intended by the manufacturer. No product can be a success if its labeling fails to meet this objective. In accomplishing its objective, a responsible party would be ill advised to ignore the relevant labeling regulations. The balance of this book is devoted to a review of these regulations in the markets that account for 94 percent of the total consumption of medical devices.

Part II

Argentina and Brazil

2 Argentina

The *República Argentina* (Argentine Republic) occupies part of the southern cone of the continent of South America. Argentina is a founding member of the Southern Cone Common Market (Mercosur). On January 1, 1995, the Mercosur member states of Argentina, Brazil, Paraguay, and Uruguay put into effect the terms of the Protocol of *Ouro Prûto*. This agreement eliminated tariffs on about 90 percent of the goods traded between the member states and erected a tariff averaging 12 percent on goods imported from elsewhere. The Mercosur alliance has created a market of 219 million people with economic activity exceeding $1 trillion US dollars in 2000. At present, Argentina, Brazil, and Uruguay have a developed system for regulating medical devices, although they have not yet put in practice the regulations necessary to harmonize product registration. Paraguay has not yet implemented a system for regulating medical devices.

BACKGROUND AND GENERAL INTENT OF THE LAW

Regulation of medical devices in Argentina derives its statutory authority from a 1985 decree of the Ministry of Health and Social Action. Decree No. 2,505/85 established the need for regulations to cover certain technical aspects that contribute to a better handling of the activities covered by the decree. In addition, the regulations are expected to enable more efficient health oversight of these activities and of the products involved. The regulations foreseen by Decree No. 2,505/85 were published as Resolution 255/95.

The Ministry of Health and Social Action has delegated the authority to promulgate and enforce the necessary regulations to the National Administration of Drugs, Foods, and Medical Technology (ANMAT).

SCOPE OF THE REGULATIONS

The Argentinean medical device regulations apply to the following biomedical products:

1. Those products that are for single use and whose labels identify them as atoxic, sterile, and free of pyrogens
2. Those products listed in Table 2.1 that may be resterilized, even if their manufacturer recommends that they are only for single use and if their labels identify them as atoxic, sterile, and free of pyrogens (The ANMAT may, from time to time, modify this list, as needs dictate.)
3. Other products included in Article 1 of the Decree No. 2,505/85 that are not included in items 1 and 2 above (These devices are listed in Table 2.2.)

The regulations cover the production, processing, sterilization, fractionation, control, storage, distribution, marketing, importation, and exportation of medical devices. They also specify the requirements and conditions for the use and reuse of covered devices.

TABLE 2.1
Products Whose Labels Identify Them as
Atoxic, Sterile, and Free of Pyrogens

Catheters for coronariography and arteriography
Counterpulsation balloons
Catheters for coronary, visceral, cerebral, or limb arteries
Metal guides
Swan-Ganz catheters with optic point
Catheters for electrophysiological studies
Carotid shunts
Retroplegia cannulas

Source: ANMAT, Resolution 255/95, Annex I.

TABLE 2.2
Products Included in Article 1 of the Degree No. 2,505/85

Material in contact with blood
Equipment and/or devices for transfusion and blood containers
Equipment and/or devices for lungs and/or for the heart
Equipment and/or devices for hemodialysis
Catheters
Needles and intravenous cannulas
Material used for the administration of fluids
Equipment and/or devices for parenteral solutions
Equipment and/or devices for peritoneal dialysis
Equipment and/or devices for irrigation
Equipment and/or devices for parenteral nutrition
Devices used for anesthesia and/or breathing
Needles for lumbar puncture
Tubes for tracheotomy
Equipment and/or devices for epidural anesthesia
Devices used for drainage and/or suction
Probes
Equipment and material for surgical drainage
Suture materials
Syringes
Gynecological implants (intrauterine contraceptive devcies [IUCD], etc.)
External medical applications.
Any other device or instrument used in medical procedures not mentioned specifically in the above mentioned points

Source: Ministry of Health and Social Action, Resolution 2,505/85, Article 1.

BRINGING DEVICES TO MARKET IN ARGENTINA

The Argentinean regulations prohibit the manufacture, distribution, importation, and exportation of medical devices that are not registered with the ANMAT. To implement this part of the regulation, the ANMAT has created and maintains a Register of Biomedical Products.

To obtain registration for a product family (each product and its variants), the manufacturer or importer must file an application providing the information listed in Annex VII of the regulations for domestically produced products or Annex VIII for imported products. The application must be signed by the head of the company or its legal representative and the company's technical director.

TABLE 2.3
List of Reference Countries

Germany
Australia
Austria
Belgium
Canada
Swiss Confederation
Denmark
Spain
United States of America
France
Israel
Italy
Japan
The Netherlands
United Kingdom
Turkey

Source: ANMAT, Resolution 255/95, Annex IX.

The ANMAT will evaluate the application within 120 days from the date of its filing. When dealing with products that are already authorized for marketing and use in any of the countries in Table 2.3, the ANMAT will evaluate the application within 40 days of filing.

For domestically fabricated products or imported products that are similar to products that are already registered and marketed in Argentina, the ANMAT has 40 days to act on the application. After that time, the company has the right to provisionally market the product until a definitive pronouncement has been made by the ANMAT. Imported products must be processed in approved plants located in any country listed in Table 2.3 or in a plant located in another country that is approved by any of the countries listed in Table 2.3 or by the ANMAT. The period of provisional sale shall be interrupted if the ANMAT requests additions or clarifications to the application. Sale of the product may not resume until the request is satisfied.

For the purposes of this part of the regulations, the word "similar," refers to that product that meets the conditions of:

- identical material, and
- equivalent clinical application and mode or form of use.

To perform the control it deems necessary, the ANMAT may require the presentation of samples of the finished product.

IMPORTING PRODUCTS

Imported products must be approved for marketing and use in the country of origin and/or the source country prior to the application for registration or importation.

When medical devices are manufactured in plants located in countries that are listed in the ANMAT's List of Reference Countries, the analytical protocols and the certificates of origin issued by the manufacturer will be accepted as valid. The ANMAT considers the countries listed in Table 2.3 to have high levels of health oversight.

If the product is manufactured in plants located in a country other than those listed in Table 2.3, but the plant is approved by the health authorities from one of these countries or by ANMAT, the analytical protocols and the certificates of origin issued by the manufacturer will be accepted as valid. However, in this case, the manufacturer must also submit duly authenticated and certified proof of the plant's approval.

The application for registration of an imported product must contain the technical details described in Annex VIII of the Argentinean regulations and the following must be enclosed:

- An authenticated copy of the Certificate of Free Sale or Certificate of Registration in the country of origin certified by an Argentina consular representative and the Ministry of Foreign Relations and Worship
- If applicable, proof(s) of commercialization on the local market of similar product(s)

The ANMAT will not certify to the National Customs Administration that an imported product is cleared to enter the market before it is registered. Any product that is not registered must be reexported or, failing that, destroyed.

TECHNICAL DIRECTOR

Companies that produce and/or import biomedical products in Argentina must be registered with the competent authorities. In order to become registered, a company must have a person on staff serving in the capacity of technical director. The technical director, along with senior management, is directly responsible for the activities or the processes required by the medical device regulations.

Only a person with certain academic credentials is considered competent to discharge the duties of technical director for an establishment devoted to the production and/or importation of medical devices. For single-use products whose labels identify them as atoxic, sterile, and free of pyrogens, the technical director must hold the degree of pharmacist. For a firm devoted to the production and/or importation of other medical devices, it is necessary for the technical director to hold the degree of pharmacist, biochemist, engineer, industrial engineer, chemical engineer, *licenciado* (bachelor) of chemistry, or a university degree associated with those mentioned that, in the opinion of the competent authority, certifies the required training.

To discharge the duties of technical director of an establishment devoted to wholesale or retail marketing of the products covered by the Argentinean regulations, it will be necessary to hold the degree of pharmacist.

In all cases, an individual seeking the position of technical director must apply to the competent authority. Only after having received authorization from the ANMAT may the individual take up his or her duties. An individual may act as a technical director in only one company at a time.

ADULTERATION AND MISBRANDING

The Argentinean regulations do not use the terms "adulteration" or "misbranding." These concepts, however, are clearly present in the regulations. Both in-country manufacturers and importers of medical devices must maintain premises that are deemed by the ANMAT to be adequate for the tasks to be performed. These premises must be adequately equipped for the production, processing, fractionation, and, if applicable, sterilization of the product. These processes include packaging and labeling of medical devices.

Those establishments involved in the production or importation of medical devices are required to maintain an archive of samples in sufficient quantity of sales units for each batch, lot, and/or series. These must be kept for six months after the date of expiration of the product. These samples permit a quality check if the ANMAT so requires. The ANMAT may, in response to a request, authorize exceptions with regard to the number of samples retained. The manufacturer must maintain the following records:

- A Fabrication and/or Fractionation Book in which are recorded the batches, lots, and/or series processed or fractionated, and the units obtained
- A Quality Control Book corresponding to the control of the raw materials, intermediate products, and finished products that records the controls performed for each production batch, lot, and/or series, and the results obtained

For a manufacturer, the Quality Control Book for raw materials must list the quantity received, the origin, and the supplier.

Importers must maintain the following records:

- A Record Book of Intake of Imported Products that reports the date of clearance to market, product type, number of batch, lot, and/or series, brand, quantity, source, and origin
- A Quality Control Book that records the controls performed for each production batch, lot, and/or series, and the results obtained

The importer's Quality Control Book must make reference, when applicable, to the appropriate protocols and certificates of origin of the device.

Those organizations that provide services such as contract sterilization to a manufacturer or importer must also maintain records that are subject to inspection by ANMAT. Depending on the nature of the service, the same records must be maintained as those for the establishment of the company to which it provides the service.

In all cases, the books must be numbered, must be approved by the ANMAT, and all the records entered in them must be signed by the technical director.

For the purposes of determining compliance with the regulations, the technical officials from the ANMAT are authorized to perform inspections with or without removal of samples from the establishments where any of the activities mentioned above are performed or presumed to be performed.

Any violation of the provisions of the regulations will be penalized in accordance with the stipulations of Law No. 16,463 and of Decree No. 341/92 without prejudice to the sanctions that may apply through the penal system.

GENERAL REQUIREMENTS FOR LABELING OF MEDICAL DEVICES

The requirements for medical device labeling in Argentina are described in Section 5 of Annex IV of the medical device regulations. This section of Annex IV sets out the requirements for packaging materials, considerations for sterile packing, the arrangement of labels, the contents of the label or labels, and the language requirements for labeling.

PACKAGING AND LABELING MATERIALS

Packaging materials shall be sufficiently strong to withstand the transport, handling, and storage conditions to which the device may be exposed. In those products that have labels bonded to the packaging, the labels may not be removable from the packaging without tearing.

CONSIDERATIONS FOR STERILE PACKING

Sterile products must be packaged in a hermetically sealed, inviolable-unit package that ensures the protection of the product against contamination and maintains sterility. When multiple-unit packages are contained in an outer package, extraction of one unit shall not affect the sterility of the remaining unit packages.

The effect of the sterilizing agent on the packaging materials, as well as on the product, must be considered when validating the sterilization process. The functional stability of the product and its packaging materials corresponds to the expiration date for the product.

ARRANGEMENT OF LABELS

Each unit package will have a printed label. The unit package may be so small that it is not practical to set out on the label attached to the unit package all of the information required by the regulations. If the unit package cannot be labeled individually, then the labeling must be placed on an outer package that contains multiple-unit packages. The label on the outer package must specify the number of unit packages contained in addition to other labeling requirements.

PACKAGE LABEL CONTENTS

The labels of the products shall include the following data:

- Full name of the product and, if applicable, the use for which it is intended
- Recommendations for use, contraindications, incompatibilities, and/or warnings, if applicable
- Production batch, lot, and/or series number
- Name and address of the manufacturer and importer, as applicable
- Name of the technical director
- Sterilization method, if applicable
- Date of sterilization and validity period, or expiration date, if applicable

The following inscriptions: "Sterile," "Single-use material," "Atoxic, sterile, and free of pyrogens," "Do not use if the packaging is not intact," or phrases of equivalent meaning, as applicable*

- Contents of the unit package
- Special storage conditions, if applicable
- Any other inscription that the ANMAT requires, considering the nature of the product and its intended use
- The inscription, "*Industria Argentina*," or "Made in [country of origin of the product]," as applicable

LANGUAGE REQUIREMENTS

For all products, the information required by the regulations shall be in Spanish in addition to any other languages.

---

* Section 3.1 of Annex VI of the Argentinean medical device regulations specifies that one of the objectives of the sterilization process for medical devices is to ensure that a sterility safety coefficient (CSE) of 10^{-6} is obtained.

DEVICE REUSE

The reuse of products marked for single use and whose labels identify them as atoxic, sterile, and free of pyrogens is prohibited by Article 2 of Argentinean medical device regulations.

Certain products, however, may be resterilized even if their manufacturer recommends that they are only for single use and if their labels identify them as atoxic, sterile, and free of pyrogens. These products are listed in Table 2.1. The ANMAT may, from time to time, modify this list, as needs dictate.

An establishment that undertakes to reuse any of the devices listed in Table 2.2 must maintain a Book of Procedures, numbered and approved by the ANMAT, where the following information is recorded:

- Patient name
- Clinical record number
- Date of procedure
- Members of surgical team
- Biomedical product(s) used with indication of type, brand, and source

The establishment where this reuse is to take place must have authorization from the ANMAT. To obtain authorization, an application signed by the medical director must be filed, demonstrating that the establishment meets all of the following requirements:

- Have adequate structure, technical capacity, and suitable personnel to comply with the conditions for the sterilization procedures established in Annex V of the medical device regulations
- Ensure that the products to be reused can be adequately decontaminated, cleaned, and sterilized, that their physical and functional characteristics are not affected, and that their use remains safe and effective
- Give written procedures for decontamination, cleaning, preparation, packaging, labeling, storage, and destruction, as well as the controls that must be performed

The labels to be placed on devices prepared for reuse must include the following data:

- Name of the institution
- Complete description of the product
- Date of sterilization

DEVICE REPROCESSING

The Argentinean medical device regulations provide that the devices discussed in the previous section may be reprocessed if it is possible to guarantee the equivalent conditions of functionality and sterility as those of the original product. However, for purposes of the medical device regulations, "reprocessing" means the process applied to an unused product whose packaging has not been opened or damaged.

THINGS TO REMEMBER

The manufacture or importation of medical devices in Argentina is governed by the regulations published as Resolution 255/95. The Argentinean Ministry of Health and Social Action has delegated the authority to promulgate and enforce the necessary regulations to the ANMAT.

The Argentinean medical device regulations apply to (1) those products that are for single use and whose labels identify them as atoxic, sterile, and free of pyrogens; (2) those products listed in Table 2.1 that may be resterilized; and (3) devices that are listed in Table 2.2. The regulations cover the production, processing, sterilization, fractionation, control, storage, distribution, marketing, importation, and exportation of medical devices. They also specify the requirements and conditions for the use and reuse of covered devices.

The Argentinean regulations prohibit the manufacture, distribution, importation, and exportation of medical devices that are not registered with the ANMAT. To obtain registration for a product family (each product and its variants), the manufacturer or importer must file an application providing the information listed in Annex VII of the regulations for domestically produced products or Annex VIII for imported products.

While the regulations do not use the terms "adulteration" or "misbranding," these concepts are clearly present in the regulations. Both in-country manufacturers and importers of medical devices must maintain premises that are adequately equipped for the production, processing, fractionation, and, if applicable, sterilization of the product. These processes include packaging and labeling of medical devices. Any violation of the provisions of the regulations will be penalized in accordance with the stipulations of the law without prejudice to the sanctions that may apply through the penal system.

The minimum requirements for medical device labeling in Argentina are described in the medical device regulations. These include requirements for packaging materials, considerations for sterile packing, the arrangement of labels, the contents of the label or labels, and language requirements for labeling. Spanish is required on all labels.

Those establishments involved in the production or importation of medical devices are required to maintain an archive of samples in sufficient quantity of sales units for each batch, lot, and/or series, to kept for six (6) months after the date of expiration of the product. These samples permit a quality check if the ANMAT so requires. The ANMAT may, in response to a request, authorize exceptions with regard to the number of samples retained.

The reuse of products marked for single use and whose labels identify them as atoxic, sterile, and free of pyrogens is prohibited. Certain products, however, may be resterilized even if their manufacturer recommends that they are only for single use. An establishment that undertakes to reuse any of these devices must obtain permission from the ANMAT and must follow the procedure set forth in the regulations. Devices prepared for reuse must bear the minimal labeling specified in the regulations.

3 Brazil

The *República Federativa do Brasil* (Federative Republic of Brazil) is the largest country in South America both in land mass and population. It is also the third largest market for medical devices in the Western hemisphere, behind the United States (US) and Canada. Brazil, along with Argentina, Paraguay, and Uruguay, is a founding member of the Southern Cone Common Market (Mercosur).

BACKGROUND AND GENERAL INTENT OF THE LAW

On December 31, 1998, the Brazilian President signed a Provisional Measure #1791, which created the *Agência Nacional de Vigilância Sanitária* (Brazilian National Health Vigilance Agency) or the ANVISA. The ANVISA replaced the National Secretary of Sanitary Vigilance (SVS), which was established in 1975 under Public Law #6360. The ANVISA has enforcement powers similar to the US Food and Drug Administration (FDA) that enable it to regulate the manufacture, import, and distribution of pharmaceuticals and medical devices. The ANVISA was created as a public company under contract to the Ministry of Health. The ANVISA Director is nominated by the President of Brazil and serves a five-year term. The law provides for a formal separation between the ANVISA, which is responsible for all sanitary and health inspection, and the Ministry of Health, which is responsible for public policy related to health issues.

As a public company, the ANVISA is expected to cover the cost of its operation through user fees. The ANVISA charges a fee for services including establishment registrations, good manufacturing practices (GMP) reviews, and product registrations. Product registrations are usually good for five years. Establishment registrations and GMP reviews are generally good for one year.

SCOPE OF THE REGULATIONS

The Brazilian law mandates registration for the following products (Brazil p.2):

- Pharmaceuticals for human use, their active ingredients, and other related materials, processes and technology
- Cleaners and sanitation products for decontamination and hygiene of hospitals, clinics, public transportation, and homes
- Diagnostic kits, reagents, and other items intended for a similar purpose
- Equipment and materials, devices for hospital, medical, dental, blood banks, laboratory use, and image diagnostics
- Immune-biological products and their active ingredients
- Blood and its derivatives
- Organs and human and veterinary tissues for transplants or reconstructive surgery
- Radioisotopes for *in vitro* diagnostics (IVDs), radio-pharmaceuticals, and radioactive products used in diagnostics and therapy
- Any and all products posing any health risks, obtained by genetic engineering, processed or submitted to radiation sources

The ANVISA also regulates the installations, equipment, technologies, environment, and procedures involved in manufacturing these items, their disposal, and respective residues.

BRINGING DEVICES TO MARKET IN BRAZIL

It is virtually impossible to bring a device to market in Brazil without a local presence — a local manufacturing unit, office, distributor, or in-country agent.

PRODUCT REGISTRATION

Product registration in Brazil can be a laborious exercise for both domestic and foreign manufacturers. For imported products, the application process must be handled by the company's local office or through an in-country agent. The registration is valid for five years and can be renewed continuously for the same period.

Manufacturers should take a series of measures in order to guarantee its rights to the registration. These include (Brazil p. 13):

- Appling for registration of the trademarks and patents with the National Industrial Property Institute (INPI) using a local law firm.
- For imported products, establishing a solid contract with the distributor to protect the manufacturer's rights, including the ownership of its registration.
- For a foreign manufacturer, establishing specific clauses in the contract to transfer the ownership of the registration from the agent to the manufacturer. A transfer of ownership can occur only if the foreign company opens an office or plant in Brazil, since no registration can be transferred overseas.

Steps to protect intellectual property are extremely important because manufacturers must disclose many of the technical details of the product to the ANVISA. For example, product drawings and lists of components must be included with the registration application (Brazil p. 13).

By law, the registration process is to be completed within 90 days after the registration is requested. However, the product registration process often takes more than one year (Brazil p. 13). Should the process take longer than three months, importers and producers are allowed to use a protocol number provided by the Brazilian authorities to distribute their products in Brazil. There is some risk of product liability claims should the manufacturer's product be found to be unsafe by the ANVISA.

The following information is required for registration of medical devices (Brazil pp. 13–14):

- Name of company
- Address
- Product name
- Product description including indications and contraindications from the instructions for use
- Final product drawings
- List of components/materials with particular attention to materials contacting tissue, including the material name, trade name, and the component(s) of the device where the material is used
- Summary of the manufacturing process (flow chart)
- Labels/directions for use
- Description of the sterilization process and sterilization parameters (if applicable)
- Quality control testing (protocols and reports covering electrical, mechanical, and electromagnetic qualification)
- Clinical publications (rationale and published literature supporting no need for a clinical trial, or the full clinical protocol and report if new features are introduced where there is no supporting data or published literature)
- Product Brochure (catalog page)

IMPORT OF MEDICAL PRODUCTS

Any product that comes in contact with the human body is controlled by the ANVISA. This includes pharmaceuticals, vitamins, cosmetics, and medical equipment/devices. These products can only be imported and sold in Brazil if (Brazil p. 12):

- The foreign manufacturer establishes a local Brazilian manufacturing unit or local office, or
- The foreign manufacturer appoints a Brazilian distributor who is licensed by the Brazilian authorities to import and distribute medical products, and
- The product is registered with the ANVISA.

All products applied to the skin, injected into the body, inserted into the eye (e.g., contact lens and cleaning liquids), or having any other medical application must be registered.

For distributors to be able to register, import, and offer for sale medical products in Brazil, they must possess the proper permits issued by the sanitary authority of the state where they conduct business as well as a similar permit issued by the federal government (Brazil p. 12). The distributor must have a contract with a qualified technician (such as a chemist, pharmacist, engineer, etc., according to different types of industry) who takes responsibility for the application. This document is called "Terms of Technical Responsibility," and is signed by the professional.

When registering a product, the in-country agent must submit the following documents (Brazil, p. 15):

- An application form obtained from the ANVISA.
- An original copy of the machine-stamped bank slip that serves as proof that the applicable registration fees have been paid.
- A copy of the distributor's *Alvará de Funcionamento*—a trading permit granted by the state sanitary authorities. This allows the company to import, distribute, store, and sell the product registered with the ANVISA.
- A copy of the distributor's *Autorização de Funcionamento*—a permit like the *Alvará de Funcionamento*, but granted by the federal government.
- A quality-control certificate issued by a recognized certification entity. (The distributor must have a contract with a local Brazilian laboratory to produce the quality-control certificate for each of the products to be registered. This laboratory must be an "OCC–*Organismo de Controle e Certificação*" [Control and Certification Laboratory]. This is an official registered certification organization registered with the Brazilian Ministry of Health.)
- A technical report on the product describing the instructions, directions, cautions, etc., and, if applicable, the components of the formula.
- A label sample, brochures, and pertinent information about the product translated into Portuguese.
- For a product not clearly mentioned in the Brazilian law, it is mandatory to provide information about its utilization in order to demonstrate its efficacy and safety.
- A notarized copy of the registration granted to the products at the country of origin (or a copy of the Free Sale Certificate).
- A copy of the legal document in which the manufacturer authorizes its distributor to trade and distribute the products
- For medical equipment, all documents demonstrating product safety, showing the country of origin, and detailing the equipment's inner structure (exploded view), as well as any user manual.

MISBRANDING

The Brazilian Customer Protection (BCP) code came into force on September 12, 1990. The BCP requires that product labeling provide the consumer with correct, clear, precise, and easily readable information about the product's quality, quantity, composition, price, guarantee, shelf life, origin, and risks to the consumer's health and safety. Failure to provide this information or providing false or misleading information is a violation of the BCP and can lead to penalties being assessed against the manufacturer or the manufacturer's in-country agent.

GENERAL LABELING PROVISIONS

Brazilian law authorizes the ANVISA to regulate the labeling of all medical devices imported into or distributed in Brazil.

LANGUAGE REQUIREMENTS

Portuguese is the official language of Brazil. For imported products, the instructions, directions, cautions, labels, brochures, and any other pertinent information about the products must be translated into Portuguese. The Portuguese translation may appear on the labels and in accompanying documents along with the same material in other languages.

MINIMUM LABELING REQUIREMENTS

The product labeling must provide the intended user with correct, clear, precise, and easily readable information about the product's quality, quantity, composition, price, guarantee, shelf life, origin, and risks to the consumer's health and safety. The instructions for use must include the indications and contraindications for use of the product.

Because metric units are the official units of measure in Brazil, products should be labeled in metric units or show a metric equivalent.

THINGS TO REMEMBER

In Brazil, medical devices must be registered with the quasi-public organization health department—the ANVISA—prior to sale. As a public company, the ANVISA is expected to cover the cost of its operation through user fees.

Brazilian law mandates registration for a wide variety of products including pharmaceutical, cleaning and sanitation products, IVD kits and reagents, equipment used in hospitals, dentistry, blood banks, and diagnostic imaging, organs and human and veterinary tissues for transplants or reconstructive surgery. The ANVISA also regulates the installations, equipment, technologies, environments and procedures involved in manufacturing these items, their disposal, and respective residues.

Product registration in Brazil can be a laborious exercise for both domestic and foreign manufacturers. For imported products, the application process must be handled by the company's local office or through an in-country agent. The registration is valid for five years and can be renewed continuously for the same period.

For distributors to be able to register, import, and offer for sale medical products in Brazil, they must possess the proper permits issued by the sanitary authority of the state where they conduct business, as well as a similar permit issued by the federal government.

To obtain registration, the manufacturer or his in-country agent must submit a dossier on the product. The manufacturer should take steps to protect its intellectual property because many of the technical details of the product must be disclosed during the registration process. For example, product drawings and lists of components must be included with the registration application.

Portuguese is the official language of Brazil. For imported products, the instructions, directions, cautions, labels, brochures, and any other pertinent information about the products must be translated into Portuguese.

Part III

Australia

4 The Therapeutic Goods Act of Australia

The regulation of medical devices in the Commonwealth of Australia dates from 1966, when the Australian Parliament passed the Therapeutic Goods Act 1966. This act placed responsibility for control of therapeutic goods under the Therapeutic Goods Administration (TGA) within the Australian Department of Health and Aging (DOHA).* On February 15, 1991, the Therapeutic Goods Act 1966 was superceded by the Therapeutic Goods Act 1989.** The Therapeutic Goods Act 1989 placed additional controls on medical devices (called therapeutic devices in the act) that are perceived to present a significant risk to the patient or user.

Under the provisions of the Therapeutic Goods Act 1989, therapeutic goods are goods that are in any way represented to be, or are likely to be used, for (Therapeutic Goods Act pp. 9–10):

- the prevention, diagnosis, cure, or alleviation of a disease, ailment, defect, or injury in a person or animal;
- influencing, inhibiting, or modifying a physiological process in a person or animal;
- testing the susceptibility of a person or animal to a disease or ailment;
- influencing, controlling, or preventing conception in a person;
- testing for pregnancy in a person; or
- the replacement or modification of parts of the anatomy in a person or animal.

Goods that are used as ingredients or components in the manufacture of therapeutic goods are included within the definition for purposes of the law. Goods used as a container, or as parts of a container, for therapeutic goods are also covered by the law.

Therapeutic devices are therapeutic goods "consisting of an instrument, apparatus, appliance, material, or other article (whether used alone or in combination), together with any accessories or software required for its proper functioning, which does not achieve its principal intended action by pharmacological, chemical, immunological, or metabolic means though it may be assisted in its function by such means" (Therapeutic Goods Act p. 9).

Devices may be excluded from regulation under the Therapeutic Goods Act 1989 by an order of the DOHA Secretary (Therapeutic Goods Act pp. 16–17). The Secretary may declare that particular devices or classes of devices are not therapeutic goods when they are (a) clearly not used as therapeutic goods, or (b) when labeled in a particular way are not therapeutic goods. The Secretary may, upon receipt of a written application, take action to exclude a particular device or classes of devices from regulation under the Therapeutic Goods Act 1989.

On the other hand, the DOHA Secretary may decide that particular devices or classes of devices are therapeutic goods when used or labeled in a particular way (Therapeutic Goods Act pp. 8–9). In either case, the Secretary makes his or her decisions known by publishing a declaration in the

---

* The responsibility for administering the act was assigned to the Australian Commonwealth Department of Community Services and Health. This department underwent several name changes and adjustments to its portfolio in the intervening years. In November 2001, the official name of the department was changed to the Department of Health and Aging (DOHA).

** Therapeutic Goods Act 1989, as amended. Commonwealth of Australia (March 21, 2002).

Commonwealth of Australia Gazette specifying the conditions (i.e., the labeling requirements) under which particular devices are, or are not, to be considered therapeutic devices.

A summary of the devices that are regulated as therapeutic devices, along with those that are exempted or are declared not to be therapeutic goods, is given in Appendix B.

BACKGROUND AND GENERAL INTENT OF THE LAW

In the 1980s, serious faults in a range of therapeutic devices—including heart valves, intraocular lenses (IOLs), intrauterine contraceptive devices (IUCDs), and powered drug-infusion systems—lead to an alarming number of deaths and serious injuries. These problems prompted the Australian Government to tighten controls over therapeutic devices. In 1987, the Australian government amended the Therapeutic Goods Act 1966 to have all therapeutic devices sold in Australia recorded on a National Register of Therapeutic Goods. Devices were classified as either "registered" or "listed" products for purposes of being recorded on the National Register of Therapeutic Goods. Products already on the market were "grandfathered" onto the register. After the grandfathering process, all products classified as registered products would undergo a premarket evaluation for quality, safety, and effectiveness. A form was to be completed on any new listed products for inclusion on the National Register of Therapeutic Goods.

Included with the 1987 amendment was the enactment of Schedule 8 Part 5(h) of the Customs Act 1901,* which called for all therapeutic devices to receive import approval (i.e., an import certificate) prior to their arrival in Australia. This regulation did not cover goods manufactured in Australia, although the Government asked for these companies to comply with the same requirements that imported devices had to meet. If a device was not listed on the National Register of Therapeutic Goods, it was held by Australian customs officials until the proper documentation was received.

All therapeutic devices must have been listed on the National Register of Therapeutic Goods unless they were registered devices or devices exempted by either the regulations or a Therapeutic Goods Order (TGO). Exempted devices were mainly nonimplantable diagnostic devices that did not require sterilization, and nonpowered surgical instruments that were supplied in a nonsterile condition.

With the passage of the Therapeutic Goods Act 1989, a second grandfathering of existing products was undertaken that effectively made import certificates obsolete, and all products again had to be recorded on a register. The new name for this was the Australian Register of Therapeutic Goods (ARTG). The process remained the same as the 1987 amendment, except that the sponsor was charged a fee to submit registered products for evaluation and incurred an annual fee to keep both "listed" and "registered" devices on the ARTG.

The Therapeutic Goods Act 1989 is intended to provide for "the establishment and maintenance of a national system of controls relating to the quality, safety, efficacy, and timely availability of therapeutic goods" (Therapeutic Goods Act p. 12). The act applies to therapeutic devices that are used in Australia, regardless whether they are produced domestically or are imported. The provisions of the Therapeutic Goods Act 1989 also apply to devices produced for export from Australia.

THE THERAPEUTIC GOODS AMENDMENT OF 2002

In June 1998, Australia signed a Mutual Recognition Agreement (MRA) with the European Union (EU) covering medical devices. On March 21, 2002, the Australian Parliament passed the Therapeutic Goods Amendment (Medical Devices) Bill 2002. This bill provides for the introduction of an

---

* Customs Act 1901, as amended. Commonwealth of Australia (August 1, 1994).

internationally harmonized framework for the regulation of medical devices in Australia. The EU-MRA came into force on October 5, 2002. In this process of preparing the necessary regulations, the TGA will complete their review of Australian and international standards and gazette the standards that demonstrate compliance with the essential principles set out in the 2002 amendment to the Therapeutic Goods Act. While the new regulations have not been issued at the time of this book's publication, Chapter 6 provides a preview based on material already published by the TGA.

SCOPE OF THE TGA REGULATIONS

Section 6 of the Therapeutic Goods Act 1989 specifies that the provisions of the Act apply to the activities of individuals or corporations in the course of, or in preparation for, trade or commerce among the states and territories and between Australia and other countries. The provisions of the Therapeutic Goods Act 1989 are intended to supercede any law of a state or territory, other than laws identified in the regulations (Therapeutic Goods Act p. 13).

Section 58 of the Therapeutic Goods Act 1989 reserves for the DOHA Secretary the power to issue export certificates for devices intended for therapeutic use in humans. A state or territory is explicitly prohibited from issuing export certificates for devices intended for therapeutic use in humans (Therapeutic Goods Act p. 116).

THE REGULATIONS

Section 63 of the Therapeutic Goods Act 1989 specifies that the Governor-General of Australia may make regulations prescribing matters necessary or convenient for carrying out the provisions of the act (Therapeutic Goods Act p. 124). Particularly with regard to labeling of therapeutic devices, the regulations may:

- prescribe requirements for informational material that is included with the therapeutic device and
- prescribe the requirements for advertising therapeutic devices.

The regulations are published in the *Commonwealth of Australia Gazette*. Once a final regulation or TGO is published in the *Gazette*, it has the force of law.

Labeling regulations for therapeutic devices promulgated by the Secretary of Health and Aging are found in Therapeutic Goods Order No. 37 (TGO 37).

LABELS AND LABELING

Section 3(1) of the Therapeutic Goods Act 1989 defines a label as a display of printed information, and includes:

- Information displayed on or attached to the therapeutic device
- Information displayed on or attached to a container or primary pack in which the therapeutic devices are supplied
- Information sealed within the container or pack, but excluding any label that is intended to be returned by the consumer to the supplier or manufacturer as a record of purchase

The primary pack is the complete package in which a device or devices and their container are supplied to the customer. The container means the vessel, bottle, tube, ampule, syringe, vial, sachet, strip pack, blister pack, wrapper, cover, or other similar article that immediately covers a therapeutic good.

The label constitutes an important element of the "presentation" of a therapeutic device. The presentation also includes the packaging and any advertising or other informational material associated with the device.

LABELING AND ADVERTISING

Advertising is defined in Section 3(1) of the Therapeutic Goods Act 1989 to include "any statement, pictorial representation or design, however made, that is intended, whether directly or indirectly, to promote the use or supply" of a therapeutic device (Therapeutic Goods Act p. 1).

Normally, the TGA does not review product information, including promotional material that is not supplied with the device. The sponsor is responsible for ensuring that this information is entirely consistent with the information supplied with the therapeutic device and that additional claims are not made (DR4 p. 58).

The TGA may, by written notice, require the sponsor of a therapeutic device to provide information concerning the advertising material relating to the device. Failure to provide the requested information in a reasonable time period or providing material that is false or misleading will subject the guilty party to a fine (Therapeutic Goods Act p. 18).

ADULTERATION AND MISBRANDING

The Therapeutic Goods Act 1989 does not use the terms "adulteration" or "misbranding." However, these concepts do appear in the Australian law. Conveying by way of sale, exchange, gift, lease, loan, hire, hire-purchase, sample, or advertising (whether free of charge or otherwise) an adulterated or misbranded therapeutic device is an offense under the Therapeutic Goods Act 1989. A person who is found guilty of "supplying" an adulterated or misbranded therapeutic device in violation of the Therapeutic Goods Act 1989 is subject to a substantial fine.

In general, a therapeutic device can be considered adulterated if the sponsor of the device knowingly or recklessly breaches any of the conditions for registration or listing of the device (Therapeutic Goods Act pp. 33–36). These include: the device not being safe for the purposes for which it is to be used, failure to comply with appropriate standards, or the device not being manufactured under acceptable quality-control procedures.

Section 3(5) of the Therapeutic Goods Act 1989 makes it a crime to supply a therapeutic device that "is capable of being misleading or confusing as to the contents or proper use" (Therapeutic Goods Act p. 11). Section 3(5) of the Therapeutic Goods Act 1989 deems a therapeutic device misbranded if:

- the label states or suggests that the device contains ingredients, components, or characteristics that it does not possess;
- The name applied to the device is the same as the name applied to another therapeutic device that is supplied in Australia and the device contains different therapeutically active ingredients;
- the label of the device does not declare the presence of a therapeutically active ingredient;
- the form of presentation of the therapeutic device may lead to unsafe use of the device, or suggests a purpose other than those for which the device has been approved in Australia; or
- the requirements that are specified in the regulations applicable to the device have not been met.

FALSE OR MISLEADING LABELING

The labeling of a therapeutic device must not be false or misleading, and labeling need not be untrue, forged, fraudulent, or deceptive to be so viewed. A word, statement, or illustration may be strictly true, yet be misleading to the customer. Misleading means labeling that leads "to unsafe use of the goods or suggests a purpose that is not in accordance with conditions applicable to the supply of the goods in Australia" (Therapeutic Goods Act pp. 11–12). A false impression may be created not only by false or deceptive statements, but also by ambiguity and misdirection. For example, labeling of a therapeutic device that states that "it is reported" that the particular device has the effect claimed may constitute misbranding when, in fact, those results were not obtained. Equivocation in the language is in itself misleading and ambiguous, even though it may not be technically false and is even, perhaps, literally true.

Failure to inform the customer of material facts may render the device misbranded (see Chapter 5) just as much as a blatantly false or exaggerated claim. A label that is silent on important considerations may be just as deceptive as one that makes exaggerated claims. To illustrate this point, consider the increased risk of toxic shock syndrome (TSS) that results from the use of menstrual tampons. Failure of a manufacturer to inform the consumer about TSS would constitute the omission of material facts about the product.

ADVERTISING AND PROMOTION

The advertising of therapeutic devices is regulated as part of the labeling of the device under the Therapeutic Goods Act 1989. Section 22 makes it a crime, punishable by a fine, to make false and misleading statements about a therapeutic device or its status under the regulations. Specifically, the sponsor of a therapeutic device may not knowingly and recklessly advertise the device for an indication other than those accepted in the application for inclusion of the device on the ARTG. In relation to a therapeutic device, indication means the specific therapeutic purpose for which it is to be used. The Therapeutic Goods Act 1989 takes a broad view by including advertising "by any means" (Therapeutic Goods Act pp. 34–35) under the control of the regulations. A person may not knowingly or recklessly represent that a therapeutic device is included in the ARTG when it is not. Likewise, a person may not represent a therapeutic device as exempt if it is not. Finally, a person may not represent a therapeutic device as being included in one part of the ARTG when it is included on another part.

In Australia, all advertising and other generic information* about therapeutic goods that is provided directly to the public must comply with provisions of the Therapeutic Goods Act 1989, the Therapeutic Goods (Medical Devices) Regulations 2002 (TGR 2002), and the Therapeutic Goods Advertising Code (TGAC). A new TGAC came into effect on April 19, 2000. The object of the TGAC is to ensure that the marketing and advertising of therapeutic goods to consumers is conducted in a manner that promotes the quality use of therapeutic goods, is socially responsible, and does not mislead or deceive the consumer (TGAC p. 1). The TGAC does not apply to advertising directed exclusively to (TGAC p. 9):

- medical practitioners, psychologists, dentists, veterinary surgeons, pharmacists, physiotherapists, dietitians, scientists working in medical laboratories, nurses;
- persons who are:

---

* Generic information, in relation to therapeutic goods, includes any statement, pictorial representation, or design, however made, about the composition, properties, or other characteristics of therapeutic goods, but does not include (a) an advertisement about the goods; (b) generic information included in, or associated (directly or indirectly) with, an advertisement about therapeutic goods; or (c) bona fide news.

- engaged in the business of wholesaling therapeutic goods or
- purchasing officers in hospitals; or
- herbalists, homoeopathic practitioners, chiropractors, naturopaths, nutritionists, practitioners of traditional Chinese medicine or osteopaths registered under a law of a state or territory of Australia.

Advertising for therapeutic goods directed exclusively to healthcare professionals are governed by industry codes of practice and are not subject to the TGAC.

Advertising for therapeutic goods to the consumer must (TGAC p. 3):

- comply with the statute and common law of the Commonwealth of Australia, its states and territories; and
- contain correct and balanced statements only and claims that the sponsor has already verified.

Advertising for therapeutic goods to the consumer must not (TGAC pp. 3–4):

- be likely to arouse unwarranted and unrealistic expectations of product effectiveness;
- be likely to lead to consumers self-diagnosing or inappropriately treating potentially serious diseases;
- mislead directly, by implication, or through emphasis, comparisons, contrasts, or omissions;
- abuse the trust or exploit the lack of knowledge of consumers or contain language that could bring about fear or distress;
- contain any matter that is likely to lead persons to believe:
  - that they are suffering from a serious ailment, or
  - that harmful consequences may result from the therapeutic good not being used (Sunscreen preparations are exempted from this requirement if the claims made in the advertising are consistent with current public health messages.);
- encourage inappropriate or excessive consumption;
- contain any claim, statement, or implication that the product is infallible, unfailing, magical, miraculous, or that it is a certain, guaranteed, or sure cure;
- contain any claim, statement, or implication that the product is effective in all cases of a condition
- contain any claim, statement or implication that the goods are safe, that their use cannot cause harm, or that they have no side-effects; or
- be directed to minors, except in certain special cases such as condoms and personal lubricants, bandages and dressings, and devices for management of chronic conditions under medical supervision.

Under Regulation 9 of the Therapeutic Goods Regulations 1990, sponsors of therapeutic goods are prohibited from making use of a restricted representation (including its use on the label of goods or in information included in the package in which the goods are contained). The diseases, conditions, ailments, and defects that are included in this restriction are listed in Table 4.1. A sponsor may apply to the DOHA Secretary for an exemption to this regulation. The sponsor must provide a justification for use of the restricted representation within their advertising material and meet the public interest criteria outline in Appendix 6 of the TGAC. If the Secretary is satisfied that the representation is accurate, balanced, and is not misleading or likely to be misleading, an exemption will be granted. The DOHA Secretary may impose conditions on the exemption (TGA, Restricted p. 2).

TABLE 4.1
Diseases, Conditions, Ailments, and Defects for Which Advertising of Serious Forms is Restricted

Cardiovascular diseases
Dental and periodontal diseases
Diseases of joint, bone, and collagen, and rheumatic disease
Diseases of the eye or ear likely to lead to blindness or deafness
Diseases of the liver, biliary system, or pancreas
Endocrine diseases and conditions including diabetes and prostatic disease
Gastrointestinal diseases or disorders
Hematological diseases
Infectious diseases
Immunological diseases
Mental disturbances
Metabolic disorders
Musculo-skeletal diseases
Nervous system diseases
Poisoning, venomous bites, and stings
Renal diseases
Respiratory diseases
Skin diseases
Substance dependence
Urogenital diseases aconditions

Source: TGA, Restricted p. 1.

Advertising for therapeutic goods must contain the following (TGAC pp. 5–6):

- The trade name of the therapeutic goods
- A reference to the indication(s) that is approved for the use of the therapeutic goods
- Where applicable, a list of ingredients or the following statement prominently displayed or communicated:

ALWAYS READ THE LABEL

except in the case of direct marketing and Internet marketing, where the catalogue or Internet communication must contain a full list of active ingredients

- Words to the following effect, prominently displayed or communicated:

USE ONLY AS DIRECTED

- When the advertising contains claims relating to symptoms of diseases or conditions:

IF SYMPTOMS PERSIST, SEE YOUR DOCTOR/HEALTHCARE PROFESSIONAL

- In the case of therapeutic goods that must be obtained directly from, or on the order of, a health professional, the following statements should be prominently displayed or communicated:

YOUR [APPROPRIATE HEALTHCARE PROFESSIONAL] WILL ADVISE YOU WHETHER THIS PREPARATION [PRODUCT NAME] IS SUITABLE FOR YOU/YOUR CONDITION

BRINGING DEVICES TO MARKET IN AUSTRALIA

Under the provisions of Section 17 of the Therapeutic Goods Act 1989, the TGA is required to maintain the ARTG, for the purpose of compiling information on therapeutic goods used on humans. The ARTG is divided into two parts—registered goods and listed goods. All therapeutic goods, including all therapeutic devices, must be listed on the ARTG unless they are exempted in the regulations or by written notice from the DOHA Secretary. The goods that are exempted under this provision are listed in Table C.3 in Appendix B. Section 20 of the Therapeutic Goods Act 1989 makes it a crime, punishable by a fine, for a person to knowingly and recklessly supply a therapeutic device for human use in Australia which is not (a) on the ARTG or (b) exempt from regulation under Section 19 of the Therapeutic Goods Act 1989.

When a therapeutic device is added to the ARTG, the TGA assigns it a unique registration number. When a group of therapeutic goods is added to the ARTG, a listing number is assigned to the group of therapeutic goods. Section 20(2) of the Therapeutic Goods Act 1989 makes it a crime, punishable by a fine, for the sponsor of therapeutic goods recorded on the ARTG to supply them in Australia unless the registration number is properly displayed on the label.

The Therapeutic Goods Act 1989 provides an exemption to this requirement for therapeutic devices that are listed on the ARTG. Under Section 20(2)(e) of the Therapeutic Goods Act 1989, therapeutic devices that are listed on the ARTG are not required to have the listing number on the label. Registered devices are always required to bear the registration number on the label.

A person may not place on the label of a therapeutic device, or on a container or package that contains a therapeutic device, a number that purports to be a registration number if that number is not the registration number assigned to that device and sponsor. To do so renders the device misbranded under Section 22(1) of the Therapeutic Goods Act 1989. A person found guilty of placing a false registration or listing number on a therapeutic device is subject to a fine.

The DOHA Secretary may, by written notice, require the sponsor of a therapeutic device to provide the following information or documents on a registered or listed device or on a device that is being considered for registration or listing (Therapeutic Goods Act p. 61):

- The formulation of the device
- The composition of the device
- The design specifications of the device
- The manufacturer who produced the device
- The presentation (labeling, packaging, advertising, etc.) of the device
- The safety and efficacy of the device for the purposes for which it is to be used
- The conformity of the device to the standards applicable to the device, or the conformity to the requirements for advertising applicable to the device
- Any additional information prescribed in the regulations, such as the commercial history of the device (DR4 p. 55)

For registered therapeutic devices, the Secretary may also require the following information or documents:

- The quality of the devices
- The method and place of manufacture or preparation of the device and the procedures employed to ensure that proper standards are maintained in the manufacture and handling of the device
- The regulatory history and status of the device in another country

Once a therapeutic device is included on the ARTG, it remains until the registration or listing is cancelled. The registration or listing can be cancelled either by the sponsor applying in writing to the ARTG, or once a year when the annual fee renewal notice is sent to the sponsor. The Secretary of Human Services and Health may cancel the registration or listing by a written order if it appears that a failure to do so would create an imminent risk of death, serious illness, or injury. The Secretary may also cancel the registration or listing of a device if the device is found to be adulterated or misbranded. If the Secretary revokes the registration or listing of a device, the sponsor must, within a reasonable time, as is specified in the order, inform the public or a specified class of persons about the cancellation. The sponsor may be required to make the notification in a specific way. The sponsor may also be required to take steps to recover any of the devices that have been distributed. A person who knowingly or recklessly refuses to comply with the order of the Secretary is guilty of an offense and is subject to a penalty.

The ARTG is not open for public inspection. However, the sponsor of a device may request a copy of its entry in the ARTG. The DOHA Secretary may, following a request from the sponsor, alter the entry for a device if the entry contains information that is incomplete or incorrect. The DOHA Secretary must alter the entry on the ARTG when requested if the only effect of the change would be to reduce the class of persons for whom the device is suitable or to add a warning or precaution. The warning or precaution may not include any comparison of the device with any other therapeutic goods by reference to quality, safety, or efficacy. The Secretary may alter the entry on the ARTG if the TGA is satisfied that the request does not indicate any reduction of the quality, safety, or efficacy of the device for the purpose for which it is intended to be used (Therapeutic Goods Act pp. 67–69).

REGISTRATION OF THERAPEUTIC DEVICES

Under the Australian system, therapeutic devices that are perceived to present a significant risk of death or serious injury for a patient or user are required to be registered with the TGA. Registration means that these devices are subject to a premarket evaluation of their quality, safety, and effectiveness; they can be sold only after they are approved by the TGA. Once approved, these devices are recorded on the registered goods part of the ARTG. At the time of this book's publication, the devices listed in Table 4.2 required registration.

The person who wishes to manufacture, import, or export a device requiring registration must make a written application to the TGA and pay an appropriate fee. The TGA evaluates the application to determine (Therapeutic Goods Act p. 41):

- Whether the quality, safety, and efficacy of the therapeutic device for the purposes for which the device is to be used have been established.
- Whether the presentation of the device is acceptable. In this case, presentation means the way the therapeutic device is presented, including the name of the device, the labeling and packaging of the device, and any advertising or other informational material associated with the device.* Specifically, the submission should include (DR4 pp. 57–58):
  - draft or sample labels complying with TGO 37;
  - user manuals, programming manuals, programming instructions or equivalents;
  - physician manuals and implanting instructions;
  - technical manuals, or similar manuals, including specifications;

---

* Product information that is to be provided separately from the product, for whatever purpose, is not reviewed during the evaluation process. The sponsor is responsible for ensuring that this information is consistent with the information approved during the evaluation.

TABLE 4.2
Registrable Therapeutic Devices

1. Active implantable Medical Devices (AIMD), including (DR4 pp. 85–93):
   - Implantable central nervous system pulse generators
   - Diaphragmatic/phrenic nerve stimulators
   - Carotid sinus stimulators
   - Intracerebellar/subcortical and deep brain stimulators
   - Cerebellar stimulators
   - Vegal nerve stimulators
   - Implantable drug infusion pumps
   - Cardiac pacing systems
   - Permanently implantable accessories such as leads and lead adaptors, extensions, caps and plugs, catheters and ports, for use with active implantable devices
2. Devices of animal origin or components of devices that are of animal origin, for use in the body or for application to broken skin, other than devices that (DR4, pp. 94–100):
   - Are manufactured using animal-derived waxes
   - Incorporate heparin, unless the heparin is being delivered as a drug
   - Are sutures conforming to a standards determined under Part 2 of the Therapeutic Goods Act 1989
   - Are made from sintered hydroxyapatite
   - Incorporate gelatin that conforms to generally accepted pharmacopoeia definition
3. Implantable breast prostheses made of materials other than water or saline that could unintentionally migrate from the implant site (DR4 pp. 101–105)
4. Powered, nonimplantable drug infusion systems that regulate the flow of liquids into the patient under positive pressure generated by a pump (DR4 pp. 106–112)
5. Extracorporeal-therapy systems that contain tissues, cell lines, or substances derived from human or animal sources, or are used as components within a system to separate, purify, or maintain a body fluid or tissue ex vivo prior to infusion, transfusion, implantation, or transplantation (DR4 pp. 113–117)
6. Heart-valve prostheses (DR4 pp. 118–130)
7. Devices of human origin—Devices that are made from or contain human tissue are classified into four separate categories based on the source of the material (DR4 pp. 131–138):
   - Commercial tissue—Commercially retrieved and processed human tissue must be registered on the ARTG.
   - Primary processed tissue—Human tissue that is procured by a tissue bank or hospital for implantation in or on the human body without any deliberate alteration to its biological or mechanical properties is exempt from registration in the ARTG. However, the supplying Tissue Bank must comply with the *Australian Code of Good Manufacturing Practice for Therapeutic Goods – Human Tissues*.
   - Secondary processed tissue (tissue engineered)—Human tissue procured by a tissue bank or hospital for implantation in or on the human body, that is processed by steps that deliberately alter the biological or mechanical properties of the tissue, must be registered on the ARTG The tissue bank must comply with the *Australian Code of Good Manufacturing Practice for Therapeutic Goods – Human Tissues*.
   - Direct donor-to-recipient transplantation (without storage)—Fresh viable human tissue (other than blood), human organs, parts of organs, human bone marrow for direct donor-to-matched-recipient transplant are not therapeutic goods under Section 7 of the Therapeutic Goods Act 1989 and are exempt from registration on the ARTG.
8. Implantable IOLs (DR4 pp. 139–147)
9. Intraocular visco-elastic fluids (DR4 pp. 148–151)
10. IUCDs unless the principal contraceptive action of the IUCD is achieved chemically through the release of a drug (in that case, the IUCD will be treated as a drug) (DR4 pp. 152–156)
11. Barrier contraceptive devices to which a Standard does not apply (e.g. female condom, cervical cap) (DR4 pp. 158–162)
12. Breast prostheses containing only saline that are manufactured using established materials and technology and are intended by the manufacturer to be left permanently in place (DR4 pp. 163–166)
13. Instrument-grade sterilants and hospital-grade or household/commercial-grade sterilants where specific claims are made must be registered on the ARTG (Hospital-grade sterilants without specific claims are listable in the ARTG.) (DR4, pp. 167–168)
14. Diagnostic kits for *in vitro* diagnosis of patients infected with the Human Immunodeficiency Virus (HIV) or with the Hepatitis C Virus (HCV) are required to be registered on the ARTG (DR4 pp. 169–176)

Source: DR4 pp. 85–175.

- copies of any information provided separately to clinical staff, including training or instructional literature, videocassettes, movies, or visual-aid materials;
    - copies of any information provided separately to patients, including warnings and cautions;
    - copies of any promotional or advertising material concerning the device, components, and accessories;
    - forms and details of any systems used to track and trace the device;
    - copies of the information and instructions for use that are given to the practitioner, including copies of any training or instructional literature, models, video cassettes, movies, or visual-aid materials, where applicable;
    - when applicable, copies of information for healthcare professionals and instructions for use by the patient that should include detailed instructions, appropriate warnings, and cautions;
    - copies of any promotional or advertising material concerning the device, its components, and accessories; and
    - copies of service manuals covering the device, components, and accessories when service is to be performed by someone other than the manufacturer.
- Whether the device conforms to any applicable standards.
- Whether any requirements in the regulations relating to advertising the device have been satisfied.
- For devices manufactured outside Australia, whether the manufacturing and quality-control procedures used in the manufacturing procedures acceptable. For those goods manufactured outside of Australia, the TGA may require evidence from a relevant overseas authority to establish that the manufacture of the goods conforms to an acceptable standard.
- For devices manufactured in Australia, whether the devices are manufactured in accordance with the manufacturing practices in Part 4 of the Therapeutic Goods Act 1989.
- Whether the device contains substances that are prohibited under the Customs Act 1901.
- Whether any other matters considered relevant by the DOHA Secretary have been satisfied.

After the therapeutic device has been evaluated, the TGA must issue a written notification to the applicant. If the decision is not to register the device, the TGA must provide the reasons for the decision. If the TGA decides to register the device, the applicant's completed registration form, submitted with the initial application, is processed. As soon as practical, the TGA issues a certificate of registration that lists the unique registration number assigned to that therapeutic device and sponsor.

In connection with an application for registration of a therapeutic device, a person must not knowingly or recklessly make a statement that is false or misleading in a material particular. To knowingly or recklessly make a false or misleading statement in an application for registration is a violation of Section 22A of the Therapeutic Goods Act 1989 and makes the guilty party subject to a substantial fine.

CLAIMING EQUIVALENCE TO A REGISTERED DEVICE

Therapeutic goods are registered subject to the condition that no changes are to be made to the data on which the decision to register the device was made (DR4 p. 21). The manufacturer of a therapeutic device is responsible for identifying those changes to features of a registered device or its components, or the differences between a new model and a registered device that may have an effect on safety and effectiveness (DR4 p. 63). There are two types of modifications that are considered as product changes (DR4 p. 21):

1. Variations to product information for registered or listed therapeutic devices that do not result in a new registration
2. Additions of products to grouped registrations

Depending on the type of modification, the sponsor must (DR4, Annex 3 p. 1):

- Obtain approval from the TGA prior to the change being made. (A sponsor who believes that the registrable device is equivalent to a predicate device that has been approved for supply in Australia may be able to lodge an abridged submission to support the application.)
- Notify the TGA as soon as practicable and not later than three months after implementation of the changes.
- Submit a notification directly to the ARTG as soon as practicable and not later than three months after implementing the change. No fee is required for change in the sponsor address or the contact person.

For these changes or differences, the manufacturer or sponsor must be able to provide data that demonstrate that the quality, safety, and effectiveness of the device will be equivalent to that of the currently registered device. Annex 3 to the Australian Medical Devcie Requirements (DR4) provides a summary of the requirements for the sponsor to notify or seek approval for changes to registered (or listed) therapeutic devices or groups of devices.

Some changes to the labeling of a therapeutic device are viewed as significant in making a determination of equivalence with the currently registered device. These include the following (DR4, Annex 3 pp. 3-6):

- A change in labeling content (e.g., sponsor details, directions, warnings) with particular attention to:
  - a change to the intended use/clinical purpose with safety ramifications, including new or extended indications for use
  - a change in claims for the device
  - a deletion from the contraindications/warnings
  - an extension of the recommended shelf life
  - a change in the product information or clinical manuals that relates to the original requirements for registration/listing
  - a change in the performance specifications that affect the intended use or the technological characteristics of the device
  - the addition of a new product within a grouped registration/listing
- A change in the manufacturing location
- A change in the sterilization technique

A second group of changes requires the sponsor to notify the TGA as soon as practicable and not later than three months after implementing the change. These include (DR4, Annex 3 pp. 3-6):

- a change to the intended use/clinical purpose with no safety ramifications, including strengthening instructions intended to enhance its safe use;
- a change to the contraindications/warnings, not concerning a deletion;
- a change in the recommended storage conditions, including new conditions;
- a change to the product name;

- a change of device storage conditions; and
- a change of the manufacturer's name. (A change of manufacturing site requires prior TGA approval and may be sufficient to require a new registration for the device.)

A third group of changes requires the sponsor to submit a notification directly to the ARTG as soon as practicable and not later than three months after implementing the change. These include (DR4, Annex 3 pp. 3-6):

- a change in the sponsor name (same sponsor),
- a change in the sponsor address,
- a change in the contact person,
- a change in the sponsor, and
- discontinuing a registered product.

Some changes to the labeling or the information supplied with the device are considered non-significant provided that they are validated against the manufacturer's specifications and the manufacturer can demonstrate that they will not have a detrimental effect on the safety and efficacy of the device. These include (DR4, Annex 3 pp. 3-6):

- a change of unit pack or pack size,
- a change in the color of the label,
- a change in the print size (having a height of 1.00 mm or more),
- a change in the layout of the labels or labeling,
- a reduction of the recommended shelf life, and
- a change in the service manuals.

Because certain types of sterilants and disinfectants are registered products in Australia, changes to such a product's use/description, claims, or indications may require TGA approval prior to the change being made. Other changes require the sponsor to submit a notification directly to the ARTG as soon as practicable and not later than three months after implementing of the change. These include (DR4, Annex 3 pp. 7-9):

- a change in the sponsor name (same sponsor)
- a change in the sponsor address
- a change in the contact person
- a change in the sponsor
- a change in the label color
- a change in the print size or typeface
- a change of manufacturer's address information (if different from the sponsor).

LISTING OF THERAPEUTIC DEVICES

The person who wishes to manufacture, import, or export a device requiring listing must make a written application to the TGA. The TGA will not refuse to list the device in relation to the person making the application unless there is evidence that (Therapeutic Goods Act pp. 45–46):

- The device is not eligible for listing (e.g., the device is required to be registered).
- The device is not safe for the purposes for which it is to be used.

- The presentation of the device is not acceptable.
- The device does not conform to the applicable standards.
- One or more requirements in the regulations relating to advertising of the device have not been satisfied.
- For a device manufactured outside Australia, the manufacturing and quality control procedures used in the manufacturing process are not acceptable. For those goods manufactured outside Australia, the TGA requires evidence from a relevant overseas authority to establish that the manufacturing practice conforms to an acceptable standard.
- For a device manufactured in Australia, the device is not produced in accordance with the manufacturing practices in Part 4 of the Therapeutic Goods Act 1989 (licensing of therapeutic goods manufacturers).
- The therapeutic device has been manufactured in Australia solely for export, and the device has been refused registration or listing for supply in Australia.
- Any other matters considered relevant by the DOHA Secretary have not been satisfied.

After considering the application, the TGA must issue a written notification to the applicant. If the decision is not to list the device, the TGA must provide the reasons for the decision. If the TGA decides to list the device, the TGA issues a certified copy of a Certificate of Listing that details the listing number assigned to that therapeutic device and sponsor.

CLAIMING EQUIVALENCE TO A LISTED DEVICE

As with registered therapeutic goods, goods are listed subject to the condition that no changes are to be made to the data on which the decision to list the device was made (DR4 p. 21). The manufacturer of a therapeutic device is responsible for identifying those changes to features of a listed device or its components, or the differences between a new model and a listed device, that may either require notification or prior approval by the TGA (DR4, Annex 3 p. 1).

For listable devices, most changes to the labeling or the information supplied with the device are considered nonsignificant provided that they are validated against the manufacturer's specifications and the manufacturer can demonstrate that they will not have a detrimental effect on the safety and efficacy of the device. A few changes require the sponsor to submit a notification directly to the ARTG as soon as practicable and not later than three months after implementing of the change. These include (DR4, Annex 3 pp. 3-6):

- a change in the sponsor name (same sponsor),
- a change in the sponsor address,
- a change in the contact person,
- a change in the sponsor, and
- discontinuing a listed product.

A few more changes require the sponsor to notify the TGA as soon as practicable and not later than three months after implementing the change. These include (DR4, Annex 3 pp. 3-6):

- a change of the manufacturer's name,
- the addition to the Pharmaceutical Benefits List for those devices that are to be supplied under the Australian Pharmaceutical Benefits Scheme (PBS) for *in vitro* diagnosis, and
- a change to the product name.

In addition, a few changes to the labeling of a listed therapeutic device are viewed as significant and require approval by the TGA before the change is made. These include (DR4, Annex 3 pp. 3-6):

- a change in the manufacturing location and
- a change in the sterilization technique used on the product.

LICENSING OF THERAPEUTIC GOODS MANUFACTURERS

The Australian manufacturer of a registered or listed therapeutic device must be appropriately licensed to carry out the manufacture of the goods or class of goods within which the device is included. This provision of the Therapeutic Goods Act 1989 includes any subcontractor or testing facility in Australia that is contracted to, or otherwise engaged to, carry out all or part of the manufacture of the registered or listed device (DR4 p. 35). Section 35 of the Therapeutic Goods Act 1989 makes it a crime for a person to knowingly or recklessly carry out a step in the manufacture of a therapeutic device requiring a license without a proper license. Violation of this provision subjects the perpetrator to a substantial fine. The holder of a license who knowingly or recklessly breaches a condition of the license is also subject to a fine.

The licensing provision of the Therapeutic Goods Act 1989 applies unless the device, or person performing the manufacturing operation, is exempted by the regulations. A list of the devices exempted from the licensing provision of the act can be found in Table B.6 in Appendix B. A list of the persons exempted from this requirement is given in Table B.7 in Appendix B.

Section 36(1) of the Therapeutic Goods Act 1989 enables the DOHA Secretary to establish written principles to be observed in the manufacture of therapeutic devices for human use. These include quality assurance (QA) and quality-control procedures and good manufacturing practices (GMPs) to be employed in the manufacture of therapeutic devices. These principles extend to proper control of labeling and packaging operations to prevent labeling mix-ups between similar products or labels. The extent of production-area controls will be determined by the likelihood that a labeling mix-up can occur.

A foreign manufacturer of therapeutic devices is required to comply with the same level of GMP that is expected of the manufacturer of similar products in Australia. A sponsor seeking registration or listing of a new therapeutic device must provide evidence of GMP compliance at the overseas manufacturing site(s). If more than one manufacturer is involved in the manufacture of a therapeutic device (e.g., assembly at one location and sterilization at another), then GMP evidence must be provided for each site (DR4 p. 36).

When the TGA considers it necessary, the sponsor of a therapeutic device must agree to pay the cost of an audit of the foreign manufacturing site(s). The TGA, however, considers that some countries have GMP audits at least equivalent to that of Australia. This list includes Japan and the United States. Australia has entered into an MRA with the European Union (EU) that enables the TGA to accept certificates issued by designated Notified Bodies under the European medical device directives.* Certification by the regulatory authorities in these countries that the manufacturer operates to a satisfactory standard will be accepted as evidence of GMP compliance (Overseas GMP p. 7).

For manufacturers of devices that are registrable if supplied in Australia, or active implantable medical devices (AIMDs), compliance with ISO 9001/EN 46001 or ISO 13485 is required. For manufacturers of other devices, compliance with ISO 9002/EN 46002 or ISO 13488 is required as a minimum as specified in Schedule 3 of the TGR 2002. A manufacturers may elect to adopt ISO 9001/EN 46001 or ISO 13485 instead of ISO 9002/EN 46002 or ISO 13488 (DR4 p. 35).

---

* An EC Type-Examination Certificate (Annex III(3)) or Design Examination Certificate (Annex II(2) Section 4) is not acceptable as evidence of an acceptable quality system.

COMMENCING THE SUPPLY OF A REGISTERED OR LISTED DEVICE

The sponsor of a registered or listed therapeutic device must provide a copy of the label(s) of the device to the TGA at the time of application. Where practical, actual labels should be provided. These labels are to be attached to the application form. Actual-size photocopies are acceptable where the label information is printed or embossed directly onto the container.

EXPORT OF THERAPEUTIC DEVICES

Unless they are exempt, therapeutic devices intended solely for export from Australia must be listed on the ARTG. Some countries that import therapeutic devices from Australia require a certificate from the TGA showing the regulatory status of the product in Australia. The required details include copies of the labeling. Certification may be issued only for products registered or listed in the ARTG (DR4 p. 10).

A registrable device that is solely for export becomes a listable device on the ARTG. No formal evaluation is undertaken because the device will never be supplied in Australia (DR4 p. 256).

Except in unusual circumstances and with the written approval of the DOHA Secretary, therapeutic devices may not be exported from Australia if the devices do not conform to the applicable standards. This provision of the Therapeutic Goods Act 1989, however, does not apply to standards that specify the labeling required on devices that are to be supplied in Australia (Therapeutic Goods Act p. 23).

IMPORT OF THERAPEUTIC DEVICES

Therapeutic devices imported into Australia must meet all of the same requirements as devices produced in Australia. Unless they are exempted by the regulations, they must be registered or listed on the ARTG. In the case of imported devices, the registration number must be added to the labeling before the devices are supplied in Australia. The sponsor of the imported device must ensure that the labeling complies with the requirements of other applicable Commonwealth legislation, such as the Commerce (Trade Descriptions) Act, which requires the name of the country in which the device was made or produced to be included (TGO 37 p. 9).

Sponsors may import therapeutic goods that are not registered or listed on the ARTG (unauthorized goods) and display them at conferences, trade fairs, and other events provided certain conditions are met. The sponsor must display unauthorized devices in a way that makes it clear that (DR4 p. 25):

- The devices are currently unauthorized in Australia.
- They are not available for supply.
- They have not been entered on the ARTG.
- Their safety, quality, and efficacy have not been established by the TGA.

Any promotional material about these products distributed at the meeting should also indicate the conditions listed above.

Until they are entered onto the ARTG, these devices must be held under the direct control of the sponsor. The sponsor may hold such unapproved devices for up to twelve months. However, the sponsor must maintain records of the source and supply of the devices. The TGA may request these records at any time. The products must be destroyed or returned to the consignor of the devices within one month of the end of that period (DR4, p. 25).

TABLE 4.3
List of Materials Designated Biological Materials by the Australian Quarantine and Inspection Service (AQIS)

Products that contain material components sourced from microorganisms
Cell lines and hybridomas
Serum
Antiserum
Enzymes
Hormones
Antibodies
Toxins
Toxoids
Tissues or tissue extracts
Secretions or exudates
Blood and blood components
Cell or microbiological culture media
Microbial fermentation products
Microbial extracts
Microbial components

Source: DR4 p. 9.

The approval of the Australian Quarantine and Inspection Service (AQIS) of the Department of Primary Industries and Energy (DPIE) is required before the importation of most therapeutic goods that contain human, animal, or plant material. The AQIS defines biological material as any products of animal and/or microbiological origin. A list of biological materials subject to quarantine and inspection are listed in Table 4.3.

The importation of these materials generally requires a "Permit to Import Quarantine Material" issued by the AQIS. However, for AQIS purposes, tissues (including organs) or fluids (including blood, serum, plasma, semen, and urine) of solely human origin (excluding feces and cell lines) that are clearly labeled or certified as such, do not require a permit. The use of this human origin material is controlled by the TGA (DR4 p 9).

STANDARDS FOR THERAPEUTIC DEVICES

The DOHA Secretary may, by publishing an order in the *Commonwealth of Australia Gazette*, establish standards for therapeutic devices (Therapeutic Goods Act p. 20). Standards under the Therapeutic Goods Act 1989 may include monographs in the *British Pharmacopoeia*, monographs in other publications approved by the DOHA, or publications of the Standards Association of Australia. Such a standard may require that a therapeutic device, or a class of therapeutic devices, be labeled or packaged in a specific way or be kept in containers that comply with the standard. The order establishing a standard may direct that additional particulars appear on the label of a therapeutic device or class of devices. The order may also specify the particular manner in which the information is to be displayed.

Except with the written consent of the DOHA Secretary, a person must not import or supply a therapeutic device for use in Australia that does not comply with the standards applicable to the device. To import or supply such a device is a violation of Section 14(1) of the Therapeutic Goods Act 1989, which subjects the perpetrator to a very substantial fine.

EXEMPTIONS FOR SPECIAL AND EXPERIMENTAL USES

The DOHA Secretary may, by written notice, grant an exemption that allows an unapproved device to be used for special purposes. This exemption allows a person to import, export, or supply a device in Australia that is not exempt, not listed, and not a registered therapeutic device recorded on the ARTG. The special purposes recognized by Section 19(1) of the Therapeutic Goods Act 1989 are:

- use by a person who is registered, in a state or internal territory, as a medical practitioner in the treatment of another person (i.e., a "custom device");
- use solely for experimental purposes in humans (i.e., an "investigational device"); and
- regulatory purposes.

Both custom and investigational devices are exempted from the general requirements, including the labeling requirements of the Therapeutic Goods Act 1989. However, they may be subject to conditions specified in the notice of approval. Because they are exempt, they do not have to comply with the requirements of TGO 37, which are described in Chapter 5 (TGO 37 p. 1).

INVESTIGATIONAL USE

The intent of the exemption for "investigational use" is to obtain reliable data on the clinical safety and effectiveness of a device when these data do not already exist (DR4 p. 43). The approval of the TGA is not required for clinical trials of devices that are currently on the ARTG.

There are two forms of application, the Clinical Trial Notification (CTN) and the Clinical Trial Exemption (CTE). Both schemes require Institutional Ethics Committee approval, and the Institutional Ethics Committee determines which scheme is appropriate (DR4 p. 43).

Both CTNs and CTEs incur a fee. However, a CTE is more expensive because of the evaluation conducted by the TGA.

In general, neither the CTN nor CTE cover the situation where a medical practitioner or related healthcare professional, in treating a particular patient, uses or modifies a registered or listed device outside the approved uses reflected in the labeling of the device. In doing so, the practitioner must consider the requirements of the appropriate Institutional Ethics Committee, informed patient consent, and his or her own legal obligations. It is a serious offense for a sponsor to knowingly supply a therapeutic device in Australia for human use unless the device is registered, listed, or exempt.

Clinical Trial Notification (CTN) Scheme

The use of devices that fall under the CTN scheme does not require the approval of the DOHA Secretary, although certain conditions must be met to qualify for the exemption (DR4, pp. 47–48). The Institutional Ethics Committee is required to review each Clinical Investigation Plan and related Informed Consent Form in conjunction with data provided by the sponsor to support the proposal. The TGA encourages discussion with other Institutional Ethics Committees currently reviewing the trial. The Institution Ethics Committee may authorize the trial or decline to accept the proposed trial under the CTN scheme and inform the sponsor that the trial should be conducted under the provisions of the CTE.

If the trial is accepted under the CTN scheme, the Institutional Ethics Committee informs the investigator and certifies this on the CTN. The CTN is forwarded to the TGA. The sponsor must ensure that the trial does not commence until approval has been obtained from the Institutional Ethics Committee, and the TGA has provided formal acknowledgment of the notification.

The TGA recommends that the Institutional Ethics Committee use the definitions and methodology prescribed in European Standard EN 540:1993, and consult the departmental publications

Guidelines for Good Clinical Research Practice 1991 and *Statement on Human Experimentation and Supplementary Notes 1992.*

The sponsor must notify the TGA in writing of severe adverse events and all adverse device effects occurring in multicenter clinical investigations within 15 days; deaths or serious injuries are to be notified within 72 hours. The sponsor must cease supplying devices for the investigation if requested by the TGA.

The device must be clearly labeled "investigational use only" or with equivalent wording. *In vitro* diagnostic (IVD) devices used to diagnose infection with Human Immunodeficiency Virus (HIV) and Hepatitis C Virus (HCV) may also require research trials before they can be registered on the ARTG. Such devices must be labeled "for research use only" or with equivalent wording.

Clinical Trial Exemption (CTE) Scheme

Although the CTN scheme is the more commonly used clinical trial process, the CTE scheme may be better suited in the instance where the experimental device introduces new technology or materials that have not previously been evaluated. Goods subject to the CTE scheme are identified as exempt goods and require approval by the DOHA Secretary prior to their use under Section 19 of the Therapeutic Goods Act 1989. A CTE application must be accompanied by the information required by the TGA (DR4 p. 45). After considering the application, the TGA must provide the applicant with a written response. If the decision is not to authorize the experimental use of the device, the TGA must include the reasons for the decision in the notification.

Consistent with the intent of the TGA to harmonize requirements for clinical trials with those of the EU, the TGA requires that data submitted for the CTE scheme must comply with European standard EN 540:1993, with reference to the TGA documents, *Guidelines for Good Clinical Research Practice 1991* and *Statement on Human Experimentation and Supplementary Notes 1992* (DR4 p. 43).

The sponsor must notify the TGA in writing of severe adverse events and all adverse device effects occurring in multicenter clinical investigations within 15 days; deaths or serious injuries are to be reported within 72 hours. The sponsor must cease supplying devices for the investigation if requested by the TGA.

The device must be clearly labeled "investigational use only" or with equivalent wording.

Custom Devices

Australia offers two schemes that enable practitioners to gain access to therapeutic devices that would otherwise not be available. They are the Individual Patient Use (IPU) scheme and the Authorized User Approval (AUA) program.

The IPU scheme enables a registered practitioner to access and use unregistered or unlisted therapeutic devices for an individual patient on a specific occasion. Approval is granted when the individual patient has a demonstrated clinical need, is likely to benefit, and there is no approved device currently marketed in Australia that is sufficient. Approval for use of a device is arranged by direct contact between that practitioner and the TGA (DR4 pp. 48–49).

The AUA program is an extension of the IPU scheme that allows authorized practitioners to access and use unregistered and unlisted devices under limited conditions without requiring a separate approval for each use. To be authorized, a practitioner has to meet the conditions established in the regulations and apply to the TGA for approval under the AUA program (DR4 pp. 50–51).

There is no requirement for specific labeling such as "custom-made device" under either the IPU or AUA programs.

GOODS EXEMPTED FOR REGULATORY PURPOSES

Part 7 of the TGR 2002 exempts certain goods from registration or listing on the ARTG. Item 1.3 of Schedule 4 specifically exempts samples of therapeutic goods imported, exported, manufactured, or supplied for (TGR 2002 p. 118):

- submission to a regulatory authority;
- developmental or quality-control procedures;
- examination, demonstration, or display; or
- analysis or laboratory testing

However, these devices may not be for supply for therapeutic use in humans.

THINGS TO REMEMBER

Medical devices in the Commonwealth of Australia are regulated under the provisions of the Therapeutic Goods Act 1989. This act of Parliament is intended to establish and maintain a national system of controls relating to the quality, safety, efficacy, and timely availability of therapeutic goods.

Under its provisions, the Governor-General of Australia may make regulations prescribing matters necessary or convenient for carrying out the terms of the act. Particularly with regard to the labeling of therapeutic devices, the regulations may both prescribe requirements for informational material that is included with the therapeutic device and set out requirements for advertising of therapeutic devices.

Although the Therapeutic Goods Act 1989 does not use the terms "adulteration" or "misbranding," these concepts do appear in the Australian law. Supplying an adulterated or misbranded therapeutic device is an offense that subjects the guilty party to a substantial fine.

A therapeutic device may be considered misbranded if its labeling or advertising is false or misleading, leads to unsafe use of the device, suggests a purpose other than those for which the device has been approved in Australia, or fails to meet the requirements specified in any regulations or standards applicable to the device. Advertising is defined in the Therapeutic Goods Act 1989 to include statements, pictorial representations, or designs that are intended, directly or indirectly, to promote the use or supply of a therapeutic device.

The TGA maintains a register, known as the ARTG, for the purpose of compiling information on therapeutic goods used on humans. The ARTG is divided into two parts—registered goods and listed goods. Under the Australian system, therapeutic devices that are perceived to present a significant risk of death or serious injury for a patient or user are required to be registered on the ARTG. Registration means that these devices are subject to premarket evaluation of their quality, safety, and effectiveness. This includes an evaluation of the labeling to be provided with the device. These devices can be sold only after being approved by the TGA and after being recorded on the registered devices section of the ARTG.

All other therapeutic devices, unless a TGO exempts them, are required to be listed on the ARTG. The person who wishes to manufacture, import, or export a device requiring listing can do so only after it has been included on the ARTG.

The DOHA Secretary may grant an exemption that allows an unapproved device to be used for special purposes. This exemption allows a person to import, export, or supply a device in Australia that is neither exempt nor a therapeutic good listed on the ARTG.

The intent of the exemption for "investigational use" is to obtain reliable data on the clinical safety and effectiveness of a device when these data do not already exist. The approval of the TGA is not required for clinical trials of devices that are currently on the ARTG.

5 General Therapeutic Device Labeling in Australia

Therapeutic devices, unless they are exempted in the regulations or by written notice from the Secretary of the Australian Department of Health and Aging (DOHA), are regulated by the Therapeutic Goods Act 1989. The act applies to therapeutic devices regardless of whether they are produced domestically or are imported. The provisions of the Therapeutic Goods Act 1989 also apply to devices produced for export from Australia.

Failure to label a device in accordance with requirements in the Therapeutic Goods Act 1989 or the regulations promulgated to carry out the provisions of the Therapeutic Goods Act 1989 renders the device misbranded. Misbranding is a violation of the letter of the law and exposes the sponsor of the device to substantial penalties. This chapter discusses the regulations that deal with misbranding of therapeutic devices.

MISBRANDING

Section 3(5) of the Therapeutic Goods Act 1989 states that the presentation of therapeutic goods is unacceptable if it (a) is misleading or confusing as to the proper use of the device, (b) may lead to the unsafe use of the device, (c) suggests a purpose for which the device has not been approved, or (d) fails to comply with the applicable regulations. For therapeutic devices, the first three points are covered in the Therapeutic Goods Act 1989 under the heading of "directions for use." The general labeling requirements established by the TGA for therapeutic devices are contained in Therapeutic Goods Order (TGO) No. 37, which is discussed in detail later in this chapter.

Failure to provide adequate directions for use or to follow or satisfy the requirements in TGO No. 37 will lead to the device being deemed misbranded. Selling or advertising for sale a misbranded device subjects the manufacturer or sponsor to a substantial fine.

ADEQUATE DIRECTIONS FOR USE

The term "directions for use" in the Therapeutic Goods Act 1989 refers to the directions under which a therapeutic device can be used safely and for the purposes for which it is intended. "Adequate directions for use" means directions sufficient for the ultimate consumer (layperson or healthcare provider) to use the device safely and for the intended purpose. This includes:

- statements of all conditions, purposes, or uses for which the device is prescribed, recommended, or suggested in its oral, written, printed, or graphic advertising;
- 'Indications," or "Indications for Use," which identifies the target population in which there is sufficient valid scientific evidence to demonstrate that the device as labeled will provide clinically significant results without presenting an unreasonable risk of illness or injury;
- the dosage for each use and the usual dosage for persons of differing ages and physical conditions;
- the frequency and duration of treatment for each indication; and
- the contraindications and appropriate warnings, precautions, and adverse reactions reasonably associated with the use of the device

GENERAL REQUIREMENTS FOR LABELS FOR THERAPEUTIC DEVICES

The Therapeuitc Goods Administration (TGA) has developed a set of labeling requirements that is applicable to most therapeutic devices. These requirements are considered to be the minimum information that should appear on the label of a therapeutic device in order to ensure the safe use of the device and to allow it to be traced to the sponsor and to a particular cycle in the manufacturing process. These requirements are contained in TGO 37. The TGA recognizes that there may be many other requirements related to the safe use of a particular class of devices that should appear on the label or be included in the package insert. TGO 37 does not attempt to address all of these requirements.

TGO 37 applies to therapeutic devices supplied in any form that is normally available to the consumer. The consumer is defined as the person, authority, or institution to which a device is supplied for use as a therapeutic device (TGO 37 p. 2). Some devices, hovever, are exempt from the requirements of TGO 37. These include (TGO 37 p. 1):

- A custom therapeutic device that, unlike any other therapeutic device, is manufactured and supplied for use by a specific individual. Unregistered and unlisted devices can be made available to a registered practitioner as custom devices (see Chapter 4). While these devices are exempt from the requirements of TGO 37, the manufacturer is responsible for labeling the device to ensure its safe use.
- A therapeutic device for dental use that is constructed externally to the mouth, and is fitted or fixed into the mouth on a temporary or permanent basis to correct irregularity or deficiency. (Devices of human or animal origin and devices implanted directly into bone or soft tissue are not exempt from this regulation.)
- A therapeutic device intended for veterinary use.
- A device intended for use as an *in vitro* diagnostic (IVD).
- A therapeutic device intended solely for export from Australia.
- A therapeutic device intended solely for the purpose of conducting clinical trials (see Chapter 4).

TGO 37 specifies requirements in the following areas:

- Presentation requirements for numbers, letters, and symbols required or permitted on labels
- General labeling requirements for all covered devices
- Labeling for sterile devices and nonsterile implantable therapeutic devices

The requirements in TGO 37 detail the minimum information set that must be included in the labeling for a therapeutic device. The following sections discuss each of these requirements in detail. The section number (e.g., §5) of the corresponding section in TGO 37 is listed with each topic for reference.

PRESENTATION REQUIREMENTS (§4(2)(C))

To satisfy the requirements of TGO 37, each number, letter, or symbol required or permitted by this order to be included on a label must appear as follows:

- Each required item must be clearly visible on the label.
- Each item must be printed in durable form. The information required by the TGO cannot be hand written.

- Each character must be legible and must be printed in a type style having a letter height of not less than 1.0 millimeters.
- Each item must be presented in the English language.
- Units of measure must be expressed in the International System of Units (SI). Units generally accepted in clinical practice may also be used. For example, units of mmHg may be used when measuring blood pressure (DR4 p. 59).

General Labeling Requirements

Part 2 of TGO No. 37 details a set of general labeling requirements that apply to all covered therapeutic devices that are not supplied sterile, or are not nonsterile implantable therapeutic devices. Every therapeutic device to which this part of the regulation applies must have a label or labels that meet the requirements of this part.

Name of the Therapeutic Device (§7(a))

The device label must contain the name of the therapeutic device. In this context, name means a name that is sufficiently descriptive to indicate the true nature of the therapeutic device to which it is applied (TGO 37 p. 2). Examples of therapeutic-device names include, for example, insulin syringe, absorbable sutures, implantable breast prostheses, and implantable cardiac pacemaker.

Name and Place of Business (§ 7(b))

The device label must contain the name and address of the manufacturer or sponsor of the therapeutic device. If the manufacturer or sponsor has a registered name, that name must appear on the label along with the city, town, or locality in which the registered office or registered place of business is situated (TGO 37 p. 2). A registered name means the name of a partnership, firm, business, company, or corporation registered or incorporated in accordance with the appropriate laws of any state or territory (TGO 37 p. 3).

For a manufacturer or sponsor that does not have a registered name, the label must include the name under which the manufacturer or sponsor conducts business and the address of the principal place of business of that manufacturer or sponsor. The address must include, as applicable, the street number or numbers, the street name, the city or town, and the state or territory in Australia. For manufacturers or sponsor whose place of business is outside Australia, the address must also include the name of the country. The label must not include a post-office, cable, telegraphic, or code address (TGO 37 p. 3). The address must be sufficiently detailed to allow an inspector from the Compliance Branch of the TGA to visit the sponsor's place of business.

For purposes of the regulations, the sponsor of a therapeutic device is the person who (TGO 37 p. 4):

- In Australia, manufacturers the devices, or arranges for another person to manufacture the devices, for supply (whether in Australia or elsewhere)
- Imports, or arranges the importation of, the devices into Australia
- Exports, or arranges the exportation of, the devices from Australia

A person is not considered the sponsor of a device if they perform any of these functions on behalf of another person who at the time is a resident of or is carrying on business in Australia. For example, a person who imports a device for sale by an Australian distributor would not be the sponsor of the device. The Australian distributor would be considered the sponsor.

In this context, "manufacture" is defined in the Therapeutic Goods Act 1989 as the production of the device, or engaging in any part of the process of producing the device, or bringing the device to its final state. Manufacturing includes engaging in the processing, assembling, packaging, labeling, storage, sterilizing, testing, or releasing for sale of the therapeutic device. Manufacturing also includes performing any of these processes on any of the components or ingredients of the device (Therapeutic Goods Act p. 6).

In Australia, a sponsor may contract with another person to manufacture the device. The contract manufacturer is not considered a sponsor of the device and does not have to be named on the label. A person who imports or exports a therapeutic device on behalf of another person who is a resident of or is doing business in Australia is also not considered the sponsor of the device.

BATCH OR SERIAL NUMBER (§7(C))

The label must include a batch or serial number, which is a characteristic marking given by the manufacturer to a particular device or to all devices in a batch for the purpose of uniquely identifying the device or batch (TGO 37 p. 1). A batch means a quantity of therapeutic devices that are uniform in composition, method of manufacture, and probability of chemical or microbial contamination. They must be made in one designated cycle of manufacture, and, in the case of sterilizing or freeze-drying, they must be sterilized or freeze-dried in one cycle. The batch or serial number must enable the device or batch to be traced through any or all of the critical stages of its manufacture and supply.

The batch number or serial number may be immediately preceded, as appropriate, by the words "Batch," "Batch Number," "Batch No.," "Lot," "Lot Number," "Lot No.," or "Lot Code," "Serial Number," "Serial No.," or by words having a similar meaning. The symbols "B," "(B)," or "Ⓑ" "SN," "S/N," or "FABR" may be used as an alternative to the word designations (TGO 37 pp 1-2). While the regulations permit a variety of designations for devices produced in batches, the preferred identification is "Batch Number," "B," "(B)," or "Ⓑ"

The date of manufacture of the therapeutic device may be used as the batch or serial number if it is clearly recognizable as a date (TGO 37 p. 2).

MULTIPLE DEVICES IN A PACKAGE (§7(D))

If two or more therapeutic devices are contained within the same package, the label must give the names and quantities of each of the goods. This provision applies if the package contains identical or different devices or consists in part of a substance that is a solid, semisolid, liquid, or gaseous element, compound, or mixture (TGO 37 p. 4). In relation to therapeutic goods, quantity means either the number of discrete units in the package or, where the goods are a substance, their nominal weight or volume (TGO 37 p. 3).

LABELING OF STERILE DEVICES AND NONSTERILE IMPLANTABLE THERAPEUTIC DEVICES

Part 3 of TGO 37 details a set of labeling requirements that apply to therapeutic devices that are supplied sterile or are non-sterile implantable therapeutic devices. Every therapeutic device to which this part of the regulation applies must have one or more labels on, or attached to, the unit packaging. If the device has an outer package, then the outer package must also have one or more labels attached that meet the requirements of Part 3.

Under TGO 37, a therapeutic device is considered implantable if it is designed to be implanted into the soft tissue or a body cavity, other than the teeth, of a person and remain there for 30 days or more (TGO 37 p. 2).

UNIT PACKAGES (§10)

For a sterile device or nonsterile implantable device, the unit package is the outermost level of packaging that maintains the sterility of the device or devices. In other cases, the unit package is the packaging closest to the therapeutic device or devices that wholly envelops all of the goods in the package (TGO 37 p. 5).

The label(s) on or attached to the unit package of a therapeutic device must contain all of the applicable information listed in this section, and the information must be visible to the consumer. The consumer is defined as the person, authority, or institution to which a device is supplied for use as a therapeutic device. The term consumer does not include a person, such as a distributor, to whom the device is supplied for the principal purpose of trade in therapeutic devices (TGO 37 p. 2). The following information is required on the unit package:

- The name of the therapeutic device must appear on the label. The requirements for the name of the device were discussed in the earlier section, "Name of the Therapeutic Device."
- The names and quantities of all devices within the unit package must appear on the label if the package contains two or more identical or different devices, or consists in part of a substance. The requirements for "Multiple Devices in a Package" are discussed in the prior section under that title.
- If the therapeutic device is manufactured in more than one size or design, the label must include the model designation—the characteristic markings or color characteristics given by the manufacturer to all devices that have identical designs (TGO 37 p. 2). If there is no model designation, the label must give the size of the device. The size is a descriptive term given by the manufacturer or sponsor to describe the size or shape of the device. It may include the physical measurements of the device or a descriptive expression such as "small," "medium," or "large" (TGO 3, p. 3).
- If the therapeutic device is intended by the manufacturer or sponsor to be used only once, the words "Single-use" or "Use only once," or words having a similar meaning may appear on the label. However, the word "disposable" must not be used to indicate the intentions of the manufacturer or sponsor.
- The label must contain the batch number or serial number. The requirements for batch numbers and serial numbers were discussed in the earlier section, "Batch or Serial Number."
- The label must include the name of the manufacturer or sponsor of the therapeutic device. A registered trademark may be substituted for the name of the manufacturer or sponsor if the trademark is registered under the Trade Marks Act 1955, and the manufacturer or sponsor is the proprietor or registered user (TGO 37 p. 3).
- If the therapeutic device is supplied in a sterile condition, the word "Sterile" must appear on the label.
- For a therapeutic device where the sterility may be qualified or limited to areas, components, or total presentation, the label must bear an additional statement to this effect. For example, this might include a caution statement advising the user that only the fluid path of a dialyzer is sterile:

CAUTION: ONLY THE FLUID PATH OF THE SET IS STERILE AND NONPYROGENIC. DO NOT USE IN A STERILE OR ASEPTIC AREA WITHOUT PROPER PRECAUTIONS.

OUTER PACKAGES (§13)

The outer package is the outermost level of packaging in which one or more therapeutic devices could normally become available to the consumer (TGO 37 p. 3). Subject to the requirements in the subsequent sections titled "Individually Wrapped Goods" and "Transparent Packages," a label on the outer package must include most of the information required on the unit package. Specifically, the label on the outer package must include all of the following:

- The name and address of the manufacturer or sponsor
- The name of the therapeutic device
- The names and quantities of all devices within the unit package, if the package contains two or more identical or different devices, or consists in part of a substance
- If the therapeutic device is manufactured in more than one size or design, the model designation must be included on the outer package label. If there is no model designation, then the label must give the size of the device.
- If the therapeutic device is intended by the manufacturer or sponsor to be used only once, the words "Single-use" or "Use only once," or words having a similar meaning must appear. However, the word "disposable" must not be used to indicate the intentions of the manufacturers or sponsors.
- The batch number or serial number of the device
- If the therapeutic device is a nonsterile implantable device, a label on the outer package must also bear the words "Nonsterile" or "Sterilize before use," or words having a similar meaning.

SMALL PACKAGES (§11)

The unit package for a therapeutic device may be so small that it is not practical to set out on the label attached to the unit package all of the information required by the regulations. If a therapeutic device is enclosed within both a unit package and an outer package, then it is sufficient that the label on the unit package contain the following information:

- The name of the therapeutic device
- The name of the manufacturer or sponsor of the therapeutic device, or the registered trademark of the manufacturer or sponsor
- If the therapeutic device is represented to be supplied in a sterile condition, the word "Sterile"

INDIVIDUALLY WRAPPED GOODS (§12)

The individual wrapping of a single-use therapeutic device such as first-aid strips, contact-lens solutions, swabs, or similar devices is considered to be the unit package for the device. If the unit package is enclosed in an outer package that is labeled as specified in the earlier section on "Outer Packaging," then it is sufficient that the label on the individual wrapper contains the following information:

- The name of the therapeutic device (e.g., first-aid strips)
- The name of the manufacturer or sponsor of the therapeutic device, or the registered trademark of the manufacturer or sponsor
- If the therapeutic device is represented to be supplied in a sterile condition, the word "Sterile"

Transparent Packages (§§14 and 15)

If the unit package containing one or more therapeutic devices is made of transparent material such that a person could clearly recognize the nature of the devices within the package, and could count the number of devices within the package, the label on the unit package need not contain:

- the name of the therapeutic devices,
- the name of the goods within the unit package, or
- the quantity of each of the goods within the unit package.

The label on the device may contain some or all of the information required on the unit or outer package. If the label on the device can be read through the packaging, then the label on the unit package or outer package need not duplicate those particulars that are clearly visible on the device label. The user must be able to read the information without the need to manipulate the device within the package. However, for sterile therapeutic devices, the word "Sterile" must still appear on the label attached to the unit package. For a therapeutic device where the sterility may be qualified in respect to components or total presentation, the label on the unit package must bear an additional statement to this effect.

IN VITRO DIAGNOSTIC (IVD) GOODS

In Australia, certain therapeutic goods, which would be devices according to the definition in the Therapeutic Goods Act 1989, have been declared to be drugs under the act. This approach is adopted for devices where safety considerations require that the device be evaluated in a manner similar to drugs. The expertise necessary for this type of evaluation resides within the Drug Safety and Evaluation Branch of the TGA (DR4 p. 2). Examples include blood and blood products, intrauterine contraceptive devices (IUCDs) containing hormones, and chemical oxygen generators. A list of devices declared to be drugs is found in Table B.2 in Appendix B.

Diagnostic goods for *in vitro* use are included in the list of devices declared to be drugs. There are, however, four groups of IVD goods that are required to be listed on the Australina Register of Therapeutic Goods (ARTG) prior to supply in Australia. The groups are (DR4 p. 222):

- IVD goods for home use,
- IVD goods supplied as a Commonwealth Pharmaceutical Benefit under the Pharmaceutical Benefit Scheme (PBS),
- IVD goods that incorporate material of human origin,
- IVD goods that are used for the diagnosis of infection with Human Immunodeficiency Virus (HIV) and Hepatitis C Virus (HCV),

All other IVD goods are exempt from registration or listing on the ARTG (see Table B.3 in Appendix B).

Some of these goods may be subject to the Quarantine Act 1901. The sponsor is responsible for obtaining quarantine permits for importation of covered devices from the Australian Quarantine and Inspection Service (AQIS) of the Department of Primary Industry and Energy (DPIE). In addition, goods that contain drug metabolites listed in the fourth schedule of the Customs (Prohibited Imports) Regulations must obtain the necessary permits to import these goods.

DEVICES SUPPLIED FOR HOME USE AND AS A COMMONWEALTH PHARMACEUTICAL BENEFIT

These IVD goods must be labeled in accordance with TGO No. 37. Information for the patient must be in plain English and must clearly describe the nature, use, and limitations of the test. All measurement must be expressed in SI units (DR4 p. 223).

For these goods, copies of the label and patient information must be submitted with the application for listing on the ARTG.

DEVICES INCORPORATING MATERIAL OF HUMAN ORIGIN

IVD goods that incorporate material of human origin must comply with the requirements in TGO 34—*Standards for Diagnostic Goods of Human Origin*. The sponsor must ensure that these goods comply with TGO 34 and that the evidence to substantiate this claim can be made available to the TGA upon request (DR4 p. 223).

DEVICES USED FOR THE DIAGNOSIS OF INFECTION WITH HUMAN IMMUNODEFICIENCY VIRUS (HIV) AND HEPATITIS C VIRUS (HCV)

Applications for registration of IVD goods for diagnosis of HIV or HCV are forwarded to the TGA's Conformity Assessment Branch (CAB). The CAB will coordinate the evaluation. The TGA will evaluate the kit integrity while the National Serological Reference Laboratory (NRL) evaluates quality and efficacy aspects.

HIV test kits are specified as being suitable for routine screening or supplemental purposes. The conditions relating to the registration will specify the appropriate category. HCV test kits are categorized as being suitable either for routine screening or for supplemental purposes only, and the condition of entry on the ARTG will specify the category of supply. Once registered, there is no restriction to the supply of HCV test kits approved for screening. However, it is Commonwealth policy that all HCV test kits approved for use as supplemental assays and those using newer technology (e.g., polymerase chain reaction and branched DNA amplification) be supplied only to laboratories approved by state/territory health authorities (DR4 pp. 169-170).

APPLICATION OF THE REGISTRATION NUMBER

Section 20(2) of the Therapeutic Goods Act 1989 requires that the registration number of a therapeutic device must be set out on the label in a prescribed manner. For imported devices, the registration number must be affixed to the device before it is supplied in Australia. The registration number is to be set out (a) on the label on the device or (b) on the label on the outermost level of the packaging in which the device is to be supplied to its user. The registration number must be written so that it is clearly visible to the user. When more than one device is packaged together for supply, the registration number must appear on the outermost label of the outermost package.

The registration number must appear on the main label or on a securely affixed sticker adjacent to the main label. The registration number is preceded by "AUST R" for registrable devices or "AUST L" for listable devices. Each character must be legible and must be printed in a type style having a letter height of not less than 1.0 millimeter.

LABELING OF COMPONENTS AND KITS

For device kits, the provisions of TGO 37 apply to the kit as an entity, but the listing numbers are not required. If the kit is the unit pack of a registered device, the registration number preceded by "AUST R" or "AUST L" as appropriate is required on the kit.

If the device components of a kit are subject to other TGOs (e.g., sutures), the labeling requirements of the specific TGO apply to that component. Exemptions from the labeling requirements of the specific TGO may be sought from the TGA where these requirements cause particular difficulties.

INFORMATION TO BE SUPPLIED FOR SPECIFIC REGISTRABLE DEVICES

Certain therapeutic devices must be registered with the TGA before they can be sold in Australia. Registration means that these devices are subject to premarket evaluation of their quality, safety, and effectiveness. These devices include:

- active implantable medical devices (e.g., cardiac-pacing systems, implantable-drug infusion pumps, implantable central-nervous system pulse generators);
- devices of animal origin;
- breast prostheses not filled with water or saline;
- powered drug-infusion systems;
- extracorporeal-therapy systems;
- heart-valve prostheses;
- devices of human origin;
- intraocular lenses (IOLs);
- intraocular visco-elastic fluids;
- IUCDs;
- barrier contraceptive devices;
- breast prostheses filled with saline;
- disinfectants and sterilants; and
- HIV/HCV IVD kits.

The TGA has established specific requirements for information that must be submitted with the registration application. These requirements are outlined in Block 2 of the *Australian Medical Device Requirements Version 4* under the Therapeutic Goods Act 1989 (DR4). In some cases, these requirements define specific information that must be included in the labeling of the devices, either directly or by reference to standards.

HEART-VALVE PROSTHESES

A heart-valve prosthesis can be either a mechanical heart valve or a biological heart valve. The labeling requirements for these devices are contained in ISO 5840:1996, *Cardiovascular implants—Cardiac Valve Prostheses*. In addition, if resterilization is recommended, the instructions supplied to the practitioner must include details of approved methods (DR4 p. 126).

As a condition of registration, the heart-valve prosthesis manufacturer may be required to maintain a register containing the names (or identifiers) of all heart-valve recipients and treating physicians. The minimum data set for the register includes the manufacturer and sponsor identifier(s), device model and catalogue numbers, type and size of the valve, device serial and batch number, patient identifier, hospital identifier, physician identifier, and procedure date. The manufacturer will need to provide a registration form to collect the required information.

ACTIVE IMPLANTABLE MEDICAL DEVICES (AIMDs)

An AIMD is any medical device that (DR4 p. 85):

- Relies on any source of energy other than that directly generated by the human body or gravity for its operation and performance;
- Is designed for implantation, in full or in part, by surgical or medical operation into the human body, or by medical intervention into a natural orifice; and
- Is intended to remain in place after the procedure.

The labeling for these devices must conform to the requirements in TGO No. 37. Detailed labeling requirements for these devices are contained in EN 45502-1, *Active Implantable Medical Devices – Part 1: General Requirements for Safety, Marking and Information to be Provided by the Manufacturer*. EN 45502-1 is the technical equivalent of ISO 14708-1.

DRUG-INFUSION SYSTEM (POWERED, NONIMPLANTABLE)

These are nonimplantable devices that are intended to regulate the flow of liquids into the patient under positive pressure generated by the pump. Labeling requirements for these devices are described in IEC 60601-1 and its applicable collateral standards, including IEC 60601-1-2 on electromagnetic compatibility and IEC 60601-1-4 on safety of programmable electronic medical systems. Particular requirements are found in IEC 60601-2-24, *Particular Requirements for Infusion Pumps and Controllers*. Labeling on pump administration sets must clearly state the device with which the set is recommended to work and the recommended change period of the set (DR4 p. 109).

BREAST PROSTHESES (NOT SALINE OR WATER)

These are implantable breast prostheses that are constructed of an outer polymeric shell and contain an inner filling material that is not water or saline (e.g., silicone gel or soya bean lipid).

The instructions for use must describe how to identify the make, model, and batch number of the implant using some imaging modality. As a condition of registration, the nonsaline breast-prosthesis manufacturer may be required to maintain a register containing the names (or identifiers) of all recipients and the device identification. The manufacturer will need to provide a registration form to collect the required information. The manufacturer should provide an implant-identification card for the patient and at least two self-adhesive labels with the implant details necessary for product identification. Advice should be given that the labels are to be attached to the hospital's and surgeon's records of the patient.

THERAPEUTIC DEVICES OF ANIMAL ORIGIN

The labeling for these devices must conform to the requirements in TGO No. 37. The outer package label should include a statement indicating the device contains material of animal origin (e.g., bovine, ovine). The product literature should include the following statement, or one of similar intent, in information provided to the patient (DR4 p. 95):

> THIS DEVICE IS DERIVED FROM ANIMAL TISSUE AND, ALTHOUGH CARE HAS BEEN EXERCISED IN ITS MANUFACTURE TO MINIMIZE PATHOGEN CONTENT, A POTENTIAL RISK IS PRESENT, AND ABSOLUTE FREEDOM FROM INFECTIVE AGENTS CANNOT BE GUARANTEED.

THERAPEUTIC DEVICES OF HUMAN ORIGIN

The labeling for these devices must conform to the requirements in TGO 37. The outer package label should include a statement indicating the device contains material of human origin. The product literature should include the following statement, or one of similar intent, in information provided to the patient (DR4 p. 133):

THIS DEVICE IS DERIVED FROM HUMAN TISSUE AND, ALTHOUGH CARE HAS BEEN EXERCISED IN ITS MANUFACTURE TO MINIMIZE HUMAN PATHOGEN CONTENT, A POTENTIAL RISK IS PRESENT, AND ABSOLUTE FREEDOM FROM INFECTIVE AGENTS CANNOT BE GUARANTEED.

The first application to supply dura mater/ophthalmic products must include a copy of all product literature for review. The package insert must include the following information for the practitioner (DR4 p. 136):

- Information from the US Center for Disease Control suggests that evidence of Creutzfeldt-Jakob disease is found in 1/10,000 autopsies.
- The inactivation procedures performed during the manufacturing process cannot be relied on to completely inactivate the agent of Creutzfeldt-Jakob disease.
- The Department of Health and Aging (DOHA) recommends that the use of dura grafts be restricted to indications where there is no satisfactory alternative.
- A list of any other virus not tested for and not destroyed by the decontamination process must be included in the package insert.

INTRAOCULAR LENSES (IOLs)

The labeling for these devices must conform to the requirements in TGO 37. Where the IOL requires specific cleaning and/or disinfection solutions or accessories, a description of containers, composition of solutions, any equipment needed, and their specifications must be included in the instructions for use.

INTRAUTERINE CONTRACEPTIVE DEVICES (IUCDs)

The prescribing information and instructions for use of an IUCD should include information on (DR4 p. 153):

- contraindications for use;
- recommended time and technique for insertion;
- use of accessory inserting device;
- trimming of withdrawal tail (if any);
- complications during insertion;
- recommended maximum in utero residence time;
- techniques for location in situ;
- techniques for removal; and
- details of shelf-life and storage conditions

All product information, promotional literature, and pre- and postinsertion leaflets for the patient must include (DR4 p. 154):

- the name and a descriptive diagram of the IUCD,
- the recommended removal time,
- details on possible side effects and complications along with recommended action to be taken; and
- instructions for checking the position of the IUCD.

Barrier Contraceptive Devices

Barrier contraceptive devices include products such as the female condom, cervical cap, an so on. In addition to the requirements in TGO 37, the label must include (DR4 p. 159):

- the date of manufacture,
- the shelf life or expiration date, and
- storage conditions.

Consumer information must be written in plain English, and must fully describe the use, effectiveness, contraindications, warnings, and method of disposal.

Saline Breast Prostheses

Breast prostheses containing only saline that are manufactured using established materials and technology and are intended by the manufacturer to be left permanently in place are registrable devices in Australia.

The instructions for use must describe how to identify the make, model, and batch number of the implant using some imaging modality. As a condition of registration, the saline breast-prosthesis manufacturer may be required to maintain a register containing the names (or identifiers) of all recipients and the device identification. The manufacturer will need to provide a registration form to collect the required information.

HIV/HVC IVD Kits

These are test kits intended for the diagnosis of patients infected with HIV or with HCV. The labeling for these test kits must include (DR4 p. 173):

- the test kit name,
- the name of all reagents,
- the sponsor's or manufacturer's name and address,
- the expiration date,
- warnings,
- the lot/batch number, and
- the AUST R number.

THINGS TO REMEMBER

The Therapeutic Goods Act 1989 established a national system of controls relating to the quality, safety, efficacy, and timely availability of therapeutic devices in Australia. The Therapeutic Goods Act 1989 applies to therapeutic goods that are used in Australia regardless of whether they are produced domestically or are imported. The provisions of the *Therapeutic Goods Act 1989* also apply to devices produced for export from Australia.

Failure to label a device in accordance with requirements in the Therapeutic Goods Act 1989 or the regulations promulgated to carry out the provisions of the Therapeutic Goods Act 1989 renders the device misbranded. Misbranding is a violation of the letter of the law, and exposes the sponsor of the device to substantial penalties.

The TGA has developed a minimum set of labeling requirements that are applicable to most therapeutic devices. The requirements are considered to be the minimum information that should appear on the label of a therapeutic device in order to ensure the safe use of the device and to allow the device to be traced to the sponsor and to a particular cycle in the manufacturing process. These requirements are contained in the Therapeutic Goods Act 1989 and in TGO 37.

In Australia, certain therapeutic goods, which would be devices according to the definition in the Therapeutic Goods Act 1989, have been declared not to be therapeutic devices under the Act. Diagnostic goods for *in vitro* use are included in the list of devices declared to be drugs. There are, however, four groups of IVD goods that are required to be listed on the ARTG prior to supply in Australia. All other IVDs are exempt.

6 The Therapeutic Goods Amendment of 2002

In June 1998, Australia signed a Mutual Recognition Agreement (MRA) with the European Union (EU). The EU MRA applies to medical devices manufactured in the EU, Australia, and New Zealand that are subject to third-party conformity assessment. Devices incorporating animal-derived tissues, radioactive materials, *in vitro* diagnostics (IVDs), and devices bearing the CE marking that are manufactured in other countries are excluded (DR4 p. 33).

The MRA authorizes the Department of Health and Aging (DOHA) to recognize that EU conformity-assessment bodies as competent to assess whether or not medical devices conform to the Australian requirements for entry on the Australian Register of Therapeutic Goods (ARTG). Conversely, the EU recognizes the competence of conformity-assessment bodies designated by the DOHA to undertake assessment of medical devices for compliance with the requirements for certification (CE marking) for entry into the EU market. At the time of publication of this book, the Therapeutic Goods Administration (TGA) is the only designated conformity-assessment body in Australia under the terms of the MRA.

This agreement opened the door for Australia to harmonize its regulatory system for medical devices by adopting the essential principles of quality, safety, performance, and vigilance requirements, and the use of international standards as recommended by the Global Harmonization Task Force (GHTF). The new system incorporates elements of the European regulatory requirements, but not the European "notified body" evaluation system.

On March 21, 2002, the Australian Parliament passed the Therapeutic Goods Amendment (Medical Devices) Bill 2002. As the name of the legislation indicates, this regulatory reform is being introduced as an amendment to the Therapeutic Goods Act 1989. The intent of the amendment is to (TGA, New Legislation p. 1):

- align Australia's system for regulating medical devices with internationally accepted best practice based on GHTF guidelines and the EU requirements;
- make no change to the Australian regulatory regime for medicines;
- retain Australia's sovereignty by providing a choice to accept or reject future changes, particularly through the recognition of new standards;
- continue to regulate the quality and safety of therapeutic goods not covered by the EU system, including tampons and disinfectants; and
- provide for a transition to the new scheme.

The Therapeutic Goods Amendment (Medical Devices) Bill 2002 received Royal Assent on April 4, 2002. The provisions of this amendment come into force six months after the Act receives Royal Assent. Consequently, the new harmonization system came into force on October 5, 2002. After that time all new medical devices were required to comply with the new harmonization requirements, except for a small group of products, such as currently exempt products, which will have until October 4, 2004 to meet the new requirements. All medical devices currently approved for use in Australia will have until October 4, 2007 (five years) to comply with the new requirements (TGA, General Information p. 2).

SCOPE OF THE NEW REGULATION

Because the new scheme is very different from the previous regime for regulating therapeutic goods, a new, separate section in the Therapeutic Goods Act 1989 has been developed. This approach extends rather than modifies the existing regulatory regime for devices. The Therapeutic Goods Act retains the old regime for regulating "therapeutic devices" (the term used in the 1989 act) while adding the new scheme for the regulation of "medical devices" (the new term as used in the EU system). For the purposes of the new regulation, a "medical device" is any instrument, apparatus, appliance, material, or other article (whether used alone or in combination, and including the software necessary for its proper application) intended, by the person under whose name it is or is to be supplied, to be used for human beings for the purpose of:

- diagnosis, prevention, monitoring, treatment or alleviation of disease;
- diagnosis, monitoring, treatment, alleviation of, or compensation for an injury or handicap;
- investigation, replacement, or modification of the anatomy or of a physiological process; and/or
- control of conception

and that does not achieve its principal intended action in or on the human body by pharmacological, immunological, or metabolic means, but may be assisted in its function by such means. An accessory to such an instrument, apparatus, appliance, material, or other article is also considered a medical device (TGA, General Information p. 1). Under this definition, medical devices include a wide range of products such as medical gloves, bandages, syringes, condoms, contact lenses, X-ray equipment, heart-rate monitors, surgical lasers, pacemakers, dialysis equipment, baby incubators, and heart valves.

THE APPROACH

The new section of the act provides a complete and integrated system for regulating medical devices. The new section sets out the "essential principles" applicable to all medical devices, and then provides the TGA with broad powers to implement the provisions of the Therapeutic Good Act. Therefore, the regulation of devices will be a two-part scheme—one part regulating therapeutic devices and the other part regulating medical devices. However, only one set of Australian regulations will apply to any one product at any one time.

This approach allows Australia to continue regulating the quality and safety of therapeutic goods not covered by the EU system. The current legislation, which deals with both medicines and therapeutic devices, will continue to function with minimal change and disruption.

DEVICE CLASSIFICATION

Under the Therapeutic Goods Amendment (Medical Devices) Bill 2002, medical devices are classified based on their "intended purpose." Intended purpose is defined as the purpose for which the manufacturer of the device intends it to be used. The manufacturer establishes the intended purpose through statements in:

- the information provided with the device,
- the instructions for use of the device, or
- any advertising material applying to the device.

Most medical devices are classified into one of five classes (I, IIa, IIb, III and active implantable medical devices [AIMDs]). The rules for classifying devices are contained in Schedule 2 of the medical device regulations. All classes will comply with a minimum requirement for safety and performance and will be included on the ARTG.

Many medical devices that are currently listable, such as thermometers, gauze dressings, and stethoscopes, are Class I devices and are subject to the lowest level of regulatory oversight. Many exempt medical devices, such as nonpowered hospital furniture and simple nonsterile, nonpowered surgical and dental instruments will also become Class I. Other listable medical devices, such as hearing aids, dental filling materials and oxygen meters, will become Class IIa devices. Higher-risk listable and registrable medical devices will become Class IIb and III devices. AIMDs are assigned their own classification group. An AIMD is any active medical device that is intended by the manufacturer:

- either to be, by surgical or medical intervention, introduced wholly or partially into the body of a human being; or to be, by medical intervention, introduced into a natural orifice in the body of a human being; and
- to remain in place after the procedure.

All classes of medical devices will be required to demonstrate their conformity with quality, safety, and performance requirements. Manufacturers of Class IIa, IIb, and III devices and AIMDs are required to have a quality system. Class III devices and AIMDs are subject to the most extensive premarket assessment in Australia's new regulatory system.

At the time of publication, the TGA was continuing to consult on the inclusion of IVD medical devices in the new regulatory system.

ESSENTIAL PRINCIPLES

The Therapeutic Goods Amendment (Medical Devices) Bill 2002 establishes a set of essential principles for safety and performance with which all medical devices must comply. The essential principles, including requirements for labels and accompanying documents, are listed in Schedule 1 in the regulations. Compliance with these requirements is mandatory and failure to conform can result in a range of administrative and legal penalties.

STANDARDS

Under the new system, the TGA will continue to list standards that are deemed to demonstrate a level of quality, safety, and performance consistent with the essential principles in the Therapeutic Goods Act in the *Commonwealth of Australia Gazette*. However, compliance with these standards will not be mandatory. Compliance with the relevant standards will lead the DOHA Secretary to presume that the medical device meets the essential principles of safety and performance.

AUSTRALIAN REGISTER OF THERAPEUTIC GOODS (ARTG)

Medical devices offered for sale in Australia will continue to be included on the ARTG. Most of the administrative requirements in Part 3 of the Therapeutic Goods Act will continue to be used for the new system. While medical devices will no longer be "registered" or "listed," the sponsor will still have to apply to the TGA for entry onto the ARTG. The cancellation of entries, recalls, and other administrative actions will be much as it is under the old system (TGA, New Legislation p. 2).

EXPORT OF MEDICAL DEVICES

Medical devices intended by their manufacturer for export only are still required to be included on the ARTG. However, under a special classification rule, these devices are all classified as Class I.

IMPORT OF MEDICAL DEVICES

Imported medical devices must have an in-country sponsor who takes responsibility for the device in Australia. The device itself must meet the same standards of quality, safety and performance as devices produced in Australia. The foreign manufacturer is required to comply with the same level of good manufacturing practice (GMP) as is expected of the manufacturer of similar products in Australia. A sponsor seeking registration or listing of a new therapeutic device must provide evidence of GMP compliance at the overseas manufacturing site(s). When the TGA considers it necessary, the sponsor of a therapeutic device must agree to pay the cost of an audit of the foreign manufacturing site(s). The TGA, however, considers that some countries have GMP audits at least equivalent to those of Australia. This list includes the EU, Switzerland, Japan, Singapore, and the United States. Certification that the manufacturer operates to a satisfactory standard by the regulatory authorities in these countries will be accepted as evidence of GMP compliance.

The sponsor who imports a medical device must ensure that the sponsor's name and address, or some other information identifying the sponsor, is provided with the device.

POSTMARKET REQUIREMENTS

The new regulatory scheme will place greater emphasis on postmarket activities such as vigilance. Under the new scheme, the TGA is proposing that sponsors must advise the admimistration of:

- serious public-health threats or concerns within 48 hours,
- incidents involving serious injury or death within 10 days, and
- nonserious incidents within 30 days.

Manufacturers must establish systems that facilitate systematic review of experience gained during the postmarketing phase and to implement any necessary corrective action. The manufacturer may be required to take action when information demonstrating a deficiency in the labeling, instructions for use, or advertising of a device becomes known (Therapeutic Goods Regulations §6.5(c)).

Manufacturers of most Class IIa, IIb, and III devices and AIMDs are required to have a certified quality system. Periodic inspection by the TGA of the manufacturer's quality system will be required. The requirement for periodic inspection of the manufacturer's quality system under the new regulatory system will be essentially the same as the current system for licensing therapeutic-goods manufacturers.

ESSENTIAL PRINCIPLES FOR LABELING

Schedule 1 of the Therapeutic Goods Amendment (Medical Devices) Bill 2002 sets out the essential principles that apply to all medical devices on the ARTG. Section 13 in Schedule 1 establishes the requirements for the information that must be provided with a medical device. These requirements include general requirements, requirements for the labels, and requirements for the instructions for use.

GENERAL REQUIREMENTS

Taking into account of the training and knowledge of the potential users of the device, the following information must be provided with every medical device:

- Information identifying the device
- Information identifying the manufacturer of the device
- Information explaining how to use the device safely

In particular, the information required by Essential Principles 13.3 must be provided with a medical device. If practical, the required information must be provided on the device itself. If it is not practicable to place all of the required information on the device itself, the information must be provided on the packaging used for the device. In the case of devices that are packaged together because individual packaging of the devices is not practicable, the required information must be placed on the outer packaging used for the devices.

If it is not practicable to place the required information on the device itself or its package labels, the information must be provided on a leaflet or other document supplied with the device (e.g., a packaging insert).

If instructions for use of the device are required, the information required by Essential Principle 13.5 must be provided in the instructions for use. Instructions for the use of a medical device need not be provided with the device, or may be abbreviated, if:

- the device is a Class I medical device or a Class IIa medical device, and
- the device can be used safely for its intended purpose without instructions.

LANGUAGE REQUIREMENT

The required information must be provided in English. However, it may also be provided in any other language.

LABELING FORMAT

The format, content, and location of the required information must be appropriate for the device and its intended purpose.

Any number, letter, symbol, or letter or number that forms part of a symbol used in the required information must be legible and at least one millimeter high. If a symbol that is not included in a recognized medical-device standard is used in the information provided with the device, or in the instructions for use of the device, the meaning of the symbol must be explained in the information provided with the device or the instructions for use of the device.

LABEL REQUIREMENTS

Where practical, the following information must be provided on the medical device itself or on its package labels:

- The manufacturer's name, or trade name, and address
- The intended purpose of the device, the intended user of the device, and the kind of patient for whom the device is intended

- Sufficient information to enable a user to identify the device or, if relevant, the contents of packaging
- Any particular handling or storage requirements applying to the device
- Any warnings or precautions that should be taken in relation to use of the device
- Any special operating instructions for the use of the device
- If applicable, an indication that the device is intended for a single use only
- If applicable, an indication that the device has been custom-made for a particular individual and is intended for use only by that individual (e.g., "Custom-Made Device")
- If applicable, an indication that the device is intended to be used only for clinical or performance investigations before being supplied (e.g., "Exclusively for Clinical Investigation")
- For a sterile device, the word "STERILE" and information about the method that was used to sterilize the device
- If applicable, a statement of the date (expressed as a month and year) up to when the device can be safely used (e.g., expiration date)
- If an expiration date is not required, then a statement of the date of manufacture of the device (This may be included in the batch code, lot number, or serial number of the device.)
- If applicable, the words "For Export Only"

In addition to the information required above, other provisions of these regulations may also require certain information to be provided with a medical device. For example, information identifying the sponsor of particular medical devices imported into Australia must be provided with the device. Also, the conformity assessment procedures may require a mark of conformity to be affixed to medical devices to which the procedures have been applied.

INSTRUCTIONS-FOR-USE REQUIREMENTS

If instructions for the use of a medical device must be provided with the device, the instructions must include the following information, as appropriate:

1. The manufacturer's name, or trade name, and address
2. The intended purpose of the device, the intended user of the device, and the kind of patient for whom the device is intended
3. Information about any risk arising because of other equipment likely to be present when the device is being used for its intended purpose (e.g., electrical interference from electrosurgical devices, or magnetic field interference from magnetic resonance images)
4. Information about the intended performance of the device and any undesirable side effects caused by use of the device
5. Any contraindications, warnings, or precautions that may apply in relation to use of the device
6. Sufficient information to enable a user to identify the device or, if relevant, the contents of packaging
7. Any particular handling or storage requirements applying to the device
8. If applicable, an indication that the device is intended for a single use only
9. If applicable, an indication that the device has been custom-made for a particular individual and is intended for use only by that individual (e.g., "Custom-Made Device")
10. If applicable, an indication that the device is intended to be used only for clinical or performance investigations before being supplied (e.g., "Exclusively for Clinical Investigation")

11. For a sterile device, the word "STERILE" and information about the method that was used to sterilize the device

12. For a device that is intended by the manufacturer to be supplied in a sterile state,
    - an indication that the device is sterile;
    - information about what to do if sterile packaging is damaged; and
    - if appropriate, instructions for resterilization of the device

13. For a medical device that is intended by the manufacturer to be sterilized before use — instructions for cleaning and sterilizing the device that, if followed, will ensure that the device continues to comply with the relevant essential principles

14. Any special operating instructions for the use of the device

15. Information to enable the user to verify whether the device is properly installed and whether it can be operated safely and correctly, including details of calibration (if any) needed to ensure that the device operates properly and safely during its intended life

16. Information about the nature and frequency of regular and preventative maintenance of the device, including information about the replacement of consumable components of the device

17. Information about any treatment or handling needed before the device can be used

18. For a device that is intended by the manufacturer to be installed with, or connected to, another medical device or other equipment so that the device can operate as required for its intended purpose, sufficient information about the device to enable the user to identify the appropriate other medical devices or equipment that will ensure a safe combination

19. For an implantable device, information about any risks associated with its implantation

20. For a reusable device,
    - information about the appropriate processes to allow reuse of the device (including information about cleaning, disinfection, packaging, and, if appropriate, resterilization of the device); and
    - an indication of the number of times the device may be reused safely

21. For a medical device that is intended by the manufacturer to emit radiation for medical purposes, details on the nature, type, intensity, and distribution of the radiation emitted

22. Information about precautions that should be taken by a patient and the user if the performance of the device changes

23. Information about precautions that should be taken by a patient and the user if it is reasonably foreseeable that use of the device will result in the patient or user being exposed to adverse environmental conditions

24. Adequate information about any medicinal product that the device is designed to administer, including any limitations on the substances that may be administered using the device

25. Information about any medicine that is incorporated into the device as an integral part of the device

26. Information about precautions that should be taken by a patient and the user if there are special or unusual risks associated with the disposal of the device

27. Information about the degree of accuracy claimed if the device has a measuring function

28. Information about any particular facilities required for use of the device or any particular training or qualifications required by the user of the device

MEDICAL DEVICES USED FOR A SPECIAL PURPOSE

The new Australian medical device regulations recognize two types of devices used for a special purpose—custom-made medical devices and medical devices intended for clinical investigations or experimental purposes. Part 7 of the medical device regulations provides special conformity assessment procedures for these devices. The manufacturer must retain the documentation relating to these devices for at least 5 years after the manufacture of the last medical devices for which the documentation applies (TGR 2002 §7.6(2)).

CUSTOM-MADE MEDICAL DEVICES

Custom-made medical devices are those manufactured to the specifications of a healthcare professional for the use of a particular individual. The manufacturer must prepare a written statement that includes the following (TGR 2002 §7.2(2)):

- the name and address of the manufacturer
- the name of the individual patient for whom the device is intended
- a description of the effect the device is intended to have on the named patient
- sufficient information to enable the user to identify the device or, if relevant, the contents of the packaging
- the name and business address of the healthcare professional who provided the specification for the device
- the particular design characteristics of the device
- a statement that the device complies with the applicable essential principles in Schedule 1, or a list of any of the applicable essential principles with which the device does not comply and the reasons for noncompliance

The statement must be signed and dated by an individual authorized by the manufacturer.

DEVICES INTENDED FOR CLINICAL INVESTIGATIONS OR EXPERIMENTAL PURPOSES

These are devices intended to investigation or experimental purposes in human subjects. The purpose of the clinical investigation is to demonstrate that, in addition to any other conformity procedures required by the regulations, the device complies with the following essential principles (TGR 2002 §3.11(1)):

- The device does not compromise the clinical condition or safety of a patient, or the safety and health of the user or any other person, when the device is used on a patient under the conditions and for the purposes for which the device was intended (Schedule 1, Section 1).
- The device is designed, produced, and packaged in a way that ensures that it is suitable for one or more of the purposes of a medical device (Schedule 1, Section 3).
- The benefits from using the medical device as intended by the manufacturer outweigh any undesirable side effects arising from its use (Schedule 1, Section 6).

The device should comply with all the applicable essential principles, including those labeling provisions in Section 13 in Schedule 1, except for the aspects under study.

SYSTEM OR PROCEDURE PACK

A system or procedure pack contains therapeutic goods, at least one of which is a medical device, that are used as a unit, either in combination as a system or in a medical or surgical procedure. The manufacturer of a system or procedure pack must prepare a declaration of conformity in relation to the system or procedure pack that includes (TGR 2002 §7.5(2)):

- the name and business address of the manufacturer of the system or procedure pack
- sufficient information for the user to identify the system or procedure pack or, if relevant, the contents of packaging
- a description of each item that is included in the package
- a statement that the manufacturer has evidence:
  - that each medical device has undergone the relevant conformity assessment procedures, and
  - that each medical device in the package complies with the applicable essential principles
- the registration or listing number for each medicine or other therapeutic goods in the package
- a statement that each medical device in the package is intended to be used for its original intended purpose, and that each medicine or other therapeutic goods in the package is intended to be used within the approved indications for use specified by their original manufacturers
- a statement that for each medical device, medicine, or other therapeutic goods in the package:
  - the mutual compatibility of all components has been verified in accordance with any instructions for use provided by the original manufacturer of each item or the approved indications for use of each item, and
  - the system or procedure pack has been manufactured in accordance with those instructions (if any) or indications
- a statement that the instructions for use of the system or procedure pack include the instructions for use provided by the original manufacturer of each item in the package
- a statement that the system or procedure pack has been manufactured, verified, and packaged under a documented system of internal control and inspection that ensures the safety, quality, performance, and effectiveness of each item in the package
- a statement that, if the system or procedure pack is supplied in a sterile state, production quality assurance procedures have been applied to the system or procedure pack in accordance with the original manufacturer's instructions for use, or the approved indications for use, of each item in the package

The declaration of conformity must be signed and dated by an individual authorized by the manufacturer of the system or procedure pack.

THINGS TO REMEMBER

On March 21, 2002, the Australian Parliament passed the Therapeutic Goods Amendment (Medical Devices) Bill 2002. The intent of this amendment to the Therapeutic Goods Act 1989 is to align Australia's system for regulating medical devices with internationally accepted best practice while retaining Australia's sovereignty over the regulation of medical devices. No change to the Australian regulatory regime for medicines is planned. Australia will continue to regulate the quality and safety of therapeutic goods not covered by the EU system, including tampons and disinfectants.

The Therapeutic Goods Act retains the old regime for regulating "therapeutic devices" (the term used in the 1989 act) while adding the new scheme for the regulation of "medical devices" (the new term as used in the EU system). This enables Australia to regulate under the previous system the quality and safety of therapeutic goods not covered by the EU system, including tampons and disinfectants. However, only one set of Australian regulations will apply to any one product at any one time.

Under the new regime, medical devices are classified based on their "intended purpose." Most medical devices will be classified into one of five classes (I, IIa, IIb, III and AIMDs). All classes will comply with a minimum requirement for safety and performance and be included on the ARTG.

The Therapeutic Goods Amendment (Medical Devices) Bill 2002 establishes a set of essential principles for safety and performance with which all medical devices must comply. The essential principles, including requirements for labels and accompanying documents, are listed in Schedule 1 in the regulations. Compliance with these requirements is mandatory, and failure to conform can result in a range of administrative and legal penalties.

Part IV

Canada and Mexico

7 The Food and Drugs Act of Canada

Canadian society places a high priority on ensuring that all of its citizens have equitable access to high-quality healthcare services. Because medical devices play an important role in the delivery of these services, ensuring that these therapeutic and diagnostic devices—most of which are imported into Canada—are safe and effective is given high priority by the Canadian government.

The Canadian government, through a federal department (Health Canada), carries substantial responsibility for the Canadian healthcare-delivery system. Established in 1944, Health Canada derives its authority from acts of Parliament passed in the late 1950s and early 1960s. In partnership with provincial and territorial governments, Health Canada provides national leadership to develop health policy, enforce health regulations, promote disease prevention, and enhance healthy living for all Canadians. The Minister of Health is responsible to Parliament for administering some 20 health-related laws and associated regulations that govern the overall programs and policies of the department.

Within the Canadian system, each of the 10 provincial governments is responsible for the administration of the healthcare-delivery system within its territory, with funding partially provided by the federal government. The federal government, however, controls the sale of drugs and medical devices under the Food and Drugs Act (F&DA) of Canada.

Under the provisions of the F&DA, a medical device is any article, instrument, apparatus, or contrivance (including any component, part, or accessory) manufactured, sold, or represented for use in (F&DA §2.):

- the diagnosis, treatment, mitigation, or prevention of a disease, disorder, or abnormal physical state, or its symptoms in human beings or animals;
- restoring, correcting, or modifying a body function or the body structure of human beings or animals;
- the diagnosis of pregnancy in human beings or animals; or
- the care of human beings or animals during pregnancy and after the birth of the offspring, including care of the offspring

This definition includes contraceptive devices that do not rely on drugs to achieve their intended purpose. The Health Canada medical device notification database contains information on almost 490,000 medical devices. Another 25,000 new devices appear on the Canadian market each year.

BACKGROUND AND GENERAL INTENT OF THE LAW

The movement to regulate medical devices in Canada began in 1973. On March 27, 1973, the Bureau of Biologics and Medical Devices was formed within the Health Protection Branch (HPB) of the then Department of National Health and Welfare. It took the Bureau of Medical Devices a little over two years to produce the draft of an initial set of medical device regulations directed primarily to the regulation of cardiac pacemakers.

The first medical device regulations were passed by the Canadian Cabinet on August 27, 1975, and were published in the *Canada Gazette* on September 24, 1975. These initial regulations established the basic structure of the regulatory system that exists in Canada today.

Initially, the premarket evaluation clause in the regulations was written to cover only cardiac pacemakers. This part of the regulation came into force on November 1, 1977. In addition to cardiac pacemakers, intrauterine contraceptive devices (IUCDs) were placed on the list of devices requiring premarket evaluation because of the widespread injuries caused by certain IUCDs. In 1981, the Director of the Bureau of Medical Devices proposed extending the breadth of premarket evaluation to cover all devices intended to be implanted for 30 days or more. In addition, menstrual tampons and prolonged-wear contact lenses were included in the list of devices requiring premarket evaluation. The expanded scope for the medical device regulations was adopted by the Canadian Cabinet on October 7, 1982, and came into force on April 1, 1983. Since then, test kits for the detection of serological markers indicative of infection with retroviruses associated with acquired immune deficiency syndrome (AIDS) were added to the list of devices requiring premarket evaluation. On January 13, 1994, prolonged-wear contact lenses were removed from the list of devices requiring premarket evaluation.

In 1985, the Bureau of Medical Devices was amalgamated with the Radiation Protection Bureau to create the Bureau of Radiation and Medical Devices. Under this combined bureau, the emphasis on premarket evaluation of certain devices with an established history, such as cardiac pacemakers, was reduced.

However, as a result of the well-publicized problem with silicone breast implants, concern about the ability of the Canadian system to ensure the safety and effectiveness of medical devices was called into question. A Royal Commission was formed to review medical device regulation in Canada. The Medical Devices Review Committee (MDRC) was established in February 1991 to formulate recommendations to the Minister of Health concerning regulation of medical devices and associated activities. Named after its chairman, the Hearn Committee delivered its report to the Minister in May 1992.

The Hearn Committee concluded, among other things, that a need existed to strengthen the organizational structure of the medical device regulatory program. As a result, a new, separate Medical Devices Bureau (MDB) was formed in 1993. The Hearn Committee recommended greater harmonization and closer collaboration with the United States (US) medical device program. They recommended that Canada adopt certain aspects of the US program, including the definition of medical devices, premarket notification, a risk-based classification system, and the concept of substantial equivalence. The committee suggested that for certain high-risk devices, Canada and the United States should agree on uniform acceptability criteria, submission protocols, and a minimum test-data package (MDRC p. 5).

Under the direction of the Minister of Health, Health Canada's Health Protection Branch (HPB) initiated a comprehensive re-engineering of the Canadian medical device regulatory system. This reengineering effort culminated in two significant changes in the Canadian system:

- the adoption of new risk-based medical devices regulations featuring a four-tiered device-classification system similar to that of the European Union (EU), and
- centralized regulation of therapeutic, preventative and diagnostic products in a single organization—the Therapeutic Products Programme (TPP) within the Health Products and Food Branch (HPFB) of Health Canada.

On April 1, 2001, the TPP was split into three distinct organizations:

- the Therapeutic Products Directorate (TPD) responsible for drugs and medical devices;
- the Biologics and Genetic Therapies Directorate (BGTD), responsible for blood, tissues, organs, biologics (including blood products, vaccines, and drugs produced by biotechnology), and radiopharmaceuticals; and
- the Bureau of Compliance and Enforcement (BCE) responsible for inspections and investigations as well as most establishment licensing and related laboratory analysis functions

Within the TPD, the MDB handles the major work on medical devices.

SCOPE OF THE THERAPEUTIC PRODUCT DIRECTORATE REGULATIONS

Regulations and regulatory technical standards for medical devices in Canada are established by Health Canada under the authority of the F&DA. Section 30 of the F&DA authorizes the Minister of Health Canada to promulgate regulations necessary for the orderly implementation of the F&DA. The intent is to initiate and implement programs to ensure that only safe and efficacious medical devices are sold in Canada. The major concerns include potential hazards, deficiencies in performance, and unsubstantiated claims made by manufacturers. With respect to labeling and packaging, the regulations are intended to prevent the purchaser or consumer from being deceived or misled about the design, construction, performance, intended use, quality, character, value, composition, merit, or safety of the medical device. In addition, the F&DA gives the Minister the responsibility for assessing the effect that the release of a device into the environment may have on the environment or on human life or health. The Minister has the power to limit the importation or sale of any devices whose release is judged to have an adverse effect on the environment or on human life or health.

The authority of the Minister to implement the provision of the F&DA is delegated to the TPD.

THE REGULATIONS

The regulations relating to medical devices are published in the *Canada Gazette,* Part II—Statutory Orders and Regulations. The Canadian "Medical Devices Regulations" (CMDR) are divided into five basic sections covering the following areas:

Part 1   General requirements applicable to all medical devices that are not subject to Part 2 or 3, including requirements for labeling of devices
Part 2   Custom-made devices and medical devices to be imported or sold for special access
Part 3   Medical devices for investigational testing involving human subjects
Part 4   Export certificates
Part 5   Transitional provisions, repeal, and coming into force

In general, products that are subject to regulation under the authority of the F&DA are exempted from regulation under other laws intended to protect health and safety.

LABELS AND LABELING

The official definition contained in Section 2 of the F&DA defines a label as "any legend, word or mark attached to, included in, belonging to or accompanying any food, drug, cosmetic, device or package" (F&DA §2). The package includes anything in which a device is wholly or partly contained, placed, or packed. Canadian regulations refer to an inner label and an outer label for regulated products. The inner label is the label on or affixed to the device or its innermost container. The outer label is the one affixed to the container in which the product is displayed or visible under normal conditions of sale (CC&CR §1). If there is only one container, then that container is considered the inner container.

The labeling for a device encompasses the labels on or affixed to the device and its packaging. It also includes package inserts and other material that provide the instructions for use (IFU) needed by the clinician and, when appropriate, the patient.

LABELING AND ADVERTISING

Advertising is defined in the F&DA to include "any representation by any means whatever for the purpose of promoting directly or indirectly the sale or disposal of any food, drug, cosmetic, or device" (F&DA §2). As part of the labeling of the device, the advertising and promotional material is subject to the provisions of the CMDR. Advertising and promotional material is not normally reviewed by the TPD. However, in unusual cases, the TPD may request copies of all promotional literature relating to the device.

ADULTERATION AND MISBRANDING

Section 20 of the F&DA makes it a crime to "label, package, treat, process, sell or advertise any device in a manner that is false, misleading, or deceptive" (i.e., misbranded) (F&DA §20(1)). Section 20 goes on to define a misbranded device as one:

- That creates an erroneous impression regarding its design, construction, performance, intended use, quantity, character, value, composition, merit, or safety;
- Is not labeled or packaged in accordance with applicable regulations; or
- Is not labeled or packaged in accordance with a standard prescribed for the device

In general, a medical device can be considered adulterated if the manufacturer knowingly or recklessly fails to provide a device that is fit for use under the conditions for which its use is recommended in the labeling. This can include providing a device that is contaminated (e.g., an *in vitro* diagnostic (IVD) reagent containing dirt or filth) or one that is unsafe or ineffective for the purpose for which the device is recommended.

For Class II, III, or IV medical devices, the device could be considered adulterated if the manufacturer fails to meet any of the conditions under which the license was issued. These conditions include the device not being safe for the purposes for which it is to be used, the failure of the device to comply with appropriate standards, or the device not being manufactured under acceptable quality-control procedures.

A person designated by the Minister of Health Canada as an inspector for the purpose of the enforcement of the F&DA may seize and detain a medical device where reasonable grounds exist to believe the device is adulterated or misbranded (F&DA §23(1)(d)). A person who contravenes the provisions of the F&DA may be subject to a fine and/or imprisonment (F&DA §31). Where a person has been convicted of a contravention of the F&DA, the court may order that the article involved is forfeited to Her Majesty and may be disposed of as the Minister of Health Canada may direct (F&DA §27(2)).

FALSE OR MISLEADING LABELING

The labeling of a medical device must not be false, misleading, or deceptive in a material particular (F&DA §20(1)). Labeling need not be untrue, forged, or fraudulent to be considered false, misleading, or deceptive. A word, statement, or illustration may be strictly true but yet be misleading to the customer. Section 20 of the F&DA defines misleading labeling as labeling that "is likely to create an erroneous impression" (F&DA §20(1)). A household air purifier, for example, may be misbranded if its labeling were to claim relief from the breathing discomfort associated with respiratory conditions such as allergies, asthma, coughs, sinus colds, and hay fever. The labeling may be considered deceptive because the layperson could be induced by these claims to buy the device in order to secure relief from the effects of these illnesses.

A false impression may be created not only by false or deceptive statements, but also by ambiguity and misdirection. Failure to inform the customer of material facts may render the device misbranded just as much as a blatantly false or exaggerated claim. A label that is silent on important considerations may be just as deceptive as one that makes exaggerated claims.

ADVERTISING AND PROMOTION

The advertising of medical devices is regulated as part of the labeling of the device under the F&DA. The F&DA allows for a broad interpretation by including as advertising "any representation by any means whatever for the purpose of promoting directly or indirectly" the sale of a medical device (F&DA §2). The F&DA prohibits the advertising of a device in a manner that is false, misleading, or deceptive. In relation to a medical device, the advertising is misleading or deceptive if it is likely to create an erroneous impression regarding the design, construction, performance, intended use, quantity, character, value, composition, merit, or safety of the device. Such advertising may render the device misbranded and the perpetrator subject to legal action.

In addition, no person may advertise to the general public any food, drug, cosmetic, or device as a treatment, preventative, or cure for any of the diseases or conditions listed in Table 7.1 (F&DA §3(1)). Further, the F&DA makes it a crime for any person to sell to the general public any food, drug, or device that has been advertised as a treatment, preventative, or cure for any of the diseases listed in Table 7.1 (F&DA §3(2)).

No person shall advertise a Class II, III, or IV medical device for the purpose of sale unless the manufacturer of the device holds a license for that device or the advertisement is placed only in a catalogue that includes a clear and visible warning that the devices advertised in the catalogue may not have been licensed in accordance with Canadian law (CMDR §27). If a licensed Class II, III or IV device had been modified to the extent that the license must be amended, advertising of the

---

TABLE 7.1
Diseases, Disorders, or Abnormal Physical States

Alcoholism	Gout
Alopecia (except hereditary androgenetic alopecia)	Heart disease
Anxiety state	Hernia
Appendicitis	Hypertension
Arteriosclerosis	Hypotension
Arthritis	Impetigo
Asthma	Kidney disease
Bladder disease	Leukemia
Cancer	Liver disease (except hepatitis)
Convulsions	Nausea and vomiting of pregnancy
Depression	Obesity
Diabetes	Pleurisy
Disease of the prostate	Rheumatic fever
Disorder of menstrual flow	Septicemia
Dysentery	Sexual impotence
Edematous state	Thrombotic and embolic disorders
Epilepsy	Thyroid disease
Gall bladder disease	Tumor
Gangrene	Ulcer of the gastrointestinal tract
Glaucoma	Venereal disease

Source: F&DA, Schedule A

device must cease until the amended license is issued by the TPD. The conditions that require the manufacturer to seek an amended medical device license are described later in this chapter.

BRINGING DEVICES TO MARKET IN CANADA

The current Canadian Medical Device Regulations (CMDR) came into force on July 1, 1998. The 1998 regulations replaced the regulatory system that had been on the books since 1975. The 1998 regulations are based on two principles: (1) the level of scrutiny afforded a device should be dependent upon the risk that the device presents; and (2) the safety and effectiveness of medical devices can best be assessed through a balance of quality-systems requirements, premarket scrutiny, and postmarket surveillance.

The CMDR sets out a system for classifying medical devices based on the perceived risk to the patient associated with the device. There are four classes, with Class I containing the lowest-risk devices and Class IV containing the highest-risk devices. The level of scrutiny a device receives will depend upon its classification.

DEVICE CLASSIFICATION

A medical device is classified into one of four classes (I to IV) using the classification rules set out in Schedule 1 of the CMDR. Schedule 1 is divided into two parts. Part 1 applies to all medical devices other that IVD devices. This part sets out 16 rules that the manufacturer must use to determine the classification for a particular device. The classification rules group devices into three broad categories: invasive, noninvasive, or active.

An invasive device is one that is intended to come into contact with the surface of the eye or to penetrate the body, either through a body orifice or through the body surface. A noninvasive device is one intended to remain entirely outside the body although it may contact injured skin or may be used for channeling or storing gases, liquids, tissues, or body fluids for the purpose of introduction into the body by means of infusion or other means of administration. The third broad category includes active devices. An active device is one that depends for its operation on a source of energy other than energy generated by the human body or gravity. Within each category is a set of rules that determines the ultimate classification based upon the intended use of the device. If a medical device can be classified into more than one class, the class representing the higher risk applies.

In Rule 16, the regulation enables the TPD to classify devices based on other issues or concerns not covered in the other classification rules. These devices and their classification are set out in a table in Rule 16. This table may be amended from time to time through the normal rulemaking process in Canada. The devices covered by this rule as of August 2002 are shown in Table 7.2.

The classification rules for IVD devices are covered in Part 2 of Schedule 1. This part sets out nine rules for classification of IVD devices. The classification rules group IVDs into three broad categories based on their association with transmissible agents, their proximity to the patient, and other special conditions.

TABLE 7.2
Medical Devices Other Than IVDs Classified According to Rule 16 of Schedule 1, Part 1

Item	Medical device	Class
1.	Breast implants	IV
2.	Tissue expanders for breast reconstruction and augmentation	IV

Source: CMDR, Schedule 1, Part 1, Rule 16

TABLE 7.3
IVDs Classified According to Rule 9 of Schedule 1, Part 2

Item	IVD Device	Class
1.	Near-patient IVD device for the detection of pregnancy or for fertility testing	II
2.	Near-patient IVD device for determining cholesterol level	II
3.	Microbiological media used to identify or infer the identity of a microorganism	I
4.	IVD device used to identify or infer the identity of a cultured microorganism	I

Source: CMDR, Schedule 1, Part 2, Rule 9

As with medical devices, the regulations enable the TPD to classify IVD devices based on issues or concerns not covered in the other classification rules. These devices and their classification are set out in a table in Part 2, Rule 9. This table may be amended from time to time through the normal rulemaking process in Canada. The devices covered by this rule as of August 2002 are shown in Table 7.3.

ESTABLISHMENT LICENSE

In Canada, a person or entity who imports or sells a medical device must hold an establishment license issued by the TPD. Certain persons or entities are exempted from this requirement. They are (CMDR §44(2)):

- a retailer;
- a healthcare facility;
- the manufacturer of a Class II, III or IV medical device; or
- the manufacturer of a Class I device, if the manufacturer imports or distributes solely through a person who holds an establishment license.

A person requiring an establishment license in order to conduct business in Canada must apply to the TPD, providing the information set out in CMDR Section 45. The application must contain a statement as to whether the activity of the establishment is importation or distribution, or both. A senior official of the establishment must attest that the establishment has documented procedures in place for handling distribution records, complaints, mandatory problem reporting, and recalls. If the establishment imports or distributes Class II, III, or IV devices, a senior official of the establishment must attest that the establishment has documented procedures in place, where applicable, for handling, storage, delivery, installation, corrective action, and servicing of these devices. As part of its cost-recovery program, the TPD will assess the applicant a fee for processing the establishment license. The fee is due and payable at the time of application.

If there are any changes to the information submitted by the holder of the establishment license, the new information must be submitted to the TPD within 15 days of the change (CMDR §46(1)). All establishment licenses expire on December 31 of each year (CMDR §46(2)). The TPD will assess a fee for processing the renewal application.

The TPD must refuse to issue an establishment license if there are reasonable grounds to believe that issuing such a license would constitute a risk to the health or safety of patients, users or other persons (CMDR §47(2)). The TPD may suspend an establishment license if there are reasonable grounds to believe that (CMDR §49):

- the licensee has made a false or misleading statement in the application;
- the licensee has contravened the regulations or any provision of the F&DA relating to medical devices including those responsibilities outlined in the following sections; or
- the failure by the TPD to suspend the establishment license would constitute a risk to the health or safety of patients, users, or other persons.

Normally, the TPD will send the licensee a written notice that sets out the reason for the proposed suspension, any corrective action required to be taken, and the time within which it must be taken. However, the TPD may suspend an establishment license without giving the licensee an opportunity to be heard by giving the licensee a notice in writing that states the reason for the suspension if doing so will prevent injury to the health or safety of patients, users, or other persons, (CMDR §§49-50).

DISTRIBUTION RECORDS

The manufacturer, importer, and distributor of a medical device shall *each* maintain a distribution record with respect to each device sold in Canada. It is important to note that if more than one entity is involved in the distribution of a device in Canada, then each is responsible for maintaining the appropriate distribution records. This provision does not apply to a retailer or a healthcare facility for a medical device that is distributed for use within that facility (CMDR §52).

The purpose of the distribution record is to permit complete and rapid withdrawal of a medical device from the market. Consequently, the record must contain sufficient information to locate devices in case of a recall, and the record must be maintained in a way that allows their timely retrieval when necessary. The manufacturer, importer, and distributor must retain the distribution record for a medical device for either the projected useful life of the device or two years after the date the device is shipped, whichever is longer (CMDR §55).

In addition to the information described above, the distribution record maintained by the manufacturer of an implanted device shall contain a record of the information received on the implant-registration card forwarded to the manufacturer from a healthcare facility (CMDR §54(1)). The implant manufacturer must update this information with any information received from the healthcare facility or the patient. The required contents of the implant-registration card are described in the next chapter.

COMPLAINT HANDLING

The manufacturer, importer and distributor of a medical device shall *each* maintain records of any reported problems relating to the performance characteristics or safety of the device, including any consumer complaints received by the manufacturer, importer, or distributor after the device was first sold in Canada. The complaint record shall detail all actions taken by the manufacturer, importer, or distributor in response to the problems reported (CMDR §57(1)). The manufacturer, importer, and distributor of a medical device must *each* establish, document, and implement procedures that will enable the manufacturer, importer, or distributor to conduct an effective and timely investigation of reported problems and, when justified, conduct an effective and timely recall of the device (CMDR, § 58). This provision does not apply to a retailer or a healthcare facility for a medical device that is distributed for use within that facility (CMDR §57(2)).

MANDATORY PROBLEM REPORTING

Both the manufacturer *and* the importer of a medical device shall make a preliminary and a final report to the TPD concerning any incident that comes to their attention involving a device that is sold in Canada that (CMDR §59):

- is related to a failure of the device, a deterioration in its effectiveness, or any inadequacy in its labeling or in its directions for use; and
- has led to the death or a serious deterioration in the state of health of a patient, user, or other person, or could do so were it to recur.

A manufacturer is not required to report an incident that occurs outside Canada unless the manufacturer has indicated to the regulatory agency of the country where the incident occurred that corrective action is planned (CMDR §59(2)). The manufacturer is required to report any incident that occurs outside Canada if the responsible regulatory agency requires the manufacturer to take corrective action.

If a reportable incident occurs in Canada, a preliminary report shall be submitted to the TPD (CMDR §60):

- within 10 days after the manufacturer or importer becomes aware of an incident that has led to the death or a serious deterioration in the state of health of a patient, user, or other person; or
- within 30 days after the manufacturer or importer becomes aware of an incident that could lead to the death or a serious deterioration in the state of health of a patient, user, or other person if it were to recur.

With regard to an incident that occurs outside Canada, the manufacturer must report the incident to the TPD as soon as possible after notifying the regulatory agency of the country where the incident occurred of their intention to take corrective action, or after the regulatory agency has required the manufacturer to take corrective action.

Implant Registration

The manufacturer of an implant shall provide, with the implant, two implant-registration cards. The minimum required information is described in the following chapter. These implant-registration cards must be printed in both official languages; however, the manufacturer may choose to provide four cards, two in English and two in French (CMDR §66). A member of the staff of the healthcare facility where an implant procedure takes place is required, as soon as possible after the completion of the procedure, to enter the required information on each implant-registration card. One card is given to the implant patient and the other is forwarded to the manufacturer of the implant or to the person designated by the manufacturer for the collection of implant-registration information (CMDR §67(1)).

The patient's name and address must not be entered on the implant-registration card forwarded to the manufacturer or person designated by the manufacturer except with the patient's written consent. The healthcare facility, the manufacturer, or the person designated by the manufacturer may not disclose the patient's name or address, or any information that might identify the patient, unless the disclosure is required by law (CMDR §§67(2-3)).

The manufacturer of an implant may apply to the TPD for authorization to use another implant-registration method. The TPD can authorize the use of an alternative implant-registration method if the TPD determines that the method will enable the manufacturer to achieve the intended purpose as effectively as the use of implant-registration cards (CMDR §68).

Safety and Effectiveness Requirements

All medical devices, regardless of their classification, shall be designed and manufactured to be safe. To this end, the manufacturer shall, in particular, take reasonable measures to (CMDR §10):

- identify the risks inherent in the device;
- eliminate these risks, if possible;
- if the risks cannot be eliminated,
  - reduce the risks to the extent possible,
  - provide for protection appropriate to those risks, including the provision of alarms, and
  - provide, with the device, information relative to the risks that remain; and
- minimize the hazard from potential failures during the projected useful life of the device.

A medical device shall not, when used for the medical conditions, purposes, or uses for which it is manufactured, sold, or represented, adversely affect the health or safety of a patient, user, or other person, except to the extent that a possible adverse effect of the device constitutes an acceptable risk when weighed against the benefits to the patient and the risk is compatible with a high level of protection of health and safety (CMDR §11).

A medical device must perform as intended by the manufacturer and must be effective for the medical conditions, purposes, and uses for which it is manufactured, sold or represented (CMDR §12).

During its projected useful life, the characteristics and performance of the medical device must not deteriorate under normal use to such a degree that the health or safety of a patient, user, or other person is adversely affected (CMDR §13).

The characteristics and performance of a medical device must not be adversely affected by the transport or storage conditions specified by the manufacturer (CMDR §14).

Reasonable measures shall be taken to ensure that every material used in the manufacture of a medical device shall be compatible with every other material with which it interacts and with material that it may come into contact with in normal use, and shall not pose any undue risk to a patient, user or other person (CMDR §15).

The design, manufacture, and packaging of a medical device shall minimize any risk to a patient, user or other person from reasonably foreseeable hazards, including (CMDR §16):

- flammability or explosion;
- the presence of a contaminant or chemical or microbial residue;
- radiation;
- electrical, mechanical, or thermal hazards; and
- fluid leaking from or entering into the device.

A medical device that is to be sold in a sterile condition shall be manufactured and sterilized under appropriately controlled conditions, and the sterilization method used shall be validated (CMDR §17).

A medical device that is part of a system shall be compatible with every other component or part of the system with which it interacts and shall not adversely affect the performance of that system (CMDR §18).

A medical device that performs a measuring function shall be designed to perform that function within tolerance limits that are appropriate for the medical conditions, purposes, and uses for which the device is manufactured, sold, or represented (CMDR §19).

If a medical device consists of or contains software, the software shall be designed to perform as intended by the manufacturer, and the performance of the software shall be validated (CMDR §20).

The medical device labeling must provide the information specified in CMDR Section 21. These requirements are described in detail in the following chapter.

Class I Medical Devices

A person holding a valid establishment license may manufacture or import for sale a Class I medical device that meets the safety and effectiveness requirements described in the previous section without prior approval of the TPD. The manufacturer must maintain objective evidence to establish that the medical device meets those requirements (CMDR§9(2)). If after reviewing a report or other information brought to their attention, the TPD has reasonable grounds for believing that a Class I medical device may not meet the safety and effectiveness requirements, it may request that the manufacturer submit, on or before a specified day, information to enable the TPD to determine whether the device meets those requirements.

The TPD may direct the manufacturer to stop the sale of a Class I medical device if the manufacturer does not provide the requested information by the date specified. The TPD may also stop the sale if the information provided indicates that the device does not meet the safety and effectiveness requirements, including the labeling requirements specified in the CMDR as well as any standards with which compliance is claimed. The TPD may lift the direction to stop the sale if the manufacturer provides the information requested and:

- corrective action has been taken to ensure that the medical device satisfies the safety and effectiveness requirements or
- the TPD's determination was unfounded.

Class II, III and IV Medical Devices

Unlike Class I medical devices, Class II, III, and IV devices cannot be manufactured or imported for sale without the prior approval of the TPD, even if the devices meet the safety and effectiveness requirements described in the previous section.

Device License

No person shall import or sell a Class II, III, or IV medical device unless the manufacturer of the device holds a valid license for the device (CMDR §27). If a system is licensed, all of its components or parts that are manufactured by the manufacturer of the system are also licensed. For IVD devices, if a test kit is licensed, all of its reagents or articles that are manufactured by the manufacturer of the test kit are also licensed.

General Requirements for a Medical Device License Application

To obtain a license for a medical device, the manufacturer must submit an application to the TPD in a format established by the TPD. The application must contain the following information (CMDR §32(1)):

- The name of the device
- The class of the device
- The identifier of the device, including the identifier of any medical device that is part of a system, test kit, medical device group, medical device family, or medical device group family
- The name and address of the manufacturer as it appears on the device label
- The name and address of the establishment where the device is being manufactured, if different from the name and address of the manufacturer that appears on the device label

For the purposed of the regulation, a device identifier is a unique series of letters or numbers, or any combination of these, or a bar code the manufacturer assigns to a medical device, that identifies it and distinguishes it from similar devices.

If the information and documents submitted with an application are insufficient to enable the TPD to determine whether a medical device meets the safety and effectiveness requirements, the TPD may request the manufacturer to submit, on or before a specified day, additional information necessary for making the determination. In the course of examining the application, the TPD may require the applicant to provide samples of the medical device (CMDR §35). The TPD will assess a fee for examining the license application. The fee is based on the classification of the device and the anticipated/actual gross revenue from the sale of the device in Canada. The fee structure is set out in Schedule 1102 of the Financial Administration Act entitled "Fee in Respect of Medical Devices Regulations."

Class II Medical Device License Application

In addition to the general information required in the preceding section, an application for a Class II medical device license must contain the following (CMDR §32(2)):

- A description of the medical conditions, purposes, and uses for which the device is manufactured, sold, or represented.
- A list of the standards with which the manufacture of the device claims compliance in order to satisfy the safety and effectiveness requirements.
- Attestations by senior officials of the manufacturer that:
  - The manufacturer has objective evidence to establish that the device meets the safety and effectiveness requirements.
  - The device label meets the applicable labeling requirements of the CMDR.
  - In the case of a near-patient IVD device, that investigational testing has been conducted on the device using human subjects representative of the intended users and under conditions similar to the conditions of use.
  - Based on an audit by an organization that performs quality-system audits, the quality system under which the device is manufactured satisfies CAN/CSA-ISO 13488-98, *Quality Systems – Medical Devices – Particular Requirements for the Application of ISO 9002*, as amended from time to time. The quality-system requirement was scheduled to come into force on July 1, 2001. However, the TPD decided to postpone the in-force date until January 1, 2003. An 18-month voluntary implementation phase commenced on July 1, 2001. Quality-system certificates could have been provided on a voluntary basis with all new device license applications and license renewals in 2001 and 2002.

Class III Medical Device License Application

In addition to the general information, an application for a Class III medical device license must contain the following (CMDR §32(3)):

- A description of the device and of the materials used in its manufacture and packaging
- A description of the features of the device that permit it to be used for the medical conditions, purposes and uses for which it is manufactured, sold, or represented
- A list of the countries other than Canada where the device has been sold, the total number of units sold in those countries, and a summary of any reported problems with the device and any recalls of the device in those countries

- A list of the standards with which the manufacturer of the device claims compliance in order to satisfy the safety and effectiveness requirements
- A summary of all studies on which the manufacturer relies to ensure that the device meets the safety and effectiveness requirements, and the conclusions drawn from those studies by the manufacturer
- In the case of a device to be sold in a sterile condition, a description of the sterilization method used
- A copy of the device labels
- In the case of a near-patient IVD device, a summary of the investigational testing conducted on the device using human subjects representative of the intended users and under conditions similar to the conditions of use
- A bibliography of all published reports dealing with the use, safety, and effectiveness of the device
- After January 1, 2003, an attestation by a senior official of the manufacturer, based on an audit by an organization that performs quality-system audits, that the quality system under which the device is designed and manufactured satisfies CAN/CSA-ISO 13485-98, *Quality Systems – Medical Devices – Particular Requirements for the Application of ISO 9001*, as amended from time to time

Class IV Medical Device License Application

In addition to the general information, an application for a Class IV medical device license must contain the following (CMDR §32(4)):

1. A description of the device and of the materials used in its manufacture and packaging
2. A description of the features of the device that permit it to be used for the medical conditions, purposes and uses for which it is manufactured, sold, or represented
3. A list of the countries other than Canada where the device has been sold, the total number of units sold in those countries, and a summary of any reported problems with the device and any recalls of the device in those countries
4. A risk assessment comprising an analysis and evaluation of the risks, and the risk-reduction measures adopted to satisfy the safety and effectiveness requirements;
5. A quality plan setting out the specific quality practices, resources, and sequence of activities relevant to the device
6. The specifications of the materials used in the manufacture and packaging of the device
7. The manufacturing process of the device
8. A list of the standards that the manufacture of the device claims compliance with in order to satisfy the safety and effectiveness requirements
9. Detailed information on all studies that the manufacturer uses to ensure that the device meets the safety and effectiveness requirements, including
   - Preclinical and clinical studies,
   - Process-validation studies,
   - If appropriate, software-validation studies, and
   - Literature studies
10. In the case of a medical device other than an IVD device manufactured from or incorporating animal or human tissue or their derivative, objective evidence of the biological safety of the device

11. In the case of a near-patient IVD device, detailed information on investigational testing conducted on the device using human subjects representative of the intended users and under conditions similar to the conditions of use

12. A summary of the studies referred to in item 8 above and the conclusions drawn from those studies by the manufacturer

13. A summary of the investigational testing referred to in item 10 above and the conclusions drawn from that testing by the manufacturer

14. A bibliography of all published reports dealing with the use, safety, and effectiveness of the device

15. A copy of the device labels

16. After January 1, 2003, an attestation by a senior official of the manufacturer, based on an audit by an organization that performs quality-system audits, that the quality system under which the device is designed and manufactured satisfies CAN/CSA-ISO 13485-98, *Quality Systems – Medical Devices – Particular Requirements for the Application of ISO 9001*, as amended from time to time

Amended Medical Device License Application

If the manufacturer proposes to make one or more of the changes described below, the manufacturer must submit to the TPD, in a format established by the TPD, an application for a medical device license amendment. The application must include the information and documents set out in the previous sections that are relevant to the change. The changes that require an amended license are (CMDR §34):

- a change that would affect the class of the device;
- a change in the name of the manufacturer;
- a change in the name of the device; or
- a change in the identifier of the device, including the identifier of any medical device that is part of a system, test kit, medical device group, medical device family, or medical device group family.

In the case of a Class II medical device, an amendment is required if there is a change in the medical conditions, purposes, or uses for which the device is manufactured, sold, or represented. In the case of a Class III or IV medical device, an amendment is required if there is any significant change in the device. A significant change means any change that could reasonably be expected to affect the safety or effectiveness of a medical device. It includes a change to any of the following (CMDR §1):

- The manufacturing process, facility, or equipment
- The manufacturing quality-control procedures, including the methods, tests, or procedures used to control the quality, purity, and sterility of the device or of the materials used in its manufacture
- The design of the device, including its performance characteristics, principles of operation and specifications of materials, energy source, software, or accessories
- The intended use of the device, including any new or extended use, any addition or deletion of a contraindication for the device and any change to the period used to establish its expiration date

Issuance of License

If the TPD is satisfied that all the conditions specified in the regulations have been met and the safety and effectiveness requirements are met, the TPD will issue a new medical device license or amend an existing license. Annually before November 1, every manufacturer of a licensed medical device must furnish the TPD with a statement signed by the manufacturer or by a person authorized to sign on the manufacturer's behalf confirming that all the information and documents supplied by the manufacturer with respect to the device are still correct or describing any change to the information and documents supplied by the manufacturer with respect to the device, other than those to be submitted pursuant to a medical device license amendment (CMDR §43(1)). If the manufacturer fails to comply with this requirement, the TPD may cancel the medical device license (CMDR §43(2)).

If the holder of a medical device license discontinues the sale of the medical device in Canada, the licensee shall inform the TPD within 30 days after the discontinuance, and the license shall be cancelled at the time that the TPD is informed (CMDR §43(3)).

Foreign Manufacturers

A manufacturer located outside Canada may be exempted from submitting the information usually required for a Class II, III, or IV medical device license application if (CMDR §33(1)):

- the applicant is governed, in that country, by a regulatory authority that is recognized by the TPD; and
- the application is accompanied by a certificate of compliance and a supporting summary report, issued by a conformity-assessment body of that country that is recognized by the Minister of Health, which certify that the medical device meets the Canadian safety and effectiveness requirements.

The TPD will, on request, make available to any interested persons the list of recognized regulatory authorities and conformity-assessment bodies of countries other than Canada.

STANDARDS FOR MEDICAL DEVICES

Under the provisions of the F&DA, the TPD may establish performance and safety standards for medical devices. When a standard has been established for a device, it is a violation of the F&DA to label, package, sell, or advertise any article in a manner that is likely to be mistaken for that device, unless the article complies with the established standard (F&DA §21).

While the law allows the TPD to establish mandatory device standards, the current regulatory approach is to use standards in a manner similar to that implemented in the European medical device directives and the US Food and Drug Admnistration (FDA) standards recognition program. Under CMDR Sections 32(2)(b), 32(3)(d), and 32(4)(h), a manufacturer must provide a list of the standards complied with in the design and manufacture of the device to demonstrate that minimum safety and effectiveness requirements have been met. This applies to applications for new and amended medical device licenses, applications for investigational-testing authorizations, and applications for custom-made devices and devices to be imported or sold for special-access authorizations (TPD Standards Policy §2).

Sections 10 through 20 of the CMDR specify safety and effectiveness requirements while Section 21 specifies general labeling requirements that all medical devices must meet. Because these requirements are stated in general terms, both manufacturers and the TPD will frequently need clearly defined criteria for determining whether a device meets the requirements. One way to provide such criteria is to make use of medical device standards issued by national or international standards-writing organizations. In the past, the TPD has informally used medical device standards, in whole or in part, in its premarket and postmarket evaluations.

The TPD has published a policy statement that establishes a procedure for the use of recognized standards for demonstrating that a medical device meets minimum safety and effectiveness requirements. The policy was published on April 12, 2002, and is available through the TPD Web site. The following discussion is based on that policy.

RECOGNITION OF STANDARDS

The TPD has no plans to recognize standards for every kind of medical device. However, where the need arises, a standard may be recognized for any type of device in any of the four classes defined in the regulations. Recognized standards may be vertical (applying to devices of a particular kind, such as latex condoms) or horizontal (applying to characteristics common to a broad range of device types, such as biocompatibility or electrical safety). When determining if a standard should be recognized, the TPD will consider if the proposed standard fulfills some or all of the following needs (TPP Draft Standards Policy §4.1.1):

- The standard sets adequate requirements for certain of the safety and effectiveness characteristics of the device below which an unacceptable hazard would exist, and the device would therefore be in violation of the regulations.
- The standard sets labeling requirements that meet the labeling requirements of the regulations.
- The standard specifies an acceptable test method for measuring the characteristics of the device and the adoption of this method would simplify the process of assessing the compliance of the device with requirements.

Any standard developed through a consensus process and that is not in conflict with any legislation, regulations, or policies under which the TPD operates, can be a candidate for recognition.

Only published standards are eligible for recognition. The TPD may reference all or part of a standard. Lists of standards being considered for recognition will be posted on the TPD Web site. The standards accepted for recognition will be published in policy documents.

Revisions to standards will not automatically be recognized until the new version has been assessed by the TPD for suitability. However, the TPD will actively monitor draft revisions to a recognized standard so that this decision can be made quickly once the new edition is final. If the TPD decides to recognize the new edition of the standard, the old edition will no longer appear on the list of TPD-recognized standards. After the effective date of the new list, compliance with the old edition of the standard will no longer be acceptable in a new license application.

If a recognized standard contains a normative reference (i.e., requires conformance with all or part of another standard), the normative reference will be recognized to the extent that it is used in the TPD-recognized standard.

The TPD may withdraw recognition of a standard once it is judged by the TPD to no longer be appropriate for meeting the safety and effectiveness requirements.

The TPD will identify eligible standards and review their suitability for recognition. Manufacturers, trade associations, or members of the public may also suggest standards for recognition by presenting a rationale to show that recognition would fulfill the purposes described above. The TPD will give prior notice of its intention to add a standard to its recognized list.

USE OF RECOGNIZED STANDARDS

The TPD will consider the requirements in a recognized standard to be a reasonable interpretation of the minimum safety and effectiveness requirements for the medical devices to which that standard applies. If a medical device complies with the standard, the TPD will consider it to have met those aspects of CMDR Sections 10 through 21 that are addressed by the standard. In cases where the standard does not address all relevant aspects of safety and effectiveness, the TPD will specify other requirements in policy documents or standard operating procedures.

Conformance with recognized standards is voluntary for manufacturers (TPD Standards Policy §4.1.1). A manufacturer may either demonstrate conformance to a recognized standard or may choose to use another method to address safety and effectiveness. For a license to be issued for a device for which a standard(s) has been recognized, a manufacturer must:

1. meet the standard,
2. meet an equivalent or better standard, or
3. provide alternate evidence of safety or efficacy.

In the case of options 2 and 3, detailed information must be submitted with the device license application.

A list of recognized standards is annexed to the TPD Standards Policy. A manufacturer should expect this list to be updated from time to time as new standards are recognized, or new amendments/editions of previously recognized standards are published.

A manufacturer who elects to demonstrate conformance with the safety and effectiveness requirements or the labeling requirements by using one or more recognized standards must submit a Declaration of Conformity. The form of the Declaration of Conformity is set out in Appendix 1 of the TPD Standards Policy.

A manufacturer may use the Declaration of Conformity to demonstrate conformance to a recognized standard in partial fulfillment of the necessary safety and effectiveness evidence in order to obtain:

- a medical device license for a Class II, III, or IV device (CMDR §32) and, if applicable, a medical device license amendment;
- an authorization for special access (CMDR §69);
- an authorization to sell or import a Class III or IV custom-made device (CMDR §70); or
- an authorization for investigational testing of a Class II, III, or IV device (CMDR §80).

The manufacturer must maintain all records and test data relating to a manufacturer's compliance and/or Declaration of Conformity with standards for a period of two years after approval of the device or for the expected design life of the device, whichever is longer (TPD Standards Policy §4.1.2).

If the TPD ceases to recognize a standard, conformance with that standard will no longer be acceptable for obtaining a new device license or an authorization. This includes standards that have been removed from the list of recognized standards because a new edition has been published. However, licenses and authorizations issued under conformance with the old standard will continue to be valid (TPD Standards Policy §4.1.2).

To obtain a device license or authorization, the manufacturer may submit a Declaration of Conformity with a recognized standard. The manufacturer has the option of proposing any alternate means of establishing safety and effectiveness, such as compliance with a standard not recognized by the TPD, or other objective evidence of safety and effectiveness. However, the manufacturer would have to show that the alternate standard or method is equivalent to or better than the recognized standard.

Compliance with a consensus standard may still require submission of data with the license application. For example, compliance with a testing standard may require submission of the test results. Details of data requirements for specific standards will be itemized in further guidance documents. The TPD may also request further information related to the use of the standard during the review of a license application.

In the Declaration of Conformity, the manufacturer must, for each standard with which the device complies (TPD Standards Policy §4.1.2.1):

- identify the recognized standard(s) that was met;
- attest that all the requirements for each standard have been met, except for requirements that do not apply or deviations, in which case the manufacturer must:
  - identify any sections or requirements of the standard that are not applicable to the device;
  - identify any ways in which the standard has been adapted for application to the device in question (e.g., by choosing one of several acceptable test methods specified in the standard); and/or
  - specify any deviations from the standard, such as deviations from an international standard necessary to meet national or provincial regulations;
- specify any differences between the device tested for conformance with the standard and the device to be marketed, and justify the use of the test results in case of differences; and
- provide the name and address of any third-part laboratory or certification body that was employed in determining conformance with the standard.

When a recognized standard describes a test method, but does not specify a performance limit, the TPD requires that the test results be submitted.

The manufacturer must be aware that the TPD review of a specific device may raise issues not addressed by recognized standards. For example, in reviewing a Class III or IV medical device, the TPD may require data from animal testing or clinical testing not addressed in recognized standards. Manufacturers must include all the information necessary to support a determination of safety and effectiveness including safety and effectiveness evidence in the application (TPD Standards Policy §4.1.3).

INVESTIGATIONAL DEVICE EXEMPTION (IDE)

The manufacturer of a medical device may be required to submit evidence of the effectiveness of the device as part of the registration application submitted to the TPD. That information can include the results of clinical trials. Under Canadian law, it is illegal to import or sell a medical device for investigational testing on human subjects except in the following instances (CMDR §80):

- For a Class I medical device, a manufacturer or importer may sell the device to a qualified investigator for the purpose of conducting investigational testing if the manufacturer or importer possesses records that contain all the information and documents required by Section 81 of the CMDR.

- For a Class II, III, or IV medical device, a manufacturer or importer may sell the device to a qualified investigator for the purpose of conducting investigational testing if the manufacturer or importer holds an authorization issued by the TPD and possesses records that contain all the information and documents required by Section 81 of the CMDR.

An investigator is a person who is a member in good standing of a professional association of persons entitled under the laws of a Canadian province to provide healthcare in the province. To be qualified, the investigator must be designated as the person to conduct the testing by the ethics committee of the healthcare facility at which investigational testing is to be conducted.

Once these criteria have been met, the manufacturer or importer may sell the device to any qualified investigator designated by the manufacturer for the purpose of obtaining the required clinical data (CMDR §83). Special requirements for labeling of investigational devices are described in the following chapter.

No person may advertise a medical device that is the subject of investigational testing unless that person holds an authorization issued by the TPD to sell or import the device and the advertisement clearly indicates that the device is the subject of investigational testing and states the purpose of the investigational testing.

SALE OF CUSTOM-MADE DEVICES AND MEDICAL DEVICES FOR SPECIAL ACCESS

The Canadian regulations allow two special types of devices to be imported or sold in Canada. They are custom-made devices and devices for special access. A custom-made device is a medical device that differs from medical devices generally available for sale and is manufactured in accordance with a healthcare professional's written direction giving its design characteristics. A custom-made device is sold for the sole use of a particular patient of that professional, or for use by that professional to meet special needs arising in the course of his or her practice. A special access device is intended for emergency use or for use where conventional therapies have failed, are unavailable, or are unsuitable.

For Class I and II medical devices, the practitioner may apply directly to the manufacturer or importer for the needed devices. For Class III and IV medical devices, the professional must apply to the TPD for an authorization that would permit the manufacturer or importer of the device to sell, or to import and sell, the device to that professional. No person may import or sell a Class III or IV custom-made device or a medical device for special access unless the TPD has issued an authorization for its sale or importation (CMDR §70). The required contents of the application are set forth in Section 71 of the CMDR.

For Class III and IV medical devices, the TPD must understand (1) the reasons the device was chosen for the diagnosis, treatment, or prevention; (2) the risks and benefits that are associated with its use; and (3) the reasons the diagnosis, treatment, or prevention could not be accomplished using a licensed device that is available for sale in Canada. If the application submitted by the practitioner is persuasive, the TPD will issue a letter of permission authorizing the manufacturer or importer to sell the device. The letter of permission will specify (CMDR §72(2)):

- the number of units of the device authorized to be imported;
- the number of units of the device authorized to be sold; and
- the name of the healthcare professional to whom the manufacturer or importer may sell the device.

Special requirements for labeling of custom-made devices or devices for special access are described in the following chapter.

EXPORT OF MEDICAL DEVICES

The F&DA does not apply to any packaged device not manufactured for consumption in Canada and not sold for consumption in Canada. Such devices are exempt from the requirements of the CMDR provided a certificate has been issued by the manufacturer that the package and its contents do not contravene any known requirement of the law of the country to which it is or is about to be consigned. The form and contents of the export certificate are prescribed in Schedule 3 of the CMDR.

It is an offense under Canadian law for a person to sign an export certificate that is false or misleading, or that contains omissions that may affect its accuracy and completeness (CMDR §90). The exporter of a device must maintain, at its principal place of business in Canada, records that contain the completed export certificates. The exporter can be required to submit the export certificates for examination by an authorized inspector. The exporter must retain the export certificate for a period of not less than five years after the date of export (CMDR §92).

The package must be marked in distinct overprinting with the word "Export" or "*Exportation.*"

THINGS TO REMEMBER

Regulations and regulatory technical standards for medical devices in Canada are established by the federal government under the authority of the F&DA. The F&DA authorizes the Minister of Health Canada to promulgate regulations necessary for the orderly implementation of the F&DA. The intent is to initiate and implement programs to ensure that only safe and efficacious medical devices are sold in Canada. The TPP is assigned responsibility for establishing performance and safety standards for medical devices. Within the TPP, the TPD is assigned responsible for review and approval of both drugs and medical devices.

When a standard has been established for a device, it is a violation of the F&DA to label, package, sell, or advertise any article in a manner that is likely to be mistaken for that device unless the article complies with the established standard.

Section 20 of the F&DA makes it a crime to label, package, treat, process, sell, or advertise any device that is adulterated or misbranded. In general, a medical device can be considered adulterated if the manufacturer knowingly or recklessly fails to provide a device that is fit for use under the conditions for which its use is recommended in the labeling. If a person has been convicted of a contravention of the F&DA, the court may order that the article involved be forfeited.

The CMDR sets out a system for classifying medical devices based on the perceived risk to the patient associated with the device. There are four classes, with Class I containing the lowest-risk devices and Class IV containing the highest-risk devices. The level of scrutiny a device receives will depend upon its classification.

In Canada, a person who imports or sells a medical device must hold an establishment license issued by the TPD. In the application for an establishment license, a senior official of the establishment must attest that the establishment has documented procedures in place for the handling of distribution records, complaints, mandatory problem reporting, and recalls. If the establishment imports or distributes Class II, III, or IV devices, a senior official of the establishment must attest that the establishment has documented procedures in place, where applicable, for handling, storage, delivery, installation, corrective action, and servicing in respect of these devices.

A person holding a valid establishment license may manufacture or import for sale a Class I medical device that meets the safety and effectiveness requirements described in this chapter without prior approval of the TPD. No person shall import or sell a Class II, III, or IV medical device unless the manufacturer of the device holds a valid license for the device.

The manufacturer or importer of a medical device may conduct investigational testing on human subjects if additional evidence of the effectiveness of the device is required. For a Class I medical device, a manufacturer or importer may sell the device to a qualified investigator for the purpose of conducting investigational testing if the manufacturer or importer possesses records required by the CMDR. For a Class II, III, or IV medical device, a manufacturer or importer may sell the device to a qualified investigator for the purpose of conducting investigational testing if the manufacturer or importer holds an authorization issued by the TPD.

A practitioner who wishes to obtain a device that could not otherwise be sold in Canada may apply to the TPD in writing for permission to obtain the needed device. The TPD, if convinced that the use of the device is based on compassionate grounds and that the benefits to the patient outweigh the risks associated with the use of the device, will issue a letter of permission authorizing the manufacturer or importer to sell the device.

8 General Medical Device Labeling in Canada

The Canadian government derives its authority to regulate medical devices from the Food and Drugs Act (F&DA). The requirements for labeling of the medical devices sold in Canada are contained in the Canadian "Medical Devices Regulations" (CMDR). These regulations are published in the *Canada Gazette*, Part II – Statutory Orders and Regulations as SOR/98-282. These general requirements may be supplemented by particular requirements in guidance documents issued by the Therapeutic Products Directorate (TPD), in performance and safety standards for medical devices established by the TPD, or in standards recognized by the TPD as demonstrating that a medical device meets minimum safety and effectiveness requirements.

A medical device that is not labeled as required by, or is labeled contrary to, the requirements in the CMDR is deemed to be misbranded under Section 20(1) of the F&DA. Misbranding is a violation of the law and subjects the perpetrator to significant penalties including seizure and forfeiture of the misbranded goods. This chapter explores the parts of the regulations that deal with misbranding of medical devices.

MISBRANDING

The TPD, which is a part of the Health Products and Food Branch (HPFB) of Health Canada, has developed a minimum set of labeling requirements that are applicable to all medical devices. These requirements are contained in Sections 21 through 23 of the CMDR, and include:

- general device requirements,
- prominence of required information,
- requirements for medical devices intended to be sold to the general public, and
- language requirements.

Failure to follow or satisfy these requirements will lead to the device being deemed as labeled in a false, misleading, or deceptive manner (i.e., misbranded). The following sections discuss each of these requirements in detail.

GENERAL LABELING REQUIREMENTS

The F&DA defines a label as any legend, word, or mark attached to, included in, belonging to, or accompanying any device or package (F&DA §2). Canadian law or regulations do not draw a distinction between labels that appear on the device or its packaging and material that accompanies the device such as package inserts, brochures, and leaflets. In this chapter, the term label will be used to describe both labels affixed to the device or its packaging and material accompanying the device.

Section 21 of the CMDR details 10 general labeling requirements. These requirements can be grouped into four basic categories:

- device identification,
- instructions for use (IFU)
- sterile devices and devices with a limited life, and
- Class III or IV devices.

The specific requirements relating to each of these categories is described in more detail in the following sections.

DEVICE IDENTIFICATION

The label of a medical device must uniquely identify the device and distinguish if from similar devices. In this context, a device includes a system, medical device group, medical device family, or medical device group family. For purposes of the CMDR, the following definitions apply (CMDR §2):

- A system is a medical device comprising a number of components or parts that are intended to be used together to fulfill some or all of the device's intended functions and are sold under a single name.
- A medical device group is a medical device comprising a collection of medical devices, such as a procedure pack or tray, that is sold under a single name.
- A medical device family is a group of medical devices that are made by the same manufacturer, have the same design and manufacturing process, and have the same intended use, and they differ only in shape, color, flavor, or size.
- A medical device group family is a collection of medical device groups that are made by the same manufacturer and have the same generic name specifying their intended use, and they differ only in the number and combination of products that comprise each group.

The device identification section of a label includes the following elements:

1. The label must include the name of the device. The name of the device must include all the information necessary for the user to identify the device and to distinguish it from similar devices. The name on the label may describe one device, an administrative grouping of devices sold for convenience under a single name, or a grouping of devices that carry the same generic name specifying the intended use of the devices.
2. The label must include the name and address of the manufacturer. For the purpose of the regulation, the manufacturer is a person who sells a medical device under his or her own name, or under a trademark, design, trade name, or other name or mark owned or controlled by that person, and who is responsible for designing, manufacturing, assembling, processing, labeling, packaging, refurbishing, or modifying the device, or for assigning to it a purpose, whether those tasks are performed by that person or on his or her behalf. The name and address should provide sufficient detail to serve as a postal address. The regulation does not preclude placing the name and address of other persons, such as an importer or distributor, on the label. However, if multiple names appear on the label, the relationship of each name to the device must be made clear. There may be private labeling agreements between a fabricator and a distributor or importer where the distributor or importer's name and product name appear on the label. Such agreements not withstanding, the sole name and address on the label is, by definition, that of the manufacturer. If a device license is required, it is issued to the manufacturer named on the label. Further, the named manufacturer is required to

satisfy the safety and effectiveness requirements for the device. The TPD does not accept qualifying the solely named manufacturer on the label with words such as "imported by" or "distributed by" (TPD Labeling Guidance p. 5).

3. In addition, the label must include the identifier of the device, including the identifier of any medical device that is part of a system, test kit, medical device group, medical device family, or medical device group family. The identifier is a unique series of letters, numbers, or a combination of letters and numbers, or a bar code, which is assigned to a medical device by its manufacturer. The identifier must be sufficiently unique that when taken with the device name, the resulting combination distinguishes it from all other devices. It may be a catalog number, model number, or a bar code.

If the contents are not readily apparent, the label must include an indication of what the package contains, expressed in terms appropriate to the device, such as the size, net weight, length, volume, or number of units. The intent is to provide sufficient information about the package contents to enable the user to make an informed choice when comparing similar devices. The information will also allow the user to pick a size suitable for his or her purposes. Units should be expressed in metric or SI (International System of Units) units. In the case of devices containing natural-rubber latex, this material should be identified (TPD Labeling Guidance p. 6).

INSTRUCTIONS FOR USE (IFU)

The label must include those instructions necessary for the safe and effective application of the device. The IFU include the indications for use of the device. Indications for use is a general description of the disease(s) or condition(s) the device will diagnose, treat, prevent, or mitigate. The indications for use should include, when applicable, a description of the patient population for which the device is intended. The indications for use are generally labeled as such, but may also be inferred from other parts of the labeling, including the directions for use, precautions, and warnings. Specifically, the IFU must include:

1. Unless it is self-evident to the intended user, the IFU must succinctly identify the medical conditions, purposes, and uses for which the device is manufactured, sold, or represented. This description must include the performance specifications of the device if those specifications are necessary for proper use.

There are some devices for which the intended user commonly understands the indications for use. For these devices, labeling the intended use may not be necessary. For example, the uses of certain surgical instruments are obvious to the intended user, such as a stainless-steel scalpels, non-medicated adhesive bandages, or tongue depressors (TPD Labeling Guidance p. 7).

The detail and level of the language used in the labeling should be appropriate to the educational level or expertise of the intended user.

The IFU must contain the device's directions for use, unless directions are obvious to the intended user and are, therefore, not required for the device to be used safely and effectively. The directions for use comprise all the necessary information about the procedures recommended for achieving the optimum performance of the device. This includes any cautions, warnings, contraindications, and possible adverse effects from application of the device. Cautions, warnings, contraindications, and possible adverse effects are important elements of the directions for use because they alert the user to important safety-related aspects of the device. They are frequently grouped together and placed near the beginning of the label (e.g., at the front of the package insert or users manual) in order to make them noticeable and easily found by the intended user. Advice on presenting cautions, warnings, and so on is located in Chapter 27. For the purposes of the CMDR, the following definitions apply:

- CAUTIONS: Sometimes referred to as "precautions," this section of the directions for use alerts the user to the need for exercising special care in order to achieve safe and effective use of the device. Cautions should be written to attract the user's attention, to inform about the seriousness of the hazard, and to recommend steps to avoid the hazard.
- WARNINGS: Warnings describe serious adverse and potential safety hazards that can occur in the proper use, or misuse, of a device, along with ways to limit the consequences in use and mitigating steps to take if they occur. The TPD suggests that when a condition or circumstance may result in death or serious injury, a succinctly worded warning enclosed within a distinctive visual box contained within the labeling should be provided (TPD Labelling Guidance, p. 8).
- CONTRAINDICATIONS: These are conditions, especially any conditions of disease, that render some particular line of treatment improper or undesirable. This section should describe situations where devices should not be used because the risk outweighs any potential therapeutic benefit. Examples could include: "Contraindicated for use in pregnancy," or "Not to be used in a patient who has an implanted Cardiac Pacemaker/Defibrillator" (TPD Labeling Guidance p. 8).
- ADVERSE EFFECTS: This section should list the adverse effects that have been reported in association with the use of the device. A description and the frequency of the most serious adverse effects should also be provided.

Although not specifically mentioned in the regulation, the TPD recommends that the labeling material include, when appropriate, a brief summary of clinical studies used to establish the safety and effectiveness of the device. This summary would describe the design of the studies, how they were conducted, and the results in support of the labeling claims. Providing this type of labeling information is most appropriate for Class III and, in particular, Class IV devices (TPD Labeling Guidance p. 8).

The directions for use should be written at a level appropriate to the training and experience of the user.

2. The label must describe any special storage conditions applicable to the device. Some devices may need to be stored and handled under certain environmental conditions in order to prevent deterioration due to temperature, humidity, atmospheric pressure, light, and so on. The user must be provided with this information in order to decide if such storage conditions are accessible or within their means. The TPD recommends that storage temperatures should be labeled in degrees Celsius (TPD Labeling Guidance p. 9).

STERILE DEVICES OR DEVICES WITH A LIMITED LIFE

For devices sterilized by the manufacturer and sold in a sterile condition, the words "Sterile" and "Stérile" must appear on the label. The absence of the words "Sterile" and "Stérile" indicates that the device is not intended to be sold to the user in a sterile condition.

When applicable, the label must contain the expiration date of the device. The expiration date is determined by the manufacturer on the basis of the component that has the shortest projected useful life. Frequently, for sterile devices, the expiration date is based on the demonstrated integrity of the sterile packing barrier. For other devices, there are other factors that determine the expiration date. For example, the expiration date for an active implantable medical device is frequently determined by the shelf life of the battery. The manufacturer must have objective evidence that demonstrates that the device will perform as intended and will meet its specifications until that date. The date should be expressed in the internationally accepted format:* year (in four digits), month (in two digits), and, when necessary, day (in two digits). The separator for the portions of the date should be a hyphen (-) (TPD Labeling Guidance p. 6).

CLASS III OR IV DEVICES

For a Class III or IV device, the label must display the manufacturer's control number. The control number is a unique series of letters, numbers, or symbols, or any combination of these, that is assigned to a medical device by its manufacturer. The control number provides traceability so that a history of the manufacture, packaging, labeling, and distribution of a unit, lot, or batch of the device can be determined. The control number encompasses both lot and serial numbers.

This is a requirement for Class III and IV devices only. Although not mandatory for Class I and II devices, a control number enhances postmarket traceability (TPD Labeling Guidance p. 5).

PROMINENCE OF REQUIRED INFORMATION

The information required by the CMDR must appear in a legible, permanent, and prominent manner on the label. It must be expressed in terms that are easily understood by the intended user (CMDR §21(2)).

REQUIREMENTS FOR MEDICAL DEVICES INTENDED TO BE SOLD TO THE GENERAL PUBLIC AT A SELF-SERVICE DISPLAY

Self-service implies the absence of an "informed intermediary" such as a physician or other health-care professional that can assist the user in the safe and effective use of the device. Self-service also implies a variety of ways of delivering the device to the customer. These include over-the-counter (OTC) displays, by catalog mail order, and via the Internet. If a medical device is to be sold to the general public without an informed intermediary, the general labeling requirements described above must be (CMDR §22(1)):

- set out on the outside of the package that contains the device, and
- be visible under normal conditions of sale.

When a package that contains a medical device is too small to display all the information on the outside of the package in a manner visible under normal conditions of sale, the IFU shall accompany the device. However, they need not be set out on the outside of the package or be visible under normal conditions of sale (CMDR §22(2)).

LANGUAGE REQUIREMENTS

The directions for use for medical devices that are sold at a self-service display in Canada must be labeled in both English and French. Other medical devices must be labeled in either English or French. If the directions for use are supplied in only one of the official languages at the time of sale, the purchaser may request the directions for use in the other official language. The manufacturer must be prepared to satisfy this request as soon as possible (CMDR §23).

---

* This date format is defined in ISO 8601:1988, *Data elements and interchange ormats–Information exchange–Representation of dates and times.*

IN VITRO DIAGNOSTIC (IVD) DEVICES

In Canada, an *in vitro* diagnostic (IVD) device is considered a medical device and is regulated under the CMDR. Therefore, the labeling for an IVD device must comply with the requirements in Sections 21 through 23 of the CMDR. These requirements are described earlier in this chapter. However, the labeling of an IVD device must convey particular information that is essential for its proper use. For example, the directions for use intended to be used by a layperson should be clearly written in a step-by-step format and include illustrations and drawings where appropriate. Both the professional and lay-user should understand what action must be taken in the case of a particular result and what is the possibility of a false positive or false negative result.

An IVD device must have a label that provides the information specified in Section 21(1) of the CMDR. This label includes, but is not limited to, the package insert, the immediate device-container label, and the reagent/component label. These labels must satisfy the same requirements for the prominence of information and language as the labeling of all other medical devices. IVD devices that are sold through self-service outlets must meet the requirements described in the previous section for medical devices intended to be sold to the general public at a self-service display.

To assist IVD device manufacturers, the TPD has prepared a draft guidance document entitled *Guidance for the Labelling of* In Vitro *Diagnostic Devices*. The material in this section was extracted from that guidance document.

PACKAGE INSERT

A package insert is essential for most IVD devices. The requirements for a package insert described in this section apply to the majority of test kits for all classes of IVD devices. The extent of the information required in the package insert may depend upon the complexity and safety considerations of the test performed using the kit (TPD IVD Guidance pp. 5–11).

Device Identification

The IVD device must be uniquely identified so that the user may distinguish it from similar device. To meet this requirement, the package insert must include:

- The name of the IVD device.
- The name and mailing address of the manufacturer.
- The identifier or catalog number.
- A list of kit contents, including quantities, descriptions, volumes, number of tests, and so on. If more than a single test may be performed using the product, any statement of the number of tests must be consistent with the IFU and the amount of material provided.

Instructions for Use (IFU)

As with other medical devices, the IFU section of an IVD device's package insert must include those instructions necessary for the safe and effective application of the IVD device. This includes:

1. The package insert must clearly indicate intended use(s) and indications for use of the IVD device. The following information should be detailed in this section (TPD IVD Guidance p. 5):
   a. The intended use (e.g. screening, monitoring, diagnostic). For a Class IV IVD device not intended for donor screening, the package insert and the device container label must indicate "**Not for donor screening**."

b. The technology employed in the IVD device (e.g. ELISA, chromatographic).

c. Whether the test is qualitative or quantitative.

d. The specific disorder, condition, or risk factor of interest for which the test is intended (i.e., the analyte to be measured).

e. A description of the patient population for which the IVD device is intended.

f. An indication of where the IVD device is to be used (e.g., in clinical laboratories, alternative-care sites, or home use). The limitations section of the package insert should include any specific training required for test performance or use.

g. The type of specimen(s) required (e.g. serum, plasma).

h. Indication if the IVD device must be used in combination with, installed with, or connected to other medical devices or equipment.

i. Specific contraindications for use (e.g. "Use of this device is contraindicated in recent influenza vaccine recipients..." when considerable cross-reactivity can be expected in recent influenza vaccine recipients).

The package insert should include a section on the performance characteristics of the IVD device and provide a summary of data from the clinical trials upon which the performance of the test is based. Performance characteristics would include such things as sensitivity, specificity, predictive values, reproducibility, repeatability, stability, limits of detection and measurement range, earliest clinical detection in comparison with tests of reference, and so on. When appropriate, the confidence intervals (i.e., 95%) associated with each of the performance characteristics should be provided (TPD IVD Guidance p. 10).

2. The package insert should contain the directions for use unless directions are not required for the device to be used safely and effectively. Most IVD devices will require directions for use. The TPD guidance document on labeling of IVD devices includes the following information under the general heading of directions for use (TPD IVD Guidance pp. 6–10):

a. Components (reagents, supplies, etc.). The description of a component must include the name of the component and a description of the contents in terms of quantity (e.g., number of vials, if applicable), mass and/or volume, or concentration. For reagents, indicate the quantity, proportion, concentration, or activity of each reactive ingredient. For biologics, indicate the source and measure of activity.

The description should include a statement indicating the presence of catalytic or non-reactive ingredients, such as buffers, preservatives, or stabilizers, where this information is needed for the safe and effective use of the test.

The maximum number of tests that can be performed with the stated contents must be specified.

The component description must contain complete directions for preparation (reconstitution, mixing, or dilution) when necessary. Storage instructions for both opened and unopened reagents are to be provided. Information regarding possible deterioration of the reagent (i.e. indicators of reagent, calibrator or quality-control material deterioration, where applicable) must also be provided.

b. Essential components and/or special equipment or instruments not provided. The directions for use must indicate any essential components and/or special equipment or instruments that are not provided with the kit. This should include details such as sizes, numbers, types, quality, and so on. Examples include: incubators, precision pipettes, calibrated thermometers, appropriate disinfectants and disinfection procedures, appropriate reaction vessels (specify glass, polystyrene, polypropylene). For instruments such as microplate readers, indicate the required specifications such as wavelength, bandwidth, absorbance, precision, filters, and so on.

If any dedicated instruments, equipment, or software is required, the directions for use must include the name of the instrument, its model number(s)/version number(s), and a brief description of use or function, performance characteristics/specifications, warnings and precautions, limitations, and so on.

c. Warnings and precautionary statements. The directions for use must indicate appropriate warnings and precautionary statements for the safe and effective use of the IVD device. The use of international symbols and signal words such as "warning" and "caution" are effective in alerting the user to a hazard.

For all classes of IVD devices, the following statement should appear in the directions for use indicate the statement (TPD IVD Guidance p. 7):

For In Vitro Diagnostic Use.

For IVD devices containing material of human or animal origin, TPD requires a disclosure statement such as (TPD IVD Guidance p. 5):

CAUTION: This device contains material of human or animal origin and should be handled as a potential carrier and transmitter of disease.

For IVD devices containing potentially infectious agents, the directions for use should indicate whether any antigens and/or control sera have been inactivated. It must provide a complete description of the tests that have been performed for Hepatitis C Virus (HCV), Heapatitis B Virsu (HBV), Human T-lymphotropic Virus (HTLV), and Human Imumunodeficient Virus (HIV), and the results obtained. If the testing revealed the presence of an infectious agent, the TPD requires a hazard statement to the effect:

HAZARD: This device may transmit [infectious agent] and should be handled with extreme caution. No known test method can offer complete assurance that products derived from human blood will not transmit infectious agents.

d. Specimen collection and handling. The directions for use must describe the specimen to be collected and indicate the necessary patient preparation, precautions, and procedure for collecting the specimen (e.g., removal of particulate matter by centrifugation). Criteria for accepting or rejecting the specimen must be described. Additives and preservatives that are added to preserve the integrity of the specimen must be described. Any special storage and handling requirements to preserve the specimen must be described. Any known interferences must be detailed.

e. Test procedure. The directions for use must provide the user with a description of the test procedure to be followed. This information should include a description of the required amounts of reagents, samples, and controls; incubation schedules; temperature; wavelengths used for measurement; and other relevant environmental conditions under which the device is to be used. Sample selection and handling procedures must to be provided along with performance and turnaround-time specifications. Calibration information, including reference samples, blanks, preparation of a standard curve, indication of the maximum and minimum levels of detection, and so on, should be provided. The quality-control procedures and materials required must be described. The IFU should indicate whether positive and negative controls are required, and what are considered to be satisfactory limits of performance. Finally, the stability of the final reaction product should be explained.

For the individual reagents, the directions for use must contain complete instructions for preparing use-dilutions or mixing of individual reagents. Test volumes and instructions for using the reagent should be included. This information could be provided in the general directions for use or in an alternate section of the package insert dealing with reagents.

f. Results. A step-by-step procedure for calculating the value of the test sample, including appropriate formulas and a sample calculation, must be provided.

g. Interpretation of results. It is important that the directions for use indicate the criteria for accepting or rejecting the test results, and indicate to the user whether further testing is required if a particular result is obtained. For example, this might include requiring duplicate tests if the initial test is reactive. When appropriate, a positive or negative result must be clearly defined with cutoff levels. If the test is designed to provide qualitative results, include an explanation of expected results. If the test requires the interpretation of "visual" results (e.g. colorimetric reactions) include a high-quality photograph or reproduction of results. Indicate to the user the significance of the results obtained. This would include information as to what degree a negative test does or does not exclude the possibility of exposure to, or infection by, the organism, and so on.

h. Limitations of the test. This section must indicate the limitations of the test. This may include an indication that results should only be used in conjunction with other clinical and laboratory data. Any patient and clinical factors that may affect marker levels and factors that should be considered when interpreting test results should be discussed. The necessary qualifications for personnel performing the test or interpreting the test results should be detailed. All known contraindications to using the test, with appropriate references, should be included here unless they are described in a separate section of the package insert.

i. Expected values. The directions for use should indicate the range of expected values based on studies of test results from various populations. Describe how the expected range was established and clearly identify the population(s) that was studied. Appropriate literature references should be provided.

j. Disposal. The directions for use should describe the appropriate decontamination and disposal procedures for used or expired kits or reagents. Disposal of all specimens and kit components must comply with all applicable waste disposal requirements. Decontamination and disposal information may also be provided in the "Warnings and Precautions" section of the package insert.

3. The package insert should indicate the storage conditions necessary to ensure the stability of the IVD device in the unopened state. This includes both the device and the individual reagents. Recommended storage-temperature intervals and other conditions for storage such as light, humidity, and so on, should be stated. Storage conditions also need to be provided for opened or reconstituted/mixed reagents (TPD IVD Guidance pp. 10–11).

Sterile Devices or Devices with a Limited Life

The CMDR requires devices sterilized by the manufacturer and sold in a sterile condition to be labeled with the words "Sterile" and "Stérile." This would apply to the IVD device as a whole or to components provided in the kit. When applicable, the label must contain the expiration date of the device. Again, this could apply to the device as a whole or to components such as reagents. This information could be provided in the package insert. However, it seems most reasonable to provide this information on the immediate container label where it is readily available to the user. The package insert could contain appropriate warnings, such as:

DO NOT USE IF THE STERILE PACKAGE HAS BEEN OPENED OR DAMAGED.

DO NOT USE THE KIT OR ANY KIT COMPONENT PAST THE INDICATED EXPIRATION DATE.

Class III or IV Devices

For Class III or IV IVD devices, the label must display the manufacturer's control number. Once again, this information could be provided in the package insert. However, it seems most reasonable for the manufacturer to provide this information on the immediate container label because package inserts would not be custom-printed for each kit. If the control number is a complex structure encoding information such as manufacturing date, lot numbers, and so on, it may be appropriate to provide the user with an explanation of this coding system in the package insert.

Other Material

The TPD recommends that the date of issue of the directions for use or of any revision should be indicated in the package insert. The TPD also recommends that the package insert include a bibliography of pertinent up-to-date references for the information cited in the text and any other references related to the subject matter (TPD IVD Guidance p. 11).

IMMEDIATE CONTAINER LABEL

The immediate container label should contain much of the same information as the package insert but in an abbreviated form.

Device Identification

The device identification section of the immediate container label should include (TPD IVD Guidance pp. 11–12):

- The name of the IVD device.
- The name and mailing address of the manufacturer.
- The identifier or catalog number.
- A list of kit contents, including quantities, descriptions, volumes, number of tests, and so on. If more than a single determination may be performed using the product, any statement of the number of tests must be consistent with IFU and the amount of material provided.

Instructions for Use (IFU)

The IFU section of the immediate container label should include a summary of the detailed material provide in the package insert. If a package insert is not provided, the immediate container label must include those instructions necessary for the safe and effective application of the IVD device. This includes (TPD IVD Guidance p. 11):

- The immediate container label must clearly and succinctly indicate the intended use(s) and indications for use of the IVDD. An example of an appropriate statement for the immediate container label is:

[ASSAY NAME] FOR THE DETECTION OF ANTIBODIES TO HUMAN IMMUNODEFICIENCY VIRUS TYPES I AND II (HIV-1/HIV-2) IN HUMAN SERUM OR PLASMA.

- Class IV IVD devices not intended for donor screening must indicate "**Not for donor screening**" on the device's immediate container label and package insert.
- When appropriate, the immediate container label should set out any necessary operating instructions. The immediate container label should also provide the user of the IVD devcie with appropriate warnings or precautions. For all IVD devices, the immediate container label should include the statement:

FOR *IN VITRO* DIAGNOSTIC USE

- For IVD devices containing potentially infectious agents, whether inactivated or not, indicate a statement such as:

HANDLE ALL THE REAGENTS AS THOUGH CAPABLE OF TRANSMITTING INFECTION.

- The immediate container label must list the necessary storage instructions, including any special storage conditions, applicable to the IVD device.

Sterile Devices or Devices with a Limited Life

The CMDR requires devices sterilized by the manufacturer and sold in a sterile condition to be labeled with the words "Sterile" and "Stérile." This would apply to the IVD device as a whole or to the components provided in the kit.

When applicable, the label must contain the expiration date of the device. The expiration date is based upon the component of the IVD device having the shortest projected useful life. Expiration dates are required for the unopened IVD device or its components (reagents, calibrators, quality control materials, etc.) and for the opened IVD device or its components if the date is different from the unopened IVD device (TPD IVD Guidance pp. 11-12).

Class III or IV Devices

A control number is required for Class III and IV IVD devices, in order to determine the complete manufacturing history of the product (TPD IVD Guidance p. 12).

REAGENT LABEL

The reagent label should contain the same basic information as the immediate container label.

Device Identification

The device-identification section of the reagent label should include (TPD IVD Guidance pp. 12–13):

- The name of the IVD device and the reagent. For reagents used within a single kit, indicate the name of the reagent and the name of the IVD device. For multipurpose reagents that can be used with a number of kits, the name of the reagent should be sufficient.
- The name and mailing address of the manufacturer.

- If applicable, the identifier or catalog number of the reagent.
- The quantity, proportion, or concentration of each reactive ingredient and, for a reagent derived from a biological material, the source and a measure of its activity.
- If more than a single test may be performed using the product, any statement of the number of tests should be consistent with the IFU and amount of material provided.

Instructions for Use (IFU)

The IFU section of the reagent label should include instructions necessary for the safe and effective application of the reagent. This includes (TPD IVD Guidance pp. 12–13):

- The IFU section of the reagent label must indicate warnings and precautions appropriate to the reagent. For all reagents, this label should include the statement:

FOR *IN VITRO* DIAGNOSTIC USE

- For reagents containing potentially infectious agents, whether inactivated or not, indicate a statement such as:

HANDLE THE REAGENT AS THOUGH IT IS CAPABLE OF TRANSMITTING INFECTION.

- The reagent label must include appropriate storage instructions adequate to ensure the stability of the product. Where applicable, the label should include information such as conditions of temperature, light, humidity, and other pertinent factors. The label for reagents that will be stored in the original bottle after being reconstituted, mixed, or otherwise processed before use should provide appropriate storage instructions for the reconstituted or mixed product.

Sterile Devices or Devices with a Limited Life

When a reagent is sterilized by the manufacturer and sold in a sterile condition, the reagent label should contain the words "Sterile" and "Stérile."

When applicable, the reagent label must contain the expiration date of the individual reagent in both its unopened and opened state (TPD IVD Guidance pp. 11-12).

Class III or IV Devices

A control number is required for Class III and IV reagents in order to determine the complete manufacturing history of the product (TPD IVD Guidance p. 13).

LABELING FOR IVD DEVICES CONTAINING EXPLOSIVE MATERIALS OR COMPONENTS

In addition to the requirements referred to in Section 21 of the CMDR, the TPD requires the label of an IVD device containing explosive material or components to have the following information (TPD IVD Guidance pp. 14–15):

- The identity of the material or the components
- The nature of the potential hazard
- The precautions that should be taken during handling, storage, or disposal of the device in order to avoid an explosion

Guidance on disclosing this type of information taken from the Canadian Consumer Chemicals and Containers Regulations (CC&CR) is presented later in this chapter.

TABLE 8.1
Medical Devices Classified as Implants

Item	Device
1	Heart valve
2	Annuloplasty ring
3	Active implantable device system
3(a)	All models of implantable pacemakers and leads
3(b)	All models of implantable defibrillators and leads
3(c)	Artificial heart
3(d)	Implantable ventricular support system
3(e)	Implantable drug infusion system
4	Devices of human origin
4(a)	Human dura mater
4(b)	Wound covering containing human cells

Source: CMDR, Schedule 2

IMPLANT-REGISTRATION CARD

The manufacturer of a medical device classified as an implant in Schedule 2 of the CMDR must provide, along with the implant itself a means to register that implant. The devices classified as an implant at the time of publication are listed in Table 8.1.

The CMDR specifies the information content of a registration card that meets the intent of the regulation. During the registration process, the manufacturer may propose an alternative method for registering an implant. If the TPD determines the proposed method enables the manufacturer to achieve the purpose established in Paragraph 66(1)(c) of the CMDR as effectively as the use of the implant card, the Minster of Health Canada can authorize the use of the alternative method.

Unless another system has been approved by the Minister of Health Canada, the manufacturer of an implant must provide two implant-registration cards. The minimum required information is (CMDR §66(1)):

- the name and address of the manufacturer;
- the name and address of a representative designated by the manufacturer for the collection of implant-registration information;
- a notice advising the patient that the purpose of the cards is to enable the manufacturer to notify the patient of new information concerning the safety, effectiveness, or performance of the implant, and any required corrective action; and
- a statement advising the patient to notify the manufacturer of any change of address.

The implant-registration card must be designed to enable a person at the healthcare facility where the implant procedure took place to record the following information (CMDR §66(2)):

- The name of the device, its control number, and its identifier, including the identifier of any medical device that is part of a system, test kit, medical device group, medical device family or medical device group family.
- The name and address of the healthcare professional who carried out the implant procedure.
- The date on which the device was implanted.

- The name and address of the healthcare facility at which the implant procedure took place.
- The patient's name and address **or** the identification number used by the healthcare facility to identify the patient. The patient's written consent is required before the patient's name and address may be entered on the implant-registration card that is forwarded to the manufacturer or the manufacturer's representative (CMDR §67(2)). The healthcare facility, the manufacturer, or the manufacturer's representative may not disclose the patient's name or address, or any information that might identify the patient, unless the disclosure is required by law (CMDR §67(3)).

The manufacturer must provide either (CMDR §66(3)):

- two implant-registration cards printed in both English and French, or
- four implant-registration cards, two in English and two in French.

As soon as possible after the completion of the implant procedure, a staff member of the healthcare facility where an implant procedure took place must enter the required information on each implant-registration card. One card is given to the implant patient and the second card is forwarded to the manufacturer of the implant or the person designated by the manufacturer to receive and record the information.

LABELING FOR INVESTIGATIONAL DEVICES

From time to time, the manufacturer or importer of a medical device may be required to submit evidence of the effectiveness of the device as part of the registration process. That information can include the results of clinical trials conducted on human subjects. Under Canadian law, a manufacturer may only sell a medical device for investigational testing on human subjects to a qualified investigator. For a Class II, III, or IV medical device, the manufacturer or importer must also possess an authorization issued by the TPD. In all cases, a device must be have a label that sets out the following (CMDR §86):

- The name of the manufacturer
- The name of the device
- The statements "Investigational Device" and *"Instrument de recherché,"* or any other statement, in English and French, that conveys that meaning
- The statements "To Be Used by Qualified Investigators Only" and *"Réservé uniquement à l'usage de chercheurs compétents,"* or any other statement, in English and French, that conveys that meaning
- In the case of an IVD device, the statements "The performance specifications of this device have not been established" and *"Les spécifications de rendement de l'instrument n'ont pas été établies,"* or any other statement, in English and French, that conveys that meaning

LABELING OF CUSTOM-MADE DEVICES AND MEDICAL DEVICES FOR SPECIAL ACCESS

Custom-made devices and medical devices imported for special purposes must have a label that (CMDR §75):

- sets out the name of the manufacturer,
- sets out the name of the device, and
- specifies whether the device is a custom-made device or is being imported or sold for special access.

EXPORT OF MEDICAL DEVICES

The F&DA does not apply to any packaged device not manufactured for consumption in Canada and not sold for consumption in Canada. The labels must conform to the requirements of the country to which it is or is about to be consigned, and the package is marked in distinct overprinting with the word "Export" or *"Exportation."*

SPECIAL LABELING REQUIREMENTS FOR SPECIFIC DEVICES

Sections 21 through 23 of the CMDR set out the basic labeling requirements applicable to all medical devices. Under the provisions of the F&DA, the TPD can publish additional guidance or requirements as needed in order to ensure safety and effectiveness. This material can appear in guidance documents such as *Guidance for the Labelling of Medical Devices* or in TPD policy documents such as the *Policy on Safety and Effectiveness Requirements for Latex Condoms.** These documents are intended to replace the Schedules and Information Letters relevant to the old Regulations.

LABELING FOR SOFT CONTACT LENSES

Contact lenses are any prosthetic device that covers the cornea, and may cover a portion of the limbus or the sclera, for the purpose of correcting refractive errors of the eye. Soft contact lenses are manufactured from a flexible polymer material. In addition to the information required by Section 21 of the CMDR, the labeling for soft contact lenses shall include the following (TPD Labeling Guidance p. 11):

1. The outer label of the package shall display the correction factor of the contact lens.
2. The outer label, or the package insert, shall contain information indicating:
   - at least two lens-care systems that are recommended by the manufacturer for the contact lens;
   - a warning statement contraindicating the use of noncompatible lens-care products, if applicable;
   - that the safety and effectiveness of contact lenses depends on proper use;
   - that an eye-care professional should be consulted regarding proper use;
   - the recommended period of continuous wear, expressed in hours, or, in the case of an extended-wear lens, in days;
   - the minimum period the contact lens should be left out of the eye before reinsertion;
   - the recommended number of times, if any, that the contact lens can be cleaned;
   - that adequate follow-up by an eye-care professional is essential for the safe use of the contact lens;
   - that infection, with possible permanent damage to vision, could result from the failure to strictly follow recommended directions for use and lens-care procedures;
   - that an eye-care professional should be consulted regarding the use of the contact lens in certain atmospheric or environmental conditions that can cause irritation to the eye;

---

* Therapeutic Products Programme.1998. *Policy on Safety and Effectiveness Requirements for Latex Condoms,* (June 30) Ottawa: Health Canada.

- that, in the event of an adverse reaction to the wearing of the contact lens, including discomfort to the eye, red eye, and blurred vision, the user should immediately remove the contact lens and consult an eye-care professional before resuming use;
- where the contact lens is a cosmetically tinted contact lens, a warning statement that the tinted contact lens can reduce visibility in low-light condition;
- where the contact lens is an extended-wear lens, a warning statement that users of extended-wear lenses have a higher risk of infection and permanent damage to their vision; and
- where the soft contact lens is not an extended-wear lens, a warning statement that the wearing of the contact lens while sleeping increases the risk of infection and permanent damage to vision.

3. Where the preceding information is displayed in a package insert, the following statement is to appear on the outer label:

ATTENTION: READ AND SAVE THE ENCLOSED INFORMATION.

MISE EN GARDE: VEUILLER LIRE ET CONSERVER LES RENSEIGNEMENTS CI-JOINTS.

LABELING FOR MENSTRUAL TAMPONS

A menstrual tampon is a medical device used to absorb menstrual flow. Menstrual tampons have been linked with an increased risk of Toxic Shock Syndrome (TSS). The higher the absorbency of the tampon, the greater the risk the user will develop TSS. The TPD-recommended labeling is intended to provide the consumer with information to help select the tampon with the minimum absorbency needed to control menstrual flow in order to reduce the risk of contracting TSS. In addition to the information required by Section 21 of the CMDR, the labeling for menstrual tampons shall include the following (TPD Labeling Guidance pp. 13–14):

1. Absorbency identification information shall appear on the display panel that is the part of the package displayed or visible under normal conditions of sale or advertisement to the consumer. This absorbency identification is found in Column II of Table 8.2 and it represents the range of absorbency of the menstrual tampon as set out in Column I of Table 8.2. The absorbency of a menstrual tampon must be measured by an accepted test method.
2. Anywhere on the outer label, the statement:

ATTENTION: TAMPONS ARE ASSOCIATED WITH TOXIC SHOCK SYNDROME (TSS). TSS IS A RARE BUT SERIOUS DISEASE THAT MAY CAUSE DEATH.

TABLE 8.2
Menstrual Tampon Absorbency Identification

Item	Column I Range of Absorbency (grams)	Column II Absorbency Identification
1.	Less than or equal to 6	Junior Absorbency
2.	Greater than 6 less than 9	Regular Absorbency
3.	Greater than 9 less than 12	Super Absorbency
4.	Greater than 12 less than 15	Super Plus Absorbency
5.	Greater than 15	Ultra Plus Absorbency

Source: TPD Labeling Guidance p. 14

Mise en garde: Les tampons hygiéniques sont associés au syndrome de choc toxique (SCT). Le SCT se manifeste rarement, mais il n'en constitue pas moins une maladie grave qui peut être mortelle.

3. Information provided on the label or in a package insert shall:
   - explain to the user the warning symptoms and risks of TSS associated with the use of menstrual tampons;
   - advise the user on the duration of use and proper hygiene during use;
   - advise the user to use menstrual tampons with the minimum absorbency needed to control menstrual flow in order to reduce the risk of contracting TSS;
   - explain to the user the various ranges of absorbency, described in Table 8.2, and the corresponding absorbency identification of menstrual tampons sold in Canada by that manufacturer;
   - describe to the user how to compare the ranges of absorbency and the corresponding absorbency identifications to select the tampon with the minimum absorbency needed to control menstrual flow in order to reduce the risk of contracting TSS;
   - advise the user to seek medical attention before using menstrual tampons again if TSS warning symptoms have occurred in the past, or if the user has any questions about TSS or tampon use; and
   - describe to the user the material composition of the tampon.
4. If the preceeding information is provided in a package insert, the following statement is to appear on the outer label:

Attention: Read and save the enclosed information.

Mise en garde: Veuillez lire et conserver les renseignements ci-joints.

Labeling for Contraceptive Devices

Section 3(1) of the F&DA prohibits the advertising to the general public of any contraceptive device except as authorized by the CMDR, Section 24(b). As detailed in Section 21(1), a device label must contain information regarding the directions for use for the device and its performance characteristics. For labeling of natural-rubber latex condoms, refer to the TPD policy document dated 30 June 1998, entitled, *Policy on Safety and Effectiveness for Latex Condoms*. The next three sections describe labeling features that should apply to all contraceptive devices. The last two sections describe labeling features recommended by the TPD to satisfy Section 21(1)(h) and (i) of the CMDR for condoms made of synthetic materials.

Individual Containers for Contraceptives Devices

Each individual contraceptive or container should bear at least the following information (TPD Labeling Guidance p. 15):

- The identity of the manufacturer or distributor (e.g. trademark, name, abbreviated name)
- The manufacturer's identifying reference for traceability (e.g., the batch/lot number)
- The expiry date recommended to be expressed as the year (four digits), month (two digits), and day (two digits) separated by a hyphen (-)

Contraceptive Effectiveness

The communication of information on pregnancy rates to users is important for the safe and effective use of barrier contraceptive devices. The manufacturer should include efficacy data resulting from clinical studies detailing the rate of observed pregnancies that may be expected during routine use of the device over a one year period (TPD Labeling Guidance p. 15).

Prophylactic Effectiveness

Users need to be informed whether or not the contraceptive device is also effective in the prevention of disease. The manufacturer should specify for the user whether or not the contraceptive device provides protection against sexually transmitted diseases (STDs) (TPD Labeling Guidance p. 15).

Individual Containers for Synthetic Condoms

The outside of the individual condom container containing the minimum unit of sale to the customer should bear at least the following information (TPD Labeling Guidance pp. 15–16):

- A description of the condom (e.g., color, tip type, ribbing, lubrication, and material of manufacture such as polyurethane)
- The number of condoms contained
- The nominal width or sizing of the condom
- The name of the manufacturer
- The expiration date indicating year and month, in an unambiguous format
- Instructions for storage
- Whether the condom is lubricated or dry, and, if lubricated, whether spermicidal and the nature of the spermicide

Instructions for Use (IFU) of a Synthetic Condom

The IFU of a synthetic condom should include (TPD Labeling Guidance p. 16):

- the need to remove the condom from the package carefully so as to avoid damage to the product;
- how to put on the condom, and that the condom should be donned prior to any sexual contact with the partner in order to effectively assist in the prevention of STDs;
- the need to withdraw the penis soon after ejaculation, while holding the condom in place at the base of the penis or while holding the collar of the condom in place;
- the advantage of concomitant use of a spermicidal preparation for prevention of pregnancy;
- instructions for the safe and sanitary disposal of the condom; and
- a statement that the condom is for single use only.

LABELING FOR MEDICAL GLOVES

In Canada, gloves used for a medical purpose are considered a medical device and are subject to the requirements of Section 21(1) of the CMDR. This section describes the minimum information the TPD considers necessary to satisfy the requirements of the regulation.

Examination Gloves Only (Sterile and Non-Sterile)

Each smallest unit for dispensing examination gloves shall be labeled with the following information (TPD Labeling Guidance p. 17):

- The brand name of the gloves, if applicable.
- The lot number.
- The number of gloves contained in the dispensing unit.
- Wording to clarify to the user that the gloves are for single use only. For example:

SINGLE USE ONLY

- "Disposable" is not acceptable labeling, because some disposable goods can be reused many times.
- Wording to clarify to the user that the gloves are for medical use. For example:

MEDICAL EXAMINATION GLOVES

- In the case of nonsterile gloves, wording to indicate that the glove is not sterile. For example:

NON-STERILE

- The glove size, using one of the following designations: extra-small, small, unisize, medium, large, extra-large.

Material of Manufacture

The TPD's expectations for the labeling of medical glove composition are described in the following sections. The TPD does not require that the exact wordings given in these examples be used. Manufacturers may choose to add additional informative labeling.

Powdered Natural Rubber Latex

- The TPD recommends that gloves that are labeled "latex," "natural-rubber latex," or "natural-latex rubber" be labeled "natural-rubber latex" for uniformity.
- The manufacturer may choose to include a warning statement regarding allergenicity.
- The manufacturer may choose to label the gloves as "powdered."
- Medical gloves labeled as "low powdered" should have test data to demonstrate less powder than the "regular" powdered gloves.

Powderless Natural Rubber Latex

- The TPD recommends that gloves that are labeled "latex," "natural-rubber latex" or "natural-latex rubber" be labeled "natural-rubber latex" for uniformity.
- The manufacturer may choose to label the gloves as "powderless."
- The label should include descriptive information about the coating or surface treatment that renders the glove "powderless" unless the coating or treatment aids only in donning the gloves and the chemical and physical properties as well as the recommended use, remain unchanged.
- Medical gloves labeled as "powderless" should have test data to demonstrate that the gloves have negligible powder levels and produce negligible airborne particle and allergen levels as compared to "powdered" gloves.

Natural-Rubber Latex with a Thin Inner Polyurethane Coating

- The TPD recommends that gloves that are labeled "latex," "natural-rubber latex", or "natural-latex rubber" be labeled "natural-rubber latex" for uniformity.
- The label should include descriptive information about the chemical and physical properties of the coating (e.g., permeability of the thin inner coating, unless the coating aids only in the donning of the glove).
- The label should include descriptive information if the thin inner coating changes the recommended use of the glove.

Bilayer Neoprene-Latex Rubber

The TPD recommends that medical gloves that are labeled "latex," "natural-rubber latex" or "natural-latex rubber" be labeled "natural-rubber latex" for uniformity, and that the gloves have a "neoprene outer layer" (TPD Labeling Guidance p. 18).

Neoprene

Medical gloves that are made of neoprene should be labeled "neoprene" (TPD Labeling Guidance p. 18).

Polyolefin or Other Thin Film Copolymer

The TPD recommends that medical gloves that are made of the hydrocarbon copolymer should be labeled "polyolefin." Medical gloves of another copolymer composition (e.g., ester copolymer) should be labeled as such to distinguish them from polyolefin (TPD Labeling Guidance p. 18).

Vinyl

The TPD recommends that medical gloves that are made of vinyl should be labeled "vinyl," "polyvinyl chloride," "PVC" or "chlorinated olefin" (TPD Labeling Guidance p. 18).

Hydrocarbon Polymer

The TPD recommends that medical gloves made of hydrocarbon polymer should be generally labeled "synthetic hydrocarbon polymer" or more precisely labeled identifying the polymer used (TPD Labeling Guidance p. 18).

LABELING FOR MEDICAL ELECTRICAL EQUIPMENT

Electromedical equipment is the category that covers most electrical instruments and apparati used in the patient environment, and those electrically powered devices having a direct effect on the safety of the patient. These are medical devices that (a) are electrically powered or generate electrical potentials, and (b) may come in contact with a patient. Implantable devices, such as cardiac pacemakers, are not considered electromedical devices. Until January 13, 1994, electromedical equipment was regulated under a performance standard in Schedule VII of the CMDR. Privy Council Order P.C. 1994-63 revoked this schedule of the CMDR. The specific requirements for electromedical equipment were removed from the CMDR because all of the provincial governments in Canada require that line-powered electromedical equipment be certified as conforming to the applicable parts of the Canadian Electric Code. Because this mandatory certification provides reasonable assurance of electrical safety, the specific requirements in Schedule VII were redundant. As medical devices, however, electromedical equipment is still subject to the general requirements of the CMDR.

Compliance with the Canadian Electric Code can be demonstrated by adhering to the requirements of CSA C22.2 No. 601.1 (CSA). The labeling requirements from Clause 6 of this standard are described here.

CSA C22.2 No. 601.1 is an adaptation of the second edition of the International Electrotechnical Commission (IEC) Standard 60601-1 as amended. The modifications to IEC 60601-1 were limited to aligning the document with essential requirements in the Canadian Electric Code and other statutory requirements. The Canadian adoption made minimal changes to Clause 6 on identification, markings, and documents.

The Canadian Standards Association (CSA) provides certification services for manufacturers who wish to obtain a license to use the appropriately registered CSA marks on their products to indicate conformity with CSA Standards. As of November 30, 1990, manufacturers could, at their discretion, use CSA C22.2 No. 601.1 for certification of electromedical equipment. Unless exempted in the standard, all electromedical equipment must have been certified to CSA C22.2 No. 601.1 as of January 1, 2000.

Markings on Electromedical Equipment

All required markings must be clearly legible and removable only with a tool or by application of considerable force. If the equipment is permanently installed, then all required markings must be visible when the equipment is mounted in its normal-use position. For transportable equipment or stationary equipment not permanently installed, it is preferred that required markings are visible during normal use. However, when necessary, the required marking must become visible when the equipment is moved or has been removed from the rack for dismountable rack units. However, warning statements and operating instructions must be affixed in a prominent location so they are legible with normal vision from the operator's position (IEC 60601-1 p. 45). Where written safety warnings appear as equipment markings, these should appear in the English and the French languages (CSA p. 20).

Markings required on the outside of the equipment depend largely on the type of power source. CSA C22.2 No. 601.1 defines three configurations. They are:

- equipment that receives its energy from the supply mains in the facility where it is used (mains powered),
- equipment that is able to operate from an internal electrical power source (internally powered), and
- equipment that is supplied by another specified power source (other than supply mains).

The markings required on the outside of the equipment are described in Table 8.3. For permanently installed equipment, the nominal supply voltage or voltage range may be marked on either the outside or the inside of the equipment enclosure. The preferred location of the markings is adjacent to the supply-connection terminals.

Electromedical equipment must have on its external surface a permanent label that sets out the equipment type as B, BF, or CF. Each equipment type has associated with it a maximum risk current that can flow through any person in contact with the equipment. A device may not exceed the maximum allowable risk current specified in the risk-current limit in Table IV in Subclause 19.3 of CSA C22.2 No. 601.1 for the labeled equipment type of the device. If the equipment has more than one part in contact with the patient, and the parts in contact with the patient have different degrees of protection, then the appropriate symbols must be clearly marked on each applied part, or on or near the relevant outlets.

TABLE 8.3
Marking on the Outside of Medical Electrical Equipment

Items to be Marked	Mains-Operated Equipment	Internally Powered Equipment[a]	Equipment Supplied from a Specified Power Source[b]
1. Name and/or trade mark and address of the manufacturer or supplier claiming that the equipment complies with CSA C22.2.601.1	*	*	*
2. Model or type identification (i.e., a combination of figures, letters, or both used to identify the equipment)	*	*	*
3. Mains supply characteristics (rated voltage/voltage range, current waveform—a.c., d.c., or dual supply, number of phases, etc.)	***		
4. Supply frequency (Hz)	***		
5. Rated power input	***		
6. Auxiliary-mains socket output	**		
7. Equipment classification[c]	**	**	**
8. Defibrillation protection[d]	**	**	**
9. Mode of operation[e]	**	**	**
10. Fuses	**	**	**
11. Output	**	**	**
12. Physiological effects[f]	**	**	**
13. Category AP/APG equipment (see Appendix C, Symbols 23 and 24)	**	**	**
14. High-voltage terminals accessible without a tool (see Appendix C, Symbol 25)	**	**	**
15. Cooling requirements	**	**	**
16. Mechanical-stability marking if special precautions are required	**	**	**
17. Protective packaging, if required during transport and storage	**	**	**
18. Earth-terminals markings[g]	**	**	**
19. Removable protective means	**	**	**
20. Marking required for medical electrical systems[h]	**	**	**
21. Other legally required markings	**	**	**

* Marking is required on all equipment
** Marking is required if applicable to the equipment.
*** Marking is not required on permanently installed equipment that is marked appropriately on the inside.
[a] Internally powered equipment is capable of operating from an internal electrical power source.
[b] The equipment is intended to be supplied from a power source other than supply mains.
[c] Classification includes the degree of protection against electrical shock (see Appendix C, Symbol 11), the protection against leakage current (see Appendix C, Symbols 17, 18, and 19), and the harmful ingress of liquids using the IP Code defined in IEC 60529.
[d] Amendment 2 introduced new symbols (see Appendix C, Symbols 20, 21, and 22) for defibrillation-proof equipment. For clear differentiation, Symbols 17 and 20 must not be applied in such a way as to give the impression of being inscribed in a square. If protection against defibrillation is partly in the patient cable, then a symbol instructing the operator to consult the accompanying documentation must be displayed near the relevant outlet (see Appendix C, Symbol 14).
[e] If equipment is unmarked, continuous operation is assumed.
[f] Equipment producing physiological effects that cause danger to the patient and/or operator must prominently display a symbol instructing the patient/operator to consult the accompanying documentations for warnings (see Appendix C, Symbol 14). If the device produces non-ionizing radiation such as high-power microwaves, then Symbol 43 in Appendix C must be displayed.
[g] Earth terminal markings include the markings for connection to a potential equalized conductor (see Appendix C, Symbol 8), a functional earth terminal (see Appendix C, Symbol 5), and a protective earth conductor (see Appendix C, Symbol 6).
[h] If the documentation accompanying the medical electrical system gives a warning related to a particular safety hazard from a nonmedical equipment component of the system, then a symbol instructing the patient/operator to consult the accompanying documentations for warnings (see Appendix C, Symbol 14) must be provided on the particular nonmedical electrical equipment or on a particular part of that equipment.

Sources: IEC 60601-1, Subclause 6.1; IEC 60601-1-1, Subclause 6.1.201

When the size of the equipment or the configuration of the enclosure makes it impractical for all of the required information in Table 8.3 to be affixed to the outside of the equipment, then at least the information in items 1, 2, 3, 7, and, if applicable, 12 should be included on the outside of the enclosure. The remaining items may then be included in the accompanying documentation. If

TABLE 8.4
Colors for Indicator Lights[a] and Push Buttons[b] and their Recommended Meanings

Color	Meaning	Required Usage	Recommended Usage
Red	Warning of danger and/or the need for urgent action	X	
Yellow	Caution or attention required		X
Green	Ready for use		X
Any other color	A meaning other than that associated with red or yellow		X

[a] Dot matrix or other alphanumeric displays are not considered to be indicator lights.
[b] An unilluminated push button may be colored red only if it is used to interrupt the function of the equipment in the event of an emergency.

Source: IEC 60601-1, Subclause 6.7 a).

it is impractical to include any markings on the outside of the enclosure, then all of the required information may be placed in the accompanying documentation.

CSA 22.2 No. 601.1 specifies requirements for marking controls and instruments on medical electrical equipment, as well as marking on the inside of the equipment. These markings on the controls must include the following information, as applicable:

- The main power switch must be clearly identified with its "ON" and "OFF" positions indicated by a symbol, adjacent indicator light, or other unambiguous means (see Appendix C, Symbols 9 and 10). Indicator lights and push buttons should follow the conventions listed in Table 8.4.
- Controls and indicators with a safety function must be clearly identified.
- The different positions of control devices and switches must be indicated by figures, letters, or other visual means (see Appendix C, Symbols 26 and 27). If the control is intended to adjust a device setting during normal operation, and changing the setting could cause a safety hazard to the patient, the control must be equipped with:
  - an associated indicating device (e.g., instruments or scale) or
  - an indication of the direction in which the magnitude of the function changes.

A label near the point where the power supply conductors connect to the equipment must show the correct method of connecting the supply connector, unless no hazard would occur if the connections were interchanged. If there is insufficient room to appropriately label the connections on the equipment, the instructions can be included in the accompanying documentation.

In permanently mounted equipment, the terminal for connection to the neutral supply conductor must be indicated by a capital letter "N."

The protective earth terminal must be marked (see Appendix C, Symbol 6) adjacent to the terminal unless the protective earth terminal is part of a detachable power cord.

With permanently connected equipment that has a terminal box that reaches a temperature of more than 75 °C when tested according to the normal temperature test specified in IEC 60601-1, the following statement must be placed near the point of connection (IEC 60601-1 pp. 54–55):

FOR SUPPLY CONNECTION, USE WIRING MATERIALS SUITABLE FOR AT LEAST ... °C.

POUR LE RACCORDEMENT L'ALIMENTATION, UTILISER DES CABLES ET CONDUCTEURS ADAPTS UNE TEMPRATURE D'AU MOINS ... °C.

The labels identifying the correct method of connecting the power supply, the supply neutral terminals, the protective earth terminal, and instructions about high temperature materials may not be affixed to any part that must be removed in order to make the supply connections. In addition, the labels must remain visible after the connections are made.

The presence of parts that carry more than 1,000 V a.c., 1,500 V d.c, or 1,500 V peak value must be clearly marked (see Appendix C, Symbol 25). The functional earth terminal must be marked with the prescribed symbol (see Appendix C, Symbol 5).

For those devices that contain heating elements or heating lamps, the maximum power loading must be permanently and indelibly marked near the heating element or lamp holder.

For devices with internal fuses accessible only with the aid of a tool, the fuse ratings and characteristics must be marked near the fuse. Information on fuses may also appear in the instruction manual accompanying the equipment. However, a marking on the equipment (e.g., a diagram number) must establish a reference to the place in the instruction manual where information on the fuses is located.

For equipment that incorporates batteries, the type of battery and mode of insertion must be marked adjacent to the battery holder.

The device may contain capacitors that remain charged after the equipment has been deenergized. If the capacitors or the circuit parts connected to them become accessible when an access cover is removed, the hazard must be clearly marked.

Documents Accompanying Electromedical Equipment

The electromedical equipment must be accompanied by documents giving the IFU, a technical description of the equipment, and an address to which the user can refer. Any markings required on the device that have not been permanently affixed to the equipment should be explained. All warning statements and explanations of warning symbols on the equipment must be provided in the accompanying documents (IEC 60601-1 pp. 59–63).

Instructions for Use (IFU)

The IFU must contain all of the information that the intended user needs in order to operate the equipment so that it achieves its intended purpose. This includes such things as (IEC 60601-1 pp. 59–61):

- the function of controls, displays, and signals;
- the sequence of operations, including:
  - preparing the device for use;
  - steps to confirm the safe operation of the device including a description of alarms and indicators;
  - operation immediately before use, paying particular attention to abnormal displays, the underlying trouble, and the interlocks to prevent hazardous output;
  - operation during use, including the sequence of operations, the methods for reading displays, and necessary adjustments for maintaining normal function;
  - measures following use to prepare for the next use of the device;
  - measures for both short- and long-term storage of the device;
- instructions for connecting and disconnecting detachable parts and accessories;
- replacement of materials consumed during operation of the equipment;
- a description of recognized accessories, detachable parts, and materials, if the use of other parts or accessories could degrade the minimum safety of the equipment;

- instructions on cleaning, preventive inspections, and maintenance to be performed, including the frequency of such maintenance;
- identification of parts on which preventive inspections, maintenance, and calibration should be performed by others—such as the manufacturer's authorized representatives; and
- the meaning of figures, symbols, warning statements, and abbreviations used on the equipment.

For equipment parts that come into contact with the patient, the IFU should contain a clear description of the methods for cleaning, disinfection, and sterilization of the device. When sterilization is required, the instructions should identify suitable sterilization agents and list temperature, pressure, humidity, and time limits that the equipment can tolerate.

A device may be connected to other equipment for the purpose of sending or receiving signals to or from other equipment (e.g., for display, recording, or data processing). If the signal outputs or inputs are intended only for connection to other equipment that meets the requirements of CSA C22.2 No. 601.1 (i.e., leakage current limits), this must be specified in the IFU.

Some medical electrical devices may be capable of operating both from power-supply mains and from an internal power source (i.e., batteries). If the internal power source is not automatically maintained at a fully usable state, the IFU must contain a warning statement describing the necessity for periodic checking or replacement of the internal power source. If this type of equipment relies on a protective earth conductor in addition to basic insulation for protection against electric shock, the instructions for use must warn the user that the equipment is to be operated from its internal power source if the integrity of the connection to the external protective earthing system is in doubt.

If the device uses primary batteries, the IFU must contain a warning to remove the batteries if the device is not likely to be used for some time.

The IFU for a device that uses rechargeable batteries must contain instructions to ensure safe use and adequate maintenance. If a specific power supply or battery charger is required to ensure compliance with CSA C22.2 No. 601.1, these accessories must be identified in the IFU.

Technical Description

The technical description provides detailed information that supports and supplements the IFU. It should include details on all the markings required on the outside of the equipment (see Table 8.3). In addition, the technical description should explain all the characteristics (or an indication of the location of specific items) that the user needs to understand for the safe operation of the device. This could include such items as:

- an explanation of the device performance, giving special attention to the physiological effect and outputs that may be hazardous to the patient or user;
- the appropriate environmental conditions required for the accurate, safe, and correct use of the device (i.e., ambient temperature, relative humidity, and atmospheric pressure);
- if applicable, a description of the required power source (i.e., voltage, frequency, power rating, and allowable ranges);
- required site preparation and installation procedures, including such items as space required for use, environmental conditions, protective barriers (radiation—ionizing and electromagnetic), and so on;
- the construction of the equipment (e.g., using photographs, block diagrams, wiring diagrams, etc.) with special attention to the interconnection of system elements;

- a description of the mechanical structure required for the safe use of the device, which includes such items as precautions to be taken with movable parts, motor circuits, radiation shields (like those on X-ray and laser equipment), and mobile equipment;
- an explanation of the activation principle for the safe and correct use of the device; and
- for portable devices, the preparation of the device for safe transport.

The technical description should contain the type and rating of fuses utilized in the supply-mains circuit. This is required if the type and rating are not apparent from the rated current and mode of operation of the device.

If the device incorporates interchangeable and/or detachable parts that are subject to deterioration in normal use, the technical description must provide the user with instructions for replacement of those parts.

As part of the technical description of the device, the manufacturer must include a statement that he or she will make information available that will assist the user's appropriately qualified technical personnel to repair those parts that the manufacturer has designated as repairable. This information could include circuit diagrams, component-parts lists, descriptions, and calibration instructions. This information could be provided in a service manual.

If the device cannot withstand the shipping and storage conditions specified in Subclause 10.1 of CSA C22.2 No. 601.1, the technical description must contain the permissible environmental conditions for transport and storage. This information must be repeated on the outermost packaging of the device.

GUIDANCE ON LABELING DEVICES IN PRESSURIZED CONTAINERS

Sections 11, 12, and 13 of the "old medical device regulations"* set out requirements for a medical device packaged in a disposable metal container that is designed to release pressurized contents by use of a manually operated valve that forms an integral part of the container (e.g., an aerosol container). These sections of the "old regulation" made reference to the applicable requirements of the Hazardous Products Act**, which in turn referred to the CC&CR. The current CMDR regulations no longer specifically refer to those requirements. Section 3(b) of the Hazardous Products Act specifically exempts any product that is a device within the meaning of the F&DA.

However, the TPD's guidance on IVD labeling describes additional information that the TPD feels is appropriate for an IVD device containing explosive material or components (TPD IVD Guidance p. 15). Therefore, the following sections are provided as guidance for those medical devices where a risk of fire or explosion may be present.

PRINCIPAL DISPLAY PANEL

The principal display panel of a regulated product is that part of the container that is displayed or visible under normal conditions of sale. For purposes of calculating various labeling requirements, the area of the display panel is defined as follows (CC&CR §1):

- In the case of a rectangular-shaped container, the total area of the side of the container containing the display panel
- In the case of a cylindrical-shaped container, the area to the top or 40 percent of the area obtained by multiplying the circumference of the container by the height of the container, whichever is larger

---

* Canadian Medical Devices Regulations, C.R.C., c. 871 (repealed June 30, 1998).
** Hazardous Products Act, R.S. 1985, c. H-3 (updated to December 31, 2000).

- In the case of a bag, the area of the largest side of the unfolded bag
- If the container is not one of the ones listed above but has an obvious display panel, then the area of the obvious display panel if that area exceeds 40 percent of the display surface
- In any other case, 40 percent of the total area of the container

The information on the container must be clear and legible and be sufficiently durable to remain legible throughout the useful life the contents. If the container is refillable, the label must remain legible during the useful life of the container under normal conditions of transportation, storage, sale, and use (CC&CR §17(b)). The color contrast between the printing and the background must be at least equal to a 70 percent screen of black on white (CC&CR §18). The printed information must the in a sans serif type that is not compressed, expanded or decorative. The type size must comply with requirements described late in this chapter where the height of the type is measured using an upper-case letter or a lower case letter that has an ascender or descender (e.g., "b" or "p") (CC&CR §19).

MANNER OF DISCLOSING REQUIRED INFORMATION

The CC&CR specify detailed requirements for the contents, size, and placement of information on the labels of products considered hazardous under the Hazardous Products Act.

The information required to be disclosed must be easily legible and in distinct contrast to other information on the container, and set out in such a way that it is sufficiently durable to remain legible under normal conditions of transportation, storage, sale, and use (CC&CR §17).

Placement of Information on Display Panels

The information required by the CC&CR is to be displayed on the principal display panel of the container immediately below the common or brand name of the chemical product (CC&CR §26(1)(a)). The information must appear in the following order (CC&CR §25(1)(a)):

- Hazard symbols
- Signal words
- Primary hazard statements, e.g., very corrosive, very flammable

The following information, when appropriate, can be displayed anywhere on the container except on the bottom of the container (CC&CR §25(1)(b)):

- Specific hazard statements, e.g., causes severe burns, contents may catch fire
- negative instruction to help avoid the hazards, e.g., do not get on skin or clothing, do not smoke
- positive instructions to help avoid the hazards, e.g., keep out of the reach of children, use only in well ventilated areas
- First-aid statements, e.g., rinse with water, induce vomiting

This information is to be placed parallel to the base of the container and centered on an imaginary vertical line bisecting the area of the display panel. The hazard symbols are located below the common name or brand name of the regulated product. The signal words are then placed below the hazard symbol. Finally, the primary hazard statements are placed immediately below the signal words (CC&CR §26). If the primary display panel is less than 10 cm high but

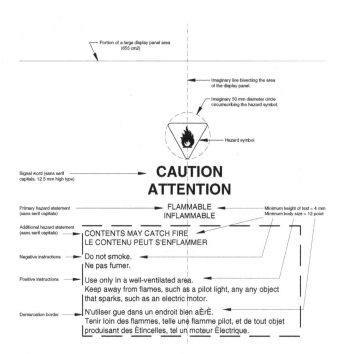

FIGURE 8.1 Example of a hazard symbol on a large label.

at least twice as wide as it is high, the signal word many be located beside the hazards symbol instead of below it (CC&CR §26(3)).

When present, the specific hazards statement, negative and positive instructions, and the first aid statements are left justified and enclosed in a border that separates this information from other information on the primary display panel. This demarcation border may be constructed as a series of dots or hatched lines, be in a color or shading, or utilize some other graphic device that distinguishes it from the background. It must be different from the label border specified in Schedule II of the CPR and from any other border on the label (CC&CR §§ 29, 30).

The size of the hazard symbols and the type style and size for the signal words and other warning statements is a function of the area of the display panel. Above a minimum threshold, the larger the area of the label, the larger the required warning information. Figure 8.1 is an example showing the approximate placement of the required information on part of a "large" display panel. In this example, large means a panel with an area of 655 cm^2. The regulations specify that the minimum size of the hazard symbol is based on the area of an imaginary circle circumscribing the symbol. The minimum area of this imaginary circle is equal to three percent of the area of the display panel or the area of a circle with a diameter of 50 mm, whichever is smaller. Three percent of an area of 655 cm^2 is equal to the area of a circle with a diameter of 50 mm. The size of the hazard symbol does not have to increase as the area of the label panel grows above 655 cm^2.

The diameter of a circle circumscribing the hazard symbol may never be less than 6 mm. The area of a circle with a diameter of 6 mm is equal to three percent of a label area of 9.5 cm^2. Figure 8.2 is an example showing the approximate positioning of the required information on a display panel with an area of 9.5 cm^2. Between these two extremes, the minimum size of the hazard symbol can be calculated based on the area of the display panel.

Where more than one hazard symbol is required on the container of a regulated product, the required hazard symbols shall be placed on the principal display panel of the container in a row such that a straight line drawn through the center of the hazard symbols is parallel to the base of the container. The hazard symbols are placed in the following order from left to right (CC&CR §25(1)):

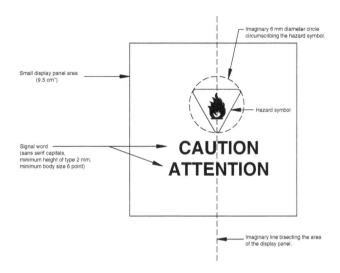

FIGURE 8.2 Example of a hazard symbol on a small label.

1. The hazard symbol that corresponds to the signal word "EXTREME DANGER"
2. The hazard symbol that corresponds to the signal word "DANGER"
3. The hazard symbol that corresponds to the signal word "CAUTION"

Where a product is packaged in a collapsible tube, the required information may be located above the common name or brand name of the product if that placement is closer to the tube opening than the placement required above (CC&CR §26(2)).

Hazard Symbols

Where a hazard symbol is required, it must be an exact reproduction (except for size and color) of that hazard symbol depicted in Schedule 2 of the CC&CR (CC&CR §21). The hazard symbol is to be displayed in a color that is not likely to be confused with other safety marks such as those required by the Transportation of Dangerous Goods Regulations.

An imaginary circle circumscribing the hazard symbol may never have a diameter less than 6 mm regardless of the size of the primary display panel. As the primary display panel increases in size, the area of the imaginary circle circumscribing the hazard symbol must increase so that it is never less than three percent of the area of the display panel. The size of the hazard symbol must continue to increase in proportion to the area of the primary display panel until the area of the imaginary circle circumscribing the hazard symbol is equal to the area of a circle with a diameter of 50 mm (CC&CR §22). The minimum sizes for the hazard symbol as a function of the area of the display panel are listed in Table 8.5.

Signal Words

The signal words, which must appear in both English and French, associated with the respective hazard symbols are "EXTREME DANGER"/*"DANGER EXTRÊME,"* "DANGER"/*"DANGER,"* and "CAUTION"/*"ATTENTION."* The signal words associated with the respective hazard symbol must be printed directly below the symbol in sans serif capitals (CC&CR §20(a)). The minimum height of the letters must be at least one-fourth of the diameter of an imaginary circle circumscribing the associated hazard symbol (see Table 8.5) (CC&CR §23).

TABLE 8.5
Minimum Hazard Symbol and Signal Word Size

Area of the Principal Display Panel (centimeters²)	Minimum Diameter of Circle Circumscribing the Hazard Symbol (millimeters)	Minimum Height of Signal Word (millimeters)
≤9.5	6	1.5
9.5 < Area < 655	$\text{Diameter} = \dfrac{\sqrt{12 \times \text{Area}(\text{cm}^2)}}{\pi}$	$\text{Height} = \dfrac{\sqrt{0.75 \times \text{Area}(\text{cm}^2)}}{\pi}$
≥655	50	12.5

Source: CC&CR §§22 and 23

TABLE 8.6
Minimum Height and Body Size of Type

Area of the Principal Display Panel (centimeters²)	Minimum Height of Type	Minimum Body Size of Type
≤100	2 mm	6 point
100 < Area ≤ 330	3 mm	8 point
Area ≤ 330	4 mm	12 point

Source: CC&CR §24

Primary Hazard Statements

When a primary hazard statement (e.g., "EXTREMELY FLAMMABLE" and *"EXTREMEMENT INFLAMMABLE"*) is required on a label, it must be printed in sans serif capitals. The minimum height of the text is dependent on the area of the primary display panel (CC&CR §24). The minimum sizes are shown in Table 8.6.

Additional Hazard Statements, Negative and Positive Instructions, and First Aid Statements

When an additional hazard statement (e.g., "CONTENTS MAY CATCH FIRE.") is required on the label, it must be printed in sans serif capitals (CC&CR §20(a)). The negative and positive instructions can be printed in upper- and lower-case sans serif type. As with the primary hazard statement, the minimum height of the additional hazard-statement text and the negative and positive instructions is dependent on the area of the primary display panel (CC&CR §24(1)). The minimum sizes are shown in Table 8.6.

When a device contains any of the chemicals regulated by the Hazardous Products Act, the label should include an appropriate first-aid statement. The first-aid statement is printed in the same type style and size as the additional hazard statement except that the phrases "FIRST AID TREAT-MENT"/*"PREMIERS SOINS"* are printed in sans serif boldface capitals (CC&CR §20(b)). The type used for the phrases "FIRST AID TREATMENT"/*"PREMIERS SOINS"* must have a minimum height of 2 mm and a minimum body size of 6 points (CC&CR §24(2)).

Small Packages

Section 25(2)(a) of the CC&CR defines a small container as one having a primary display panel with an area of less than 35 mm². The label of such a container is only required to display the hazard symbol and the associated signal word (see Figure 8.2). A container with a primary display panel that is greater than or equal to 35 mm² but less than 70 mm² may omit the negative and positive instructions other than the positive instructions required if the container has a child-proof closure when more than one hazard symbol is required (CC&CR §25(2)(b)).

LABELING FOR HOME-USE DEVICES

Home-use devices are intended to be used by laypersons having little knowledge of medicine or the technology employed in the device. The IFU must be described in language that is correct and succinct and makes clear use of words. See Chapter 27 and Chapter 28 for recommendations for preparing the IFU.

Although not required by the CMDR, the manufacturer of a home-use device that contains a substance that is characterized as "toxic," "harmful," "very corrosive," "corrosive," or an "irritant" must be packaged in a child-resistant container. A child-resistant container is one that meets the performance requirements listed in Sections 9 to 11 of the CC&CR.

A container that is child-resistant by virtue of requiring the operation or removal of a functional part of the container with a tool that is not supplied with the container does not need special labeling. Other child-resistant containers should bear instructions for opening and, where applicable, for closing the container so that it retains its child-resistant characteristics. The labeling must contain either text in English and French, or diagrams or symbols that are self-explanatory (CC&CR §11).

It is appropriate to sell a regulated product in a container other than a child-resistant container provided it has the following characteristics (CC&CR §40(b)):

- The outlet of the container is designed to dispense the regulated product in single drops.
- The primary display panel of the container discloses the following "primary hazard statement" in addition to any other labeling required on the container.

THIS CONTAINER IS NOT CHILD-RESISTANT. KEEP OUT OF REACH OF CHILDREN.

CE CONTENANT N'EST PAS UN CONTENANT PROTÈGE-ENFANTS. TENIR HORS DE LEUR PORTÉE DES ENFANTS.

A regulated product may be sold in a container that does not remain child-resistant once opened if it meets the following requirements (CC&CR §13):

- The contents of the container are intended to be used immediately after the container is opened.
- The container meets the requirements for a child-resistant container except that it cannot be reclosed.
- The primary display panel of the container discloses the following "primary hazard statement" in addition to any other labeling required on the container.

USE ENTIRE CONTENTS ON OPENING. THIS CONTAINER IS NOT CHILD-RESISTANT ONCE OPENED.

UTILISER LA TOTALITÉ DU CONTENU APRÈS OUVERTURE. UNE FOIS OUVERT, LE CONTENANT N'EST PLUS UN CONTENANT PROTÈGE-ENFANTS.

The requirement for child-resistant containers does not apply to containers with a capacity greater than 5 L unless the contents are classified as very toxic or very corrosive (CC&CR §14).

DEVICES WITH CONTENTS UNDER PRESSURE

Unless exempted under the provisions for small-size packages described in a later section, the principal panel of the inner and outer label of a medical device that is contained in a pressurized

This figure is prepared for convenience of reference
only and has no official sanction. The reader is
referred to Schedule 2 of the CC&CR.

Source: CC&CR, Schedule

FIGURE 8.3 Explosive symbol.

disposable metal container, must include the explosive symbol superimposed on the caution symbol
(see Figure 8.3). The symbols must comply with the requirements in Sections 17 through 31 of the
CC&CR. These sections of the CC&CR prescribe in detail the size and type requirements for the
symbols and accompanying signal words and text. Of course, all text must appear in English and
in French.

The explosive symbol must be accompanied by the signal words "CAUTION"/"*ATTENTION,*"
and the nature of the primary hazard presented by the container expressed in the following wording
(CC&CR §59):

CONTENTS UNDER PRESSURE

CONTENU SOUS PRESSION

The label must include the following specific hazard statement:

CONTAINER MAY EXPLODE IF HEATED

CE CONTENANT PEUT EXPLOSER S'IL EST CHAUFFÉ

Also required are the following negative and positive instructions:

DO NOT PUNCTURE.   NE PAS PERFORER.

DO NOT BURN.   NE PAS BRÛLER.

STORE AWAY FROM HEAT   CONSERVER LOIN DES SOURCES DE
                                         CHALEUR

LABELING FOR FLAMMABLE PRODUCTS

Unless exempted under the provision for small-size packages described earlier in this chapter, a
device that can produce a flame projection or a flashback must bear a flammable warning. If the
substance is packaged in a pressurized container, it may also be required to display the explosion
warning described in the previous section. The required symbols and statements must appear on the
principal display panel of the inner and outer labels of the device (CC&CR §§48–56). The symbols
must comply with the requirements in Sections 21 and 22 of the CC&CR. The required warning is
dependent on the specific hazard as determined by the test described in Schedule I of the CC&CR.

Flammable products are classified as "combustible," "spontaneously combustible," "flammable,"
or "very flammable" using the table in Section 49 of the CC&CR. The information required on the

This figure is prepared for convenience of reference only and has no official sanction. The reader is referred to Schedule 2 of the CC&CR.

Source: CC&CR, Schedule 2

FIGURE 8.4 Flammable symbol.

principle display panel depends on the classification (CC&CR §54). All flammable products must display the flammable symbol in Figure 8.4.

Devices with Combustible or Spontaneously Combustible Contents

A container whose contents are classified as combustible or spontaneously combustible must bear on the principal panel of the inner and outer label the flammable symbol (see Figure 8.4). The symbol must be accompanied by the signal words "CAUTION"/"*ATTENTION*" and the nature of the primary hazard presented by the container must be expressed in the following wording:

READ INSTRUCTIONS BEFORE USING.

LIRE LES INSTRUCTIONS AVANT USAGE.

Both the inner label and outer label must display on one of their panels the additional hazard statement:

DANGER OF COMBUSTION

DANGER DE COMBUSTION

Also required are the following positive instructions:

KEEP AWAY FROM FLAMES OR SPARKS

TENIR LOIN DES FLAMMES ET DES ÉTIN-CELLES.

If the contents are classified as spontaneously combustible, the positive instructions are replaced by:

MATERIALS SUCH AS RAGS USED WITH THIS PRODUCT MAY BEGIN TO BURN BY THEMSELVES.

LES MATÉRIAUX UTILISÉS AVEC CE PRODUIT, TELS LES CHIFFONS, PEUVENT S'ENFLAMMER SPONTANÉMENT.

AFTER USE, PUT RAGS IN WATER OR LAY FLAT TO DRY, THEN DISCARD.

APRÈS UTILISATION METTRE LES CHIFFONS DANS L'EAU OU LES SÉCHER À PLAT, PUIS LES JETER.

Devices with Flammable Contents

A container whose contents are classified as flammable must bear on the principal panel of the inner and outer label the flammable symbol (see Figure 8.4). The symbol must be accompanied by the signal words "DANGER"/"*DANGER*" and the nature of the primary hazard presented by the container must be expressed in the following wording:

FLAMMABLE

INFLAMMABLE

Both the inner label and outer label must display on one of their panels the additional hazard statement:

CONTENTS MAY CATCH FIRE.

LE CONTENAU PEUT S'ENFLAMMER.

If a vapor or fume posses a hazards, then the following text is substituted for the additional hazard statement:

FUMES MAY CATCH FIRE.

LES ÉMANATIONS PEUVENT S'ENFLAMMER.

Also required are the following negative and positive instructions:

DO NOT SMOKE.

USE ONLY IN WELL VENTILATED AREAS.

KEEP AWAY FROM FLAMES, SUCH AS A PILOT LIGHT, AND ANY OBJECT THAT SPARKS, SUCH AS AN ELECTRIC MOTOR.

NE PAS FUMER.

N'UTILISER QUE DANS UN ENDROIT BIEN AÉRÉ.

TENIR LOIN DES FLAMMES, TELLE UNE FLAMME PILOTE, ET DE TOUR OBJET PRODUISANT DES ÉTINCELLES, TEL UN MOTEUR ÉLECTRIQUE.

Devices with Very Flammable Contents

A container whose contents are classified as very flammable must bear on the principal panel of the inner and outer label the flammable symbol (see Figure 8.4). The symbol must be accompanied by the signal words "EXTREME DANGER"/"*DANGER EXTRÊME*" and the nature of the primary hazard presented by the container must be expressed in the following wording:

VERY FLAMMABLE

TRÈS INFLAMMABLE

Both the inner label and outer label must display on one of their panels the additional hazard statement:

CONTENTS MAY CATCH FIRE.

LE CONTENAU PEUT S'ENFLAMMER.

If a vapor or fume posses a hazards, then the following text is substituted for the additional hazard statement:

FUMES MAY CATCH FIRE.

LES ÉMANATIONS PEUVENT S'ENFLAMMER.

Also required are the following negative and positive instructions:

DO NOT SMOKE.

USE ONLY IN WELL VENTILATED AREAS.

KEEP AWAY FROM FLAMES, SUCH AS A PILOT LIGHT, AND ANY OBJECT THAT SPARKS, SUCH AS AN ELECTRIC MOTOR.

NE PAS FUMER.

N'UTILISER QUE DANS UN ENDROIT BIEN AÉRÉ.

TENIR LOIN DES FLAMMES, TELLE UNE FLAMME PILOTE, ET DE TOUR OBJET PRODUISANT DES ÉTINCELLES, TEL UN MOTEUR ÉLECTRIQUE.

LABELING FOR CORROSIVE PRODUCTS

Unless exempted under the provision for small-size packages described earlier in this chapter, a device that contains a substance that is classified as a corrosive substance in Section 41 of the CC&CR must bear a corrosive warning. If the substance is packaged in a pressurized container, it may also be required to display the explosion warning described in the previous section. The required symbols and statements must appear on the principal display panel of the inner and outer labels of the device (CC&CR §39). The symbols must comply with the requirements in Sections 21 and 22 of the CC&CR. Most corrosive products must display the corrosive symbol in Figure 8.5

Devices with Corrosive Contents

For devices with substances that are classified as corrosive in Section 41 of the CC&CR, the corrosive symbol must be accompanied by the signal words "DANGER"/"*DANGER*," and the nature of the primary hazard presented by the container expressed in the following wording (CC&CR §46(2)):

CORROSIVE

CORROSIF

This figure is prepared for convenience of reference only and has no official sanction. The reader is referred to Schedule 2 of the CC&CR.

Source: CC&CR, Schedule 2

FIGURE 8.5 Corrosive symbol.

TABLE 8.7
Required Information for Devices with Substances Classified as Corrosive

Type of Information	Route of Exposure	English Information	French Information
Specific hazard statement	(a) All	CAUSES BURNS	PROVOQUE DES BRÛLURES
	(b) Inhalation	DANGEROUS FUMES FORM WHEN MIXED WITH OTHER PRODUCTS	DÉGAGE DES ÉMANATIONS DANGEREUSES LORSQUE MÉLANGÉ AVEC D'AUTRES PRODUITS
Negative instructions	(a) All	When appropriate and before the other negative instructions: Do not mix with [*Insert description of other products that react with the chemical product, such as toilet bowl or drain cleaners, bleach or ammonia.*].	When appropriate and before the other negative instructions: Ne pas mélanger avec [*Insert description of other products that react with the chemical product, such as des nettoyants pour cuvettes de toilette ou tuyaux d'évacuation, des agents de blanchiment ou de l'ammoniaque.*].
	(b) Oral	Do not swallow.	Ne pas avaler.
	(c) Eyes	Do not get in eyes.	Éviter tout contact avec les yeux.
	(d) Dermal	Do not get on skin or clothing.	Éviter tout contact avec la peau ou les vêtements.
	(e) Inhalation	Do not breathe fumes.	Ne pas respirer les émanations.
Positive instructions	(a) All	Handle with care. Keep out of reach of children.	Manipuler avec soin. Tenir hors de la portée des enfants.
	(b) Oral	Wear [*Insert description of the specific safety equipment relevant to the hazard, e.g., a mask.*].	Porter [*Insert description of the specific safety equipment relevant to the hazard, e.g., un masque.*].
	(c) Dermal	Wear [*Insert description of the specific safety equipment relevant to the hazard, e.g., rubber gloves, safety glasses.*].	Porter [*Insert description of the specific safety equipment relevant to the hazard, e.g., des gants de caoutchouc, des lunettes de sécurité.*].
	(d) Inhalation	Use only in a well-ventilated area. Wear [*Insert description of the specific safety equipment relevant to the hazard, e.g., a mask, a respirator.*].	N'utiliser que dans un endroit bien aéré. Porter [*Insert description of the specific safety equipment relevant to the hazard, e.g., un masque, un respirateur.*].
First-aid Statement	(a) All	FIRST AID TREATMENT Contains [*name of hazardous ingredients in descending order of proportion*]. If swallowed, call a Poison Control Centre or doctor immediately. Do not induce vomiting.	PREMIERS SOINS Contient [*name of hazardous ingredients in descending order of proportion*]. En cas d'ingestion, appeler immédiatement un centre antipoison ou un médecin. Ne pas provoquer le vomissement.
	(b) Eyes	If in eyes, rinse with water for [*Insert appropriate period of time.*].	En cas de contact avec les yeux, rincer avec de l'eau pendant [*Insert appropriate period of time.*].
	(c) Dermal	If on skin, rinse well with water. If on clothes, remove clothes.	En cas de contact avec la peau, bien rincer avec de l'eau. En cas de contact avec les vêtements, enlever ceux-ci.
	(d) Inhalation	If breathed in, move person to fresh air.	En cas d'inhalation, transporter à l'air frais la personne exposée.

Source: CC&CR §46(2)

The labels of corrosive products must include particular specific hazard statements, negative and positive instructions, and first-aid information depending on the route of exposure. These markings are shown in Table 8.7.

Devices with Very Corrosive Contents

For devices with substances that are classified as very corrosive in Section 41 of the CC&CR, the corrosive symbol must be accompanied by the signal words "EXTREME DANGER"/"*EXTRÊME DANGER,*" and the nature of the primary hazard presented by the container expressed in the following wording (CC&CR §46(1)):

VERY CORROSIVE

TRÈS CORROSIF

The labels of very corrosive products must include particular specific hazard statements, negative and positive instructions, and first-aid information depending on the route of exposure. These marking are shown in Table 8.8.

Devices with Contents Classified as an Irritant

Devices with substances that are classified as an irritant in Section 41 of the CC&CR are not required to display the corrosive symbol. They are required to display the signal words "CAUTION"/"*ATTEN-TION,*" and the nature of the primary hazard presented by the container expressed in the following wording (CC&CR §46(3)):

IRRITANT

IRRITANT

The labels of products classified as an irritant must include particular specific hazard statements, negative and positive instructions, and first-aid information depending on the route of exposure. These marking are shown in Table 8.9.

LABELING FOR TOXIC PRODUCTS

Unless exempted under the provision for small-size packages described earlier in this chapter, a device that contains a substance that is classified as toxic using the procedure in Section 35 of the CC&CR must bear a toxicity warning. If the substance is packaged in a pressurized container, it may also be required to display the explosion warning described in the previous section. The required symbols and statements must appear on the principal display panel of the inner and outer labels of the device (CC&CR §39). The symbols must comply with the requirements in Sections 21 and 22 of the CC&CR. All toxic products must display the toxic symbol in Figure 8.6.

Devices with Contents Classified as Toxic

For devices with substances that are classified as toxic in Section 33 of the CC&CR, the toxic symbol must be accompanied by the signal words "DANGER"/"*DANGER,*" and the nature of the primary hazard presented by the container expressed in the following wording (CC&CR §39(1)):

POISON

POISON

The labels of toxic products must include particular specific hazard statements, negative and positive instructions, and first-aid information depending on the route of exposure. These marking are shown in Table 8.10.

TABLE 8.8
Required Information for Devices with Substances Classified as Very Corrosive

Type of Information	Route of Exposure	English Information	French Information
Specific hazard statement	(a) All	CAUSES SEVERE BURNS	PROVOQUE DES GRAVES BRÛLURES
	(b) Dermal and contains a concentration of 0.5 percent or more of available floride ions.	SYMPTOMS MAY NOT APPEAR IMMEDIATELY	LES SYMPTOMES PEUVENT NE PAS SE MANIFESTER IMMÉDIATEMENT
	(c) Inhalation	DANGEROUS FUMES FORM WHEN MIXED WITH OTHER PRODUCTS	DÉGAGE DES ÉMANATIONS DANGEREUSES LORSQUE MÉLANGÉ AVEC D'AUTRES PRODUITS
Negative instructions	(a) All	When appropriate and before the other negative instructions: Do not mix with [*Insert description of other products that react with the chemical product, such as toilet bowl or drain cleaners, bleach or ammonia.*].	When appropriate and before the other negative instructions: Ne pas mélanger avec [*Insert description of other products that react with the chemical product, such as des nettoyants pour cuvettes de toilette ou tuyaux d'évacuation, des agents de blanchiment ou de l'ammoniaque.*].
	(b) Oral	Do not swallow.	Ne pas avaler.
	(c) Eyes	Do not get in eyes.	Éviter tout contact avec les yeux.
	(d) Dermal	Do not get on skin or clothing.	Éviter tout contact avec la peau ou les vêtements.
	(e) Inhalation	Do not breathe fumes.	Ne pas respirer les émanations.
Positive instructions	(a) All	Handle with care. Keep out of reach of children.	Manipuler avec soin. Tenir hors de la portée des enfants.
	(b) Oral	Wear [*Insert description of the specific safety equipment relevant to the hazard, e.g., a mask.*].	Porter [*Insert description of the specific safety equipment relevant to the hazard, e.g., un masque.*].
	(c) Dermal	Wear [*Insert description of the specific safety equipment relevant to the hazard, e.g., rubber gloves, safety glasses.*].	Porter [*Insert description of the specific safety equipment relevant to the hazard, e.g., des gants de caoutchouc, des lunettes de sécurité.*].
	(d) Inhalation	Use only in a well-ventilated area. Wear [*Insert description of the specific safety equipment relevant to the hazard, e.g., a mask, a respirator.*].	N'utiliser que dans un endroit bien aéré. Porter [*Insert description of the specific safety equipment relevant to the hazard, e.g., un masque, un respirateur.*].
First-aid Statement	(a) All	FIRST AID TREATMENT Contains [*name of hazardous ingredients in descending order of proportion*]. If swallowed, call a Poison Control Centre or doctor immediately. Do not induce vomiting.	PREMIERS SOINS Contient [*name of hazardous ingredients in descending order of proportion*]. En cas d'ingestion, appeler immédiatement un centre antipoison ou un médecin. Ne pas provoquer le vomissement.
	(b) Eyes	If in eyes, rinse with water for [*Insert appropriate period of time.*].	En cas de contact avec les yeux, rincer avec de l'eau pendant [*Insert appropriate period of time.*].
	(c) Dermal	If on skin, rinse well with water. If on clothes, remove clothes.	En cas de contact avec la peau, bien rincer avec de l'eau. En cas de contact avec les vêtements, enlever ceux-ci.
	(d) Inhalation	If breathed in, move person to fresh air.	En cas d'inhalation, transporter à l'air frais la personne exposée.

Source: CC&CR §4 t6(1)

TABLE 8.9
Required Information for Devices with Substances Classified as an Irritant

Type of Information	Route of Exposure	English Information	French Information
Specific hazard statement	(a) All	CAUSES SEVERE BURNS	PEUT IRRITER LES YEUX
	(b) Dermal	MAY IRRITATE SKIN	PEUT IRRITER LA PEAU
	(c) Inhalation	DANGEROUS FUMES FORM WHEN MIXED WITH OTHER PRODUCTS	DÉGAGE DES ÉMANATIONS DANGEREUSES LORSQUE MÉLANGÉ AVEC D'AUTRES PRODUITS
Negative instructions	(a) All	When appropriate and before the other negative instructions: Do not mix with *[Insert description of other products that react with the chemical product, such as toilet bowl or drain cleaners, bleach or ammonia.]*.	When appropriate and before the other negative instructions: Ne pas mélanger avec [Insert description of other products that react with the chemical product, such as des nettoyants pour cuvettes de toilette ou tuyaux d'évacuation, des agents de blanchiment ou de l'ammoniaque.].
	(b) Eyes	Do not get in eyes.	Éviter tout contact avec les yeux.
	(c) Dermal	Do not get on skin or clothing.	Éviter tout contact avec la peau ou les vêtements.
	(d) Inhalation	Do not breathe fumes.	Ne pas respirer les émanations.
Positive instructions	(a) All	Keep out of reach of children.	Tenir hors de la portée des enfants.
First-aid Statement	(a) All	FIRST AID TREATMENT Contains *[name of hazardous ingredients in descending order of proportion]*. If swallowed, call a Poison Control Centre or doctor immediately. Do not induce vomiting.	PREMIERS SOINS Contient [*name of hazardous ingredients in descending order of proportion*]. En cas d'ingestion, appeler immédiatement un centre antipoison ou un médecin. Ne pas provoquer le vomissement.
	(b) Eyes	If in eyes, rinse with water for *[Insert appropriate period of time.]*.	En cas de contact avec les yeux, rincer avec de l'eau pendant [*Insert appropriate period of time.*].
	(c) Dermal	If on skin, rinse well with water.	En cas de contact avec la peau, bien rincer avec de l'eau.

Source: CC&CR §46(3)

This figure is prepared for convenience of reference only and has no official sanction. The reader is referred to Schedule 2 of the CC&CR.

Source: CC&CR, Schedule 2

FIGURE 8.6 Toxic symbol.

TABLE 8.10
Required Information for Devices with Substances Classified as Toxic

Type of Information	Route of Exposure	English Information	French Information
Specific hazard statement	(a) Oral or aspiration	CONTENTS HARMFUL	CONTENU NOCIF
	(b) Dermal	CONTENTS HARMFUL	CONTENU NOCIF
	(b) Inhalation	CONTENTS HARMFUL or, if only the vapour or fume poses a hazard: FUMES HARMFUL	CONTENU NOCIF or, if only the vapour or fume poses a hazard: ÉMANATIONS NOCIVES
Negative instructions	(a) Oral or aspiration	Do not swallow.	Ne pas avaler.
	(b) Oral and contains 1 percent or more methyl alcohol and a total quantity of 5 mL or more	May cause blindness if swallowed.	L'ingestion peut causer la cécité.
	(b) Dermal	Do not get in eyes or on skin or clothing.	Éviter tout contact avec les yeux, la peau et les vêtements.
	(b) Inhalation	Do not breathe fumes.	Ne pas respirer les émanations.
Positive instructions	(a) All	Keep out of reach of children.	Tenir hors de la portée des enfants.
	(b) Oral or aspiration	Wear [*Insert description of the specific safety equipment relevant to the hazard, e.g., a mask.*].	Porter [*Insert description of the specific safety equipment relevant to the hazard, e.g., un masque.*].
	(c) Dermal	Wear [*Insert description of the specific safety equipment relevant to the hazard, e.g., rubber gloves, safety glasses.*].	Porter [*Insert description of the specific safety equipment relevant to the hazard, e.g., des gants de caoutchouc, des lunettes de sécurité.*].
	(d) Inhalation	Use only in a well-ventilated area. Wear [*Insert description of the specific safety equipment relevant to the hazard, e.g., a mask, a respirator.*].	N'utiliser que dans un endroit bien aéré. Porter [*Insert description of the specific safety equipment relevant to the hazard, e.g., un masque, un respirateur.*].
First-aid Statement	(a) All	FIRST AID TREATMENT Contains [*name of hazardous ingredients in descending order of proportion*]. If swallowed, call a Poison Control Centre or doctor immediately.	PREMIERS SOINS Contient [*name of hazardous ingredients in descending order of proportion*]. En cas d'ingestion, appeler immédiatement un centre antipoison ou un médecin.
	(b) Oral or aspiration	If person is [*Insert instructions for administering first aid, e.g., for methyl alcohol: If person is alert, induce vomiting.*]	Si la personne est [*Insert instructions for administering first aid, e.g., for methyl alcohol: Si la personne est consciente, provoquer le vomissement.*].
	(c) Dermal	If in eyes or on skin, rinse well with water. If on clothes, remove clothes.	En cas de contact avec les yeux ou la peau, bien rincer avec de l'eau. En cas de contact avec les vêtements, enlever ceux-ci.
	(d) Inhalation	If breathed in, move person to fresh air.	En cas d'inhalation, transporter à l'air frais la personne exposée.

Source: CC&CR §39(1)

Devices with Contents Classified as Harmful

For devices with substances that are classified as harmful in Section 33 of the CC&CR, the toxic symbol must be accompanied by the signal words "CAUTION"/"*ATTENTION*," and the nature of the primary hazard presented by the container expressed in the following wording (CC&CR §39(2)):

TABLE 8.11

Required Information for Devices with Substances Classified as Harmful

Type of Information	Route of Exposure	English Information	French Information
Specific hazard statement	(a) Oral or aspiration	CONTENTS MAY BE HARMFUL	LE CONTENU PEUT ÊTRE NOCIF
	(b) Dermal	CONTENTS MAY BE HARMFUL	LE CONTENU PEUT ÊTRE NOCIF
	(b) Inhalation	CONTENTS MAY BE HARMFUL or, if only the vapour or fume poses a hazard: FUMES MAY BE HARMFUL	LE CONTENU PEUT ÊTRE NOCIF or, if only the vapour or fume poses a hazard: LES ÉMANATIONS PEUVENT ÊTRE NOCIVES
Negative instructions	(a) Oral or aspiration	Do not swallow.	Ne pas avaler.
	(b) Dermal	Do not get in eyes or on skin or clothing.	Éviter tout contact avec les yeux, la peau et les vêtements.
	(b) Inhalation	Do not breathe fumes.	Ne pas respirer les émanations.
Positive instructions	(a) All	Keep out of reach of children.	Tenir hors de la portée des enfants.
	(b) Oral or aspiration	Wear [*Insert description of the specific safety equipment relevant to the hazard, e.g., a mask.*].	Porter [*Insert description of the specific safety equipment relevant to the hazard, e.g., un masque.*].
	(c) Dermal	Wear [*Insert description of the specific safety equipment relevant to the hazard, e.g., rubber gloves, safety glasses.*].	Porter [*Insert description of the specific safety equipment relevant to the hazard, e.g., des gants de caoutchouc, des lunettes de sécurité.*].
	(d) Inhalation	Use only in a well-ventilated area. Wear [*Insert description of the specific safety equipment relevant to the hazard, e.g., a mask, a respirator.*].	N'utiliser que dans un endroit bien aéré. Porter [*Insert description of the specific safety equipment relevant to the hazard, e.g., un masque, un respirateur.*].
First-aid Statement	(a) All	FIRST AID TREATMENT Contains [*name of hazardous ingredients in descending order of proportion*]. If swallowed, call a Poison Control Centre or doctor immediately.	PREMIERS SOINS Contient [*name of hazardous ingredients in descending order of proportion*]. En cas d'ingestion, appeler immédiatement un centre antipoison ou un médecin.
	(b) Oral or aspiration	If person is [*Insert instructions for administering first aid, e.g., for methyl alcohol: If person is alert, induce vomiting.*]	Si la personne est [*Insert instructions for administering first aid, e.g., for methyl alcohol: Si la personne est consciente, provoquer le vomissement.*].
	(c) Dermal	If in eyes or on skin, rinse well with water. If on clothes, remove clothes.	En cas de contact avec les yeux ou la peau, bien rincer avec de l'eau. En cas de contact avec les vêtements, enlever ceux-ci.
	(d) Inhalation	If breathed in, move person to fresh air.	En cas d'inhalation, transporter à l'air frais la personne exposée.

Source: CC&CR §39(2)v

POISON

POISON

The labels of harmful products must include particular specific hazard statements, negative and positive instructions, and first-aid information depending on the route of exposure. These marking are shown in Table 8.11.

THINGS TO REMEMBER

Under the authority of the F&DA, the Canadian government established a national system of regulation of medical devices. The medical device regulation program is intended to ensure that therapeutic and diagnostic devices are safe and effective. The requirements for labeling of these devices in Canada are contained in the CMDR, which are published in the *Canada Gazette*, Part II–Statutory Orders and Regulations.

A medical device that is not labeled as required by, or is labeled contrary to, the CMDR is deemed to be misbranded. Misbranding is a violation of the law and subjects the perpetrator to significant penalties including seizure and forfeiture of the misbranded goods. This chapter explores the parts of the regulations that deal with misbranding of medical devices.

The TPD, which is a part of the HPFB of Health Canada, has developed a minimum set of labeling requirements that are applicable to all medical devices. These requirements are contained in Sections 21 through 23 of the CMDR. To assist manufacturers in interpreting the regulations, the TPD has developed guidance documents on labeling of medical devices and labeling of IVD devices.

Under the provisions of the F&DA, the TPD can publish additional guidance or requirements as needed in order to ensure safety and effectiveness. This material can appear in guidance documents or in TPD policy documents. At the time of publication, additional labeling requirements and guidance had been developed for prolonged-wear contact lenses, menstrual tampons, contraceptive devices, and medical gloves.

Electromedical devices as a group are regulated by the federal government as medical devices and by the provincial governments in Canada under the applicable parts of the Canadian Electric Code. Compliance with the Canadian Electric Code can be demonstrated by adhering to the requirements of C22.2 No. 601.1. The labeling requirements from Clause 6 of that standard are described in this chapter.

Sections 11, 12, and 13 of the "old medical device regulations" set out requirements for a medical device packaged in a disposable metal container which is designed to release pressurized contents by use of a manually operated valve that forms an integral part of the container (e.g., an aerosol container). The current CMDR regulations no longer specifically refer to these requirements. However, the TPD's guidance on IVD labeling describes additional information that the TPD feels is appropriate for an IVD containing explosive material or components. Some guidance extracted from the CC&CR is provided for those medical devices where a risk of fire or explosion may be present.

9 Radiation-Emitting Device Labeling in Canada

In Canada, the responsibility for regulating radiation-emitting devices is assigned to the Minister of Health under the provisions of the Radiation Emitting Devices Act (REDA). The act defines radiation as energy in the form of electromagnetic waves or acoustical waves (REDA §2). Therefore, a radiation-emitting device is (REDA §2):

- any device that is capable of producing and emitting radiation, and
- any component of or accessory to such a device.

The Minister of Health has assigned responsibility for regulating these devices to the Radiation Protection Bureau (RPB). The RPB is an administrative division within the Healthy Environments and Consumer Safety sector of Health Canada.

BACKGROUND AND GENERAL INTENT OF THE LAW

Regulation of radiation-emitting devices in Canada dates from the early 1980's. The purpose and intent of this regulatory program is to assess, monitor, and assist in reducing the risks associated with radiation-emitting devices and other sources of radiation. Other sources of radiation include naturally occurring sources or ionizing radiation appearing in the environment such as radon gas, radionuclides used in industrial and medical applications, and the harmful ultraviolet (UV) radiation from sunlight. The REDA focuses on X-ray and non-ionizing radiation. Radiodiagnostic and radiotherapeutic agents are regulated as drugs under the Food and Drugs Act (F&DA) and its associated regulations.

THE REGULATIONS

Section 13 of the REDA authorizes the development of regulations that:

- prescribe classes of radiation emitting devices for the purposes of this act;
- prescribe standards regulating the design, construction, and functioning of any designated classes of radiation-emitting devices to protect persons against injury or death from radiation;
- exempt any radiation-emitting device or class of radiation-emitting device from the application of provisions of the REDA; and
- prescribe the information that must be present on any label or package and the manner in which that information must be shown.

The regulations relating to radiation-emitting devices are published in the *Canada Gazette* Part II—Statutory Orders and Regulations. The Canadian "Radiation Emitting Devices Regulations" (REDR) covers a range of products used in medical, industrial, commercial, consumer, and educational activities. Schedule I of the REDR defines the following classes of devices that are covered by the regulation:

- television receivers
- dental X-ray equipment with an extra-oral source
- microwave ovens
- baggage inspection X-ray devices
- demonstration-type gas discharge devices (used to demonstrate the production, properties or effects of glow discharges or X-rays, or the flow of electrons or ions)
- photofluorographic X-ray equipment
- laser scanners (used to read or generate codes represented by drawn or printed geometrical patterns)
- demonstration lasers (used for demonstrating the principles of optics in educational institutions)
- low-energy electron microscopes
- high intensity mercury vapor discharge lamps (including mercury vapor lamps, metal halide lamps, and self-ballasted lamps)
- sunlamps (not used to produce therapeutic effects for medical purposes)
- diagnostic X-ray equipment
- ultrasound therapy devices
- analytical X-ray equipment (used to identify elemental composition or examine the micro-structure of material)
- cabinet X-ray equipment (designed primarily for the examination of material)

This chapter will be restricted to those devices with a medical application.

LABELS AND LABELING

The REDA defines a label as "any legend, word, or mark attached to, included in, belonging to, or accompanying any radiation-emitting device or package" (REDA §2). With respect to the emission of radiation, the REDA prohibits the labeling, packaging, or advertising of a radiation-emitting device "in a manner that is false, misleading or deceptive, or is likely to create an erroneous impression regarding its design, construction, performance, intended use, character, value, composition, merit or safety" (REDA §5). A person who knowingly violates the provisions of the REDA may be subject to penalties ranging from seizure of the equipment to fines prescribes in the act.

LABELING REQUIREMENTS FOR RADIATION-EMITTING PRODUCTS

Schedule II of the REDR defines the requirements for each of the covered classes of radiation emitting products. The REDR specifies design standards, construction standards, functional standards, and labeling and information requirements for each of the covered classes of products. This chapter focuses on the labeling and information requirements.

DENTAL X-RAY EQUIPMENT WITH AN EXTRA-ORAL SOURCE

Dental X-ray equipment with an extra-oral source is X-ray-generating equipment that is primarily intended for the examination of dental structures in humans and that has an X-ray generating tube designed to be used outside the mouth.

This figure is prepared for convenience of reference only and has no official sanction. The reader is referred to Schedule II, Part II, Section 5 of the REDR.
Source: REDR, Schedule II, Part II §5(1)(vi)

FIGURE 9.1 X-ray warning symbol.

Labeling

Dental X-ray equipment with an extra-oral source must bear the X-ray warning symbol in Figure 9.1. The label bearing the symbol must (REDR §II(5)(1)(a)):

- be securely affixed to the equipment control panel;
- appear in two contrasting colors;
- be clearly visible and clearly readable from a distance of 1 m;
- have no outer dimension that is less than 2 cm; and
- bear the following words:

CAUTION: X-RAYS – ATTENTION: RAYONS X.

A warning sign that is clearly visible to the operator must indicate that the possibility of hazardous radiation will be emitted when the equipment is in operation and state that any unauthorized use is prohibited (REDR §II(5)(1)(b)).

A mark or label must appear on the external surface of the equipment control panel that indicates (REDR §II(5)(1)(c)):

- the name of the equipment's manufacturer,
- the model designation,
- the serial number,
- the date of manufacture, and
- the country of manufacture.

This mark or label must be permanent, clearly visible, and readily discernable to the operator.

The external surface of the X-ray-tube assembly must also have a clearly visible and readily discernable permanent mark or label that indicates (REDR §II(5)(1)(d)):

- the name of the manufacturer of the X-ray-tube assembly,
- the model designation,
- the serial number,
- the date of installation of the X-ray tube in the X-ray tube housing, and
- the country of manufacture.

Instructions for Use (IFU)

The manufacturer of dental X-ray equipment with an extra-oral source must provide instructions for use (IFU) that provides the following minimum information (REDR §II(5)(2)):

- operating instructions that provide the information necessary for the safe and proper operation of the equipment; and
- the following information about the functioning of the equipment:
  - the maximum allowable deviation from the specified X-ray tube current and voltage,
  - the accuracy of the controlling timer, and
  - the specific conditions on which the information described above are based.

PHOTOFLUOROGRAPHIC X-RAY EQUIPMENT

Photofluorographic X-ray equipment is an X-ray-generating appliance designed primarily for examining the human chest and the photographic recording of the image produced on a fluorescent screen.

Labeling

Photofluorographic X-ray equipment must be designed so that all controls, meters, lights, or other indicators are clearly labeled to indicate their function. The control panel must bear the X-ray warning symbol in Figure 9.1. The label bearing the symbol must (REDR §VI(6)(1)(a)):

- be securely affixed to the equipment control panel;
- appear in two contrasting colors;
- be clearly visible and clearly readable from a distance of 1 m;
- have no outer dimension that is less than 2 cm; and
- bear the following words:

CAUTION: X-RAYS – ATTENTION: RAYONS X.

A warning sign that is clearly visible to the operator must indicate that the possibility of hazardous radiation will be emitted when the equipment is in operation and state that any unauthorized use is prohibited. This warning sign must be permanently affixed to the control panel (REDR §VI(3)(b)).

The equipment must bear a replaceable label mounted on the camera hood in clear view of the operator indicating the type and date of insertion of the fluorescent screen (REDR §VI(4)(1)(i).

A mark or label must appear on the external surface of the equipment control panel that indicates (REDR §VI(4)(2)):

- the name of the equipment's manufacturer,
- the model designation,
- the serial number, and
- the date and place of manufacture.

This mark or label must be permanent, clearly visible, and readily discernable to the operator.

The external surface of the X-ray-tube housing must also have a clearly visible and readily discernable permanent mark or label that indicates (REDR §VI(4)(2)):

- the name of the manufacturer of the X-ray tube,
- the model designation,

- the serial number, and
- the date and place of manufacture.

DIAGNOSTIC X-RAY EQUIPMENT

Diagnostic X-ray equipment is an X-ray device that is used for examining the human body. Not included in this class are dental X-ray equipment with an extra-oral source covered by Part II of the REDR, photofluorographic X-ray equipment covered by Part VI of the REDR, radiation-therapy simulators, and computer-assisted tomographic equipment.

Labeling

Diagnostic X-ray equipment must be designed so that all controls, meters, warning lights and other indicators required by this part must be clearly labeled as to their function (REDR §XII(5)). The equipment must display all required information in a legible, permanent, and visible manner on the specified surfaces of the equipment.

The main control panel must bear the X-ray warning symbol in Figure 9.1. At the manufacturer's option, the symbol in Figure 9.2 may be used as an alternative. The label bearing the symbol must (REDR §XII(4)):

- be securely affixed to the equipment control panel;
- appear in two contrasting colors;
- be clearly visible and clearly readable from a distance of 1 m;
- have no outer dimension that is less than 2 cm; and
- bear the following words:

CAUTION: X-RAYS – ATTENTION: RAYONS X.

A warning sign that is clearly visible to the operator must indicate that the possibility of hazardous radiation will be emitted when the equipment is in operation and state that any unauthorized use is prohibited (REDR §XII(3)(a)).

The external surface of the X-ray-tube housing must also have a clearly visible and readily discernable permanent mark or label that indicates (REDR §XII(3)(b)):

- the name of the manufacturer of the X-ray-tube assembly,
- the model designation,

This figure is prepared for convenience of reference only and has no official sanction. The reader is referred to Schedule II, Part II, Section 5 of the REDR.
Source: REDR, Schedule II, Part XII ß4(e)(ii)

FIGURE 9.2 Alternate x-ray warning symbol.

- the serial number,
- the date of installation of the X-ray tube in the X-ray-tube housing,
- the country of manufacture, and
- the minimum permanent inherent filtration of the X-ray beam emitted from the X-ray-tube assembly, expressed in millimeters of aluminum equivalent at a specified X-ray-tube voltage.

The external surface of the X-ray-tube housing or another suitable structure permanently attached to the X-ray-tube housing must bear (REDR §II(5)(1)(c)):

- an indicator that enables the focal spot to image receptor distance to be determined to within 2 percent of that distance, and
- if the X-ray tube and the X-ray generator are not located within a common enclosure, marks that clearly indicate the anode and cathode terminals on the X-ray-tube housing and on the high-voltage generator.

If a beam-limiting device is installed on the equipment that adds filtration to the X-ray beam, the external surface of the beam-limiting device must be marked with the total permanent filtration deliverable by the beam-limiting device, expressed in millimeters of aluminum equivalent at a specified X-ray-tube voltage (REDR §II(5)(1)(d)).

Instructions for Use (IFU)

The manufacturer must provide the following information in the IFU provided with each piece of diagnostic X-ray equipment (REDR §XII(2)):

- the name and address of the manufacturer;
- the installation instructions;
- any radiological safety procedures and additional precautions that are necessary because of any unique features of the equipment;
- the maintenance instructions necessary to keep the equipment in compliance with the requirements of the REDR;
- the rated line voltage, the maximum line current and the line voltage regulation for the operation of the equipment at the maximum line current;
- the loading factors that constitute the maximum line current condition for the X-ray generator;
- the duty cycles, rectification type; and generator rating of the equipment; and
- if the equipment is battery powered, the minimum state of charge necessary for it to operate.

The manufacturer must also include in the IFU:

- the operating range of X-ray-tube voltages and the maximum deviation for any selected X-ray-tube voltage within that range of values;
- if the equipment is not operated exclusively in automatic exposure control mode, the accuracy limits of:
  - the controlling timer,
  - the X-ray-tube current;
  - the current time product; and
- where the equipment operates under automatic exposure control, the accuracy limits of that control; and
- the conditions under which the information provided in the preceding three bullets are valid.

For each X-ray-tube assembly, the manufacturer must provide (REDR §XII(2)(g)):

- the nominal focal spot sizes and the method of their determination;
- the cooling curves for the anode and for the X-ray-tube housing;
- the X-ray-tube rating charts; and
- the method by which the focal spot to image receptor distance can be determined using the indicator specified in Section 3(c)(i) of the REDR.

ULTRASOUND-THERAPY DEVICES

An ultrasound-therapy device is designed to generate and emit ultrasonic power at acoustic frequencies above 20 kHz for use in physical therapy.

Labeling

Every ultrasound-therapy device must be designed so that all marks, labels, and signs are permanently affixed and clearly visible. All user controls, meters, lights, or other indicators must be clearly visible, readily discernible, and clearly labeled to indicate their function. Every ultrasound-therapy device must bear on the external surface of its housing (REDR §XIII(2)(3)(a)):

- the name and address of the manufacturer;
- the name and address of the distributor, if the distributor is not the manufacturer;
- the type and model designation;
- the serial number;
- the month and year that the equipment was manufactured;
- the ultrasonic frequencies in kilohertz (kHz) or megahertz (mHz);
- a statement indicating if the wave produced by the device is a continuous wave or an amplitude modulated wave; and
- the line voltage used for normal operation.

For an ultrasound-therapy device that produces an amplitude-modulated wave, the external surface of the device housing must also be marked with:

- the pulse repetition rate,
- the pulse duration,
- the ratio of the temporal maximum effective ultrasonic intensity to the temporal average effective ultrasonic intensity, and
- a description of the wave shape.

If these parameters vary depending on the output power, then the parameters are to be stated at the temporal maximum ultrasonic power.

The external surface must also be marked with the ultrasound radiation-warning symbol in Figure 9.4. The label bearing the symbol must (REDR §XIII(4)(1)(a)):

- be shown in two contrasting colors;
- be clearly visible and identifiable from a distance of 1 m;
- have no outer dimensions less than 2 cm; and
- bear the following words:

This figure is prepared for convenience of reference
only and has no official sanction. The reader is
referred to Schedule II, Part XIII, Section 4 of the
REDR.
Source: REDR, Schedule II, Part XIII ß4(e)

FIGURE 9.3 Ultrasonic radiation warning symbol.

CAUTION-ULTRASOUND, ATTENTION-ULTRASONS.

The external surface of each ultrasound-therapy applicator housing must bear the following information (REDR §XIII(2)(3)(b)):

- the type and model of the ultrasound-therapy device for which the applicator is designed;
- if it is a focusing type applicator, the focal length and the focal area;
- a unique serial number or other unique identification; and
- the effective radiating area in square centimeters (cm^2).

THINGS TO REMEMBER

The Canadian Minister of Health has responsibility for regulating radiation-emitting devices under the provisions of the REDA. This includes any devices that emit radiation as energy in the form of electromagnetic waves or acoustical waves. This can include a broad range of medical and nonmedical devices. However, the regulations issued under the authority of the REDA focus on devices where the radiation could pose a direct health hazard. The REDR covers devices such as television receivers, microwave ovens, and laser scanners as well as medical products such as diagnostic X-ray and ultrasound-therapy devices. The Minister of Health has assigned responsibility for regulating these devices to the RPB.

Schedule I of the REDR defines the following classes of devices with a medical application that are covered by the regulation:

- dental X-ray equipment with an extra-oral source
- photofluorographic X-ray equipment
- diagnostic X-ray equipment
- ultrasound therapy devices

Schedule II of the REDR defines the requirements for each of the covered classes of radiation emitting products. The REDR specifies design standards, construction standards, functional standards, and labeling and information requirements for each of the covered classes of products.

10 Mexico

Mexico has one of the largest and fastest growing markets for medical devices among the Latin American countries. The approval of the North American Free Trade Agreement (NAFTA) in 1994 was a major milestone in economic relations between Mexico, the United States (US) and Canada. NAFTA has spurred hefty increase in both import and export trade with its North American trading partners. The Mexican market is expected to continue to grow rapidly.

Mexico is vigorously pursuing free-trade agreements with many other countries as a way of expanding the benefits from trade liberalization and attempting to lessen its dependence on the US market. Prior to 2000, Mexico had free-trade agreements in effect with Chile, Costa Rica, Bolivia, Colombia, Venezuela, and Nicaragua. In 2000, Mexico signed a free-trade agreement with the European Union (EU). Similar to NAFTA in its coverage, this agreement went into effect on July 1, 2000. Mexico also signed trade agreements with Israel and with the Central American "Northern Triangle" (Guatemala, El Salvador, and Honduras). Free-trade negotiations are now underway with Panama and with the European Free Trade Association (EFTA) (Norway, Switzerland, Iceland, and Liechtenstein), and Mexico continues to push Japan to begin similar negotiations. In addition to these free-trade efforts, Mexico renewed a bilateral accord with Uruguay within the framework of *Asociacion Lationamericana de Integracion* (ALADI) and has since begun negotiations on a similar preference agreement with Brazil. Mexico hopes that these accords can become stepping-stones for an eventual free-trade agreement with Mercosur (Argentina, Brazil, Paraguay, and Uruguay). (DoC, 2001 Chap. 1).

A growing affluence coupled with an expanding middle class is expected to increase the demand for healthcare services. The flow of medical devices should continue to grow rapidly as NAFTA and other free-trade agreements eliminate tariffs on most devices. NAFTA will ensure that the important Mexican public-sector market, which accounts for the bulk of medical device purchases, remains open to US and Canadian manufacturers by requiring transparent and nondiscriminatory government practices.

The manufacture and sale of medical goods in Mexico is regulated under provisions of the *Ley General de Salud* (General Health Law).* Medical devices are also covered under provisions of the Health Goods Ruling** and the Federal Consumer Protection Law.*** Under the law, medical goods are any instruments, devices, implements, machines, designs, reagents for *in vitro* usage, cultivation medias, contrast medias, healing materials, hygiene products, dental materials, prostheses and orthesis, susceptibility test for antimicrobians, reagents for *in vivo* usage, allergenics, antiserums, antigens, calipers, verifiers, or any other similar competent products, parts or accessories that (NOM-137 p 49):

- are listed on National Pharmacopoeia or its supplements;
- are designed for use in the diagnosis of an illness or other condition, in healing, mitigation, treatment or prevention of a human disease; and
- are designed to modify the structure or any function of human body.

---

* *Ley General de Salud, Secretaria de Salud, México, D.F.,*1993.
** *Reglamento de Insumos para la Salud, México, D.F.,*1998.
*** *Ley Federal de Protetión al Consumidor, publicada en el* Diario Oficial de la Federación *en 24 de diciembre de 1992.*

Medical equipment (devices) are any apparatus, accessories, and instruments, the final intended use of which is medical attention, surgery or exploration procedures, diagnosis, treatment and patient rehabilitation, as well as for performing biomedical research activities (NOM-137 p. 49).

BACKGROUND AND GENERAL INTENT OF THE LAW

Compared to other countries in Latin America and on the Pacific Rim, Mexico has a rather long history of regulating medical devices. The practice of regulating medical devices dates from well before the adoption of the current General Health Law in 1992. Medical devices are also covered under provisions of the Health Goods Ruling and the Federal Consumer Protection Law.

The cornerstone of conformity assessment in Mexico is the 1992 Federal Law of Metrology and Standardization (as amended on December 24, 1996 and May 20, 1997), which provides for greater transparency and access by the public and interested parties to the mandatory standard-formulation process. Another key document is the regulations to the Federal Law of Metrology published on January 14, 1999. Under the provisions of these laws, products must conform to all applicable Mexican standards, known as *Normas Oficiales Mexicanas* (NOMs). In addition to mandatory standards (NOMs), there are numerous voluntary standards known as *Normas Mexicanas* (NMXs). As of June 2000, there were over 6,200 valid NMXs in Mexico. NMXs become mandatory for importation purposes when they are referred to in a NOM. Many NOMs include labeling requirements that are applicable to the products covered by those NOMs. The official labeling guide for NOM-050-SCFI-1994, which covers labeling for most commercial products, lists 37 other labeling NOMs that are applicable to specific products. Medical devices are covered by the NOM-137-SSA1-1995 (NOM-137).

The May 20, 1997, revision to Mexico's Federal Law on Metrology and Standardization requires that all NOMs be reviewed at least every five years. Around every April 15, Mexico publishes, in the *Diario Oficial*, its annual standardization program, and a subsequent amendment is published in the fall of the same year. This annual program lists all committees working on the development, update, or cancellation of NOMs and NMXs for the year.

SCOPE OF THE REGULATIONS

Regulations promulgated under the provisions of the General Health Law are applicable to domestically produced and imported medical goods. NOM-137 establishes minimal general requirements for sanitary and commercial information that should be included in labeling for healing materials, diagnostic agents, dental supplies, prostheses, ortheses, functional aids, hygiene products and medical equipment that will be addressed to consumers. NOM-137 is obligatory for all the industries, laboratories and establishments working on manufacturing, processing, importing and distributing of medical goods.

Certain medical goods are excluded from the scope of NOM-137. These include (NOM-137 p. 48):

- Highly specialized medical devices that are products intended for the special use of the importer or a third fiscal or moral person. The product is acquired under a purchase-sale contract and meets the specifications of the purchaser who assumes the risk of the acquired product. These products are not sold to the general public.
- Medical devices that are not intended for commercial, industrial, or service processes and are not intended for sale directly to the consumer. The consumer is defined as the fiscal or moral person who acquires, realizes, or uses the goods, products, or services.

- Medical devices that are imported by fiscal or moral people for their own use.
- Medical devices imported by educational and scientific institutions.
- Medical devices that because of their nature or size cannot have labels or devices with labels too small to contain all the data required by the regulation. These devices are subject to the disposition of the Ministry of Health–the *Secretaria de Salud* (SSA).
- Products sold in bulk.
- Products subject to NOM-137 when imported for the purpose of obtaining the certification under the Official Mexican Rule.
- Products intended only for export.
- Any other device specifically excluded from the scope of the regulation by the Ministry of Health.

BRINGING DEVICES TO MARKET IN MEXICO

Like many of the countries in Latin America and on the Pacific Rim, Mexico requires that medical devices be reviewed by the SSA prior to sale. Although the domestic production of medical devices has markedly increased in recent years, Mexico still imports a significant percentage of the needed medical devices from the United States, Japan, and the EU. For these products, the Mexican government relies heavily on the product approvals from the regulatory entity in the country where the device was manufactured.

Product Registration

A person who proposes to bring a medical device to market in Mexico is required to register the device with the SSA. To obtain registration, the manufacturer must submit a dossier on the product. A nominal fee is charged for processing each submission. The dossier must contain the following information, as applicable to the product in question:

1. A description of the product, its purpose, a summary of the intended use(s), and engineering drawings. Published product literature may fulfill this requirement. The information, however, must be submitted in Spanish.
2. A copy of the package insert or product manual that describes product performance, handling, warnings, contraindications, precautions, and so forth is required.
3. Clinical abstracts discussing the clinical experience with the device. These documents must show clearly the source of the clinical data (i.e., the institutions where the clinical work was done).
4. A description of the materials used in the product, emphasizing those materials in contact with the body tissue. The information should stress body compatibility, corrosion resistance, and so on. This description must include evidence of the testing that was performed from the laboratory responsible for the report.
5. Evidence of the testing (functional tests) to which the product has been exposed. The evidence must include test procedures and results, and the reports must be on the letterhead of the organization performing the testing and signed by the person responsible for conducting the testing. The required battery of tests includes tests for toxicity, pyrogenicity, hemolysis, biocompatibility, hermeticity, and sterility.
6. A description of the packaging system and packaging testing that has been performed. The description should include such items as the packaging configuration (c.g., sterile package inside a sales package), the physical characteristics of the packages, and the materials used

in the package. Evidence of package stability and integrity testing must be included. When applicable, use-before-date information and testing is required.

7. When applicable, details on the sterilization methods and the expected results are to be provided.

8. Device-specific information requested by the SSA. For example, the dossiers for tissue heart valves and tissue conduits should include documentation on the following:
    - Individual certificates of results from biological tests made on low-porosity Dacron grafts, intracutaneous toxicity, implantation tests, cytotoxicity, and hemolysis tests
    - Quality controls implemented for polyester-based duct, water permeability, adaptability, wall width, internal diameter, strain force, radial resistance, Dacron-adhesion resistance, suture retention, final elongation, and spiral-adhesion resistance
    - Metallic-chemical composition of Haynes Alloy No. 25 metallic ring, as well as corrosion resistance and Rockwell hardness tests
    - Certificate of quality for raw material in heart valves of porcine or any other tissue origin
    - Certificate of biologic and immunologic tests undertaken with heart valves of porcine origin
    - Finished-product quality control showing mechanical and physical tests undertaken (e.g., hydrodynamic tests, structural design, valve/duct separation resistance)
    - Finished-product analysis certificate for 2 percent glutaraldehyde solution
    - Stability-test certificates indicating dates of initial and subsequent analyses, temperature studies, and results obtained, as well as valve/duct functionality tests

9. A certification that the product is manufactured under appropriate manufacturing controls (e.g., US Good Manufacturing Practices [GMP]) to ensure the quality of the end product.

10. A certificate showing the address of the manufacturer. This information may appear on some other required document such as the US Food and Drug Administration (FDA) Certificate for Foreign Government (CFG).

11. For imported products, evidence of approval of the product from the regulatory entity in the country where the device was manufactured is required. This could take the form of a Japanese shonin (device approval), a US FDA CFG, a US FDA Premarket Approval Letter or 510(k) Approval Letter, or a CE Conformity Certificate.

12. For imported products, a letter addressed "to whom it may concern" naming the distributor in Mexico and granting an exclusive license for distribution of the product in Mexico. This may not be required with each dossier if the distributor has a properly notarized and legalized letter on file granting the distributor an exclusive license to distribute all products manufactured by all of the manufacturer's facilities. Because Mexico is member of the Haya Convention, the legalization takes the form of an apostille or marginal annotation or note such as a notary seal so that consularization is no longer necessary.

13. A Certificate of Free Sale completes the dossier. This document is issued by a Ministry of Health and means the product is freely sold among the citizens of the country where the goods are manufactured. For goods manufactured in the United States, the Certificate of Free Sale takes the form of a CFG.

Some of the documents, such as the Free Sale Certificate and the Letter of Exclusive Distributor, must be notarized. The notarization gives the document legal status in the eyes of the SSA. Any documents required by the SSA in the form of a letter should appear on letterhead and be signed by a responsible official of the organization. Original documents are preferred, but if these are not available, notarized copes will be sufficient.

In Mexico, the Approval Letter "belongs" to the SSA. This means that even if the file is submitted by a distributor and the Approval Letter grants the distributor the rights to import, distribute, or export the goods, the manufacturer will always have the rights over its goods.

All of the items listed above are important to getting a product registered in Mexico. However, the SSA places special emphasis on (a) the certificate showing the address of the manufacturer (item 10), (b) the Approval Letter or certificate from the Ministry of Health where the product was manufactured (item 11), (c) the description of the materials in contact with the body (item 4), and (d) the description of testing (item 5).

Once a product has been accepted for registration by the SSA, a Certificate of Registration will be issued. This allows the distributor to sell the device on the open market. For imported products, the SSA registration also enables some products to be cleared by Mexican customs officials at the entry point into Mexico. Implanted goods require an Import Permit filed with a copy of the Approval Letter. However, for eligibility to sell products to government institutions, further approvals are required.

QUALIFYING FOR SALE TO THE MEXICAN GOVERNMENT

Having a product approved for sale to the healthcare agencies of the Mexican government is important because government purchases account for more than 75 percent of the Mexican market. Mexican citizens are covered by one of five health plans operated by the Mexican Government. The largest and best known is the *Instituto Mexicano del Seguro Social* (Social Security or IMSS). This is a group of hospitals that are managed by the government and provide services to employees in the private sector. The second is the *Instituto de Seguridad Social al Servicio de los Trabajadores del Estado* (Mexican Institute for Healthcare for Government Employees or ISSSTE). The ISSSTE provides services for all nonmilitary government employees. The Army and Navy are covered by their own healthcare system. The fourth plan covers national institutes whose goals include investigations. Finally, there is the system administered by the SSA for unemployed persons.

All public hospitals have to go out for bids for products listed in a catalog called *Cuadro Básico del Sector Salud*. This catalog describes in a generic form all types of goods that can be acquired by the system. Effective in 2001, hospitals are requiring goods to be quality tested in order to participate in bids. So far, the IMSS is the only system that has the capability to perform the required testing.

From the general catalog, IMSS has chosen the goods they require and have listed them in a special catalog known as *Cuadro Básico Institucional*. If a product fits into a description in the catalog, a manufacturer or distributor can request approval of its product. In order for a product to be qualified for government purchase, that product must be evaluated by an authorized third party, which could be:

- the National Polytechnic Institute;
- the *Secretaria de Salud* Laboratory; or
- the IMSS Quality Assurance (QA) Department.

To be qualified to perform this testing, the laboratory must be accredited by the *Entidad Mexicana de Acreditación* (Mexican Accreditation Entity or EMA) or the *Sistema Nacional de Acreditamiento de Laboratorios de Prueba* (National Accreditation System for Testing Laboratories or SINLAP).

The laboratories charge a nominal fee for the evaluation, and the manufacturer or distributor is usually required to provide samples for evaluation. Effective in 2000, the number of samples provided depends on the type of good. For example, pacemakers require four samples, while PTA catheters require an average of 15 samples. Sterile products must be provided in a sterile state and are usually not returned, as they will most likely be used on a patient.

Evaluation of a product by the IMSS QA Department is typical of the process that is followed. The manufacturer begins by notifying the IMSS Acquisition Department of the product from the list of products accepted in government tenders (bids) for which it is a potential supplier. Then the manufacturer will request in writing that the IMSS QA Department evaluate the product. The letter should be on letterhead, signed by a responsible person, and have the name and address of the manufacturer/distributor, the brand name, and country of origin. The letter should include a list of the samples and lot numbers submitted for analysis and establish the equivalent between the manufacturer's catalog number and the IMSS code. The dossier and the samples must be submitted with the request for testing. If anything is missing, the application will not be accepted.

In addition to the requirements listed above, the submission should include:

1. A technical dossier on the product including the following, as applicable:
   • The technical specifications for the product, including tolerance levels where appropriate.
   • The operation, technical, and service manual(s) for the product when applicable.
   • A copy of the manufacturing standards under which the product has been built, and the appropriate process validation.
   • The protocol of the QA test to which the product has been subjected during its manufacture and the laboratory test results. This should be submitted in the form of a Quality Control Release document.
   • When appropriate, certificates of the sterilization, pyrogenicity, and toxicity of the device issued by the manufacturer. (They may all be in the same certificate.)
   • An explanation of the lot or serial-number system used to identify the product.
   • A product catalog.
   • Publications that show the behavior of the product submitted for evaluation.
   • A list of countries and hospitals where the product has been used and number of cases.
2. Device-specific information requested by the IMSS QA Department. For example, the technical dossier for a tissue heart valve should include documentation on the following:
   • Washing protocol
   • Microscopic tissue control
   • Electronic microscopy investigation
   • Chemical features for fixing and preservation
   • Gradient table
   • Hemodynamic image information
   • *In vitro* investigation reports
3. Official certificates giving the regulatory status of the device, including the following:
   • A copy of the SSA Approval Letter along with the original to check/verify the copy
   • For imported products, the following certificates should be notarized and have an Apostille:
     • Certificate of Free Sale showing that the product is freely sold among the people of the country where the device was manufactured. This certificate should be issued by the Ministry of Health (e.g. a US FDA CFG).
     • GMP Certificate issued by the Ministry of Health of the country where the device was manufactured.
4. A specified number of saleable products for evaluation in the field (i.e., a hospital) and in the IMSS QA Department laboratory. If the device is labeled with an expiration date, the

sample devices should have at least 18 months before they expire (i.e., the IMSS QA Department wants fresh devices and is not willing to accept obsolete products). The IMSS will not accept imported devices labeled as being only for export from their country of origin (e.g., "Export Only," "Investigation Only," "Pilot Lot," "Experimentation Phase," or any similar labeling from the US or any other country). If support instruments (e.g., a pacemaker programmer) are required during part of the testing, the manufacturer must provide these instruments. They are returned after the evaluation is completed.

5. Where applicable, the name, address, and telephone number of a person in Mexico who will provide technical assistance (i.e., a technical representative).

The IMSS QA Department will evaluate the submitted material. If the device is acceptable, the IMSS will issue an official announcement that the product can be accepted to compete in a tender.

MISBRANDING

The requirements described in the following sections are mandatory under Mexican law. The SSA is charged with responsibility for surveillance to ensure that regulatory requirements are met. Its personnel will perform the verification and surveillance that is required. Penalties can be assessed under the provisions of the General Health Law.

GENERAL LABELING PROVISIONS

The common feature of most labeling requirements in Mexico is that package labels must carry commercial information in Spanish. The content should be easy to understand and read. If the required information appears in a language other than Spanish, then this information must appear in Spanish in the same size font, format, and clarity as it appears in the other language. Consequently, many of the labels added to packages in the past for the Mexican market are no longer acceptable. However, placing stickers over information preprinted on a package or label is permitted as long as the resulting label complies with the labeling requirements. Secondary or additional labeling should not cover original labeling. Additionally, labels can be applied or modified in Mexico provided that the process complies with Mexico's verification procedures.

Additionally, a comma must be used as a decimal point in the quantity declaration on packages, as required by NOM-008-SCFI-1993. Imported products using a period as a decimal point are likely to be rejected by Mexican customs officials. Accredited verification units exist, where companies can obtain an evaluation of their labels prior to export to Mexico.

Within the scope of NOM-137, a label is any printed bill, stamp, inscription, image, or any other descriptive or graphic, written, printed, outlined, marked, engraved in high- or low-raised work or attached form to the container or packaging.

Types of Packaging

NOM-137 defines a container or packaging for a medical good as any receptacle that guarantees the full preservation of product quality. Several layers of packaging are described in NOM-137. They are (NOM-137 p. 49):

- Primary packaging—Any container that is in direct contact with the product, with the intended purpose of preserving its physical, chemical, and microbiological integrity;
- Secondary packaging—Any container in which the primary packaging resides;

- Collective packaging—Any container or wrapping in which two or more varieties of prepacked products reside, as a mean of presentation to be sold to the consumer; and
- Multiple packaging—Any container or wrapping in which two or more primary or secondary packages reside.

Minimum Labeling Requirements

NOM-137 specifies the minimum information that must appear on the labeling of medical goods. The minimum information required in order to comply with the regulation includes (NOM-137 pp. 51–52):

1. The product's commercial name. This is the only required item that can be stated in a language other than Spanish.
2. The brand, logotype, social reason (name given in Mexico to the legal name of a company) and the commercial address of the manufacturer or distributor that is registered with the SSA. The distributor is defined as the fiscal or moral person who imports, conditions, distributes, or commercializes goods, products, or services inside Mexico. For the purpose of this regulation, a distributor is anyone who is not involved in the manufacturing process and is dedicated to the commercialization of a product.
3. For imported goods, the name and address of the importer.
4. The country of origin.
5. The approval number given by the SSA.
6. For sterile products, the following should be indicated on the label: "Sterile Product. Product sterility is not guaranteed in case primary packaging shows previous rupture/breaking signs."*
7. For sterile products, the expiration or "use by" date in those cases when the manufacturer does not guarantee at least 5 years of sterile shelf life. The expiration date (in Spanish, *fecha de caducidad*) is displayed on the primary and secondary (if applicable) packaging and delineates the useful life of the product. It is calculated from the manufacturing or sterilization date. The shelf life (in Spanish, *periodo de caducidad*) is the estimated time during which the product remains inside specifications when preserved under normal or particular storage conditions. Normal storage conditions are defined in NOM-137 as storage in a dry place (not more than 65% relative humidity), properly ventilated, room temperature between 288 K and 303 K (15 °C to 30 °C), protected from intense light and strange odors or any other form of contamination. Particular storage conditions are any conditions other than normal storage conditions. Particular storage conditions must be shown on the product label.
8. When special storage temperatures are required, these should be expressed as ___K (°C) to ___K (°C) or similarly.
9. The lot or serial number. NOM-137 defines a lot as the quantity of a product manufactured in one single process with the equipment and required substances in the same period of time to guarantee its homogeneity.
10. The manufacturing date. This can be part of the lot number.
11. The contents of the package, giving a description of the product and indicating the number of units, the volume, or the weight of the contents.
12. The nominal dimensions (if applicable).

---

* Subclause 4.1.1.11 of NOM-137 specifies the following statement in Spanish, "*Producto estéril, 'No se garantiza la esterilida del producto en caso de que el primario tenga señales de haber sufrido ruptura previa'.*"

13. When the use, handling, and conservation of the product is not obvious, this information should be stated in the instructions for use (IFU) with a statement on the label saying "Read attached instructions" or similar wording.*
14. Any warnings or precautions regarding the use of a product.
15. Any adverse secondary effect caused by the use of the product should be detailed on the label.
16. A statement that the product is atoxic, pyrogen free, or similar (if applicable).
17. An indication if the product is disposable (if applicable).
18. Precise instructions for reuse, disposal, or destruction of empty containers (if applicable).
19. For every product that is composed or made up of several ingredients, the label should list the qualitative and quantitative formula of its active ingredients (if applicable).

When the products are imported goods, they should have the required labeling in Spanish. The data required by NOM-137 can be added to imported goods inside Mexico after clearing customs but before commercialization or supply to the public.

Products that, because of their nature or the size of the units in which they are sold or supplied, cannot have labels or, because of the size, cannot have all the data required by NOM-137, will be subjected to the discretion of the SSA. In these cases, the SSA must be contacted for information. Customs brokers and agents generally have the information about NOMs that must be complied with to import into Mexico, as well as other regulatory requirements for importation.

Bulk product is only required to have original labeling on the collective packaging. Bulk product is defined in NOM-137 as product placed in any type of container where the contents may vary and have to be weighed, counted, or measured at the moment of its conditioning and later sale.

The nature of the product, its formula, composition, distinctive name or brand, generic or specific designation, and labeling should match with the authorized specifications approved by the SSA in the applicable certificate and cannot be modified.

THINGS TO REMEMBER

In Mexico, as with most other countries in Latin America and on the Pacific Rim, medical devices must be registered with the public health department—the *Secretaria de Salud*—prior to sale. To obtain registration, the manufacturer must submit a dossier on the product. Registration allows the product to be sold on the open market.

However, in order to be eligible for sale to the IMSS and/or other government institutions, the product must be evaluated by an authorized third party. This evaluation includes the review of a technical dossier on the product, and may involve laboratory testing of devices.

The minimum information that must appear on the labeling of medical goods is specified in Official Mexican Rule NOM-137-SSA1-1995. The common feature of most labeling requirements in Mexico is that package labels must carry commercial information in Spanish. If the required information appears in a language other than Spanish, then this information must appear in Spanish in the same size font, format, and clarity as it appears in the other language. Placing stickers over information preprinted on a package or label is permitted as long as the resulting label complies with the labeling requirements. Placing secondary or additional labeling over the original labeling is considered misbranding and therefore it is not permitted.**

The labeling of a product must match the authorized specifications approved by SSA in the applicable certificate and cannot be modified.

---

* Subclause 4.1.1.10 of NOM-137 provides the following example: "Léase Instructivo Anexo."
** Subclause 4.1 of NOM-137 states that additional labels should not cover the original labels.

Part V

China, Korea, and Thailand

11 People's Republic of China

The People's Republic of China (PRC) is the worlds largest potential market for medical devices. In the last two decades of the twentieth Century, China witnessed a dramatic growth in its population, rapid urbanization, the transition from a planned toward a market economy, and its integration into the global economy. Following drawn-out preliminary negotiations begun in 1985, China became the 143rd member of the World Trade Organization (WTO) in November 2001.

China's initial attempt to regulate medical devices was characterized by a good deal of overlap and competition for preeminence between agencies of the central government. The current system rationalized some of the responsibilities although overlapping requirements and redundant testing for some medical devices by different agencies remains. The framework for medical device regulation in China was established by an order of the State Council of the PRC. The Regulation on Supervision and Administration of Medical Devices (China MDR) was adopted in council on December 18, 1999, and became effective on April 1, 2000.

Under the provisions of the China MDR, a medical device is defined as "any instrument, equipment, apparatus, appliance, material, or other article (including necessary software) that is used alone or in combination on human bodies. The application of the medical device is intended to achieve:

- prevention, diagnosis, treatment, monitoring, or alleviation of diseases;
- diagnosis, treatment, monitoring, treatment, alleviation, or reparation of injuries or the handicapped;
- research, replacement, or modification of anatomy or certain physiological process:
- control of pregnancy.

The function of a medical device on or in the human body is not achieved by pharmacological, immunological, or metabolic means even though these processes may be involved and play a supporting role in the performance of the medical device (China MDR §3).

BACKGROUND AND GENERAL INTENT OF THE LAW

The Chinese system for medical device regulation began in 1994 and was administered by the State Pharmaceutical Administration of China (SPAC). Prior to April 1, 2000, SPAC oversaw the registration of medical products and the development of the medical products industry. SPAC has been succeeded by the State Drug Administration (SDA) as the central government agency in charge of drug and medical device regulation. The SDA has established supplemental regulations and orders to facilitate the regulation and supervision of medical devices.

Although the SDA retains overall authority, responsibility for supervision and administration of certain provisions of the China MDR are delegated to other levels of government within the PRC. Responsibility for dealing with domestically produced, low-risk medical devices is assigned to municipal governments. Responsibility for domestically produced, medium-risk devices is assigned to the government of the provinces, autonomous regions, and municipalities directly under the central government. The SDA retains responsibility for domestically produced high-risk devices and for registration of all imported products. The SDA has the authority to review any registration granted by another government entity and can order it to correct its mistakes if the SDA concludes that it has wrongly granted a registration certificate. If that government entity fails to correct its error within a specified time, the SDA can revoke its registration certificate by making a public announcement.

The SDA does not have exclusive jurisdiction over all aspects of the approval process, especially for imported medical devices. Depending on the product, there can be several Chinese agencies involved in granting approval.

Beginning in 1989, the State Administration for Entry/Exit Inspection and Quarantine (SAIQ)* was charged with safety licensing for a wide range of imported consumer products. In 1997, the SAIQ extended the coverage of its licensing power to include six categories of medical devices (ITA, China, p. 2). These products were required to bear the China Commodity Inspection Bureau (CCIB) mark when imported into China.

Imported medical devices that incorporate a pressure vessel were also required to have a certificate issued by the Safety Quality Licensing Office of Boiler and Pressure Vessels (SQLO), a division of the State Bureau of Quality and Technical Supervisions (SBTS), before Chinese Customs would allow the product into the country. Also, measuring and weighing devices or equipment that contain a major measurement component, such as thermometers, blood-pressure meters, syringes, biochemical analyzers, blood gas analyzers, and so on, require the approval of SBTS. Testing of these devices could be performed by SBTS or by the end-users on behalf of SBTS (ITA China, p. 2).

On December 3, 2001, the Chinese government established a Compulsory Product Certification System (CPCS). This system is jointly administered by the State General Administration for Quality Supervision and Inspection and Quarantine (AQSIQ) and the Certification and Accreditation Administration (CNCA) of the PRC. It replaces the Safety License System for Imported Commodities administered by SAIQ and Compulsory Supervisions System for Product Safety Certification administered by SBTS.

The system came into force on May 1, 2002, for products listed in the CPCS product catalog. For medical devices, the following categories are listed in the first edition of the CPCS catalog (CPCS p. 5):

- Medical diagnostic X-ray equipment
- Hemodialysis equipment
- Hollow fiber dialysis equipment
- Blood purification equipment
- Electrocardiographs
- Implantable cardiac pacemakers
- Ultrasound equipment

These devices required an Import Safety License and a CCIB mark issued by SAIQ prior to the enactment of the new regulation. An example of the CCIB mark is found in Figure 11.1.

Prior to April 30, 2003, products may be certified under either system. As of May 1, 2003, products listed in the CPCS catalog must have a new certificate and the new mark (CNCA pp. 2–3). See the section on the Compulsory Product Certification System later in this chapter.

SCOPE OF THE REGULATIONS

Article 5 of the China MDR specifies the establishment of a three-tier classification system to facilitate the management of medical devices in China. The three classes are:

---

* At one time, the State Administration for Entry/Exit Inspection and Quarantine (SAIQ) was known as the State Administration of Import and Export Commodity Inspection of China (SAIC).

Source: CPS Reg p. 2.

FIGURE 11.1 CCIB safety certification mark.

- Class I — Medical devices for which conventional management is sufficient to ensure safety and effectiveness
- Class II — Medical devices for which certain controlling measures must be applied to ensure safety and effectiveness
- Class III — Medical devices that are implanted into the human body, that are life supporting, or that may impose potential risks to humans for life support and require strict controls to ensure safety and effectiveness

To standardize the classification of medical devices, the SDA has established a Regulation for Medical Device Classification. This regulation specifies that the criteria for classifying a medical device are to be based on the intended purpose and the form of operation of a medical device. Devices intended for laboratory use and *in vitro* diagnostic (IVD) reagents are classified as medical devices. Should one medical device fit in two classes at the same time, the higher one must be used (China Classification pp. 1–3).

For those medical devices to be used in combination with other devices, the classification of each part should be determined separately. Accessories for a medical device are classified independently from the master device based on the accessories' own characteristics.

For those medical devices to be used on particular parts of the human body, the classification should be determined on the basis of the risks involved with the intended purpose of a device and its form of operation.

Software that controls the functions of a medical device should be designated to the same class as its associated medical device. Those products that are designed to monitor or affect the major functions of a medical device should be designated to the same class as the device being monitored or affected.

Devices intended for measuring and weighing or equipment that contains a major measurement component must comply with the relevant provisions of the Metrology Law of the People's Republic of China (China MDR §6).

MISBRANDING

The SDA has the right to withdraw a product registration for a device if the applicant has made false statements in the registration dossier (China MDR §40). Since the product manual is provided as part of the registration, false or misleading statements in the labeling could be viewed by the SDA as false or forged statements in the product registration dossier. Likewise, lack of evidence to support a contention in the labeling could be viewed as a false statement, even if the statement was strictly true. Loss of the product registration would result in the manufacturer being prohibited from manufacturing or marketing the product in China.

BRINGING DEVICES TO MARKET IN CHINA

Any firm that intends to manufacture or import a medical device for sale in China should apply for product registration with a competent department of the Chinese government; products that are not registered cannot be sold (China MDR §35). The basic procedures and requirements for registering a product are outlined in the following sections.

MEDICAL DEVICES MANUFACTURED IN CHINA

Medical devices manufactured in China are subject to regulation by the SDA or by an authorized agency of the municipal or provincial government* where the manufacturer is located. The governmental agency responsible for processing the registration application depends on the classification of the device. For devices in the Class III (highest) category, examination of the application is performed by the SDA. For devices in the Class II category, the application is examined and approved by the drug and medical device administration affiliated with the provincial government where the manufacturer is located. For Class I (lowest) category devices, the application is examined and approved by the drug and medical device administration affiliated with the municipal government where the manufacturer is located. For purposes of the regulation, a medical device whose final production procedures are completed within China is considered a domestically produced device. Devices produced in Taiwan, Hong Kong, and Macao are examined by the SDA regardless of class (China Registration §3).

Medical Device Registration

A manufacturer initiates the registration procedure for any product by submitting the proper registration application forms with the required accompanying documents to the local drug and medical device administration that has jurisdiction in the area where the manufacturer is located. The requirements for registration are covered in the SDA regulation, "Management Provisions on Registration of Medical Devices" (China Registration). The steps in processing the application depend on the classification of the device.

The SDA has the authority to cancel the product registration certificate for a device under the following conditions (China Registration §§25–29):

- It is determined that any information in the registration dossier is false or misleading.
- Labeling is changed to expand the scope of application and indications for the device without prior approval.
- The safety and effectiveness of a device cannot be guaranteed because an imported device fails an SAIQ import inspection.
- The SDA has concluded that another government entity has wrongly granted a registration certificate.

If a manufacturer ceases production of a medical device for more than two consecutive years, the registration certificate is automatically invalidated. The manufacturer must apply for a new registration before reentering the market (China Registration §21).

Class I Devices

For Class I devices, the application for registration is filed with the local drug and medical device administration. The application must include (China Registration §5):

---

* Provincial government includes the governments of provinces, autonomous regions, and municipalities under the direct control of the central government.

- the manufacturer's qualification certificate,
- specifications for the medical device with related notes and explanations,
- a self-test report on the performance of the device,
- a description of the manufacturer's production quality-management system,
- the instruction manual of the product, and
- a statement of guarantee on the authenticity of the materials submitted.

The local drug and medical device administration reviews the application and issues a registration certificate that is sufficient to allow sale of the device in China. The registration certificate is valid for a period of four years. Six months before the expiration of the certificate, the manufacturer can apply for a four-year extension.

Class II and III Devices

For Class II and III devices, the application for registration is filed with the local drug and medical device administration. For Class II devices, the application is sent to the drug and medical device administration associated with the provincial government. For Class III devices, the application is forwarded to the SDA.

Class I devices may proceed directly from registration to full production. However, domestically produced Class II and III devices must undergo a two-step procedure before full production is authorized. The steps are:

1. Trial production registration indicates that the basic safety and effectiveness of the product have been established (see "Investigational Use" later in this chapter). The trial production registration certificate is valid for two years. During this time, the manufacturer will gather feedback from the product users.
2. Formal production registration indicates that (a) the safety and effectiveness of the product have been established, (b) the manufacturer's quality system has been inspected, and (c) the product has achieved an acceptable level of field performance as measured by the users' quality feedback.

Trial production registrations expire naturally at the end of the two-year term. The SDA has the right to terminate a trial production registration or a production permission registration at any time based on (a) postmarket surveillance of product quality, (b) user complaints, or (c) a determination that the applicant made false statements in the registration application or product dossier.

The trial production application must include (China Registration §6):

- The manufacturer's qualification certificate.
- A technical report on the product.
- An analysis on risks to safety presented by the product.
- Specifications on the medical device with related notes and explanations.
- A self-test report on the performance of the product.
- A test report on the medical device issued within one year (half a year for biological materials) by a testing institution certified by the SDA.
- Clinical trial reports from more than two clinical-trial bases. These reports must be provided in the manner stipulated by the Provisions on Report Items of Clinical Trials for Medical Device Registration.* The clinical trials must be conducted in compliance with the Provisions on Clinical Trials for Medical Device Products.

---

* *Provisions on Report Items of Clinical Trials for Medical Device Registration* appears as an appendix to State Drug Administration Decree No. 16 (China Registration).

- The instruction manual for the product.
- A statement of guarantee on the authenticity of the materials submitted.

After seven months of trial production, the manufacturer can apply for formal production registration (China Registration §4(I)). The formal production application must include (China Registration §7):

- the manufacturer's qualification certificate,
- copies of the registration certificate for trial production,
- specifications for the medical device,
- a report on improvements made during the trial production,
- a valid certificate of inspection of the manufacturer's quality system,
- a test report on the medical device issued within one year by a testing institution certified by the SDA,
- a quality-tracking (postmarket surveillance) report on the product, and
- a statement of guarantee on the authenticity of the materials submitted.

If granted, the formal production registration is valid for four years. Six months before the expiration of the certificate, the manufacturer can apply for a four-year extension.

Under certain circumstances, a manufacturer may not have to wait seven months before applying for registration for formal production. After applying for a trial production registration, the manufacturer can immediately apply for formal production registration if the manufacturer has a quality system certified by an agency designated by the SDA, and the medical device is equivalent to one already registered for full production, or the differences in structure or function of the device have no material impact on the safety or effectiveness of the device as compared to a device already registered for formal production (China Registration §10).

When applying for renewal of any registration, the manufacture must provide a quality-tracking report on the product detailing postmarket-vigilance experience with the product. For Class II and III products, the manufacturer must also provide a test report issued within one year by a testing institution certified by the SDA and a valid certificate for inspection of the manufacturer's quality system (China Registration §8).

Drug/Device Combinations

A product that employs a drug in an ancillary role is registered as a medical device. However, the drug itself must be approved by the appropriate government agency and have a registration number before the drug/device combination is registered with the SDA.

MEDICAL DEVICES IMPORTED INTO CHINA

Imported medical devices are examined and approved by the SDA regardless of the classification. An imported medical device is one whose final production procedures are completed outside of China.

Unlike other Asian countries such as Japan, Korea, or Thailand, China issues registration certification and licenses in the name of the manufacturer, not to the local agent or distributor. An in-country agent is not required to register a product in China. However, foreign manufacturers have found it difficult to conduct registration from overseas due to communications problems and other requirements. A locally based representative, distributor, or a specialized agency firm to manage product registration on behalf of the manufacturer is recommended (ITA, China p. 7).

The application for registration is filed with the local drug and medical device administration with jurisdiction over the locality where the manufacturer's in-country agent is located. The application must be submitted in Chinese. Following some initial review, the application is forwarded to the SDA. The application must include the following information (China Registration §11):

1. Documentation demonstrating the manufacturer's business qualification. (This could include incorporation documents, manufacturing permits, and other legal documents that prove that the manufacturer is qualified to conduct business.)
2. The business license, incorporation documents, or other legally binding documents that prove the qualification of the manufacturer's in-country agent to do business and represent the manufacturer.
3. A certificate issued by the country (region) of origin indication that the medical device was approved or that permits the medical device to enter the market of that country (region) (e.g., US FDA Certificate for Foreign Government [CFG] or CE mark Certificate).
4. The technical specifications for the device (i.e., the requirements on its safety and technical performance) and tests that the device must pass before shipment from the factory. The technical specifications would include such items as:
   • a declaration of compliance with adopted standards (see the section on Medical Product Standards later in this chapter); and
   • the product characteristics as appropriate, such as material composition, physical and chemical characteristics, sterility, toxicity, biocompatibility, packaging and transport requirements, markings, and so on.
   For Class III devices, two copies of such technical specifications shall be provided.
5. The instruction manual.
6. A test report on the medical device issued within one year (half a year for biological materials) by a testing institution certified by the SDA.
7. Clinical-trial reports from more than two clinical-trial centers. These reports shall be provided in the manner stipulated by the Provisions on Report Items of Clinical Trials for Medical Device Registration. The clinical trials must be conducted in compliance with the Provisions on Clinical Trials for Medical Device Products. In most cases, the report of clinical trials conducted in the country of origin will be sufficient. However, the SDA can require on-shore clinical trials if there is reason to believe that the device may operate differently in the Chinese population or if quality problems have been identified by market vigilance.
8. A statement issued by the manufacturer, guaranteeing that the quality of the product to be registered and sold in China will have the same quality as the product sold in the country (region) of origin.
9. A letter of authorization that designates after-sales service agencies in China and a letter from each agency demonstrating that it accepts the responsibility, along with legal documents demonstrating the qualifications of the after-sales service agencies (i.e., their business license).
10. A statement of guarantee on the authenticity of the materials submitted.

These documents must be in Chinese or be accompanied by a Chinese version. The documents mentioned in items 1, 2, and 3 may be submitted as photocopies, but they must be signed and sealed by the original issuing authorities or notarized by local notary officer. Other documents mentioned above must be submitted as originals that are signed or sealed by legal representatives.

Once the SDA is satisfied that the device meets the intended level of safety and clinical effectiveness, a product registration certificate is issued. An application for import may be submitted to the custom service only after the registration certificate for the product has been issued by the SDA (China MDR §11).

The registration certificate is valid for four years. An applicant can apply for renewal of the registration within six months of the expiration of the currently valid registration certificate. The information required to accompany the application for renewal is essentially the same as that which must accompany the original applications except:

- A copy of the original registration certificate must accompany the application.
- A new test report on the medical device issued within one year by a testing institution certified by the SDA is required.
- A quality-tracking report on the product must be included.

A clinical report and the documentation on the qualifications of the manufacturer's in-country agent are not required in the renewal application.

BUSINESS LICENSING AND QUALITY-SYSTEM APPROVAL

Enterprises dealing with medical devices must have a license issued by the drug and medical device administration affiliated with the government of the province, autonomous region, or a municipality directly under the central government. Two types of licenses are issued: (a) a production license, and (b) an operation license. Both licenses are valid for five years and can be renewed upon expiration. Medical institutions are required to purchase medical devices from enterprises that have a valid production or operation license (China MDR §26).

Production License

A production license is necessary for a medical device manufacturer in China. The Chinese Administration of Industry and Commerce will not issue a business license for the firm without this production license.

The applicant for a production license must demonstrate through its application that it has (China MDR §19):

- qualified employees trained in the production of the device,
- facilities appropriate for the production of the device,
- appropriate equipment, and
- qualified institutions or personnel and relevant equipment to carry out quality inspections of the manufactured devices.

For Class I medical device manufacturers, notification of the appropriate authority is required. For Class II and III device manufacturers, the application must be examined and approved by the appropriate authority.

Operation License

Other businesses, such as importers and distributors, must have an operation license. The applicant for an operation license must demonstrate thorough its application that it has (China MDR §23):

- facilities appropriate for conducting business and handling the medical devices,
- qualified personnel to carry out quality inspections of the devices, and
- the capacity to provide after-sales services such as technical training and maintenance.

For enterprises handling Class I devices, notification of the appropriate authority is required. For enterprises handling Class II and III devices, the application must be examined and approved by the appropriate authority.

Quality System

Since July 1995, manufacturers of Class II and III medical devices must undergo review and approval of their quality systems before they can complete registration of their devices. Examination and approval of the manufacturer's quality system are the responsibility of the certification agencies on quality systems designated by the SDA.

For Class II and III devices manufactured in China, the manufacturer must complete the examination of its quality system before applying for formal production registration (China Registration §9). For imported Class III devices, the foreign manufacturer's facility will need to complete an on-site inspection of its production quality system as part of the registration process. An on-site inspection will be required every four years. If the same type of device is covered by a quality system that has passed an on-site inspection within four years, a second inspection is not required (China Registration §13).

The manufacturer of a Class II or III medical device can apply for an exemption from the quality-system inspection requirement if (China Registration §15):

- a domestic manufacturer has a valid "Certificate of GB/T19001+YY/T0287" or "GB/T19002+YY/T0288" issued by a certification agency on quality systems designated by the SDA, and the medical device being registered is covered by the certified system; or
- a foreign manufacturer has been approved by the responsible regulatory authority in the country of origin and its quality system has been certified to ISO 9000 (or equivalent) within the term of validity of the regulatory approval.

REGISTRATION CERTIFICATE ALTERATION

If there is a change in the contents of a registration certificate, the manufacturer must apply to the authority issuing the registration for modification or reregistration within thirty days of the change in circumstances (China MDR §13). Four circumstances that would require modification or reregistration are (China Registration §19):

1. A manufacturer changes the business name on a registration certificate due to renaming or merger. The manufacturer must apply for alteration of the registration certificate by submitting an application report, its new business license, and a certificate issued by the local drug and medical device administration.
2. A new product name is to be used for a registered product. The manufacturer must apply for alteration of the registration certificate by submitting an application report to the relevant authorities.

3. A lost or damaged registration certificate. The manufacturer must reapply for the same registration certificate by submitting to the appropriate authority an application report and a statement to assume relevant legal liabilities.
4. A change of production location. The manufacturer must apply for a new registration certificate. A valid certificate for the manufacturer's quality system at the new location must be submitted with the registration.

The original serial number assigned for the product is still valid, but a *gèng* (更), meaning altered, is added at the end of the original serial number. The issue date of the new registration certificate is the date when the alteration was approved. However, the expiration date is the same as the original certificate. The original certificate is revoked when the renewal certificate is issued.

COMPULSORY PRODUCT CERTIFICATION SYSTEM (CPCS)

In addition to registering a product, a foreign manufacturer of any of the seven categories of medical devices listed in the CPCS catalog must obtain a safety certificate. The CPCS is jointly administered by the SAIQ, who formulates the needed regulations, and the CNCA, who supervises the certification process. As part of its responsibilities, the CNCA selects and accredits the Designated Certification Bodies (DCB) who perform the actual testing and certification of products. Another of CNCA's responsibilities is to specify the certification model applicable for any product in the CPCS catalog (CPS Reg §§6–7). The certification models identified in the regulation are (CPS Reg §12):

- design appraisal,
- type testing,
- testing or inspection of samples taken from factories,
- testing or inspection of samples taken from the market,
- assessment of the manufacturer's quality-assurance system (QA), and
- follow-up inspection of certified products.

The manufacturer, importer, wholesaler, or retailer can apply to a DCB to have a product certified. When the importer, wholesaler, or retailer acts as an applicant, they should provide a copy of the contract with the manufacturer, along with the application forms, required technical documents, and samples (CPS Reg §§6–7). The cost incurred for the registration is borne by the applicant based on a fee chart for the services required (CNCA p. 4).

The product certification requires all or part of the following steps to be taken (CPS Reg §12):

- Acceptance of the application by the DCB
- Type testing
- Factory inspection
- Sampling and testing
- Evaluation of the certification results and approval of certification
- Follow-up inspection

Once issued, the certificate serves as valid documentation to indicate that the product meets requirements and the China Compulsory Certification (CCC) mark can be applied (CPS Reg §16). The CCC mark exists in two forms (CCC Mark §7):

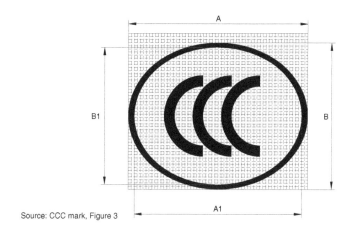

Source: CCC mark, Figure 3

FIGURE 11.2 Basic CCC mark.

1. The basic or core design is illustrated in Figure 11.2. The design can appear in one of five standard dimensions. These are given in Table 11.1. When needed, an irregular-size mark may be used as long as it is proportional to the standard size.
2. The core design with a small capital letter(s) to the right as illustrated in Figure 11.3. The letter(s) reflects the type of certification. For example, "S" represents safety. The CNCA will design and announce additional letters as required.

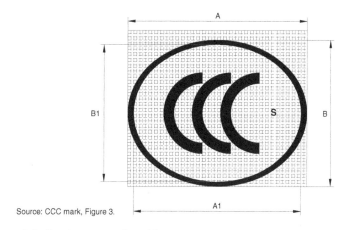

Source: CCC mark, Figure 3.

FIGURE 11.3 Basic CCC mark indicating type of certificate.

TABLE 11.1
Specification for Standard-Size CCC Mark

	Standard-Size Categories (Dimensions in mm)				
Size	1	2	3	4	5
A	8	15	30	45	60
A1	7.5	14	28	42	56
B	6.3	11.8	23.5	35.3	47
B1	5.8	10.8	21.5	32.3	43

Source: CCC mark, Table 1.

Standard-size CCC marks are printed only by the CNCA-authorized printing houses and are purchased by the manufacturer for application to the outer body of the certified product. If it is not feasible to apply the CCC mark to the body of the product, it must be applied to the smallest packaging of the product(s) and indicated in the accompanying documents(s) (CCC Mark §11). The standard-size CCC mark is a black design on a white background (CCC Mark §10).

The CCC mark may be printed, pressed, molded, screen-printed, painted, etched, carved, stamped, or sealed on a product or the nameplate of the product. A manufacturer who wishes to take this approach must submit the design to a CNCA-designated agency and can only produce the mark with the final approval of the CNCA (CCC Mark §15). The applicant must pay the administrative charges if the CCC mark is printed, pressed, molded, screen-printed, painted, etched, carved, stamped, or sealed on a product or the nameplate of the product. If the manufacturer chooses this option, the color of the background and that of the design can be reasonably altered to match the appearance of the product or the nameplate of the product (CCC Mark §10).

Domestic manufacturers must apply the CCC mark after receiving the certification but before the product is allowed out of the factory. For imported products, the CCC mark must be applied after the certification is received but before they are imported into China.

MEDICAL PRODUCT STANDARDS

The Chinese system assigns a significant role to standards as an element of the overall regulatory process. In China, medical product standards are classified into national standards, industrial standards, and registered-product standards. National or industrial standards are those that are adopted by the state; they unify technical requirements nationwide. Registered-product standards are those formulated by the manufacturer to ensure the safety of products. The process for managing the process of developing national and industrial standards and reviewing product standards is set out in a decree from the SDA entitled, Measures for the Management of Medical Appliance Standards (China Standards).

The SDA is responsibile for organizing the drafting of national standards for medical devices and for organizing the development and promulgation of the industrial standards for medical devices. The SDA also is responsibile for examining and approving registered-product standards for imported medical devices as well as the registered-product standards for Class III medical devices manufactured within the territories of the PRC. To discharge this responsibility, the SDA will organize and supervise standardization technical committees in the medical device specialties, oversee the transformation of international standards, and conduct external exchange on standardization efforts (China Standards §5).

The drug and medical device administration associated with the government of the provinces, autonomous regions, and municipalities directly under the central government are responsible for supervising the implementation of medical device standards within the administrative regions. In addition, they are responsible for examining registered-product standards for Class II devices manufactured within their jurisdiction and for the preliminary review of the registered-product standards for Class III devices manufactured within their jurisdiction. The municipal drug and medical device administration is responsible for examining registered-product standards for Class I devices manufactured within their jurisdiction.

A registered-product standard must meet the requirements of the relevant national standard(s) and industrial standard(s), and the related laws, rules, and regulations. The manufacturer should submit the text of the registered-product standard as well as an explanation of the following points when applying for product registration (China Standards §14):

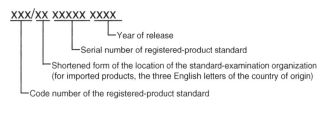

Source: China Standards §17

FIGURE 11.4 Registered-product standard code number layout.

- Have the materials in touch with the human body been used clinically and has their safety and reliability been proven?
- Provide a list of the related standards and materials cited or referred to in the product standard.
- Describe the basis for determination of the device's classification.
- Provide a product overview and basis for determining major technical clauses.
- Include a product self-testing report.
- Include the background of other contents in the standard that may need more explanation.

Once the examination is complete, the reviewing authority assigns a code number. The code number of a registered-product standard is composed of the code name of the registered-product standard, the shortened form of the location of the standard-examination organization (nationality for imported products), a serial number, and the year of registration.

For medical devices manufactured within the territories of the PRC, the shortened form of the location of the standard-examination organization will be one or two Chinese characters. These characters encode (a) the country, province, autonomous region, and municipality directly under the central authority, or (b) the province or autonomous region plus the municipality. For imported products, the shortened form of nationality is presented in three English letters, corresponding to the country of origin of the imported medical device. Figure 11.4 illustrates the layout of a registered-product standard code number.

POSTMARKET SURVEILLANCE

China has established systems for reporting and publicizing accidents caused by medical devices of poor quality. The provisions of these systems are to be jointly established by the SDA and the public health administration and family planning administration affiliated with the state council.

MEDICAL DEVICES EXPORTED FROM CHINA

A party who desires to export a medical device manufactured in China must apply for a Certificate of Free Sales. The manufacturer should provide the following documents when applying for a Certificate of Free Sales:

- An application for export approval
- The Chinese and English names of the product for export and its manufacturer
- A brief introduction to the product
- Documents submitted for domestic-market-clearance registration or the relevant certificate for a product production license
- Reports of product testing performed within one year of the application

Once these documents have been examined and verified, the Certificate of Free Sales will be issued. The SDA has the right to cancel a Certificate of Free Sales based on complaints from abroad about the device. Evidence of false or misleading statements in the application constitutes reasonable justification for cancellation of a Certificate of Free Sales.

INVESTIGATIONAL USE

Before a Class II or III medical device enters the market, it is appropriate for the manufacturer to demonstrate that the device achieves the desired safety and clinical effectiveness under normal conditions of operation. The Chinese government considers clinical investigations an important factor in the process of granting market clearance for a medical device.

The clinical investigation study and reporting requirements for Class II and various types of Class III devices are described in the Provisions on Report Items of Clinical Trials for Medical Device Registration. For imported devices, the clinical studies used to support approval in the country of origin may be accepted. However, a panel of specialists selected by the Chinese government will review the report.

ADVERTISING AND PROMOTION

The Chinese government exercises control over the advertising of medical devices under the provisions of Article 34 of the Regulation on Supervision and Administration of Medical Devices. In particular, the SDA is concerned about companies mislabeling or incorrectly identifying products in order to circumvent government regulations. An example that is frequently cited is a company that attempts to circumvent the government limit on the number of Computerized Axial Tomography (CAT) scanners imported into China. Companies have mislabeled products or shipped components under different names in order to exceed that limit.

Advertising is broadly defined as the release of information about a medical device using radio, film, television, newspapers, periodicals, and other media. Advertisements for medical devices must be examined and approved by the drug and medical device administrations affiliated with the governments above the provincial level. Advertisements without such approval may not be published, broadcast, circulated, or posted.

The content of medical device advertising must be strictly in compliance with the relevant instruction manuals approved by the SDA or the drug and medical device administrations affiliated with the governments of relevant provinces, autonomous regions, and municipalities directly under the central government.

GENERAL LABELING PROVISIONS

When examining the product labeling, the SDA will pay particular attention to the clinical verification of the indications for use (IFU). The SDA will also look for evidence that the technical performance of the device has been verified. Finally, the SDA will check to make certain the IFU contain explicit directions, precautions, and warnings so that the device can be used safely for its intended purpose.

PRODUCT MANUAL

The labeling provisions of the medical device regulations focus on the content of the technical manual. In the guide for product registration, the Certification Commission for Medical Devices (CMD) identified 11 topics that should be covered in the product manual. They are listed below.

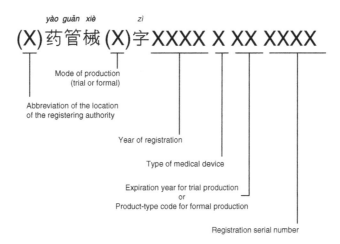

FIGURE 11.5 Form of the registration number for medical products produced in China.

- A description of the technical performance of the device
- The principles of operation of the device
- The method of application
- The indications for use
- A description of clinical effectiveness
- The period of validity (i.e., "use before" date), when appropriate (see "The Chinese Calendar" later in this chapter)
- Contraindications describing when the device should not be used
- Appropriate warnings and precautions describing serious adverse reactions, potential safety hazards, and any special care to be exercised by the practitioner
- Any instructions for maintenance and repair, if appropriate
- Any storage and transport requirements for the device
- Packaging and accessories

THE ACKNOWLEDGMENT SYMBOL FOR MEDICAL EQUIPMENT

When the registration process has been completed, the reviewing authority issues a registration number. The form of the registration number depends on whether the product is domestically produced or imported. For domestic products, the form of the registration number is illustrated in Figure 11.5, and for imported products in Figure 11.6.

The registration number is to be placed on the medical device and its external package. When marking on the device is not practical, for example, on an implanted device, it may be acceptable to place the registration number on the package label and in the instruction manual.

The Chinese have also developed a product mark known as the "CMD" symbol after the China Certification Commission for Medical Devices (CMD) (see Figure 11.7). The symbol indicates that the manufacturer's QA system has been examined and approved. At the time this chapter was written, a CMD mark was not required in China. There is no minimum size requirement for the CMD symbol.

THE CHINESE LANGUAGE

Chinese is the principal language of eastern Asia and is spoken by more people than any other primary language in the world. There are a number of major dialects. However, the northern or

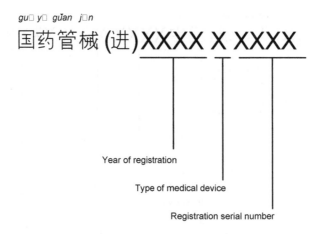

Source: China Registration p. 2

FIGURE 11.6 Form of the registration number for imported products.

Source: Xi Xianmin to Charles Sidebottom via Terri Zavada, Embassy of the United States of America, Beijing, China, facsimile transmittion of March 6, 1996.

FIGURE 11.7 China certification commission for medical devices (CMD) mark.

"Mandarin" dialect is spoken by over 70 percent of Chinese people. This national language is known in China as *putonghua*, or "common speech." *Putonghua* is taught in schools and used in universities and colleges throughout China. The majority of television, radio, and motion picture programs are made in *putonghua* (Scurfield p. viii).

CHINESE WRITING

Chinese is the oldest written language in common use today. The language is essentially monosyllabic with one character representing one idea. As the number of ideas that needed to be expressed expanded, so did the number of characters in the written language. Beginning in the last century, various Chinese governments have attempted to reduce the number of characters in common use and to simplify the remaining characters. This effort has continued under the present government, which has simplified more than 2,000 of the characters in general use.

Until the founding of the PRC in 1949, the structure of the Chinese characters had remained essentially unchanged for almost 2,000 years. Characters have changed mainly in the style of strokes, without altering the basic structure. The most commonly used form is the *k'ai* or "regular form," in which most books are printed. The running hand and cursive hand are used for personal notes and calligraphic purposes (Britannica, Chinese p. 633).

The written language does not have a phonetic alphabet but various systems have been devised for transcribing Chinese sounds into Latin script. The form in use today is known as *pinyin*, or "spell sounds." The PRC officially adopted *pinyin* in 1958 (Scurfield p. ix).

Source: Scurfield p. 160

FIGURE 11.8 Examples of Chinese dates.

THE CHINESE CALENDAR

Traditionally, China has used the lunar calendar, which is divided into twelve months of either 29 or 30 days. The alternating cycle compensates for the lunar month's mean duration of 29 days, 12 hours, and 44.05 minutes. The calendar is synchronized to the solar calendar by the addition of extra months at fixed intervals.

The beginning of the lunar year is fixed at the second new moon of the winter solstice as seen in China. Therefore, the new year begins between January 21 and February 19 of the Gregorian Calendar. For example, the year 2002 is known as the Year of the Horse (Lunar Year 4700) and began on February 12, 2002.

Both the Western (Gregorian) Calendar and the traditional lunar calendars are used publicly in China (World Almanac p. 312). However, while both calendars are in common use, the order of the date is the reverse of that used in the United States: year, month, day. In Chinese, you move from the general to the particular.

The year is read as individual numbers followed by the word *nián* (年), "year." The month is read as a number followed by the word *yuè* (月), "month." The day is read as a number followed by the word *rì* (日), "day." *Rì* is used in formal written Chinese, whereas *hào* (号) is commonly used in the spoken language (Scurfield p. 160). Therefore, a date such as August 18, 2002, would be written in Chinese as illustrated in Figure 11.8. Figure 11.8 also illustrates a mixed form of Arabic and Chinese characters that is sometimes used.

THINGS TO REMEMBER

The SDA is the agency of the Chinese central government that is responsible for the registration of medical products and the development of the medical products industry. The SDA does not have exclusive jurisdiction over all aspects of the approval process. Depending on the product, there can be several other Chinese agencies involved in granting approval. Certain medical devices are subject to the CPCS and are required to bear CCC mark. The CPCS is jointly administered by the AQSIQ and the CNCA.

Although the SDA retains overall authority, responsibility for supervision and administration of certain provisions of the China MDR are delegated to other levels of government within the PRC. Responsibility for dealing with domestically produced, low-risk medical devices is assigned to municipal governments. Responsibility for domestically produced, medium-risk devices is assigned to the government of the provinces, autonomous regions, and municipalities directly under the central government. The SDA retains responsibility for domestically produced, high-risk devices and for registration of all imported products.

The SDA derives its authority from an order of the State Council of the PRC. The Regulation on Supervision and Administration of Medical Devices (China MDR) was adopted in council on December 18, 1999, and became effective on April 1, 2000.

China maintains a three-tier classification system for medical devices that is similar to the risk-based systems in other countries. The organization responsible for supervision of the device will depend on its classification. The need for third-party product testing, quality-system review, and clinical data to support the registration application also depends on the classification. Products that apply pharmacological principals through the use of drugs, even though the drugs do not play the principal role, are considered medical devices in China. Devices intended for laboratory use and IVD reagents are classified as medical devices. Should one medical device fit in two classes at the same time, the higher one must be used.

At this time, China's interest in product labeling is focused primarily on the product manual. The manufacturer is required to submit the product manual as part of the registration application. When examining the product labeling, the authorities will pay particular attention to the clinical verification of the indications for use. The CMD will also look for evidence that the technical performance of the device has been verified by the required testing. Finally, the authorities will check that the IFU contain explicit directions, precautions, and warnings so that the device can be used safely for the intended purpose.

The SDA has the right to withdraw a product registration for a device if the applicant has made false statements in the registration dossier. Since the product manual is provided as part of the registration, false or misleading statements in the labeling could be viewed by the SDA as a false or forged statement in the product-registration dossier.

The Chinese government exercises control over the advertising of medical devices under the provisions of Article 34 of the Regulation on Supervision and Administration of Medical Devices. Advertising is broadly defined as the release of information about a medical device using radio, film, television, newspapers, periodicals, and other media. Other media are interpreted by the SDA to include product manuals and brochures.

12 Republic of Korea

The medical device market in the Republic of Korea (South Korea) is growing at a rate of 10–15 percent annually. With a medical device market of over one billion US dollars, South Korea is one of the most rapidly growing markets for medical technology in all of Asia. South Korea is seeing growth in both domestic production of medical devices and the import of medical technology.

BACKGROUND AND GENERAL INTENT OF THE LAW

Until 1995, the Korean Ministry of Health and Welfare (MOHW) managed the requirements for evaluation and testing of medical devices prior to entry into the market under the pharmaceutical law. In 1993, Dr. Hwal SUH, a researcher in medical materials, petitioned the National Assembly of Korea to establish a new Special Regulatory System for Testing and Evaluation for Implantable Devices. According to Dr. SUH, a new regulatory system was needed because the pharmaceutical law was not suitable to regulate the currently developed biochemical and physiologically functioning implantable devices. In 1994, the MOHW launched a special committee to prepare regulations for medical devices separate from the pharmaceutical law. These regulations were modeled on the system in the United States and emphasize local testing of products. After approval at the National Assembly, the first medical device regulations became effective on January 1, 1995. The regulations are intended to screen new products for safety and efficacy before they are allowed on the Korean market. The increased regulation of medical devices was prompted by recent concerns about certain medical devices combined with increased consumer pressure.

Almost immediately after implementing the regulations, the South Korean government began a process to restructure the approach to regulating medical devices. The regulations have been revised twice. The first revision occurred in 1997. This restructuring was intended to achieve two objectives: (1) to bring the system closer into conformity with international practice for regulating medical devices; and (2) to lay the foundations for improving quality and enhancing the international competitiveness of the Korean medical device industry.

The 1997 revisions were intended to improve the process in four areas:

1. Introduce a three-tiered classification system for medical devices based on the degree of risk to the patient. The approval and management procedures would be different for each classification. Class I products would involve the lowest level of risk and would only require that the MOHW be notified before the product could be manufactured or imported and sold. Class II and Class III devices would be subject to review and approval before they could be sold in Korea. Only for Class III products would safety and effectiveness data be submitted with the premarket approval application.
2. Establish and implement Good Manufacturing Practice (GMP) guidelines based on the ISO 9000 quality system. Initially, GMP would be voluntary, but eventually all domestic and foreign manufacturers would be required to comply.
3. Streamline the testing procedures for foreign manufacturers that satisfy ISO 9000 or equivalent quality-system standards.
4. Supplement the current postmarketing surveillance system.

In 1997, South Korea became a voting member of a number of the International Standards Organization (ISO) Technical Committees that deal with medical products. South Korea hopes to use these standards as part of its testing protocols to help in harmonizing the Korean regulations with those of other major markets.

In 1998, the Korean Food and Drug Administration (KFDA) was made an independent executive agency of the government. The KFDA is the principal government agency whose mission is to ensure that foods are safe, sound, wholesome, and well labeled and that medicines are safe and effective with all possible side effects detected. The KFDA is also responsible for regulating cosmetics, vaccines, blood products, medical devices, and radiation-emitting products. The KFDA has the authority to require that the manufacturer present both preclinical and clinical data establishing the safety and quality of any medical device sold in South Korea. This requirement applies to both products manufactured in South Korea and those imported into the country. Preclinical data must be provided according to the Guideline of Testing and Evaluation for Medical Devices. Only products that have been approved through this procedure may be used in clinical testing that follows the Guideline of Clinical Trials for Medical Device.

In 2000, a new law entitled Act for Medical Device Regulatory System replaced the medical device regulations. The new law was intended to further streamline the process and provide a base for new guidelines. By 2001, five regulatory guidelines related to the licensing of medical devices in South Korea had been published. They are:

- Guideline of the Appointment for Medical Devices, which deals with classification of the devices;
- Guideline of the Safety and Effect Assessment for Medical Devices dealing with documentation and the process necessary to register;
- Guideline of Permission for Medical Devices, which includes conditions for registration and licensing of the devices, and for registration of the testing agencies;
- Guideline of the General Standard of Biological Safety for Medical Devices, which deals with testing Class II and III devices that directly contact the body; and
- Guideline of Clinical Trials for Medical Devices, which regulates clinical trials.

SCOPE OF THE REGULATIONS

The South Korean medical device regulations are intended to regulate the sale of products manufactured in or imported into the Republic of Korea. According to the Guideline of the Appointment for Medical Devices, all the devices are classified into one of three classes (see Table D.1 in Appendix D).

Because of their simplicity or low risk to the patient, devices in Class I are exempted from the necessity of product testing and are subject to premarket registration only. Any medical device that is of the same type, or is substantially equivalent to the listed device, is also subject to premarket registration, according to the Guideline of the Safety and Effect Assessment for Medical Devices and the Guideline of Permission for Medical Devices.

Devices in Classes II and III are subject to testing before the product is initially placed on the market. In addition, imported products in these categories are subject to testing on each shipment entering the country. The requirements for testing of Class II and III devices are set in the Guideline of the General Standard of Biological Safety for Medical Devices and the Guideline of Clinical Trials for Medical Devices. Under the Guideline of Clinical Trials for Medical Devices, any Class II or III medical device that is directly connected to the body or creates a possibility of hazard to the body must be tested under a study in more than two registered testing hospitals and supervised by an Institutional Research Board. The clinical trial must adhere to the Guideline for Good Clinical Practice (GCP), which is governed by the Central Committee of Pharmaceutical Affairs, before the device can be approval by the KFDA.

LABELING AND ADVERTISING

At the present time, the advertising and promotion of medical devices is strictly regulated under the provisions of the medical device regulations. Advertising can promote only the indications registered and approved by the KFDA on the label.

BRINGING DEVICES TO MARKET IN KOREA

South Korean regulations require that the person holding the registration for a medical device must be a legal resident of Korea. For imported products, it is the local distributor, not the manufacturer, who applies for and serves as the legal holder of the medical device registration. Multiple distributors can hold approvals for a single product. Each distributor must undergo the full product-approval process and bear the full cost of the process. Alternatively, a "principal distributor" or an independent consultant holds the approval and the other distributors "share" the registration. The manufacturer can choose the approach that best fits its business plan.

Imported medical devices must be approved for sale in the country where they are manufactured before the KFDA will consider them for registration (ITA, Korea p. 1). A Certificate of Free Sale or Certificate of Product for Export from the country where the device was manufactured must be submitted as part of the registration package.

A manufacturer or importer that wishes to bring a product to market in Korea must register the product with the KFDA. There are two pathways to obtaining marketing clearance.

PREMARKET NOTIFICATION

Registration of Class I devices requires that the manufacturer or importer submit an application to the KFDA district office of the region where the manufacturer's or importer's main office is located. If the KFDA district office decides to accept the registration application, a certificate of completed notification is issued. This certificate allows the product to be distributed in South Korea (ITA, Korea p. 3).

PREMARKET APPROVAL

Class II and III devices are subject to premarket approval by the KFDA. The premarket approval process involves a more rigorous documentary review followed by a product "type test" in a KFDA-authorized laboratory (ITA, Korea p. 4).

The manufacturer or importer begins the process by filing an application in Korean with the KFDA's Medical Device Evaluation Department (MDED). The MDED reviews the application to evaluate the manufacturer's quality standards for materials or finished products. Most products that are evaluated by the MDED are essentially the same as those already on the market. These products do not require a safety and effectiveness review, so once the MDED completes the quality review, the product can undergo a product type test.

When products with new-to-market features in terms of materials, mechanism of action, usage, or effectiveness are submitted for registration, the KFDA may require a safety and effectiveness review. The safety and effectiveness review looks at a variety of data including information about the device, including the structure, physiochemical and biological properties, toxicity data, basic-

TABLE 12.1
KFDA Authorized Testing Agencies

Authorized Testing Agency	Device Type
Korea Testing Laboratory (KTL)	Qualified to test all medical devices
Korea Testing & Research Institute for Chemical Industry (KOTRIC)	14 product groups, including implants and supplies
Korea Electric Testing Institute (KETI)	27 product groups, including electric/electronic devices
Korea Merchandise Testing and Research Institute (KOMTRI)	14 product groups, including medical supplies
Seoul National University Hospital Clinical Trial Center	11 product groups, including implanted devices and supplies
Yonsei University Hospital Medical Technology Evaluation Center	11 product groups, including implanted devices and supplies
Yonsei University Dental College, Dental Products Testing & Evaluation Center	Dental devices only
Kyung-hee Univ. Dental College, Open Laboratories for Dental Products	Dental devices only

Source: ITA, Korea

safety data, clinical-study data, an so on. When the MDED completes both the quality and the safety and effectiveness review, the product can undergo a product type test.

Testing of Class II and III medical devices is mandatory before a product is approved. The manufacturer or importer prepares a product dossier in Korean. The dossier is submitted to a KFDA-authorized testing laboratory. The KFDA has eight authorized laboratories to test medical devices. The laboratories are listed in Table 12.1 along with the device types for which they are qualified.

For imported products, the test laboratory reviews any foreign test data that may be available. At its discretion, the test laboratory may decide to recognize foreign test reports and to forego local type testing of imported products. Once the review process has been completed, the testing agency issues a Certificate of Confirmation and a Certificate of Confirmation of Standards and Testing Methods. The agency sends the dossier to the KFDA for final review and approval. The KFDA has the authority to approve or to reject the application and ask the manufacturer to have additional tests by the test laboratory.

CLINICAL TRIALS

When clinical trails are required, The KFDA prefers trials conducted at one or more of the South Korean institutions listed in Table 12.2. Foreign clinical studies may be accepted in lieu of local trials. The KFDA requires that such studies be conducted according to Good Clinical Practice (GCP). The KFDA prefers that the final study report be published by a reputable professional journal listed in the Science Citation Index. The KFDA will consider unpublished study reports to determine acceptability, depending on GCP compliance and the reputation of the research organization.

IMPORTING PRODUCTS

For products imported into South Korea, additional steps are required. The manufacturer, or the manufacturer's in-country agent must submit an Import Notification to the Korean Medical Instrument and Industry Corporation (KMIIC), a local trade organization. Originally, the regulations called for the import notification to include quantity and unit-price information. This provision was stricken from the final regulations. The Certificate of Confirmation and the Certificate of Confirmation of Standards and Testing Methods must accompany the Import Notification. Once the KMIIC returns stamped certificates to the person submitting the application, the product may be imported into South Korea.

However, before the imported product can be sold, it may need to undergo further testing. The Class I product groups listed in Table D.1 in Appendix D are exempt from product testing. However, each shipment of products in the Class II and III product groups listed in Table D.1 in Appendix D

TABLE 12.2
KFDA Clinical Trial Agencies

Yonsei MTEC, Yonsei University, Seoul, Korea
Yeongdong Severance Hospital, Yonsei University, Seoul Korea
Seoul National University Hospital, Seoul, Korea
Catholic Medical Center, Catholic University, Seoul, Korea
Korea University Hospital, Seoul, Korea
Samsung Medical Center, Seoul, Korea
Seoul Chung-Ang Medical Center, Seoul, Korea
Chung-Ang University Hospital, Seoul, Korea
Hanlim University Medical Center, Seoul, Korea
Cha Medical Center, Seoul, Korea
Inha University Hospital, Gyeongki-do, Korea
Sooncheonhyang University Medical Center, Chungcheongnam-do, Korea
Dankook University Hospital, Chungcheongnam-do, Korea
Chungnam National University Hospital, Chungcheongnam-do, Korea
Chonbuk National University Hospital, Cheollabuk-do, Korea
Chonnam National University Hospital, Cheollanam-do, Korea
Kyeongbuk National University Hospital, Daegu, Korea
Yeongnam University Medical Center, Daegu, Korea
Busan National University Hospital, Busan, Korea
Inje University Hospital, Keongsangnam-do, Korea

Source: Dr. Hwal SUH.

must undergo testing on each shipment that enters Korea. All other products must be tested on the first shipment into South Korea, but are exempted on subsequent shipments.

Once the product shipment clears customs, it must be quarantined until testing is completed by the appropriate KFDA-authorized testing agency. The report of the testing agency is submitted to and approved by the KFDA. The approval of quality from the KFDA is submitted to the KMIIC. Once confirmation has been received from the KMIIC, the product can be released from quarantine and is available for distribution.

The content of the Korean labeling, including package inserts, sample stickers, or draft labels, is to be included with the report. For the second and later imports, this requirement is waived.

GRANDFATHERED PRODUCTS

Products that were sold in South Korea before January 1, 1995, are "grandfathered" under the provisions of the Korean medical device regulations. The manufacturer or the manufacturer's in-country agent can register these products by submitting to the appropriate KFDA-authorized testing agency the following (Larkin pp. 6–7):

- A Certificate of Free Sale or Certificate of Product for Export from the country where the device was manufactured.
- A certificate indicating that the product has been sold for more than three years, or at least 50 pieces have been imported into South Korea.
- For implanted products, records showing that the product has been sold to two or more medical institutions in South Korea plus evidence demonstrating the safety of the products.

The Certificate of Free Sale or Certificate of Product for Export must be notarized and consularized by the South Korean Council in the territory of the exporter's home country. For grandfathered products, the local testing requirement is waived.

GENERAL REQUIREMENTS FOR LABELING OF MEDICAL DEVICES

The common feature of most labeling requirements in South Korea is that key information must be provided in Korean. This information includes the directions for use and sufficient information to enable the user to identify the device, its manufacturer, importer, and/or distributor, and, when applicable, the device's expiration date.

ADEQUATE DIRECTIONS FOR USE

Medical devices must be labeled to provide a statement of the purposes for which the device is intended by the manufacturer. This statement must include the directions, dosage, or other information necessary for the use and handling of the device. It should also include an explanation of the device performance, as well as any warnings, cautions, contraindications, and possible side effects.

PACKAGE LABEL

According to Guideline of Permission for Medical Devices, the unit, shelf, or shipper package label must include the following information:

- Description of the device
- Name of the manufacturer
- Directions for use
- Lot or serial number of the product
- Date of manufacture
- Expiration date, when applicable
- Sterilization date, when applicable
- Description of package contents, expressed as weight, volume, or packaging unit

For imported products, the description of the device, the name and nationality of the manufacturer, the name and address of the importer, the name and address of the distributor, the directions for use, and the description of package contents must appear on the label in Korean. This information must be printed by the manufacturer or stickered by the importer in characters sufficiently large to be easily read by the consumer. If the container or package is too small to include all these items, the directions for use, the date of manufacture, and the description of the package contents may be deleted from the on-product label and placed in a separate package insert. If the product is very small and sold without packaging, the entire contents of the Korean label can be placed on a separate tag or leaflet that is provided along with each product.

Separate from the KFDA regulations, South Korean commercial regulations require that the country of origin be marked on the product and the package. In case of a device that bears no space to mark the country of origin or if the marking itself disturbs the biocompatibility of the product, especially in implants, marking only on the package and label is permitted.

THINGS TO REMEMBER

In 1995, South Korea joined the growing number of nations that require evaluation and testing of some medical devices by a competent authority in the country before products are allowed on the market. The South Korean medical device regulations are intended to regulate the sale of products manufactured in or imported into the Republic of Korea after January 1, 1995.

The new regulations emphasize local testing of products. Class I product groups, because of simplicity or low risk to the patient, are exempted from the necessity of product testing. However, the devices in Class II and III product categories listed in Table D.1 in Appendix D are subject to testing before the product is initially placed on the market. In addition, imported products in these categories are subject to testing on each shipment entering the country.

At the present time, the advertising and promotion of medical devices is firmly regulated under the provisions of the medical device regulations.

A manufacturer or importer that wishes to bring a product to market in Korea must register the product and have it examined by a testing agency authorized by the KFDA. For products imported into South Korea, additional steps are required. The manufacturer or the manufacturer's in-country agent must submit an Import Notification to the KMIIC. However, before the imported product can be sold, it may need to undergo testing. The Class I product groups listed in Table D.1 in Appendix D are exempted from product testing. Each shipment of products in the Class II and III product groups in Table D.1 in Appendix D must undergo testing on each shipment that enters Korea. All other products must be tested on the first shipment into South Korea, but are exempted on subsequent shipments.

Products that were sold in South Korea before January 1, 1995, are "grandfathered" under the provisions of the Korean medical device regulations.

Medical devices must be labeled to provide a statement of the purposes for which the device is intended by the manufacturer. This statement must include the directions, dosage, or other information necessary for the use and handling of the device. The Korean regulations specify requirements for the unit, shelf, or shipper package label. Some of this information must appear in Korean.

Separate from the MOHW regulations, the South Korean commercial regulations require that the country of origin be marked on the product and package.

13 Thailand

The Kingdom of Thailand is one of the fastest growing markets for medical devices in Southeast Asia. Thailand has created a strong export economy and a positive balance of trade that has continued to grow despite the general Asian economic downturn of the late 1990s. The growth in the economy has enabled the Thai government to increase spending on public health. In 2000, the Thai government was spending 9.2 percent of its national budget on public health (Britannica, 2002 p. 722).

The Thai Ministry of Public Health has the primary responsibility for ensuring that food, drugs, narcotic substances, medical devices, and household hazardous substances available to consumers are of standard quality, efficacy, and safety. The Thai government controls the sale of medical devices under the Medical Device Act of Thailand (Thai MDA).

Under the provisions of the Medical Device Act, a medical device is (Thai MDA §3):

1. equipment, products, or articles used by the medical profession, in the profession of nursing and midwifery, in the clinical practice of medicine, or in veterinary practice as prescribed by the legislation concerned;
2. equipment, products, or articles that have effects on the health, the structure, or any function of the human or animal body;
3. constituents, components, accessories, or parts of the equipment, products, or articles under (1) or (2); or
4. other equipment, products, or articles prescribed by the Ministry as medical devices by publication in the *Thailand Government Gazette*.

Under this definition, 8165 products in 33 product categories were registered in Thailand by May 2001 (Thai Medical Device pp. 11–13).

BACKGROUND AND GENERAL INTENT OF THE LAW

Thailand has a long history of regulating food and drugs to protect consumers from the adulteration of these products. The Thai Ministry of Public Health can trace its regulatory roots to a 1922 act of the Thai Parliament. The Thai Food and Drug Administration (TFDA) was established in its current form in 1985. In 1988, the Thai Parliament enacted the Medical Device Act, which gives the TFDA the statutory authority to regulate the manufacture, import, and sale of medical devices. A major responsibility of the TFDA is to ensure that medical devices are of standard quality, efficacy, and safety. The TFDA monitors both pre- and postmarket phases of the manufacture, import, transport, storage, and sale of medical devices (Thai FDA §1).

SCOPE OF THE REGULATIONS

For purposes of regulation, the TFDA divides medical device products sold in Thailand into three categories. These categories are listed below (Thai Medical Device pp. 2–5):

Class I   This category includes devices that require full product registration, including condoms, examination gloves, surgical gloves, hypodermic syringes, insulin syringes, and Human Immunodeficiency Virus (HIV) test kits for diagnostic use. A license issued by the TFDA is required before any of these devices can be manufactured or imported and marketed in Thailand.

Class II   Devices in this category do not require a license, but they must be registered with the TFDA. This category includes HIV test kits for research investigation, physical-therapy devices, alcohol-detection devices, and implanted silicone breast prostheses.

Class III This final category covers devices for which only a Certificate of Free Sale or export certificate by the competent authority of the country of manufacture (i.e., the United States [US] Food and Drug Administration[FDA]) is required for registration.

In contrast to other classification systems, Class I devices are the most regulated, while Class III devices are the least regulated. Unlike systems in the United States, Canada, the European Union (EU), and Japan, the classification system is not primarily risk based.

Thai law prohibits the importation of used or refurbished medical devices. In addition, products that cannot be marketed or sold in the country of origin (as certified by a Certificate of Free Sale, such as the US FDA's Certificate for Foreign Government[CFG]) cannot be legally imported.

Production, importation, or distribution of a medical device without the proper registration and, in the case of Class I devices, a proper license is a violation of the Medical Device Act. Failure to comply with the law can subject the offender to a substantial fine and/or a term of imprisonment.

ADULTERATION AND MISBRANDING

Although the Medical Device Act does not use the terms "adulteration" or "misbranding," these concepts do appear in the Thai law. Manufacturing, importing, or distributing a counterfeit, sub-standard, deteriorated, or unsafe medical device is a criminal offense. A person found guilty of a violation is subject to a term of imprisonment or a substantial fine, or both.

Chapter VI of the Medical Device Act defines the conditions under which the device could be considered adulterated or misbranded. These conditions include:

- A counterfeit medical device—a medical device that:
  - is wholly or partially made in imitation to be fraudulently or deceptively represented as genuine;
  - bears a name, category, type, or characteristics that are different from those granted by the TFDA (Class I devices) or those in the list of particulars submitted to the TFDA (Class II devices); or
  - bears false statements of the manufacturer's name, the source of production, the date of production, or the place of production.
- A substandard medical device—a medical device whose quality or standard is below that specified in a Ministerial order published in the *Thailand Government Gazette*.
- A deteriorated medical device—a medical device whose condition has deteriorated so that its quality is below the standard specified for the device or it is beyond its prescribed expiration date.
- An unsafe medical device—a device that:
  - is already used,
  - was manufactured or stored in unhygienic conditions,

- is contaminated by foreign or potentially health-hazardous substances,
- contains degradable substances that may be toxic to the user, or
- is a medical device whose effectiveness is still in doubt.

BRINGING DEVICES TO MARKET IN THAILAND

The Medical Device Act recognizes that medical devices vary widely in their complexity and the risks and benefits that they offer. They do not all need the same level of regulation in order to protect public health. For the purpose of product registration, the TFDA places all medical devices into one of the three categories described in the previous section. The level of regulatory review required before a product can be placed on the market is dependent on its classification.

DEVICE REGISTRATION

Full registration is required for all Class I medical devices. The registration process requires that the manufacturer demonstrate that a device complies with the applicable Thai Industrial Standards Institute standards, which are based on ISO standards. In addition, some products must undergo testing by the TFDA.

In Thailand, the registration is the property of the applicant. It is normal for foreign manufacturers to register through an agent or distributor, because the TFDA requires the person holding the registration to be a locally based legal entity. A product registration is valid for five years. Should a foreign manufacturer change the Thai agent/distributor within this five-year period, the new Thai representative must apply for a new registration (Thai Medical Device p. 2).

The registration process can be expected to take three months. When a product registration is granted for an imported product, the TFDA notifies the Customs Office to permit clearance for the product to enter the country.

The Medical Device Act requires that the manufacturer (importer) and distributor of Class I medical devices must have a license issued by the TFDA. The requirements for an applicant for a license to manufacturer, import, or distribute a Class I medical device are set out in Section 14 of the Medical Device Act. Additional requirements and the form of the application for a license are set out in Ministerial Regulations adopted under various provisions of the Medical Device Act. Ministerial Regulation No. 1* deals with the requirements for the production of medical devices in Thailand. The regulation is heavily weighted toward requirements for maintaining the hygiene and cleanliness of areas for production, packaging, and storage of manufactured products and raw materials. Ministerial Regulation No. 2** establishes the requirements for importers of medical devices. Again the requirements of the importer regulation focus on maintaining the proper facilities to protect the integrity of the imported devices. Thailand requires that a person who distributes a Class I medical device must also hold a license from the TFDA. The Medical Device Act defines distribute as "sell, dispense, dispose of, trade or transfer the right or possession to other persons for commercial purpose, including having in possession for sale" (Thai MDA §3). The requirements for a distributor are set out in Ministerial Regulation No. 3.***

---

* *Ministerial Regulation No. 1. 1990. Thailand Government Gazette*, 107, part 28 (February 19 B.E. 2533).
** *Ministerial Regulation No. 2. 1990. Thailand Government Gazette*, 107, part 28 (February 19 B.E. 2533).
*** *Ministerial Regulation No. 3. 1990. Thailand Government Gazette*, 107, part 28 (February 19 B.E. 2533).

DEVICE NOTIFICATION

Class II devices are those for which general controls alone are insufficient to assure safety and effectiveness. These products must be registered with the TFDA before being placed on the market and the manufacturer, importer, or distributor must submit a list of particulars to the TFDA in a form specified in Ministerial Regulation No. 4.* The list of particulars include a description of the device, its physical properties, contents, production process, and copies of the labeling that contain the indications for use, instructions for use (IFU), storage requirements, shelf life (if any), and the name and address for the manufacturer or importer. The notification process can be expected to take approximately fifteen days if there are no issues with the submission.

In addition to the material described in the previous paragraph, the application for an imported device must be accompanied by a Certificate of Free Sales issued by the health authority (or a related government body) in the country of origin. The TFDA accepts devices that meet the requirements in the following jurisdictions (Thai Medical Device p. 5):

- United States (US FDA)
- European Union (CE mark)
- Japan (Pharmaceutical and Medical Safety Bureau)
- Australia (Therapeutic Goods Administration)
- People's Republic of China (State Drug Administration)

A CFG issued by the US FDA must be "consularized" either by the Thai Consulate in the United States (in Washington, DC) or by the Commercial Section of the US Embassy in Bangkok.

GENERAL MEDICAL DEVICES

Most medical devices fall into Class III and are subject to the lowest level of regulatory controls. Class III devices may be imported as long as it can be demonstrated that the devices are freely marketed and sold in the manufacturing country. As with Class II devices, a Certificate of Free Sales issued by the health authority (or a related government body) in the country of origin is required to register the product. For these devices, the Certificate of Free Sales is the only document that must accompany the registration application. The registration process can be expected to take approximately 10 days if there are no issues with the submission.

REQUIRED POSTMARKET REPORTING

The Medical Device Act requires that manufacturers, importers, and distributors of Class I and Class II medical devices periodically provide the TFDA with statistics on the number of devices produced and/or sold in Thailand. The content and frequency of these reports is specified in Ministerial Regulation No. 5.** Manufacturers, importers, and distributors of Class I or Class II medical devices are also required to report when they become aware of adverse events associated with the device. An adverse event is defined in Ministerial Regulation No. 5 as one resulting in abnormal symptoms or health hazards including a death or serious injury. Events resulting from deficiencies or problems with the labeling should be considered as reportable events. A written report is to be submitted within 15 days of the manufacturer, importer, or distributor becoming aware of the adverse event. In the case of a death or serious injury, the Secretary-General of the TFDA must be notified within

---

* *Ministerial Regulation No. 4.* 1990. *Thailand Government Gazette*, 107, part 28 (February 19 B.E. 2533).
** *Ministerial Regulation No. 5.* 1990. *Thailand Government Gazette*, 107, part 28 (February 19 B.E. 2533).

24 hours. A follow-up written report must be submitted within 15 days. The forms to be used and the required follow-up on the initial report are specified in Ministerial Regulation No. 5.

THAI LABELING REQUIREMENTS

Labeling of medical devices is regulated by the TFDA under the provision of Chapter V of the Medical Device Act. The Medical Device Act considers labeling of medical devices to include labels and accompanying documents. A label includes any image, design, symbol, or statement displayed on a medical device, its container, or packaging. Accompanying documents include material (paper or otherwise) on which information about the medical device is displayed by an image, design, symbol, or statement. Accompanying documents are those inserted or included in the container or package of the medical device including the manual (Thai MDA, §3). The TFDA also regulates advertising of medical devices. Advertising is material intended for commercial purposes that does not accompany the medical device. Advertising requirements are discussed later in this chapter.

LABELS

The minimum requirements for the labels of all medical devices are set out in Section 33 of the Medical Device Act. Section 33 specifies the information that must appear on the medical device's labels. It includes:

- The name, category, and type of the medical device.
- The name and address of the manufacturer or importer of the device.
- For imported medical devices, the name of the manufacturer and the country where the device was manufactured.
- The contents of the package.
- The lot or serial number.
- The intended use of the medical device (when feasible).
- The IFU of the device (when feasible).
- The instructions for storage and maintenance of the medical device (when feasible).
- For a disposable medical device, the words "for single use" in RED must be clearly displayed.
- Warnings and precautions for use of the device when required under Section 35(5) of the Medical Device Act.
- The expiration date for the device where this information is required under Section 35(8) of the Medical Device Act.
- Other information required by the TFDA under a Ministry Announcement published in the *Thailand Government Gazette*.

ACCOMPANYING DOCUMENTS

Section 34 of the Medical Device Act specifies that the following information must appear in the accompanying documents:

1. The intended use of the medical device
2. The IFU of the device
3. The instructions for storage and maintenance of the medical device
4. Warnings and precautions for use of the device when required under Section 35(5) of the Medical Device Act

When there are accompanying documents, items 1, 2, or 3 above need not be displayed on the label of the medical device.

LANGUAGE REQUIREMENTS

The required information on the labels and the accompanying documents must appear in Thai. The labeling may bear information in languages other than Thai. However, the information content must be the same as that appearing in Thai. The type size of the Thai labeling cannot be smaller than the type size of the other languages.

Thai, the standard spoken and literary language used throughout Thailand, is specifically the dialect of Bangkok and its environs. The Thai alphabet is derived from the Devanagari of South India. Writing proceeds from left to right, and spaces indicate punctuation but not word division. The written alphabet contains 42 consonant signs, 5 tonal markers, and 41 vowels and special combinations.

EXPIRATION DATES

Expiration dates are expressed in the Buddhist Era (B.E.). The Buddhist Era begins in 544 B.C. Dates in the Christian Era are converted to the Buddhist Era by adding 543. (For example, October 1990 becomes 10/2533. A factor of 543 is used because chronologist admit no year zero between 1 B.C and A.D. 1.)

ADVERTISING

Advertising of medical devices is regulated by the TFDA under the provision of Chapter VII of the Medical Device Act. Falsely or fraudulently advertising the benefits, quality, quantity, standard, or source of medical devices is a violation of Section 41 of the Medical Device Act, and can subject the violator to a substantial fine.

Advertising of medical devices for commercial purposes must have the approval of the TFDA. This approval covers any statements, audio, or video aspects of the advertising. The advertising must be in accordance with the conditions prescribed by the TFDA (Thai MDA §42). The TFDA has the authority to issue a written order suspending any advertising that it considers violates or fails to comply with the Medical Device Act (Thai MDA §43).

IN VITRO DIAGNOSTIC (IVD) DEVICES

In Thailand, *in vitro* diagnostic (IVD) devices are treated as medical devices. In fact, HIV test kits for diagnostic use are designated Class I and HIV test kits for research investigation are designated Class II by the TFDA. In addition to meeting the requirements of the Medical Devices Act, any medical devices, including IVD devices, that contain hazardous substances must comply with the provisions of Thailand's Hazardous Substances Act (Thai HSA). The Hazardous Substances Act defines a hazardous substance as an explosive, flammable substance, oxidizing agent and peroxide, corrosive substance, irritating substance, radioactive substance, toxic substance, substance causing disease, substance causing mutation, and any other substance, either chemical or otherwise, that may cause injury to persons, animals, plants, property, or the environment (Thai HSA §4). The Minister of Public Health is one of the government officials charged with applying this law. Within the ministry, this responsibility is delegated to the TFDA.

The Hazardous Substance Act classifies hazardous substances according to the need for control. There are four classes (Thai HSA § 18):

Class IA hazardous substance where the person who manufactures, imports, exports, or possesses the substance must comply with criteria and procedures prescribed under the authority of the Hazardous Substance Act.

Class IIA hazardous substance where the person who manufactures, imports, exports, or possesses the substance must first notify the responsible authority in addition to complying with criteria and procedures prescribed under the authority of the Hazardous Substance Act.

Class IIIA hazardous substance where the person who manufactures, imports, exports, or possesses the substance must obtain a permit from the responsible authority.

Class IVA hazardous substance where the manufacture, import, export, or possession of the substance is prohibited.

Section 20 of the Hazardous Substance Act gives the TFDA the power to adopt regulations covering the composition, qualification and mixtures, containers, methods of examining and testing containers, labels, production, import, export, sales, transport, storage, disposal, destruction, treatment of hazardous-substance containers, and any other matters relating to hazardous substances.

Requirements for the registration of Class II and Class III hazardous substances that pose a risk to public health are found in a 1995 notification from the Ministry of Public Health. * This notification sets out the procedure, specifies the forms, and details the material, including samples and labeling, that must be submitted with the application.

THINGS TO REMEMBER

In Thailand, medical devices are controlled under the authority of the Medical Device Act, which gives the TFDA the statutory authority to regulate the manufacture, import, and sale of medical devices. The TFDA monitors both pre- and postmarket phases of the manufacture, import, transport, storage, and sale of medical devices.

For purposes of regulation, the TFDA divides medical device products sold in Thailand into three categories—Class I, Class II, and Class III—with Class I devices being the most regulated and Class III devices being the least regulated. Production, importation, or distribution of a medical device without the proper registration and, in the case of Class I devices, a proper license is a violation of the Medical Device Act. Failure to comply with the law can subject the offender to a substantial fine and/or a term of imprisonment.

The Medical Device Act specifies minimum labeling requirements for both the labels and accompanying documents associated with the medical devices. The required information must appear in the Thai language and dates must be expressed in the Buddhist Era (b.e.).

Advertising of medical devices is also regulated by the TFDA. The law states that advertising of medical devices for commercial purposes must have the approval of the TFDA. The TFDA has the authority to suspend any advertising that it considers violates or fails to comply with the Medical Device Act.

In Thailand, IVD devices are treated as medical devices. In addition to meeting the requirements of the Medical Devices Act, any medical devices, including IVD devices, that contain hazardous substances must comply with the provisions of Thailand's Hazardous Substance Act.

---

* Notification of the Ministry of Public Health. 1995. *Registration of Household or Public Health Hazardous Substances. Thailand Government Gazette*, 112, part 43 (May 18 b.e. 2538).

The Hazardous Substance Act classifies hazardous substances into one of four classes according to the need for regulatory control. Class II and Class III hazardous substances that pose a risk to public health must be registered with the TFDA. In addition, the organizations dealing with Class III hazardous substances must have a permit also issued by the TFDA.

Part VI

European Union

Part VI

European Union

14 The European Medical Device Directives

Long encumbered by separate regulatory systems, the European Union (EU, formerly the European Community) is now in the process of unifying its approach toward regulation of a variety of industrial products, including medical devices. The intent of the EU is to create an internal market that is open to the free movement of goods, persons, services, and capital.

For a number of product classes, the laws and regulations of the Member States of the EU that regulated these products were different. In addition, the certification and inspection procedures for such devices differed from one member state to another. Together, these disparities constituted a barrier to trade within the EU. To establish a single market for these products, the EU has created a series of European technical harmonization directives. A directive is enacted by the EU Council of Ministers and requires the member states to harmonize their national laws, regulations, and administrative procedures with the provisions of the directive within a certain time period specified in the directive.

For the medical device sector, three technical harmonization directives have been adopted by the EU. The Medical Device Directive (MDD) (Directive 93/42/EEC) was not the first of the three directives adopted by the EU. However, because it encompasses the bulk of medical devices and their accessories, the MDD will be considered first.

The MDD defines a medical device as any instrument, apparatus, appliance, material, or other article, whether used alone or in combination, including the software necessary for its proper application, intended by the manufacturer to be used for human beings for the purpose of (Directive 93/42/EEC §2(a)):

- diagnosis, prevention, monitoring, treatment, or alleviation of disease;
- diagnosis, monitoring, treatment, alleviation of, or compensation for an injury or handicap;
- investigation, replacement, or modification of the anatomy, or of a physiological process of the body; or
- control of conception in humans.

By definition, a medical device does not achieve its principal intended action by pharmacological, immunological, or metabolic means, although it may be assisted in its function by such means.

The second of the technical harmonization directives is the Active Implantable Medical Device Directive (AIMDD) (Directive 90/385/EEC). The AIMDD covers a subset of medical devices that are intended to be totally or partially introduced, surgically or medically, into the human body or by medical intervention into a natural orifice. Once introduced, the device is intended to remain in the body after the procedure. In addition, an active implantable medical device (AIMD) relies on a source of electrical energy or power other than that generated directly by the human body or gravity (Directive 90/385/EEC §2(c)).

The third and final member of the medical device directives triad covers *in vitro* diagnostic (IVD) medical devices. The *In vitro* Diagnostic Devices Directive (IVDD) (Directive 98/79/EC) defines an IVD medical device as any medical device that is a reagent, reagent product, calibrator, control material, kit, instrument, apparatus, equipment, or system, whether used alone or in combination, intended by the manufacturer to be used *in vitro* for the examination of specimens, including blood and tissue donations, derived from the human body, solely or principally for the purpose of providing information (Directive 98/79/EC §2(b)):

- concerning a physiological or pathological state,
- concerning a congenital abnormality,
- to determine the safety and compatibility with potential recipients, or
- to monitor therapeutic measures.

For the purposes of the IVDD, containers intended by their manufacturer to hold specimens are considered to be IVD medical devices. Specimen receptacles are those devices, whether vacuum-type or not, specifically intended by their manufacturers for the primary containment and preservation of specimens derived from the human body for the purpose of IVD examination. Products for general laboratory use are not IVD medical devices unless, considering their characteristics, their manufacturer specifically intendeds them to be used for IVD examination.

BACKGROUND AND GENERAL INTENT OF THE LAW

The European Economic Community (EEC) was established on January 1, 1958, by the Treaty of Rome. Originally made up of France, West Germany, Italy, Belgium, The Netherlands, and Luxembourg, the EEC was created to improve economic cooperation among the member states. The treaty aimed at establishing a general common market within a 12- to 15-year period. One of the problems faced by the EEC was its relationship with its European neighbors. Other European countries, principally the United Kingdom, sought to trade freely with the EEC member states without joining in a full economic or political union. Negotiations to resolve this issue in 1958 failed to reach a satisfactory agreement.

One result of this failure was the formation of the European Free Trade Association (EFTA) by the United Kingdom, Austria, Denmark, Sweden, Norway, Switzerland, and Portugal. The EFTA came into effect in May 1960 on the basis of a compact signed in Stockholm in November 1959. Finland, Iceland, and Liechtenstein subsequently joined the EFTA. The aim of the association was the establishment, through the gradual reduction in tariffs, of an industrial free-trade area by 1970.

In the intervening years, the cooperation between EEC and the EFTA increased. Eventually, the United Kingdom, Denmark, and Portugal, along with Ireland, Spain, and Greece, became full members of the EEC—or simply the European Communities (EC).

The Treaty of Maastricht, which was signed in February 1992, proposed the creation of the EU as the successor to the EEC, the European Coal and Steel Community, and Euratom. After an intense political battle in several of the EC member states, the Maastricht Treaty was approved in 1993. On November 1, 1993, the Maastricht Treaty came into force, creating the EU. In parallel with the approval of the Maastricht Treaty, the EC entered into an agreement with the remaining seven EFTA countries (Austria, Norway, Sweden, Finland, Switzerland, Iceland, and Liechtenstein) to create a European Economic Area (EEA) (Palmer p. 419). Four of the EFTA countries (Austria, Norway, Sweden, and Finland) applied for full membership in the EU beginning in 1995 (Schöpflin p. 420). Austria, Sweden, and Finland voted to join the EU and became full members on January 1, 1995. Norwegian voters narrowly rejected membership in the EU, choosing to remain outside the expanding political coalition.

The EU has moved to create better economic relations with the countries of central Europe. This effort will culminate on May 1, 2004 when the Czech Republic, Estonia, Hungary, Latvia, Lithuania,

Poland, Slovakia, and Slovenia, along with Cyprus and Malta, will joint the EU. This will create a 25-nation market for all CE-marked devices (EU Expands p. 5).

As a condition for participating in the EEA, the EFTA countries agreed to assume existing EC rules governing matters ranging from antitrust to the environment (Treverton p. 422). This agreement includes the products covered by the three medical device directives. Device manufacturers have access to a 19-nation market based on one certification (Active p. 5).

There are various types of decisions issued by the EU (e.g., directives, regulations, decisions, recommendations, or opinions) all having different degrees of enforcement and applicability. The directive is the type of mechanism used most frequently in the legislation forming the basis of the internal market. A directive is enacted by the Council of Ministers of the EU. As a consequence, the member states are obligated to harmonize their national laws, regulations, and administrative measures with the requirements of the directive within a certain time period.

A proposal for a directive is prepared by the Commission, the civil service of the EU. The Commission will normally consult interested parties such as trade associations, the professions, and consumers while preparing a proposal. The Commission's proposal goes through a complicated process of consultation with the European Parliament, the Economic and Social Committee (an advisory body representing trade unions, employers, consumers, and professional interests) and the Council of Ministers of the EU.

The Council of Ministers is the real decision-making body of the EU. It is made up of the Ministers responsible for the subsector under consideration. The Council votes on the final adoption of the directive.

To facilitate the development of the internal market, the EU has developed a new approach to technical harmonization and standards. Established by the Council Resolution of May 7, 1985, the new approach defines the principles that will govern EU legislation on industrial products, including medical devices. These principles are:

- The directives will not contain detailed technical provisions. Instead, they will list essential requirements (ERs), which must create legally binding and enforceable obligations when transposed into national law.
- The detailed technical provisions will be contained in "harmonized" standards adopted by recognized European standards organizations such as the European Committee for Standardization (CEN) and the European Committee for Electrotechnical Standardization (CENELEC). Conformity to harmonized standards will be presumed to imply compliance with the ERs of the applicable directive.
- The standards themselves are voluntary, not mandatory. Only the ERs are a part of the law. However, the harmonized technical standards may take on a quasi-obligatory status because more effort may be required to demonstrate compliance with the ERs by other means.
- A product that complies with the requirements of the directive and bears the officially recognized conformity marking (i.e., the initials "CE"), may circulate freely among the member states of the EU (and the EEA).

MUTUAL RECOGNITION AGREEMENTS

A number of countries and regions have developed formal mutual recognition agreements (MRA) with the EU covering medical devices. These include Australia, Canada, Switzerland, and the United States (US). The MRAs range from accepting some aspect of the medical device regulatory process such as Good Manufacturing Practice (GMP) inspections to full bilateral acceptance of product approvals.

AUSTRALIA

In June 1998, Australia signed an MRA with the EU. This MRA applies to medical devices manufactured in the European Union and in Australia and New Zealand that are subject to third party conformity assessment. The new harmonization system went into force on October 5, 2002. For more information on this MRA, see Chapter 6.

SWITZERLAND

Although geographically in the center of the EU, Switzerland is not a full member of the EU or the EEA. In order to enjoy bilateral free trade in medical devices, a MRA was required. Although in process for many years, the MRA for products covered by the MDD and the AMID did not come into force until June 1, 2002. An MRA between Switzerland and other EFTA countries, Iceland, Liechtenstein, and Norway, also came into force on June 1, 2002. Negotiations to extend the MRA to products covered by the IVDD continue although informal agreement has been reported in the trade press (Swiss MRA, p.2).

The MRA is seen as largely benefiting Swiss manufacturers as Switzerland has accepted CE marked medical devices since 1996. The MRA will enable the Swiss Agency for Therapeutic Goods (Swissmedic) to appoint Swiss conformity assessment bodies (Notified Bodies). Heretofore, Swiss manufacturers have had to use notified bodies in the EU in order to place the CE mark on their products.

EU manufacturers are no longer obliged to place the address of a Swiss representative on their labeling (Swiss MRA, p.2).

SCOPE OF THE EUROPEAN UNION REGULATION

The EU regulations for medical devices are contained in three technical harmonization directives—the MDD, the AIMDD, and the IVDD. In 1999, a fourth directive came into force in Europe that has a direct bearing on the regulation of medical devices. This is the Radio and Telecommunications Terminal Equipment Directive (R&TTED) (Directive 1999/5/EC).

A critical feature of the definition of a medical device in these directives is that it depends on the intended use of the device and on its principal intended action. Medical devices are defined as articles that are intended to be used for a medical purpose. The use of the term "intended" allows some opportunity for manufacturers to include or exclude their product from the scope of the directives. For instance, a piece of general-purpose laboratory equipment is not a medical device even though it may be used to examine human specimens. However, labeling the same piece of equipment as being "suitable for medical use" changes the classification. The piece of equipment with this labeling change satisfies the definition of a medical device falling within the scope of the IVDD (Higson, p. 143).

The protection ensured by the directives becomes valid when they are supplied to the final user. Taking this concept to its logical conclusion, raw materials, components, or intermediate products are normally not medical devices. However, raw materials, components, or intermediate products may need to present properties or characteristics that contribute to the safety and quality of finished devices. Therefore, the manufacturer of the finished device is responsible for the selection and control of raw materials, components, or intermediate products.

Spare parts supplied for replacement of existing components of a device for which conformity with a directive has been established are not medical devices. However, if the spare parts change significantly the characteristics or performance of a device for which conformity to a directive has been established, then the spare parts are to be considered as devices in their own right (CEC, Definitions §I.1.1(a)).

Software that influences the proper functioning of a medical device may be part of a device or a device in its own right if it is placed on the market separately from the related device. Software that is used with multipurpose informatics equipment may be a medical device if it has a proper medical purpose. Examples of software with a proper medical purpose include (CEC, Definitions §I.1.1(f)):

- software for calculation of anatomical sites of the body,
- image-enhancing software intended for diagnostic purposes, and
- software for programming a medical device.

There is no medical purpose for software used for administration of general patient data.

The definition of a medical device does not encompass products that are principally used for a toiletry or cosmetic purpose even though they may be used to prevent disease. Examples of products for which a medical purpose normally cannot be established are (CEC, Definitions §I.1.1(d)):

- toothbrushes, dental sticks, dental floss;
- baby diapers, hygiene tampons;
- contact lenses without corrective function, intended to provide another color to the eye;
- bleaching products for teeth; and
- instruments for tattooing

THE MEDICAL DEVICE DIRECTIVE (MDD)

The MDD came into force within the EU on January 1, 1995. The MDD applies to all medical devices and their accessories, unless they are covered by the AIMDD or the IVDD. There are certain other products that are not considered medical devices by Article 1 of the MDD. The specific exemptions are listed in Chapter 15.

ACTIVE IMPLANTABLE MEDICAL DEVICE DIRECTIVE (AIMDD)

The AIMDD came into force on January 1, 1993. The labeling requirements peculiar to AIMDs are described in Chapter 17.

IN VITRO DIAGNOSTIC DEVICE DIRECTIVE (IVDD)

The IVDD came into force on June 7, 2000. Until June 6, 2005, Member States must accept devices placed on the market that conform to the rules in force in their territory on June 7, 2000. For an additional period of two years, these devices may be put into service. The labeling requirements described in the directive that are peculiar to IVD devices are described in Chapter 19.

RADIO EQUIPMENT AND TELECOMMUNICATIONS TERMINAL EQUIPMENT DIRECTIVE (R&TTED)

New approach directives cover a wide range of products and hazards that both overlay and complement each other. As a result, several directives may have to be taken into consideration for one product, since placing a product on the market and putting it into service can only occur when the product complies with all applicable provisions. Directives may make direct reference to other directives. For example, the MDD makes reference to Directive 65/65/EEC for certain aspects of devices that administer medicinal substances and to Directive 80/836/Euratom regarding the basic safety standards for the health protection of the general public and workers against the dangers of ionizing radiation. However, for the most part, compliance with the medical device directives is sufficient to allow the manufacturer to place that device on the market. As an example, medical devices were specifically excluded from the provisions of the Electromagnetic Compatibility Directive.

On March 7, 1999, the EU adopted a new approach directive dealing with radio equipment and telecommunication terminal equipment. The R&TTED came into force on April 8, 2000. The purpose for this directive is to create an open competitive market for telecommunications equipment within Europe.

The R&TTED defines radio equipment as a product, or a component of a product, that is capable of communicating by means of the emission or reception of radio waves utilizing the spectrum allocated to terrestrial/space radio communications (Directive 1999/5/EC §2(c)). Telecommunication terminal equipment is a product, or relevant component of a product, intended to connect directly or indirectly by any means to interfaces of public telecommunications networks. A public telecommunication network is one used wholly or partly to provide publicly available telecommunications services (Directive 1999/5/EC §2(b)).

Article 1 of the R&TTED explicitly applies the provisions of this directive to medical devices within the meaning of the MDD and to AIMDs within the meaning of the AIMDD (Directive 1999/5/EC §1(1)). IVD equipment is not mentioned in the unamended version of the R&TTED.

Article 3 of the R&TTED sets out several essential requirements including compliance with the safety requirements of the Low Voltage Directive and the protection requirements of the Electromagnetic Compatibility Directive (EMCD).* The R&TTED involves the same basic conformity assessment modules as the medical device directives. These procedures require the interaction of a properly authorized Notified Body in the testing of the product or the certification of the manufacturer's quality system.

The R&TTED is also concerned about the use of the radio-frequency (RF) spectrum, particularly the use of the frequency spectrum that is not allocated for a given purpose throughout the EC. For radio equipment using frequency bands whose use is not harmonized throughout the EC the manufacturer, its authorized representative, or the person responsible for placing the equipment on the market must notify the national authority in the relevant member state responsible for spectrum management of its intention to place such equipment on the member state's national market. This notification must be given no less than four weeks in advance of placing the device on the market and shall provide information about the radio characteristics of the equipment (in particular, frequency bands, channel spacing, type of modulation, and RF-power) and the identification number of the Notified Body responsible for conformity assessment (Directive 1999/5/EC §6(4)).

LABELS AND LABELING

For the purposes of the European medical device directives, a label is defined as all written, graphic, or printed matter that relates to the identification, technical description, and use of the medical device (EN 46001 §3.10). This includes all written, graphic, or printed matter (a) affixed to a medical device or any of its containers or wrappers, or (b) accompanying a medical device. The shipping documents accompanying a medical device are not considered a label.

The directives make extensive reference to, and sets specific requirements for, the instructions for use (IFU). A natural outcome of the organization of the ERs is to begin to think of the IFU as being synonymous with a package insert or some other accompanying document. For many devices,

---

* Council Directive 89/336/EEC of 3 May 1989 on the approximation of the laws of the member states relating to electromagnetic compatibility. 1989. *Official Journal of the European Communities*, 32, no. L139 (May 23).

this will be the case. However, it is not required by the directives. When practical, the preferred location for the information needed for using the device safely and for its intended purpose is on the device itself. If placing the IFU on the device is impractical (e.g., on an implantable hip joint), the next best location is on the package, and the remaining alternative is in an instruction leaflet (Directive 93/42/EEC §13.1).

In this and subsequent chapters on the EU, the term "label" will be used to refer to the written, graphic, or printed matter on a device or any of its containers or wrappers. The term "accompanying documents" will be used to refer to the material, such as the instruction leaflet, package insert, operator's manual, service manual, and so on that accompanies the medical device.

LANGUAGE REQUIREMENTS

Article 4 of each of the directives permits the member states to require that product labels and the IFU be in the member's national language or languages, or in another community's language at the time the device reaches the final user. This requirement is independent of whether or not the final user is a healthcare professional or a layperson. Therefore, the language requirement is contained in the national legislation that transposes the directive into local law. Several of the member states, such as Germany and France, were vigorously enforcing a requirement for local language before the directive came into force. Not surprisingly, virtually all of the member states included a provision requiring their local language(s) in the legislation implementing the directives. The official languages of the member states are listed in Table 14.1.

There is no requirement regarding the use of single- or multiple-language labeling. The manufacturer is free to choose the type of labeling system that works best for a particular product.

STANDARDS

The concept of harmonized standards plays an important role in the framework of the new-approach directives. According to new-approach directives, conformity with national standards that have transposed harmonized standards confers a presumption of conformity with the ERs covered by the harmonized standard.

TABLE 14.1
Official Languages of the European Union

Member State	Official Language
Austria	German
Belgium	Dutch, French, and German
Denmark	Danish
Finland	Finnish
France	French
Germany	German
Greece	Greek
Ireland	English
Italy	Italian
Luxembourg	French or German
Netherlands	Dutch
Portugal	Portuguese
Spain	Spanish
Sweden	Swedish
United Kingdom	English

A harmonized standard is produced by European standards organizations at the instigation of the European Commission. After consultation with the member states, the European Commission may invite the European standards organizations to present harmonized standards within the meaning of each directive. This invitation is often referred to as a mandate. The documents that the standards organizations are invited to present are European Norms (ENs) or Harmonization Documents (HDs). In presenting harmonized standards, the standards organizations are not constrained to present newly developed standards. They may also identify existing standards that they judge, after examination, to meet the terms of the mandate. The standards organizations may modify an existing EN or HD to meet the terms of the mandate. In the same way, they may identify national or international standards and adopt them as European standards. The resulting European standard will then qualify as a harmonized standard (CEC, Guide p. 28).

The presumption of conformity associated with a harmonized standard depends on (a) the publication of the reference in the *Official Journal of the European Communities*, and (b) the transposition of the European standard into a national standard (CEC, Guide p. 29).

Without publication of the reference by the Commission in the *Official Journal of the European Communities*, use of the standard will not give rise to the presumption of conformity. Because of the legal implications, publication in the *Official Journal* is required so that a clear date is set from which presumption of conformity can take effect.

The presumption of conformity is also dependent on the transposition of the European standard into a national standard. This means that no presumption exists unless the European standard has been transposed, even if a reference has been published in the *Official Journal*. However, it is not necessary for transposition to take place in all member states before a harmonized standard can be used to demonstrate conformity with the ERs of a directive (CEC, Guide p. 29). The European Commission maintains a summary list of titles and references of harmonized standards relating to all new-approach directives on their Web site. The URL can be found in Appendix H.

Both the CEN and the CENELEC have active work programs underway to develop and maintain the harmonized standards needed to implement the medical device directives. A summary of the key top-level standards that specify labeling requirements is given in Table 14.2.

BRINGING A DEVICE TO MARKET IN THE EUROPEAN UNION

The medical device directives recognize only three categories of devices that may be "placed on the market" within the EU. They are (a) devices conforming to the ERs of the relevant directive (and bearing the CE conformity marking), (b) devices intended for clinical investigations, and (c) custom-made devices. Each of these categories is discussed briefly in the following sections.

The following actions are not considered placing a product on the market (CEC, Guide p. 18):

- Transfer of the product from a manufacturer located outside the EU to its authorized EU representative whom the manufacturer has made responsible for completing the procedures required to ensure that the product conforms to the applicable directive(s) in order to place it on the market
- Import into the EU with a view to reexport (e.g., under a processing arrangement)
- Transfer of a product manufactured in the EU with a view to exporting the product to a country outside the EU
- Display of the product at fairs and expositions

Only the manufacturer or its authorized representative may place a device on the market under the manufacturer's own name. A manufacturer who, under its own name, intends to place devices

TABLE 14.2
Key Technical Harmonization Standards for Medical Device Labeling

Subject	Responsibility/Key Documents	Status
Quality Systems	CEN/CENELEC Coordination Working Group on Quality Supplements	
	Quality systems –Medical devices – Particular requirements for the application of EN ISO 9001	EN 46001:1996 [a]
	Quality systems –Medical devices – Particular requirements for the application of EN ISO 9001 (revision of EN 46001) (identical to ISO 13485:1996)	EN ISO 13485:2000 [b]
	Quality systems – Medical Devices – particular requirements for the application of EN ISO 9002	EN 46002:1996 [a]
	Quality systems – Medical devices – Particular requirements for the application of EN ISO 9002 (revision of EN 46001) (identical to ISO 13488:1996)	EN ISO 13488:2000 [b]
	Quality systems – medical devices – particular requirements for the application of EN ISO 9003	EN 46003: 1999 [a]
AIMDs	CEN/CENELEC Joint Working Group	
	Active implantable medical devices – Part 1: General requirements for safety, marking and information to be provided by the manufacturer	EN 45502-1:1997 [a]
IVD Systems	CEN Technical Committee (TC) 140 and CENELEC TC 66	
	Requirements for labelling of *in vitro* diagnostic reagents for professional use	EN 375:2001
	Requirements for labelling of *in vitro* diagnostic reagents for self-testing	EN 376:2002
	Requirements for marking of *in vitro* diagnostic instruments	EN 1658:1996 [a]
	Safety requirements for external equipment, control and laboratory use	EN 61010-1:2001 [a]
Labeling	CEN TC 257	
	Graphical symbols for use in labelling of medical devices	EN 980:2001 [a]
	Information supplied by the manufacturer with medical devices	EN 1041:1998 [a]
Alarm Signals	CEN TC 259	
	Medical devices. Electrically generated alarm signals	EN 475: 1995 [a]
Nonactive Surgical Implants	CEN TC 285	
	Nonactive surgical implants – General requirements	EN ISO 14630:1997 [a]
Electromedical Equipment	CENELEC TC 62	
	Medical electrical equipment – Part 1: General requirements for safety	EN 60601-1:1996 [a]
	Medical electrical equipment. Part 1: General requirements for safety – 1. Collateral standard: Safety requirements for medical electrical systems	EN 60601-1-1:2001 [a]
	Medical electrical equipment - Part 1: General requirements for safety – 2. Collateral standard: Electromagnetic compatibility – Requirements and tests	EN 60601-1-2:2001 [a]
Sterile Devices	CEN TC 204	
	Sterilization of medical devices – Requirements for medical devices to be designated "sterile" – Part 1: Requirements for terminally sterilized medical devices	EN 556-1:2001 [a]

[a] Harmonized standard, a reference to which has been published in the *Official Journal of the European Communities*.

[b] EN ISO 13485 and EN ISO 13488 will replace EN 46001 and EN 46002, but during a transition period both may be used to comply with the directives.

on the market in the EU must inform the competent authority of the member state in which the registered place of business is located of the address of that place of business.

Under the medical device directives, a manufacturer that does not have a registered place of business within the EU may designate one or more authorized representatives who have registered places of business within the EU. The manufacturer delegates tasks in writing to the authorized representative(s), spelling out the manufacturer's obligations under the directives for which it is delegating responsibility to the authorized representative(s). The manufacturer is responsible for actions by the authorized representative(s).

The manufacturer's representative(s) must provide the business address to the competent authority of the member state in which the registered place of business is located.

An importer is any person who places on the market a product from outside the EU that is covered by a directive. Unlike the authorized representative, the importer has no preferential relationship with the manufacturer.

Therefore, if neither the manufacturer nor the manufacturer's authorized representative is based in the EU, the importer is deemed responsible under the terms of the directives for placing the imported product on the EU market. In this capacity, the importer must maintain the technical file and the manufacturer's declaration of conformity available for examination by the competent authorities (CEC, Guide p. 23).

Devices Conforming to the Essential Requirements

Each of the directives stipulates that member states may not create any obstacle to the free movement within their territories of a device that complies with the requirements of the applicable directive. The CE conformity marking, when properly affixed, indicates that the device has been the subject of an assessment of its conformity with the essential requirements.

Conformity Assessment

Conformity assessment is a process established in the directives that is intended to demonstrate that (a) a device complies with all of the applicable ERs and (b) the manufacturer has taken all necessary steps to ensure that the manufacturing processes produce devices that conform to the appropriate design documentation. Depending on the type of device, the conformity-assessment process can vary from a self-assessment done by the manufacturer to a complex design-dossier examination and quality-system certification performed by one or more external assessors (i.e., Notified Bodies). Regardless of the extent of the conformity-assessment process that is required for a particular device, it is important to remember that every device must fully conform to all of the applicable ERs in the directive. This includes the labeling requirements in the appropriate directive, as well as any labeling requirements in standards used to demonstrate compliance with the ERs.

Notified Bodies

New-approach directives have moved away from the traditional process whereby conformity assessment was the sole responsibility of the authorities within the member states. In the traditional system, the member states often delegated the technical work involved in conformity assessment to entities unknown to the other member states. Such practices make it difficult to operate a system that is based on the trust that national authorities are prepared to place in each other. The new approach replaced the traditional system with one oriented toward assured technical competence, objectivity, and transparency based on technical criteria documented in the directives themselves and in the appropriate European standards.

Under this system, member states are invited to notify the Commission of those bodies that they consider competent to carry out the conformity-assessment tasks described in a directive. This process of notification has given its name to the conformity-assessment organizations, which are known collectively as "Notified Bodies."

The member states are responsible to the other member states and to the Commission for the bodies they notify. The member state is responsible for ensuring that the Notified Bodies implement fully and at all times the conditions under which they were notified. Should a body notified by a member state cease to fulfill these conditions, the member state must inform the other member states and the Commission that notification has been withdrawn. Therefore, the Notified Body must come under the jurisdiction of the member state. This entails legal jurisdiction and means that member states can only notify bodies established on their territory (CEC, Guide p. 36).

A Notified Body may be privately owned or state-owned. The legal status is irrelevant as long as the body can demonstrate that it meets the legally binding criteria set out in the directives in relation to the conformity assessment activities (CEC, Guide pp. 39–40).

The bodies that are notified are free to offer the conformity-assessment services for which they are notified to any interested party established either inside the EU or in another country. They may carry out these activities within the territory of the member states or in another country. A Notified Body may even subcontract part of its work to another body within clearly specified limits. The subcontractor may or may not be located within the territory of the EU. The Notified Body must establish and monitor on a regular basis the technical competence of the subcontractor. Ultimately, the Notified Body remains entirely responsible for the work carried out by the subcontractor. A Notified Body cannot subcontract assessment and appraisal activities, which are the essential tasks for which it was notified (CEC, Guide p. 42). Certificates are always issued by, and in the name of, the body notified and not in the name of any subsidiary or subcontractor.

Competent Authorities

The new approach to technical harmonization places a good deal of responsibility for technical assessment on the manufacturer and Notified Bodies. The member states, however, retain the responsibility for protecting the health and safety of their citizens.

One of the elements of the legislation that transposes the requirements of a directive into national law is the naming of a "competent authority" within the member state—usually the Ministry of Health or one of its bureaus or divisions (e.g., the Medical Devices Agency within the UK Department of Health). This competent authority is charged with discharging the responsibilities of the member state under the directive. These responsibilities include:

- Adopting and publishing laws, regulations, and administrative procedures to implement the directives.
- Selecting and supervising the Notified Bodies, possibly in conjunction with official accreditation organizations within the member state.
- Reviewing clinical notifications when required by Article 10 of the AIMDD or Article 15 of the MDD.
- Taking the necessary steps under the postmarket vigilance provisions in Article 2 of the AIMDD, Article 10 of the MDD, and in Article 11 of the IVDD.
- Exercising the safeguard clause leading to the removal of a device bearing the CE conformity marking from the market when it is determined that a device is in noncompliance with the respective directive because (Directive 90/385/EEC §7; Directive 93/42/EEC §8; Directive 98/79/EC, § 8):
  - The device failed to meet the ERs in Annex I of the directive.
  - The device failed to comply with the harmonized standards with which compliance was claimed by the manufacturer.
  - There are shortcomings in the harmonized standards themselves.
- Other activities required by the directives of the authorities in the member states (i.e., receive notifications concerning clinical investigations).

DEVICES FOR CLINICAL EVALUATIONS

The directives specify that a device bearing the CE conformity marking must achieve the characteristics and performances intended by the manufacturer, and that it be suitable for one or more of the functions of a medical device. As part of the conformity-assessment process, the manufacturer must be able to document that the device, in fact, meets these requirements. This may necessitate that a clinical evaluation be carried out. A clinical evaluation means that the available evidence supports the assertion of safety and performance in clinical use. In the case of existing, established devices, published scientific literature may be adequate to establish the required level of confidence. In other cases, a clinical investigation (i.e., clinical trial) may be necessary to obtain the needed evidence.

A device intended for use in clinical investigations cannot bear the CE conformity marking because, by definition, there is insufficient evidence to demonstrate that the device meets all of the ERs of the applicable directive. Therefore, the directives provided a special mark that allows the clinical device to be "put into service" for its intended purpose. This mark is described later in this chapter.

It is important to remember that the manufacturer must attest to the fact that the investigational device in question "conforms to the ERs apart from the aspects covered by the investigation and that, with regard to these aspects, every precaution has been taken to protect the health and safety of the patient" (Directive 90/385/EEC Annex 6(2.2); Directive 93/42/EEC Annex VIII(2.2)). This includes the labeling requirements in Annex I of the directives.

DEVICES FOR PERFORMANCE EVALUATION OR REEVALUATION

For IVD medical devices, the IVDD requires that the manufacturer possess adequate performance-evaluation data to support the performances claimed by the manufacturer. The claims must be supported by a reference measurement system (when available), with information on the reference methods, the reference materials, the known reference values, the accuracy and measurement units used. This data should originate from studies in a clinical or other appropriate environment, or result from relevant biographical references.

In order to gather the necessary data, a manufacturer may need to conduct one or more performance-evaluation studies in laboratories for medical analyses or in other appropriate environments outside the manufacturer's premises. A device undergoing such an evaluation cannot bear the CE marking of conformity because, by definition, there is insufficient evidence that the device meets all of the ERs of the IVDD. Therefore, the IVDD provides a special mark that allows the investigation device to be put into service for its intended purpose. This mark is described later in this chapter. This provision does not affect national regulations relating to the ethical aspects of carrying out performance-evaluation studies using tissues or substances of human origin.

The manufacturer must prepare a statement that the device in question conforms to the requirements of the IVDD, apart from the aspects covered by the evaluation and apart from those specifically itemized in the statement, and that every precaution has been taken to protect the health and safety of the patient, user, and other persons.

CUSTOM-MADE DEVICES

A custom-made device is one that is "specifically made in accordance with a duly qualified medical practitioner's written prescription which gives, under his responsibility, specific design characteristics and is intended for the sole use of a particular patient" (Directive 90/385/EEC §1(2(d)); Directive 93/42/EEC §1(2(d))).

Like a device intended for use in clinical investigations, the custom-made device cannot bear the CE conformity marking. The directives provide a special mark that allows the custom-made device to be put into service for the sole use of a particular patient. This mark is described later in this chapter.

It is important to remember that the manufacturer must attest to the fact that the custom-made device in question "conforms to the ERs set out in Annex I [of the applicable directive] and, where applicable, indicate which ERs have not been fully met, together with the grounds" (Directive 90/385/EEC Annex 6(2.1); Directive 93/42/EEC Annex VIII(2.1)). This includes the labeling requirements in Annex I of the directives.

By definition, custom-made devices (such as dental appliances and hearing-aid inserts) are usually one-off devices that are subject to special procedures under the directives and a special conformity mark. However, intermediate products specifically intended for use in making custom-made devices may be considered as medical devices in their own right. For example, dental alloys, dental ceramics, and modular components for prostheses may be considered as medical devices if the intended use is for a medical purpose (CEC, Definitions §I.1.1(c)).

A device that must undergo further processing after reaching the final user is not automatically considered a custom-made device. Examples of final-user processing include (CEC, Definitions §I.1.1(c)):

- sterilization of medical devices supplied in a nonsterile state,
- assembling of systems,
- configuration of electronic equipment,
- preparation of a dental filling,
- fitting of contact lenses, and
- adaptation of a prosthesis to the needs of the patient.

These activities are not considered to be part of the manufacturing process if they are carried out by the healthcare professional as part of his or her professional activity. If, however, a specialist in such processing carries out these steps, they may be considered manufacturing or assembly activities within the meaning of the directives.

Devices for Trade Fairs, Exhibitions, and Demonstrations

Member states may not create obstacles to the display of devices that do not conform to the directives at trade fairs, exhibitions, demonstrations, and so on provided these devices are properly marked. A clearly visible sign must indicate that the device in question cannot be marketed or put into service until it complies with the applicable directive(s) (Directive 90/385/EEC § 4(3); Directive 93/42/EEC §4(3); Directive 98/79/EC, § 4(3)).

POSTMARKET SURVEILLANCE AND VIGILANCE

A manufacturer's responsibility does not end with the sale of a medical device. The manufacturer's responsibility continues through the useful life of the device in the form of an obligation to operate a system for obtaining feedback from the market (postmarket surveillance) and reporting serious incidents to the competent authority (vigilance).

Postmarket Surveillance

The conformity-assessment annexes of the directives include a requirement that the manufacturer maintain a postmarket surveillance system for devices bearing the CE mark of conformity. This requirement is most clearly stated in the conformity-assessment annexes of the MDD. The manufacturer is required to "initiate and keep up to date a systematic procedure to review experience gained from devices in the post-production phase and to implement appropriate means to apply any necessary corrective action" (Directive 93/42/EEC Annex II(3.1)).

The manufacturer is required to establish a system that collects data from a variety of sources, which can be as diverse as (a) customer complaints, (b) service/maintenance records, (c) patient registries, and (d) ongoing follow-up of patients involved in clinical investigations. These data must be periodically reviewed, looking for trends and seeking to identify systematic problems. When necessary, corrective action must be initiated. If an incident is serious enough, it must be reported to the competent authorities under the vigilance requirement of the directives.

VIGILANCE

The purpose of the vigilance system is to improve the protection of patients, users, and others. This improvement is to be achieved by reducing the likelihood of the same type of adverse event being repeated in different places at different times. Reported incidents are to be evaluated and, when appropriate, information that could be used to prevent repetition of the incident, or to alleviate the consequences of such an incident, is to be disseminated to all the member states. The vigilance system is intended to allow data to be correlated between competent authorities and manufacturers. It is anticipated that this sharing of information will facilitate implementation of corrective action earlier than would be the case if data on adverse incidents were collected, analyzed, and action taken on a state-by-state basis (CEC, Vigilance §3).

Incidents that need to be reported are defined in the directives as (Directive 90/385/EEC §8(1); Directive 93/42/EEC §10(1); Directive 98/79/EC §11(1)):

- those that lead to a death;
- those that lead to a serious deterioration in the state of health of a patient or user, including:
  - life-threatening illness or injury,
  - permanent impairment of a body function or permanent damage to a body structure,
  - a condition that requires medical or surgical intervention to prevent permanent impairment of a body function or permanent damage to a body structure, and
- any technical or medical reason relating to the characteristics or performance of a device that, for the reasons listed above, has lead to a systematic recall of devices of the same type by the manufacturer.

The Commission recognizes that interpretation of the term "serious" is not easy, and that it should be made in consultation with a medical practitioner whenever possible (CEC, Vigilance §5.4.2). The Commission has suggested that many points need consideration, such as:

- whether a risk was foreseeable and clinically acceptable in view of the potential patient benefit
- whether the outcome was adversely affected by a preexisting condition of the patient.

In cases of doubt, the Commission counsels that there should be a predisposition to report an incident, rather than to not report (CEC, Vigilance §5.4.2).

Just because a death or serious deterioration in the health of a patient or user did not occur is not justification for failing to report the near incident. The directives state that an incident with the real potential of causing a death or serious deterioration in the state of health is reportable. It may be that good fortune or the intervention of healthcare professionals prevented serious or fatal consequences in the incident. However, it is sufficient that an incident associated with the device occurred and that the incident was such that if it occurred again, it might lead to death or serious deterioration in health.

The manufacturer should also report if an examination of that device or the information supplied with the device indicates some factor (e.g., some inadequacy in the labeling) that could lead to an incident involving death or serious deterioration in health. Inadequacies in the labeling would include significant omissions and deficiencies, as well as inaccuracies. For example:

- The omission of important safety-related information would be reportable. An example would be the failure to include a warning of a side effect that may be produced by the device while working within specifications. However, the absence of information that should generally be known by the intended user in not considered an omission (CEC, Vigilance §5.5.3).
- A deficiency would exist if there were a lack of clarity in the instructions that leads, or could lead, to an injury.
- Inaccuracies are errors of fact that caused, or could cause, misuse or incorrect maintenance or adjustment of the device.

In Appendix 4 of the Guidelines on a Medical Device Vigilance System, the Commission provided some examples of incidents that should (or should not) be reported. These include:

- *An infusion pump delivers the wrong dose because of an incompatibility between the pump and the infusion set used.* If the combination of pump and the infusion set used was in accordance with the IFU for either the pump or the infusion set, then the incident should be reported. If the combination was used against the IFU for both pump and the infusion set, then the incident should not be reported.
- *An aortic balloon catheter leaked because of inappropriate handling of the device in use, causing a situation that was potentially dangerous to the patient.* If the inappropriate handling was in any way due to inadequacies in the labeling, then the incident should be reported as a near incident. If the labeling clearly indicated that such handling was inappropriate, the incident need not be reported.

The Safeguard Clause

Article 4 in each of the directives precludes the member states from creating obstacles to placing on the market or putting into service a device that bears the CE conformity marking. This provision of the directives notwithstanding, the member states have an overriding right and obligation to protect the health and safety of their citizens. This power and obligation is the basis of the Safeguard Clause (Article 7 of the AIMDD; Article 8 of the MDD; Article 8 of the IVDD).

Under the Safeguard Clause, a member state is required to take action if it is demonstrated that a device bearing the CE conformity marking and used for its intended purpose could endanger the health and/or safety of persons, animals, or goods. The member state must take all necessary measures to restrict or forbid the placing of the device on the market or to have the product withdrawn from the market. The Safeguard Clause is the last resort in the overall system of market vigilance and postmarket surveillance required by the directives (CEC, Guide p. 53).

The member state must establish that the CE conformity marking has been improperly affixed to the device by making a finding that is objective and based on verifiable evidence. The assessment of the risk to persons, animals, or goods is, however, the full responsibility of the member states. They bear responsibility for evaluating whether there is a foreseeable potential danger likely to have serious consequences (CEC, Guide p. 53).

The member states administer the Safeguard Clause within their own territories. It is up to the Commission, however, to manage the Safeguard Clause at the Community level and ensure that it is applied to the whole Community as soon as possible (CEC, Guide p. 54).

In theory, any noncompliance with the ERs, including noncompliance with the labeling requirements, that poses a foreseeable potential danger likely to have serious consequences, could be justification for a member state to activate the Safeguard Clause. However, activation of the Safeguard Clause is seen by the Commission as a last resort if other corrective actions, taken as part of the vigilance process, fail to alleviate the problem. The Commission is concerned that the public authorities in the member states not abuse the Safeguard Clause.

CE MARKING OF CONFORMITY

The purpose of the CE conformity marking is to symbolize to all interested parties that a product conforms to the provisions of the technical harmonization directives relevant to the product. It also indicates that the economic operator responsible has undergone all of the evaluation procedures required by EU law with respect to his or her product.

The CE conformity marking consists of the initials "CE" taking the form shown in Figure 14.1. The CE conformity marking may be reduced or enlarged, provided the proportions given in the graduated drawing in Figure 14.1 are respected. The minimum height of the CE conformity marking (dimension C in Figure 14.1) is 5 mm. However, a waver of the minimum dimension may be obtained for very small devices (Directive 93/42/EEC Annex XII).

The CE conformity marking must be affixed by the manufacturer or the manufacturer's duly authorized representative within the EU at the end of the production control phase. The CE conformity marking must be visibly, legibly, and indelibly affixed to the product or to its data plate. However, where this is not possible or is not warranted because of the nature of the product (i.e., in the case of AIMDs), it must be affixed to the packaging, if any, and to the accompanying documents. Affixing any other marking that is liable to deceive third parties as to the meaning of the CE conformity marking is prohibited (Decision 93/465/EEC Annex I(i)).

Other markings (e.g., marks indicating conformity to national or European standards) may be affixed to a product, provided such marks are not liable to be confused with the CE conformity marking. These marks may be affixed to the product, its packaging, and the accompanying docu-

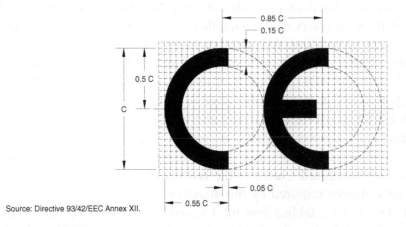

Source: Directive 93/42/EEC Annex XII.

FIGURE 14.1 CE conformity marking.

mentation provided such marks do not compromise the legibility and visibility of the CE conformity marking (Decision 93/465/EEC Annex I(j)).

When a Notified Body is required to be involved in the conformity-assessment procedure, the identification number of the Notified Body must follow the CE conformity marking (Directive 93/42/EEC §17(2); Directive 90/385/EEC §12(2); Directive 98/79/EC §16(2)). Notified Body identification numbers are assigned by the European Commission as part of the body notification procedure. Notified body identification numbers are published periodically in the *Official Journal of the European Communities*. Each Notified Body will have a single identification number no matter how many directives under which it is notified (Decision 93/465/EEC Annex I(g)).

The CE conformity marking is the only marking that certifies that a product conforms to the applicable directive. The member states must refrain from introducing into their national regulations any reference to a conformity marking other than the CE conformity marking in connection with a technical harmonization directive. Member States must make provision in their national law to exclude any possibility of confusions and to prevent abuse of the CE conformity marking. If it is determined that the CE conformity marking has been affixed unduly, the manufacturer, or the manufacturer's agent, is obligated to make the product comply under the conditions imposed by the member state. If the noncompliance continues, the member state must take all appropriate measures to restrict or prohibit the placing of the product on the market, or to ensure that the product is withdrawn from the market in accordance with the procedures in the Safeguard Clause of the relevant directive.

CE Conformity Marking on Devices Covered by the MDD

Medical devices covered by the MDD, other than custom-made devices or devices intended for clinical investigations, that meet the ERs that apply to the device in Annex I of the MDD must bear the CE conformity marking. The CE conformity marking must appear in a visible, legible, and indelible form in the following locations (Directive 93/42/EEC §17(2)):

- The CE conformity marking must appear on the instruction leaflet.
- Where practical and appropriate, the CE conformity marking must appear on the device or its sterile package.
- If a separate sales package is provided, the CE conformity marking must appear on this layer of packaging.

If, for a particular medical device, the participation of a Notified Body in the conformity assessment procedure is required, the CE conformity marking must be accompanied by the identification number of the responsible Notified Body (Directive 93/42/EEC §17(3)).

CE Conformity Marking on Devices Covered by the AIMDD

AIMDs other than those devices that are custom-made or are intended for clinical investigations, that meet the ERs in Annex I of the AIMDD must bear the CE conformity marking. The CE conformity marking must appear in a visible, legible, and indelible form in the following locations (Directive 90/385/EEC §12(2)):

- The CE conformity marking must appear on the sterile package.
- If a separate sales package is provided, the CE conformity marking must appear on this layer of packaging.
- The CE conformity marking must also appear on the instruction leaflet.

There is no requirement in the AIMDD that the CE conformity marking appear on the AIMD itself.

As all AIMDs require the participation of a Notified Body in the conformity assessment procedure, the CE conformity marking is always accompanied by the identification number of the responsible Notified Body.

CE Conformity Marking on Devices Covered by the IVDD

As with the devices covered by the other directives, IVD devices, other than those intended for performance evaluation, that meet the ERs in Annex I for the IVDD must bear the CE conformity marking. The CE conformity marking must appear in a visible, legible, and indelible form in the following locations (Directive 98/78/EC §16(2)):

- The CE conformity marking must appear on the IFU.
- The CE conformity marking must also appear on the sales packaging.
- Where practical and appropriate, the CE conformity marking must appear on the device itself.

When a Notified Body is involved in one of the conformity-assessment procedures set out in Annexes III, IV, VI, and VII of the IVDD, the identification number of that Notified Body must accompany the CE marking.

CE Conformity Marking on Devices Covered by the R&TTED

The R&TTED requires that all equipment complying with the essential requirements set out in Article 3 must bear the CE conformity marking. The CE conformity marking must appear in a visible, legible, and indelible form in the following locations (Directive 1999/5/EC Annex VII(3)):

- On the product or its data plate
- On the packaging
- In the IFU*

When the conformity-assessment procedures identified in Annex III, IV, or V of the R&TTED are used, the CE conformity marking must be accompanied by the identification number of the responsible Notified Body (Directive 1999/5/EC §12(1)). If more than one Notified Body is involved in the conformity-assessment process for a device, then all the numbers need to appear in association with the CE conformity marking.

In addition to placing the CE conformity marking on the product, the R&TTED requires that the manufacturer or the person responsible for placing the equipment on the market provide the user with the following information that shall be prominently displayed (Directive 1999/5/EC §6(3)):

- The intended use of the apparatus with respect to the provisions of the R&TTED.
- A declaration of conformity to the essential requirements of the R&TTED.

The Telecommunication Conformity Assessment and Market Surveillance Committee (TCAM)** has agreed that the following statement is sufficient to satisfy this requirement (CEC, R&TTED FAQ pp. 5–6):

---

* At some points, the R&TTED refers to accompanying documents. At others, it mentions the IFU. There appears to be no significant differences so the author has used the term IFU to be consistent with the medical device directives.
** In implementing the R&TTED, the Commission is assisted by a committee, the TCAM, composed of representatives of the member states and chaired by a representative of the Commission.

HEREBY, *[NAME OF MANUFACTURER]*, DECLARES THAT THIS *[TYPE OF EQUIPMENT]* IS IN COMPLIANCE WITH THE ESSENTIAL REQUIREMENTS AND OTHER RELEVANT PROVISIONS OF DIRECTIVE 1999/5/EC.

- The manufacturer's name and a type, batch, or serial number (Directive 1999/5/EC §12(4)).

For telecommunications terminal equipment, this information must be sufficient to identify interfaces of the public telecommunications networks to which the equipment is intended to be connected.

For radio equipment, such information must identify the member states or the geographical area within a member state where the equipment is intended to be used. It is sufficient to place this information on the packaging and the IFU of the apparatus. In addition, the manufacturer must alert the user to potential restrictions or requirements for authorization of use of the radio equipment in certain member states by placing an equipment-class identifier in proximity to the CE conformity marking when such an equipment-class identifier has been assigned. Equipment-class identifiers are assigned by the European Commission drawing on the expertise of CEPT/ERC and of the relevant European standards bodies in radio matters. The equipment-class identifier must take a form to be decided by the Commission, the details of which shall be published in the *Official Journal of the European Communities*.

Radio equipment that uses a frequency band that is not harmonized throughout the community must also bear the "alert sign" in close proximity to the CE conformity marking. The alert sign should be the same size as the CE conformity marking. The alert sign is shown in Figure 14.2 and is drawn on the same reference grid as the CE conformity marking in Figure 14.1. The combination is sufficient to inform the user that the radio equipment is subject to restrictions established by one or more member states. These member states may, under the provisions of the Treaty of Rome, prohibit or restrict the placing on the market, or require the withdrawal from the market, radio equipment that has caused or which can reasonably be expected to cause harmful interference with existing or planned services on nationally allocated frequency bands (Directive 1999/5/EC §9(5)).

CE CONFORMITY MARKING ON DEVICES COVERED BY OVERLAPPING DIRECTIVES

Placing the CE conformity marking on a product indicates that this product complies with all applicable directives that provide for affixing of the CE conformity marking (Directive 93/42/EEC §4(5)). However, from time to time a product may not qualify for a CE conformity marking under all of the applicable directives. For example, an investigational medical device may not bear the CE

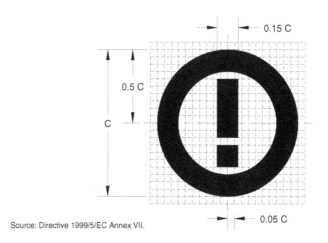

Source: Directive 1999/5/EC Annex VII.

FIGURE 14.2 Alert sign.

conformity marking under any of the medical device directive. However, it may be required to have a CE conformity marking under the R&TTED.

To address this problem, a paragraph was added to Article 4 of the MDD. Under the terms of this paragraph, the manufacturer may choose, during the transitional period, the arrangement of directives that apply. During this period, the CE conformity marking indicates that the device fulfills only the provisions of the directives applied by the manufacturer. The manufacturer must clearly identify in a prominent place in the package insert/IFU the directives with which the product complies (Directive 93/42/EEC §4(5)). This declaration must include the number and title of the directive as it appeared in the *Official Journal of the European Communities* (CEC, Demarcation §3.2.3). Such a statement might take the following form:

THIS DEVICE COMPLIES WITH THE REQUIREMENTS OF DIRECTIVE 1999/5/EC ON RADIO EQUIPMENT AND TELECOMMUNICATIONS TERMINAL EQUIPMENT AND THE MUTUAL RECOGNITION OF THEIR CONFORMITY.

INVESTIGATIONAL DEVICE MARKING

A device intended for use in clinical investigations cannot bear the CE conformity marking because, by definition, there is insufficient evidence to demonstrate that the device meets all of the ERs of the applicable directive. Therefore, the directives provide a special mark that allows the clinical device to be put into service for its intended purpose. The mark also indicates that the economic operator responsible for the product has undergone all of the evaluation procedures required by EU law in respect to his or her product. If the device is intended for clinical investigations, its labeling must bear the phrase:

EXCLUSIVELY FOR CLINICAL INVESTIGATION

This phrase must appear in the same places on the labeling as the CE conformity marking.

PERFORMANCE-EVALUATION DEVICE MARKING

An IVD device undergoing a performance evaluation cannot bear the CE marking of conformity because, by definition, there is insufficient evidence that the device meets all of the ERs of the IVDD. Therefore, the IVDD provides a special mark that allows the investigation device to be put into service for its intended purpose. The mark also indicates that the economic operator responsible for the product has undergone all of the evaluation procedures required by EU law in respect to his or her product. If the device is intended for performance evaluation, its labeling must bear the phrase:

FOR PERFORMANCE EVALUATION ONLY

This phrase must appear in the same places on the labeling as the CE conformity marking.

CUSTOM-MADE DEVICE MARKING

Like a device intended for use in clinical investigations, the custom-made device cannot bear the CE conformity marking. Therefore, the directives provide a special mark which allows the custom made device to be put into service for the sole use of a particular patient. The mark also indicates that the economic operator responsible for the product has undergone all of the evaluation procedures required by EU directives in respect to the product. If the device meets the definition of a custom-made device, its labeling must bear the phrase:

CUSTOM-MADE DEVICE

This phrase must appear in the same places on the labeling as the CE conformity marking.

OTHER DIRECTIVES OF INTEREST

Besides the medical device directives, several other EU directives may have an impact on the labeling of medical devices. The manufacturer is responsible for identifying the applicable directives and complying with their provisions. Some of the major ones are described in the following sections.

PROPRIETARY MEDICINAL PRODUCTS

A drug and its delivery system that are placed on the market as a single unit (e.g., a prefilled syringe) is covered by the Proprietary Medicinal Products Directive.* However, the syringe must also conform to the applicable ERs of the MDD. The labeling, however, should comply with the requirements applicable to medicinal products.

LOW-VOLTAGE EQUIPMENT

The Low-Voltage Directive** as amended in 1993, requires electrical equipment operating at 50-1000 V a.c. or 75-1000 V d.c. to comply with the safety requirements listed in Annex I of the directive. From January 1, 1997, equipment operating in this range must bear the CE conformity marking. Medical electrical equipment is excluded from the scope of this directive. However, ancillary equipment that is provided with a medical device (e.g., personal computers) may be subject to this directive.

MACHINERY

The Machinery Directive*** came into full force on January 1, 1995. This directive covers equipment with power-operated moving parts. Medical machinery in direct contact with the patient is excluded.

DANGEROUS SUBSTANCES

Devices that incorporate dangerous substances are subject to regulation under the Classification, Packaging and Labelling of Dangerous Substances Directive (DSD).**** Such devices must be labeled according to DSD Annex II. This directive is applicable to IVDs; its requirements are discussed in Chapter 19.

---

* Council Directive 65/65/EEC of 26 January 1965 on the approximation of provisions laid down by law, regulation, or administrative action relating to proprietary medicinal products. 1965. *Official Journal of the European Communities*, 8, no. L22 (February 9).

** Council Directive 73/23/EEC on the harmonization of the laws of the Member States relating to electrical equipment designed for use within certain voltage limits. 1973. *Official Journal of the European Communities*, 15, no. L77 (March 26) as amended by Council Directive 93/68/EEC of 22 July 1993 amending Directives 87/404/EEC (simple pressure vessels), 88/378/EEC (safety of toys), 89/106/EEC (construction products), 89/336/EEC (electromagnetic compatibility), 89/392/EEC (machinery), 89/686/EEC (personal protective equipment), 90/384/EEC (nonautomatic weighing instruments), 90/385/EEC (active implantable medicinal devices), 90/396/EEC (appliances burning gaseous fuels), 91/263/EEC (telecommunications terminal equipment), 92/42/EEC (new hot-water boilers fired with liquid or gaseous fuels) and 73/23/EEC (electrical equipment designed for use within certain voltage limits). 1993. *Official Journal of the European Communities*, 36, no. L 220 (August 30).

*** Council Directive 98/37/EC of the European Parliament and of the Council of 22 June 1998 on the approximation of the laws of the member states relating to machinery. 1998. *Official Journal of the European Communities*, 41, no. L 207 (July 23).

**** Council Directive 91/325/EEC of 1 March 1991 adapting to technical progress for the twelfth time Council Directive 67/548/EEC on the approximation of the laws, regulations and administrative provisions relating to the classification, packaging and labelling of dangerous substances. 1991. *Official Journal of the European Communities*, 34, no. L 180 (July 7).

PACKAGING AND PACKAGING WASTE

A directive covering all packaging placed on the market was adopted by the European Parliament on December 20, 1994. The Packaging and Packaging Waste Directive (Directive 94/62/EEC) came into force on December 31, 1994. This directive aims to harmonize the national laws concerning the management of packaging and packaging waste in order to reduce their impact on the environment. This directive covers all packaging placed on the market in the EU, and all packaging waste, whether it is used or released at industrial, commercial, office, shop, service, household, or any other level, regardless of the material used. Member states were required to bring into force the laws, regulations, and administrative provisions necessary to comply with this directive before June 30, 1996.

The directive establishes targets that the Member States are required to meet. By June 30, 2001, 50 to 65 percent by weight of the packaging waste must be recovered. Within the same time period, 25 to 45 percent by weight of the total packaging waste must be recycled. In addition, a minimum of 15 percent by weight of each packaging material must be recycled. Within 10 years, the Council of the European Union will establish new requirements for recovery and recycling of packaging waste with a view to substantially increasing these targets (Directive 94/62/EEC p. 365/14). Greece, Ireland, and Portugal may because of special circumstances, attain a lower initial target. However, these countries must attain the same level as the other Member States by December 31, 2005.

The directive also establishes the maximum concentration of heavy metals (i.e., lead, cadmium, mercury, and hexavalent chromium) that may be present in packaging or packaging components (Directive 94/62/EEC p. 365/15). By June 30, 1998, the sum of concentration of heavy metals could not exceed 600 ppm by weight. By June 30, 1999, this was reduced to 250 ppm by weight. By June 30, 2001, the sum of concentration levels could not exceed 100 ppm by weight. The European Commission will determine (a) the conditions under which these concentration levels will not apply to recycled materials or closed product loops and (b) the types of packaging that are exempted from the 100 ppm requirement (Directive 94/62/EEC p. 365/16).

To facilitate collection, reuse, and recovery including recycling, packaging shall indicate the nature of the packaging materials used. By December 31, 1995, the European Commission was to establish a system of numbering and abbreviations, and determine which materials shall be subject to the identification system. The system utilizes the following numbering: from 1 to 19 for plastic, from 20 to 39 for paper and cardboard, from 40 to 49 for metal, from 50 to 59 for wood, from 60 to 69 for textiles, and from 70 to 79 for glass. The identification system may also use abbreviations for the relevant materials. The identification marks will appear in the center of or below the graphical marking indicating the reusable or recoverable nature of the packaging (Directive 94/62/EEC p. 365/18).

The Packaging and Packaging Waste Directive does not establish the graphical markings to indicate the reusable or recoverable nature of the packaging. The marking shown in Table 14.3 are the ones contained in the amended proposal (CEC, Amended Proposal p. 14). However, this table was removed from the final directive, apparently because the member states were unable to agree on the symbols. Rather, the directive specifies that these markings were to be established by the European Commission before December 31, 1996 (Directive 94/62/EEC p. 365/14). At the time of publication, this task has not been completed.

The directive does require that the markings, once they are established, must appear either on the packaging itself or on a label. The marking must be clearly visible and easily legible. The marking must be appropriately durable and lasting even when the packaging is opened (Directive 94/62/EEC p. 365/15).

TABLE 14.3
Markings to Facilitate Reuse and Recovery of Packaging and Packaging Waste

No.	Symbol	Description
1	⇄	Reusable packaging
2	♻	Recoverable packaging (alternative #1)
3	△	Recoverable packaging (alternative #2)
4		Identification system: Plastic shall use a numbering from 1 to 19; paper and cardboard from 20 to 39; metal from 40 to 49; wood from 50 to 59; textiles from 60 to 69; glass from 70 to 79. The identification may also be done by using the abbreviation of the material(s) used (e.g., HDPE: high-density polyethylene). Numbering or abbreviations or both may be used to identify materials. The said identification methods shall be located in the center of or below the graphical marking indicating reusable or recoverable nature of the package.
5	⟳	Packaging made partly or entirely of recycled materials X % = Percentage of recycled material used in the manufacture of the product.

Source: CEC, Amended Proposal p. 23.

LOCAL REQUIREMENTS

The technical harmonization directives are a major step toward creating an internal market that is open to the free movement of medical devices. They do not, however, entirely eliminate the power of the member states to place requirements on medical devices. The authority of the member states to require local-language labeling is an obvious example. Member States may set other general requirements such as the German law, which makes the manufacturer responsible for recovering waste packaging. Compliance with such a law may require that package labels bear certain symbols or other information unrelated to the medical purpose of the device. The manufacturer is responsible for identifying any relevant local laws in the countries where it plans to do business and for complying with their provisions.

THINGS TO REMEMBER

The EU regulations for medical devices are contained in three technical harmonization directives—the MDD, the AIMDD, and the IVDD. The purpose of the directives is to ensure the free movement of medical devices within the internal market.

Devices that conform to the requirements of a directive may bear the CE conformity marking. The purpose of the CE conformity marking is to symbolize to all interested parties that a product conforms to the provisions of the technical harmonization directives relevant to the product. It also indicates that the economic operator responsible for the product has undergone all of the evaluation procedures required by EU law with respect to his or her product.

The directives also recognize two other product classes that may move freely in the internal market. These are custom-made devices and devices intended for clinical investigations. These products bear special inscriptions in lieu of the CE conformity marking. As with the CE conformity marking, these inscriptions indicate that the manufacturer certifies under its own responsibility that the devices meet the requirements of the directive.

A manufacturer's responsibility does not end with the sale of a medical device. The manufacturer's responsibility continues through the useful life of the device in the form of an obligation to operate a system of obtaining feedback from the market (postmarket surveillance) and reporting serious incidents to the competent authority (vigilance).

Article 4 of each of the directives permits the Member States to require that product labels and the IFU be in their national language or languages, or in another community language, at the time the device reaches the final user. This requirement is independent of whether the final user is a healthcare professional or a lay person.

In addition to the requirements of the medical device directives, a product may be subject to other EU directives such as those covering dangerous substances. The member states may impose general requirements such as the German law, which makes the manufacturer responsible for recovering waste packaging. The manufacturer is responsible for identifying and complying with any relevant EU directives and local laws in the countries where it plans to do business.

15 The Medical Device Directive (MDD)

The Medical Device Directive (MDD) is the most extensive of the three directives covering the medical device sector. It applies to all medical devices and their accessories, unless they are covered by the Active Implantable Medical Device Directive (AIMDD) or the *In Vitro* Diagnostic Device Directive (IVDD). Other excluded products are:

- medicinal products covered by the Proprietary Medicinal Products Directive (MPD);
- cosmetic products covered by the Cosmetic Products Directive;*
- devices that incorporate human blood, human blood products, human plasma, or blood cells of human origin;
- transplants, tissues, or cells of human origin, and devices incorporating or derived from tissues or cells of human origin;
- transplants, tissues, or cells of animal origin, unless a device is manufactured utilizing animal tissues that have been rendered nonviable or nonviable products derived from animal tissues; and
- radiation-emitting equipment (i.e., x-ray equipment) covered by Directives 80/836/EURATOM** and 84/446/EURATOM.***

The MDD does not apply to personal protective equipment that is covered by the Personal Protective Equipment Directive.**** In deciding whether a product is a medical device or personal protective equipment, the intended purpose for the product as expressed in the labeling should be considered. The device would be considered a medical device if the product is intended to be used in a medical context with the aim to provide protection for the health and safety of the patient, regardless whether the product also simultaneously protects the user. Examples of protective equipment classified as medical devices or as personal protective equipment are given in Table 15.1.

Some devices are on the borderline between the MDD and the MPD. As a general rule, a relevant product is regulated either by the MDD or the MPD. Normally the procedures of both directives do not apply cumulatively (CEC, Guidelines 65/65/EEC §A.2).

The principal intended action is critical in making the determination of which directive applies. A medical device typically fulfills its intended purpose by physical means (e.g., mechanical action, physical barrier, replacement of or support to organs or body functions). A medicinal product produces its intended action by pharmacological, immunological, or metabolic means. A medical device may be assisted in its function by pharmacological, immunological, or metabolic means. However, as soon as these means cease being ancillary to the principal purpose of the product, the product is no longer a device but becomes a medicinal product (CEC, Guidelines 65/65/EEC §A.2).

---

* Council Directive 76/768/EEC of 27 July 1976 on the approximation of the laws of the Member States relating to cosmetic products. 1976. *Official Journal of the European Communities*, 19, no. L 262 (September 27).

** Council Directive 80/836/EURATOM of 15 July 1980 amending the Directives laying down the basic safety standards for the health protection of the general public and workers against the dangers of ionizing radiation. 1980. *Official Journal of the European Communities*, 23, no. L 246 (September 17).

*** Council Directive 84/466/EURATOM 3 September 1984 laying down basic measures for the radiation protection of persons undergoing medical examination or treatment. 1984. *Official Journal of the European Communities*, 27, no. L 265 (October 5).

**** Council Directive 89/686/EEC of 21 December 1989 on the approximation of the laws of the Member States relating to personal protective equipment. 1989. *Official Journal of the European Communities*, 32, no. L 399 (December 30).

TABLE 15.1
Examples of Protective-Equipment Classification

Protective equipment classified as medical devices:
- Surgical gloves, examination gloves
- Face masks
- Corrective glasses (including those intended at the same time for sun protection)
- Surgeons' gowns and hats

Protective equipment classified as personal protective equipment:
- Protective gloves (e.g., for use in a medical laboratory)
- Clothing for protection against ionizing radiation
- Sun glasses
- Eye-protection devices for professional use (e.g., for welders, regardless of whether or not they contain corrective glasses adapted to the needs of the user)
- Gum shields for boxers

Source: CEC, Definitions §3.3

The claims made about the product in the labeling will represent an important factor in the classification of the product as a medical device or a medicinal product.

In the draft guidance document on the demarcation between the MDD and the MPD, the European Commission offers examples of products classified as medical devices or medicinal products. These examples can be found in Table 15.2.

Some medical devices (e.g., infusion pumps, iontophoresis devices, nebulizers) are designed to deliver a medicinal substance. These products are regulated as medical devices. The medicinal substance(s) that the device is intended to deliver is approved separately following the normal procedures for medicinal products. However, if the device and the medicinal substance form a single integral product that is not reusable, the single product is regulated as a medicinal product (CEC, Guidelines 65/65/EEC §6.2). Examples include a prefilled syringe, a nebulizer precharged with a specific medicinal substance, and a patch for transdermal drug delivery. In such cases, the essential requirements (ERs) of the MDD apply as far as the device elements (i.e., safety and mechanical features), but the labeling should comply with the requirements applicable to the medicinal substance.

A medical device may incorporate substances as an integral part that, if used separately, may be considered medicinal substances. Such products are classified as devices as long at the action of the medicinal substance is ancillary to that of the device. A catheter coated with heparin or a condom coated with a spermicide are examples of devices incorporating a medicinal substance. However, merely coating a product with a chemical does not imply that the chemical is a medicinal substance. Coatings that are in use and that are not medicinal substances are hydromers and phosphorylcholines (CEC, Guidelines 65/65/EEC §A.5).

The MDD defines an accessory as "an article which whilst not being a device is intended specifically by its manufacturer to be used together with a device to enable it to be used in accordance with the use of the device intended by the manufacturer of the device" (Directive 93/42/EEC §1(2(b))). Because accessories are regulated as devices, it does not matter if an article is a device or an accessory. However, it does matter whether the article is an accessory or a general-purpose article. A screwdriver used to install X-ray equipment is a general-purpose device and is not regulated. A sterilizable screwdriver designed for use with bone screws is an accessory that is subject to the requirements of the directive (Higson p. 11).

Software that is intended to control the function of a medical device is regulated as a medical device.

The MDD is a "specific directive" with regard to the Electromagnetic Compatibility Directive (EMCD). As a specific directive, the MDD covers all aspects related to electromagnetic compatibility

TABLE 15.2
Examples of Medical Devices/Medicinal Products

Examples of products classified as medical devices:
- Bone cement
- Dental filling materials
- Tissue adhesives
- Resorbable osteosynthesis materials (e.g., polyglycolic acid)
- Sutures, absorbable sutures
- Hard-tissue scaffolds and fillers (e.g., collagen, calcium phosphate, bioglas)
- Intrauterine devices (e.g., copper, silver)
- Blood bags

Examples of products classified as medical devices because they are accessories to medical devices:
- Contact-lens care products (disinfecting, cleaning, rinsing, and hydrating solutions)
- Disinfectants specifically intended for use with medical devices (e.g., endoscopes) [a]
- Lubricants specifically intended for use together with medical devices (e.g., for gloves, endoscopes, condoms)
- Skin-barrier powders and pastes or other skin-care products specifically intended for use together with ostomy bags

Examples of medicinal products:
- Injectable X-ray contrast media, nuclear magnetic resonance enhancing agents
- Water for injection and intravenous fluids
- Anesthetic gases
- Artificial tears
- Topical disinfectants
- Solutions for peritoneal dialysis and hemodialysis
- Agents for preservation of organs intended for transplantation

Examples of medical devices incorporating a medicinal substance with ancillary action:
- Catheters coated [b] with heparin or an antibiotic agent
- Bone cement containing antibiotic
- Blood bags containing anticoagulant
- Hemostatic devices enhanced by incorporation of collagen
- Condoms coated with spermicides
- Electrodes with steroid-coated tip
- Wound dressings, surgical, or barrier drapes with antimicrobial agent

[a] Multipurpose disinfectants or sterilization agents are not covered by the MDD; they will be covered in the future by the forthcoming directive on biocides.

[b] Merely coating a product with a chemical does not imply that the chemical is a medicinal substance. For example, hydroxyapatite, frequently used as coating for orthopedic and dental implants, is not considered a medicinal substance.

Source: CEC, Demarcation §§A.3, A.4, and A.5

(EMC) (immunity and electromagnetic interference [EMI]) of medical devices. However, in 1999, the Radio Equipment and Telecommunications Terminal Equipment Directive (R&TTED) came into force. The R&TTED applies to any equipment that is capable of communicating by means of the emission or reception of radio waves utilizing the spectrum allocated to terrestrial/space radio communications or is intended to connect directly or indirectly by any means to interfaces of public telecommunications networks. Medical devices that also fall within the scope of the R&TTED are now subject to the protection requirements with respect to EMC contained in the EMCD.

DEVICE CLASSIFICATION

Because the MDD covers such a broad range of devices, the Commission determined that the same conformity-assessment procedure would be inappropriate for all of the devices covered by the directive. It would not be feasible, economical, nor justifiable to subject all covered devices to the most rigorous conformity-assessment procedures available. A graduated system of controls is more appropriate. A graduated system provides a level of control that corresponds to the level of potential hazard inherent in the type of device concerned (CEC, Classification p. 3).

The MDD introduced a classification system that places all covered devices into one of four classes—Class I, IIa, IIb, and III. The classification of a device determines the basic process that the manufacturer must follow to demonstrate that the device complies with all of the applicable ERs and that the manufacturer has taken all necessary steps to ensure that the manufacturing processes produce devices that conform to the design documentation. For many Class I devices the manufacturer may self-certify conformity with the MDD. Class III devices require an examination of technical documentation (i.e., the design dossier) and quality-system certification by one or more Notified Bodies. Class IIa and IIb devices fall between the two extremes.

Regardless of the extent of the conformity-assessment process that is required for a particular device class, it is important to remember that every device must fully conform to all of the applicable ERs in the directive. This includes the labeling requirements in the directive, as well as any labeling requirements in standards used to demonstrate compliance with the ERs. There is one notable exception where the extent to which an ER is applicable is dependent on the device classification. This exception has to do with the instructions for use (IFU) of the device and is described later in this chapter in the section, "Instructions For Use."

MISBRANDING

The information provided by the manufacturer that accompanies a medical device is covered by the ERs of the MDD. Annex I of the MDD includes specific minimal requirements for information that must be on the label and the IFU. These requirements include (Directive 93/42/EEC Annex I(13)):

- general labeling provisions,
- devices with a measuring function,
- particulars to be present on the label,
- IFU, and
- patient information.

Failure to follow or satisfy these requirements may lead to the device being deemed improperly labeled (i.e., misbranded). Misbranding would be grounds for a Notified Body to refuse to give the manufacturer authorization to affix the CE conformity marking. Discovery of improper labeling of a device that bears the CE conformity marking would be a reportable incident under the vigilance system (CEC, Vigilance §5.5.3). In the extreme case, the failure to properly label the device could cause a competent authority to activate the Safeguard Clause, leading to the removal of the device from the market. The following sections discuss each of these five areas in detail.

As an aid to cross-referencing to the MDD, the number of the ER in Annex I (e.g., §12.9) is listed with the corresponding topic.

GENERAL LABELING PROVISIONS

The MDD requires that each device must be accompanied by the information needed to use it safely and to identify the manufacturer. In determining the information required, the manufacturer must take into account the training and knowledge of the potential users. As far as practical and appropriate, the information needed to use the device safely should be placed on the device itself and/or on the packaging of each unit or, where appropriate, on the sales packaging. If individual packaging of each device is impractical, then the information must be set out in the documentation supplied with one or more devices.

The MDD encourages the use of symbols to convey the required information. Symbols simplify labeling and reduce the need for multiple translations of words into national languages. When a manufacturer wishes to use symbols and colors to convey the required information, it must use the symbols and identification colors defined in the harmonized standards. For example, a manufacturer who wishes to use symbols to identify the "on" and "off" positions of a power switch should use Symbols 9 and 10 in Appendix C.

A European standard (EN 980) has been drafted to reduce the need for multiple translations of words into national languages, to simplify labeling where possible, and to prevent the separate development of different symbols to convey the same information. These symbols are shown in Table 15.3. This standard was drafted under a mandate from the European Commission, and is a harmonized standard under the MDD, AIMDD, and the IVDD. Colors and minimum dimensions are not specified in the standard. However, all symbols must be legible when viewed under an

TABLE 15.3
Graphical Symbols for Use in Labeling Medical Devices

No.	Symbol [a]	CEN Publication	Description
1	⊗	980 §4.1	Do not reuse
2	⧗	980 §4.2	Use by date [b]
3	LOT	980 §4.3	Batch code
4	SN	980 §4.4	Serial number
5	⋏⋏	980 §4.5	Date of manufacture [c]
6	STERILE	980 §4.6	Sterile
7	STERILE EO	980 §4.7.1	Method of sterilization using ethylene oxide
8	STERILE R	980 §4.7.2	Method of sterilization using irradiation
9	STERILE	980 §4.7.3	Method of sterilization using steam or dry heat
10	REF	980 §4.8	Catalog number
11	⚠	980 §4.9	Attention, see IFU
12	STERILE A	980 §4.10	Method of sterilization using aseptic technique
13	▪	980 §5.2	Manufacturer
14	EC REP	980 §5.3	Authorized representative in the European Community
15	∑	980 §5.4	Contains sufficient for <n> tests

[a] The symbols in this table are for convenience of reference only and have no official sanction. The reader is referred to CEN EN 980.
[b] This symbol is accompanied by a date expressed as four digits for the year, two digits for the month, and, when appropriate, two digits for the day.
[c] This symbol is accompanied by a date expressed as four digits for the year and two digits for the month.

Source: EN 980

TABLE 15.3 (CONTINUED)
Graphical Symbols for Use in Labeling Medical Devices

No.	Symbol	CEN Publication	Description
16		980 §5.5	For IVD performance evaluation only
17	IVD	980 §5.6	In vitro diagnostic medical device
18		980 §5.7.1	Upper limit of temperature
19		980 §5.7.2	Lower limit of temperature
20		980 §5.7.3	Temperature limitation
21		980 §5.8	Consult IFU
22		980 §5.9	Biological risks

a) The symbols in this table are for convenience of reference only and have no official sanction. The reader is referred to CEN EN 980.
Source: EN 980

illumination of 215 lux using normal vision, corrected if necessary, at a distance that takes into account the specifics and size of the individual device (EN 980 §3).

If a needed symbol is not defined in a harmonized standard, then the manufacturer is free to use symbols defined in other recognized sources such as the International Organization for Standardization (ISO) publication 15223 or the International Electrotechnical Commission (IEC) publication 60878. If necessary, the manufacturer may develop special symbols, remembering that, to be effective, a symbol should be clear when viewed in context of the device. If a symbol or color is not defined in a harmonized standard, it must be described in the documentation accompanying the device. Unless a symbol that appears in a harmonized standard is very well known through long use, it is a good idea to explain the symbol in the accompanying documentation.

CONTROLS AND DISPLAYS (§12.9)

The function of the user-adjustable controls and visual displays must be clearly specified on the device and/or in the IFU (EN 1041 p. 9).

INTENDED PURPOSE (§13.4)

The MDD requires that the manufacturer must state the intended purpose of the device clearly on the label and in the IFU unless the intended purpose is obvious to the user, taking into account the potential user's training and knowledge.

The formulation of the statement of intended purpose(s) for a device is important for two reasons. First, ER number 3 requires that the manufacturer must be able to demonstrate, with clinical data if necessary, that the device achieves the performances claimed in the labeling. Secondly, the statement of intended purpose will have an effect on the classification of the device and, hence, the conformity-assessment process that must be followed before the manufacturer can affix the CE conformity marking. Phrases that are used in the classification rules in Annex IX of the MDD should be included, usually in the negative sense (Higson p. 118). For example:

NOT INTENDED FOR CONTINUOUS USE FOR MORE THAN 30 DAYS.

NOT FOR INTERNAL USE.

NOT FOR USE ON THE CENTRAL CIRCULATORY SYSTEM.

DEVICES WITH A MEASURING FUNCTION

The MDD places specific requirements on devices that incorporating a measuring function. The MDD requires these devices to be designed and manufactured in such a way that the measuring function provides sufficient accurate and stability to meet the intended purpose of the device. The manufacturer must state the limits of accuracy provided by the device.

MEASUREMENT, MONITORING, AND DISPLAY SCALES (§10.2)

The measurement, monitoring, or display scales must be designed and manufactured in line with accepted ergonomic principles, taking into account the intended purpose of the device.

UNITS OF MEASURE (§10.3)

Devices that incorporate a measuring function must express the result of the measurement in legal units conforming to the provisions of Council Directive 80/181/EEC. The basic units are those of the International System of Units (SI), which is described in ISO 1000 (IEC, A2 p. 17). Units outside the SI that can be used on medical equipment are listed in Table 15.4.

PARTICULARS ON THE LABEL

ER 13.3 of the MDD specifies the minimum information that must be included on the label of a covered medical device. Although the requirements described in the following sections appear to be comprehensive, closer examination will reveal that they have been constructed around sterile medical devices. Manufacturers of other devices need to pay attention to the requirements of key product standards such as EN 60601-1 (IEC 60601-1) for medical electrical equipment.

TABLE 15.4
Units Outside the International System that Can be Used on Equipment

Plane Angle Units:
- revolution
- grade
- degree
- minute of angle
- second of angle

Time Units:
- minute
- hour
- day

Energy Units:
- electron volt
- pressure of Blood and Other Body Fluids
- millimeters of mercury

Source: IEC, A2 pp. 12–13

MANUFACTURER IDENTIFICATION (§13.3(A))

The label must bear the name or trade name of the manufacturer. A well-known trademark or logo may be sufficient to specify the manufacturer. The label must also bear the address of the manufacturer's registered place of business. The full postal address may not be required if the address is of sufficient detail to enable the manufacturer to be contacted (e.g., postcode and country) (EN 1041 p. 10).

If the manufacturer's registered place of business in not in the EU, the label, the outer packaging, or the instructions for use must also bear the name and address of the manufacturer's authorized representative in the European Union (EU) (EN 1041 p. 10).

If neither the manufacturer nor the manufacturer's authorized representative is based in the EU, the importer is deemed responsible under the terms of the directive for placing the imported product on the EU market. In this case, the label, the outer packaging, or the IFU also must bear the name and address of the importer.

IDENTITY OF THE DEVICE (§13.3(B))

The label must provide sufficient detail to allow the intended user to identify the device. For many medical devices, the identity will be obvious to the intended user from inspection of the device itself. Unpacked devices or those provided only with a storage or transport package may not require further identification. Similarly, transparent packaging may reduce the need for a detailed description on the label by allowing the intended user to see the device while it remains in the package. For more complex devices, the identity can be indicated on the device itself, on the packaging, or in the accompanying documentation.

It may be appropriate to list the contents and give the quantity of each item in the package. This would be particularly important if there are accessories that are required for use of the device that are not included in the package. If it is not practical to list the contents on the package label (i.e., because of space limitations), the information should be given in the accompanying documentation.

STERILE DEVICE MARKING (§§13.3(C), 13.3(M), AND 8.7)

If a device, or a part of a device, is provided in a sterile condition, the label must bear the word "STERILE." To avoid the need for translation, the word sterile has been turned into a symbol (see Symbol 6 in Table 15.3).

The sterile symbol should be prominently displayed on the label. If only part of the device is sterile, this should be stated on the label (e.g., "sterile fluid path") (EN 1041 p. 11).

When applicable, the label must bear the method of sterilization. Symbols have been developed that combine the method of sterilization with the word "STERILE" (see Symbols 7, 8, and 9 in Table 15.3). In this way, the single symbol satisfies the requirements in both ERs 13.3(c) and 13.3(m). In determining when the particulars of the method of original sterilization are required on the label, a prime consideration is the need for those storing, handling, and using the device to know the method of sterilization (EUCOMED p. 5).

A product that is labeled sterile in Europe must conform to the definition in Subclause 3.4 of EN 556-1:2001. To be considered sterile, a product must be put through a process that achieves a Sterility Assurance Level (SAL) of 10^{-6} (EN 556-1 §4.1).

Occasionally, a device may be available from a manufacturer in both sterile and nonsterile configurations. If the packaging of the two configurations is similar, the user may be confused and mistakenly treat a nonsterile device as if the manufacturer had sterilized it. To avoid confusion, the manufacturer should provide a prominent warning of the nonsterile nature of the device (EN 1041 p. 9).

Product Identification (§§13.3(d) and 13.5)

The MDD requires that, whenever reasonable and practical, devices and their detachable components must be identified to allow action to be taken in case of any potential risk. Such identification will enable the manufacturer to recall the device and/or its detachable components, if necessary. To satisfy this requirement, the device should bear a batch or serial number. In addition, detachable components that are intended by the manufacturer to be used separately from the original device are to be identified with a batch or serial number (EN 1041 §4.1.8).

To distinguish the identification code on the label, the batch code is to be preceded by the word "LOT." To avoid the need for translation, the word lot has been turned into a symbol (see Symbol 3 in Table 15.3). A serial number may be identified using the serial-number symbol (see Symbol 4 in Table 15.3). In either case, the use of the symbol precludes the need for translation.

Expiration Dating (§13.3(e))

Typically, the expiration date reflects either a time-related deterioration in safety, or the extent of the time period for which the manufacturer accepts responsibility for safe use. When a device, for whatever reason, has an expiration date beyond which it may not be used in complete safety, this "use by" date must appear on the label. The expiration date may be identified using the "use by" symbol (see Symbol 2 in Table 15.3).

The MDD requires that the expiration date must be expressed as a four-digit year and a two-digit month (i.e., yyyy-mm). Some devices require more precision in specifying the expiration date. For these devices, a two-digit day may be added (i.e., yyyy-mm-dd).

Year of Manufacture (§13.3(l))

If the device does not require an expiration date, the label must include the four-digit year of manufacture (i.e., yyyy). The year of manufacture may be included in the lot or serial number (e.g., 19940001234). If the year of manufacture is included as a separate piece of information on the label, it should be identified with the appropriate text or the "date of manufacture" symbol (see symbol 5 in Table 15.3).

Single-Use Devices (§13.3(f))

When a device is intended by its manufacturer to be used only once, its label must bear an indication that the device is for single use. Phrases such as "for single use only," "do not reuse," and "use only once" on the label are sufficient to convey this information. To avoid translation, the manufacturer may use the "do not reuse" symbol (see Symbol 1 in Table 15.3) to satisfy this requirement.

Storage and Handling Conditions (§13.3(i))

If there are special storage, handling, and transport conditions that are critical for the safe and proper functioning of the device, these must be provided on the outermost label on the device (i.e., on the shipping package). Otherwise, the manufacturer must prepare the device to survive the storage and handling conditions normally expected by the user for devices of the type in question. It would be generally understood that the devices should be protected from extremes of temperature, weather, and electromagnetic radiation. If, however, the device needs to be maintained within a particular range of temperature and relative humidity, this information should be indicated on the outer label (EUCOMED p. 3).

Special Operating Instructions (§13.3(j))

The manufacturer must take into account the technical and clinical knowledge and the skill of the intended user when determining the type and extent of the operating instructions that must be included on the label for the safe use of the device. In many cases, operating instructions are required on the label only if the mode of operation of the device is novel or unfamiliar and would not be self-evident to the intended user (EN 1041 p. 12).

Warnings and Precautions (§13.3(k))

It is impractical to provide particulars on all warnings and/or precautions that should be in effect with a particular device. In addition, the larger the amount of text, the greater the risk that important information will be missed in the volume of words. The manufacturer should focus on those warnings and/or precautions associated with novel or unfamiliar features that would not be self-evident to the intended user (EUCOMED p. 4).

Markings for Special-Purpose Devices (§§13.3(g) and 13.3(h))

For the special markings on investigational and custom-made devices, see the appropriate sections in Chapter 14.

Devices Incorporating Stable Human Blood Derivatives (§13.3(n))

A device that incorporates as an integral part a substance that, if used separately, might be considered a medicinal product or a constituent of a medicinal product that is derived from human blood or human plasma that acts upon the human body with an action ancillary to that of the device is subject to additional approval and labeling requirements (Directive 2000/70/EC §2(a)). In addition to meeting the requirements of Directive 93/42/EEC, such a device must be assessed and authorized using the procedure in Directive 2000/70/EC.

When a device incorporates as an integral part a human blood derivative, the Notified Body must seek a scientific opinion from the European Agency for the Evaluation of Medicinal Products (EMEA) on the quality and safety of the derivative. In particular, the provisions of Directives 75/318/EEC* and 89/381/EEC** are to be considered. The usefulness of the derivative as a part of the medical device must be verified, taking account of the intended purpose of the device. In accordance with Article 4(3) of Directive 89/381/EEC, a state laboratory designated by a member state shall test a sample from each batch of bulk and/or finished product of the human blood derivative.

The label of such a device must bear an indication that the device contains a human blood derivative.

INSTRUCTIONS FOR USE (IFU)

The MDD requires that IFU must be on the device itself and/or on the packaging for each unit. When appropriate, the information needed to use the device safely may be placed on the sales package. When individual packaging of each unit is not practical, the IFU must be set out on a leaflet supplied with one or more devices.

---

* Council Directive 75/318/EEC of 20 May 1975 on the approximation of the laws of member states relating to analytical, pharmaco-toxicological and clinical standards and protocols in respect of the testing of proprietary medicinal products. 1975. *Official Journal of the European Communities*, 18, no. L 147 (June 9).

** Council Directive 89/381/EEC of 14 June 1989 extending the scope of Directives 65/65/EEC and 75/319/EEC on the approximation of provisions laid down by law, regulation or administrative action relating to proprietary medicinal products and laying down special provisions for medicinal products derived from human blood or human plasma. 1989. *Official Journal of the European Communities*, 32, no. L 181 (June 28).

The IFU must be included in the packaging for every device. An exception to this requirement has been established for devices that fall into Class I or IIa. As mentioned earlier, this is one point where the classification of a device affects the application of the ERs. For devices that fall into Class I or IIa, no IFU are needed if these devices can be used safely without such instructions. These would be devices where the user, because of special training and experience, would be familiar with the proper use of the device.

ER 13.6 of the MDD specifies the minimum information that, when appropriate, must be included in the IFU for a covered medical device.

PARTICULARS FROM THE LABEL (§13.6(A))

The IFU must contain all of the applicable information required on the label of the device except:

- the lot or serial number of the device and
- the expiration date or year of manufacture.

PERFORMANCE INTENDED BY THE MANUFACTURER (§13.6(B))

The IFU must set out the performance claimed by the manufacturer. Care should be taken to describe the performance of the device in clear and objective terms. ER number 3 requires that the manufacturer must be able to verify any performance claims.

The manufacturer must warn the user of the undesirable side effects reasonably associated with the use of the device. Undesirable side effects should be listed in descending order based on their clinical significance as determined by their severity and frequency. The labeling should provide frequency data from adequately reported clinical studies when the data are not well known to the intended user and/or when this information is needed in deciding between the use of the device and an alternative procedure or approach.

CONNECTION TO OTHER MEDICAL DEVICES (§§9.1 AND 13.6(C))

If a device must be installed with or connected to other medical devices or equipment in order to operate as required for its intended purpose, the IFU must describe the characteristics of the device in sufficient detail so the user can obtain a safe combination. The extent of the information provided should take into account the training and experience of the anticipated users. Material must be provided if establishing a safe connection is not self-evident or requires knowledge that is outside the normal and expected training and experience of the user. Sufficient detail about the characteristics (e.g., connections) can be provided by referencing relevant published standards that provide the characteristics (EN 1041 p. 13).

INSTALLATION AND MAINTENANCE (§13.6(D))

The IFU should provide information that allows the user to verify that equipment has been properly installed and can be operated properly and safely. The IFU do not have to explain all of the steps necessary for proper installation. Similarly, the IFU must alert the user to the nature and frequency of maintenance and calibration required to ensure that the device operates properly. The step-by-step procedures for maintaining and calibrating the device do not need to be included in the IFU. If installation and/or maintenance of the device is not performed by the manufacturer or the manufacturer's agent, then required installation and maintenance procedures should be available in a separate document (i.e., a service manual) (EUCOMED p. 6).

Reciprocal Interference (§13.6(f))

Medical devices may interact with one another or with other medical treatments in ways that harm the patient, damage the device, or render one or the other of the treatments ineffective. The IFU must identify the risks of reciprocal interference. An example is damage to an external pacemaker caused by defibrillation of a patient. Another example would be an adverse reaction caused by use of certain medications in a patient who is connected to a heart-lung machine that incorporates anticoagulant coatings in the blood circuit.

Sterile Packaging (§13.6(g))

If the device is provided in a sterile condition, the IFU must include instructions to the user in the event of damage to the sterile package. These instructions could range from returning the device to the manufacturer to procedures for resterilizing the device. If resterilization is recommended, then the IFU must provide the details of the appropriate methods of sterilization. The IFU should also contain warnings about inappropriate sterilization (e.g., "Do not autoclave").

Reusable Devices (§13.6(h))

If the manufacturer intends that the device is reusable, then the IFU must describe the appropriate processes to allow the device to be prepared for reuse. This would include cleaning, disinfecting, and packaging of the device, and, if appropriate, details on the method of resterilizing it. The instructions should clearly state any restrictions on the number of reuses.

Device Preparation (§13.6(i))

Some devices may require that the user alter the characteristics of the device (e.g., final assembly, sterilization) before use. In such cases, the IFU must provide sufficient detail to enable the intended user to prepare the device. If the device is to be sterilized before use, the instructions for cleaning and sterilization must be such that, if they are correctly followed, the device will be suitable for the intended purpose.

The IFU does not have to contain instructions for handling that are implicit in normal use by the intended user. For example, it is not necessary to recommend to a physician that a sterile device be removed aseptically from the packaging.

Radiation-Emitting Devices (§§11.4.1 and 13.6(j))

Devices that emit radiation (ionizing, laser, infrared, etc.) for medical purposes must include in the IFU a detailed description of the nature, type, intensity, and distribution of the radiation. In addition to providing information on the nature of the radiation, the IFU must describe means for protecting the patient and the user, ways to avoid mishandling, and the risks inherent in installation.

Implantable Devices (§13.6(e))

If the device is implantable, the IFU must include any appropriate operating instructions, warnings, and precautions to avoid risks uniquely associated with the implantation of the device. These risks could include such things as body rejection phenomena, erosion through the skin, embolism, and infections. This category should include those risks that are recognizable and foreseeable as opposed to those that are unknown and/or improbable (EUCOMED p. 7).

PATIENT INFORMATION

The information described in the previous sections is directed at the user of the device. It is primarily intended to enable the device to be used safely and for the purpose intended. In contrast, the information in this section is intended to enable qualified medical staff to brief the patient on contraindications or precautions that the patient needs to take into account. Because a qualified professional will interpret the information, it need not be written in a manner that can be read and understood by the patient. If the information is not needed to brief the patient, or is obvious, it need not be included (EUCOMED p. 8).

When appropriate, the IFU should contain information allowing the medical staff to brief the patient on the topics described in the following sections.

CHANGES IN PERFORMANCE (§13.6(K))

The IFU should explain the precautions that the patient should take in the event of changes in the performance of the device. This could include instructions on what signals to look for that would indicate a change in performance, and what action (i.e., contact the healthcare provider immediately) is appropriate if any signals are observed.

EXPOSURE TO ENVIRONMENTAL CONDITIONS (§13.6(L))

Some devices may be affected by exposure to reasonably foreseeable environmental conditions. The MDD lists magnetic fields, external electrical influences, electrostatic discharge, pressure or variations in pressure, acceleration, and thermal ignition sources as examples. If there are reasonably foreseeable environmental conditions that the patient should avoid, the IFU should provide the necessary briefing material to the medical staff. For example, if the device is susceptible to interference from high-powered electromagnetic fields, the instructions leaflet might tell the medical staff to instruct the patient to avoid work environments where arc welders, induction furnaces, resistance welders, radio, or microwave-frequency transmitters are in use.

ADMINISTRATION OF MEDICINAL PRODUCTS (§13.6(M))

The IFU must provide adequate information on the medicinal product or products that the device is designed to administer. This includes any restrictions in the choice of substances to be delivered by the device.

DISPOSAL OF THE DEVICE (§13.6(N))

If there are unusual risks or special precautions that the patient should take when disposing of the device, these need to be coveresd in the IFU. For example, the IFU might remind the medical staff to instruct the patient to dispose of a device incorporating a needle in a container designed for sharp objects.

MEDICINAL SUBSTANCES INCORPORATED INTO THE DEVICE (§13.6(O))

If a medicinal substance is incorporated as an integral part of a device, the IFU must enable the medical staff to brief the patient on any precautions to be taken. These precautions could include avoiding ingesting certain substances (e.g., foods, alcohol, drugs) while using the device.

MEASURING ACCURACY (§13.6(P))

For devices that incorporate a measuring function, the IFU must provide any information that should be communicated to the patient regarding the accuracy claimed by the manufacturer.

SPECIAL LABELING REQUIREMENTS FOR SPECIFIC DEVICES

Many medical devices have specific labeling requirements that go beyond the minimums described in the MDD. For specific devices, the manufacturer should pay careful attention to the content of the device standards used to demonstrate compliance with the ERs of the directive. However, the existence of multiple standards may lead to conflicting or ambiguous requirements—including labeling requirements. When this happens, the manufacturer must avoid conflicting, contradictory, or ambiguous labeling. When resolving these inconsistencies, the following approach is recommended:

- The labeling must satisfy all of the applicable requirements in Annex I of the MDD. These requirements specify the minimum content of the labeling and must be met.
- Labeling or marking requirements specific to any standards that the manufacturer uses to demonstrate compliance with the ERs of the directive are added.
- To these requirements, the manufacturer must add the information that he or she considers necessary for the safe use of the device.
- If any of these sources introduce conflicting labeling requirements in either content or presentation, these should be resolved as follows:
  - Requirements of the directive override all others;
  - Requirements in harmonized standards override requirements in non-harmonized standards; and
  - Specific standards override general standards.

A key standard that manufacturers of active medical devices should take into account in this regard is EN 60601-1 on medical electrical equipment. Medical electrical equipment is the category that covers most electrical instruments and apparatus used in the patient environment, and those electrically powered devices having a direct effect on the safety of the patient. They are medical devices that make physical or electrical contact with the patient, and/or transfer energy to or from the patient, and/or detect such energy transfer. Implantable devices, such as cardiac pacemakers, are not considered medical electrical devices and are covered by the AIMDD.

EN 60601-1 is essentially identical with the second edition of the IEC Standard 60601-1. In 1995, the IEC amended IEC 60601-1. In parallel, this amendment was adopted by the European Committee for Electrotechnical Standardization (CENELEC) as an amendment to EN 60601-1. One of the purposes of this amendment is to fully align EN 60601-1 with the ERs of the MDD. The requirements described in the remainder of this section were extracted from the amended Clause 6 of EN 60601-1.

MARKINGS ON MEDICAL ELECTRICAL EQUIPMENT

All required markings must be clearly legible and removable only with a tool or by application of considerable force. If the equipment is permanently installed, then all required markings must be visible when the equipment is mounted in its normal use position. For transportable equipment or stationary equipment that is not permanently installed, it is preferred that required markings are visible during normal use. However, when necessary, the required marking must become visible when the equipment is moved or has been removed from the rack in the case of dismountable rack units. However, warning statements and operating instructions must be affixed in a prominent location so they are legible, with normal vision, from the operator's position.

Markings required on the outside of the equipment depend largely on the type of power source. EN 60601-1 defines three configurations. They are:

- Equipment that receives its energy from the supply mains in the facility where it is used (mains powered).
- Equipment that is able to operate from an internal electrical power source (internally powered).
- Equipment that is supplied by another specified power source (other than supply mains). Such equipment must be isolated from the supply mains.

The markings required on the outside of the equipment are described in Table 15.5. For permanently installed equipment, the nominal supply voltage or voltage range may be marked on either the outside or inside of the equipment enclosure. The preferred location of the markings is adjacent to the supply-connection terminals.

Medical electrical equipment must have on its external surface a permanent label that sets out the equipment type as B, BF, or CF. Each equipment type has associated with it a maximum risk current that can flow through any person upon contact with the equipment. A device may not exceed the maximum allowable risk current specified in the risk-current limit in Table IV in Subclause 19.3 of EN 60601-1 for the labeled-equipment type of the device. If the equipment has more than one part in contact with the patient, and the parts in contact with the patient have different degrees of protection, then the appropriate symbols must be clearly marked on each applied part, or on or near the relevant outlets.

Special markings are required on equipment that has been protected against the effects of the discharge of a cardiac defibrillator. These markings are described in Table 15.5.

The packaging of equipment or accessories that are provided in a sterile form must be marked with the word "STERILE." As an alternative, the sterile symbol may be used (see Symbol 6 in Table 15.3).

When the size of the equipment or the configuration of the enclosure makes it impractical for all of the required information in Table 15.5 to be affixed to the outside of the equipment, then at least the information in items 1, 2, 3, 7, and, if applicable, 12 should be included on the outside of the enclosure. The remaining items may then be included in the accompanying documentation. If it is impractical to include any markings on the outside of the enclosures, then all of the required information may be placed in the accompanying documentation.

EN 60601-1 specifies requirements for marking controls and instruments on medical electrical equipment, as well as for marking on the inside of the equipment. These markings on the controls must include the following information, as applicable:

- The main power switch must be clearly identified with its "ON" and "OFF" positions indicated by a symbol, adjacent indicator light, or other unambiguous means (see Appendix C, Symbols 9 and 10). Indicator lights and push buttons should follow the conventions listed in Table 15.6.
- EN 60601-1 requires that all operator controls and indicators must be clearly identified.
- The different positions of control devices and switches must be indicated by figures, letters, or other visual means (see Appendix C, Symbols 26 and 27). If the control is intended to adjust a device setting during normal operation, and changing the setting could cause a safety hazard to the patient, the control must be equipped with:
  - an associated indicating device (e.g., instruments or scale) or

TABLE 15.5
Marking on the Outside of Medical Electrical Equipment

Items to be marked	Mains Operated Equipment	Internally Powered Equipment [a]	Equipment Supplied from a Specified Power Source [b]
1. Name and/or trademark and address of the manufacturer or supplier claiming that the equipment complies with EN 60601-1	*	*	*
2. Model or type identification (i.e., a combination of figures, letters, or both used to identify the equipment)	*	*	*
3. Mains-supply characteristics (rated voltage/voltage range, current waveform—a.c., d.c., or dual supply, number of phases, etc.)	***		
4. Supply frequency (Hz)	***		
5. Rated power input	***		
6. Auxiliary mains socket output	**		
7. Equipment classification [c]	**	**	**
8. Defibrillation protection [d]	**	**	**
9. Mode of operation [e]	**	**	**
10. Fuses	**	**	**
11. Output	**	**	**
12. Physiological effects [f]	**	**	**
13. Category AP/APG equipment (see Appendix C, Symbols 23 and 24)	**	**	**
14. High-voltage terminals accessible without a tool (see Appendix C, Symbol 25)	**	**	**
15. Cooling requirements	**	**	**
16. Mechanical-stability marking if special precautions are required	**	**	**
17. Protective packaging, if required during transport and storage	**	**	**
18. Earth-terminals markings [g]	**	**	**
19. Removable protective means	**	**	**
20. Marking required for medical electrical systems [h]	**	**	**
21. Other legally required markings	**	**	**

* Marking is required on all equipment.

** Marking is required if applicable to the equipment.

*** Marking is not required on permanently installed equipment that is marked appropriately on the inside.

[a] Internally powered equipment is capable of operating from an internal electrical power source.

[b] The equipment is intended to be supplied from a power source other than supply mains.

[c] Classification includes the degree of protection against electrical shock (see Appendix C, symbol 11), the protection against leakage current (see Appendix C, Symbols 17, 18, and 19), and the harmful ingress of liquids using the IP Code defined in IEC 60529.

[d] Amendment 2 introduced new symbols (see Appendix C, Symbols 20, 21, and 22) for defibrillation-proof equipment. For clear differentiation, Symbols 17 and 20 must not be applied in such a way as to give the impression of being inscribed in a square. If protection against defibrillation is partly in the patient cable, then a symbol instructing the operator to consult the accompanying documentation must be displayed near the relevant outlet (see Appendix C, Symbol 14).

[e] If equipment is unmarked, continuous operation is assumed.

[f] Equipment producing physiological effects that cause danger to the patient and/or operator must prominently display a symbol instructing the patient/operator to consult the accompanying documentations for warnings (see Appendix C, Symbol 14). If the device produces nonionizing radiation such as high-power microwaves, then Symbol 43 in Appendix C must be displayed.

[g] Earth terminal markings include the markings for connection to a potential equalized conductor (see Appendix C, Symbol 8), a functional earth terminal (see Appendix C, Symbol 5), and a protective earth conductor (see Appendix C, Symbol 6).

[h] If the documentation accompanying the medical electrical system gives a warning related to a particular safety hazard from a nonmedical equipment component of the system, then a symbol instructing the patient/operator to consult the accompanying documentations for warnings (see Appendix C, Symbol 14) must be provided on the particular nonmedical electrical equipment or on a particular part of that equipment.

Sources: EN 60601-1; EN 60601-1-1

TABLE 15.6
Colors for Indicator Lights[a] and Push Buttons[b] and their Recommended Meanings

Color	Meaning	Required Usage	Recommended Usage
Red	Warning of danger and/or the need for urgent action	X	
Yellow	Caution or attention required		X
Green	Ready for use		X
Any other color	A meaning other than that associated with red or yellow		X

[a] Dot matrix or other alphanumeric displays are not considered to be indicator lights.

[b] An unilluminated push button may be colored red only if it is used to interrupt the function of the equipment in the event of an emergency.

Source: EN 60601-1

- an indication of the direction in which the magnitude of the function changes.
- Amendment 2 requires that numeric indications must be expressed in SI units. S.I. units are described in ISO 1000. Units outside the International System that can be used on medical equipment are listed in Table 15.4.

A label near the point where the power-supply conductors connect to the equipment must show the correct method of connecting the supply connector, unless no hazard would occur if the connections were interchanged. If there is insufficient room to appropriately label the connections on the equipment, the instructions can be included in the accompanying documentation.

In permanently mounted equipment, the terminal for connection to the neutral-supply conductor must be indicated by a capital letter "N." The protective earth terminal must be marked (see Appendix C, Symbol 6) adjacent to the terminal unless the protective earth terminal is part of a detachable power cord.

For permanently connected equipment having a terminal box that reaches a temperature of more than 75°C when tested according to the normal temperature test specified in EN 60601-1, the following statement must be placed near the point of connection (IEC 60601-1 pp. 54–55):

FOR SUPPLY CONNECTION, USE WIRING MATERIALS SUITABLE FOR AT LEAST __°C.

The labels identifying the correct method of connecting the supply, the supply-neutral terminals, the protective earth terminal, and instructions about parts that reach high temperatures may not be affixed to any part that must be removed in order to make the supply connections. In addition, the labels must remain visible after the connections are made.

The presence of parts that carry more than 1,000 V a.c., 1,500 V d.c., or 1,500 V peak value must be clearly marked (see Appendix C, Symbol 25).

The functional earth terminal must be marked with the prescribed symbol (see Appendix C, Symbol 5).

For those devices that contain heating elements or heating lamps, the maximum power loading must be permanently and indelibly marked near the heating element or lamp holder.

For devices with internal fuses accessible only with the aid of a tool, the fuse ratings and characteristics must be marked near the fuse. Information on fuses may also appear in the instruction manual accompanying the equipment. However, a marking on the equipment (e.g., a diagram number) must establish a reference to the place in the instruction manual where information on the fuses is located.

Equipment that incorporates batteries must have the type of battery and mode of insertion marked adjacent to the battery holder. The required information must be placed in the accompanying documentation if (a) the operator is not intended to change the batteries, and (b) changing the batteries requires the aid of a tool. A mark (see Appendix C, Symbol 14) referring the operator to the accompanying documents is to be placed adjacent to the battery compartment.

The device may contain capacitors that remain charged after the equipment has been de-energized. If the capacitors or the circuit parts connected to them become accessible when an access cover is removed, the hazard must be clearly marked.

Documents Accompanying Medical Electrical Equipment

Medical electrical equipment must be accompanied by documents giving the IFU a technical description of the equipment, and an address to which the user can refer. Any markings required on the device that have not been permanently affixed to the equipment should be explained. All warning statements and explanations of warning symbols on the equipment must be provided in the accompanying documents (IEC 60601-1 pp. 59–63).

Instructions for Use (IFU)

The IFU must contain all of the information that the intended user needs to operate the equipment so that it achieves its intended purpose. This includes such things as:

- the function and intended application of the equipment;
- the function of controls, displays, and signals;
- the sequence of operations, including:
  - preparing the device for use;
  - steps to confirm the safe operation of the device including a description of alarms and indicators;
  - operation immediately before use, paying particular attention to abnormal displays, the underlying trouble, and the interlocks to prevent hazardous output;
  - operation during use, including the sequence of operations, the methods for reading displays, and necessary adjustments for maintaining normal function;
  - measures following use to prepare for the next use of the device; and
  - measures for both short- and long-term storage of the device;
- information about potential EMI or other interference between the equipment and other medical devices, together with advice on avoiding such interference;
- instructions for connecting and disconnecting detachable parts and accessories;
- replacement of materials consumed during operation of the equipment;
- a description of recognized accessories, detachable parts, and materials, if the use of other parts or accessories could degrade the minimum safety of the equipment;
- instructions for cleaning, preventive inspections, and maintenance to be performed including the frequency of such maintenance;
- identification of parts on which preventive inspections, maintenance, and calibration should be performed by others such as the manufacturer's authorized representatives; and
- the meaning of figures, symbols, warning statements, and abbreviations used on the equipment.

For equipment parts that come into contact with the patient, the IFU should contain a clear description of the methods for cleaning, disinfection, and sterilization of the device. When sterilization is required, the instructions should identify suitable sterilization agents and list temperature, pressure, humidity, and time limits that the equipment can tolerate.

A device may be connected to other equipment for the purpose of sending or receiving signals to or from other equipment (e.g., for display, recording, or data processing). If the signal outputs or inputs are intended only for connection to other equipment that meets the requirements of EN 60601-1 (i.e., leakage-current limits), this must be specified in the IFU.

Some medical electrical devices may be capable of operating both from power-supply mains and from an internal power source (i.e., batteries). If the internal power source is not automatically maintained at a fully usable state, the IFU must contain a warning statement describing the necessity for periodic checking or replacement of the internal power source.

If this type of equipment relies on a protective earth conductor in addition to basic insulation for protection against electric shock, the IFU must warn the user that the equipment must be operated from its internal power source if the integrity of the connection to the external protective earthing system is in doubt. Amendment 2 expanded the requirement to include not only concern about the integrity of the protective earth conductor, but also its arrangement.

If the device uses primary batteries, the IFU must contain a warning to remove the batteries when the device is not likely to be used for some time if a safety hazard can arise from leaving the batteries in the equipment.

The IFU of a device that uses rechargeable batteries must contain instructions to ensure safe use and adequate maintenance. If a specific power supply or battery charger is required to ensure compliance with EN 60601-1, these accessories must be identified in the IFU.

To complete the alignment with the MDD, Amendment 2 added a requirement that the IFU identify any risks associated with disposal of waste products, residues, and so on, and of the equipment and accessories at the end of their useful lives. The manufacturer should provide advice on minimizing these risks.

Technical Description

The technical description provides detailed information that supports and supplements the IFU. It should include details on all the markings required on the outside of the equipment (see Table 15.5). In addition, the technical description should explain all the characteristics (or an indication of where they may be found) that the user needs to understand for safe operation of the device. This could include such items as:

- an explanation of the device performance, giving special attention to the physiological effect and outputs that may be hazardous to the patient or user;
- the appropriate environmental conditions required for the accurate, safe, and correct use of the device (i.e., ambient temperature, relative humidity, and atmospheric pressure);
- if applicable, a description of the required power source (i.e., voltage, frequency, power rating, and allowable ranges);
- required site-preparation and installation procedures, including such items as the space required for use, environmental conditions, protective barriers (radiation, ionizing, and electromagnetic), and so on;
- the construction of the equipment (for example, using photographs, block diagrams, wiring diagrams, etc.) with special attention to the interconnection of system elements;
- a description of the mechanical structure required for the safe use of the device, which includes such items as precautions to be taken with movable parts, motor circuits, radiation shields (like those on X-ray and laser equipment), and mobile equipment;
- an explanation of the activation principle for the safe and correct use of the device; and
- for portable devices, the preparation of the device for safe transport.

Amendment 2 draws special attention to providing the range(s), accuracy, and precision of the displayed values or an indication of where this information can be found.

The technical description should contain the type and rating of fuses utilized in the supply-mains circuit. This is required if the type and rating is not apparent from the rated current and mode of operation of the device.

If the device incorporates interchangeable and/or detachable parts that are subject to deterioration in normal use, the technical description must provide the user with instructions for replacement of those parts.

As part of the technical description of the device, the manufacturer must include a statement that it will make available information that will assist the user's appropriately qualified technical personnel to repair those parts that the manufacturer has designated as repairable. This information could include circuit diagrams, component-parts lists, descriptions, and calibration instructions. This information could be provided in a service manual.

If the device cannot withstand the shipping and storage conditions specified in Subclause 10.1 of EN 60601-1, the technical description must contain the permissible environmental conditions for transport and storage. This information must be repeated on the outermost packaging of the device. Amendment 2 made this requirement more rigorous by requiring that the permissible environmental conditions for transport and storage be provided at all times.

Medical Electrical Systems

Medical electrical equipment may be integrated with nonmedical electrical equipment to form a medical electrical system. In this case, the system must be accompanied by documents containing all data necessary for its safe and reliable use. This may include information that is in addition to the documentation accompanying all of the medical electrical equipment and other nonmedical electrical equipment, as follows (EN 60601-1-1 pp. 15–16):

- Information for cleaning and, where applicable, for sterilization and disinfection of equipment used in the system
- Additional safety measures that are to be applied following installation of the system
- An indication of which parts of the system are specified to be suitable for use within the patient environment (e.g., by markings)
- Additional measures that should be applied during preventative maintenance
- An instruction that multiple portable socket-outlets provided with the system are to only be used to supply power to equipment that is intended to form part of the system
- A warning that the multiple portable socket-outlets are not to be placed on the floor
- A warning that the additional multiple portable socket-outlets or extension cords are not to be connected to the system
- A warning not to connect items that are not specified as part of the system
- The maximum permitted load for any multiple portable socket-outlet(s) used with the system
- An explanation of the risks of connecting nonmedical electrical equipment that has been supplied as a part of the system directly to a wall outlet when the nonmedical electrical equipment is intended to be supplied via a multiple portable socket-outlet with a separating transformer
- An explanation of the risks of connecting electrical equipment that has not been supplied as a part of the system to the multiple portable socket-outlet
- Any restrictions in the environmental conditions to ensure safety
- Instructions to the operator not to touch simultaneously the patient and parts of nonmedical electrical equipment that are accessible without the use of a tool during routine maintenance, calibration, and so on

- Advice to:
  - The installer, recommending that the system be installed in a way that enables the user to achieve optimal use
  - The user, to carry out all required cleaning, adjustment, sterilization, and disinfection procedures

THINGS TO REMEMBER

The MDD is the most extensive of the three directives covering the medical device sector. It applies to all medical devices and their accessories, unless they are covered by the AIMDD or by the IVDD.

The information provided with a medical device by the manufacturer is covered by the ERs in Annex I of the MDD. Failure to follow or satisfy these requirements may lead to the device being deemed improperly labeled (i.e., misbranded).

The labeling requirements are contained in ER 13 in Annex I. The directive specifies the contents of the label (ER 13.3) and the instructions for use (ER 13.6). The labeling must satisfy all of the applicable requirements in Annex I of the MDD. These requirements specify the minimum content of the labeling and must be met.

Many medical devices have specific labeling requirements that go beyond the minimums described in the MDD. For specific devices, the manufacturer should pay careful attention to the content of the device standards used to demonstrate compliance with the ERs of the directive. A key standard that manufacturers of active medical devices should take into account in this regard is EN 60601-1 on medical electrical equipment.

16 The Active Implantable Medical Device Directive (AIMDD)

The Active Implantable Medical Device Directive (AIMDD) covers the placing on the market and putting into service of a subset of products in the medical device sector. To qualify under the AIMDD, a product must be a medical device that is, at the same time, both "active" and "implantable." The "device" definition relates to a product intended by the manufacturer for a medical purpose "whether used alone or in combination, together with any accessories or software for its proper functioning" (Directive 90/385/EEC §1(2(a))).

The medical purpose may be achieved by a stand-alone device or as a result of several devices being used as a system. When the medical purpose is achieved by a system, each part of the system must be regarded as a medical device. As a consequence, the device definition may be applied to the system or to interchangeable parts intended, together with other devices, to form a system. Each part of the system is covered by the AIMDD regardless of whether the part by itself is "active" or "implantable" (CEC, Field, § 2.1.1). Examples of active implantable medical devices (AIMDs) include such products as an implantable pacemaker with or without electrodes, leads (electrodes) for an implantable pacemaker, or a pacemaker programmer.

The AIMDD specifies that a device is active if it "relies for its functioning on a source of electrical energy other than that generated by the human body or gravity" (Directive 90/385/EEC §1(2(b))). For example, a hydrocephalus pressure relief that allows the release of cerebrospinal fluid when a build-up of pressure overcomes a spring is not "active." Even if the device can be adjusted by an external "programmer," the device is not an AIMD because the medical purpose of the device is to relieve pressure, not to be adjusted (CEC, Field §2.1.2). In another example, a cochlear implant activated by an external power transmitter is considered an "active" implant. The implanted component depends on the external power source to do useful work, that is, to carry out its primary function of converting the power it receives into signals that stimulate the appropriate sensory channels of the brain (CEC, Field §2.1.2). A device that merely transmits heat, light, pressure, or vibration into the body is not automatically considered "active."

To be considered "implantable," a device must be "totally or partly introduced, surgically or medically, into the human body or by medical intervention into a natural orifice, and which is intended to remain after the procedure" (Directive 90/385/EEC §1(2(c))). The AIMDD is concerned about potential risks that may be inherent due to the impossibility of maintenance, calibration, or control, and problems relating to the aging of material in long-term implants. An external drug-infusion pump, although active and intended for long-term use, is not considered an AIMD even though it is connected to a catheter "partially introduced" into the body. Such a device does not present the hazards associated with "implantable" devices (CEC, Field §2.1.3).

"Accessories" to AIMDs are by definition AIMDs in their own right. This does not mean that the definition presupposes that the attributes "active" and "implantable" must necessarily be met by a product described by the manufacturer as an accessory. All that is necessary is that the article enables the AIMD to be used for its intended purpose. A catheter used with an implantable drug pump is not "active," but it is clearly necessary for the functioning of the AIMD and is, therefore, also an AIMD. Following this same logic, a programmer or external activator for controlling an AIMD is covered by the definition of "active implantable medical device" (CEC, Field §2.2).

TABLE 16.1
Examples of Active Implantable Medical Devices

1. Implantable cardiac pacemakers
2. Implantable defibrillators
3. Leads, electrodes, and adapters for 1 and 2
4. Implantable nerve stimulators
5. Bladder stimulators
6. Sphincter stimulators
7. Diaphragm stimulators
8. Cochlear implants
9. Implantable active drug-administration devices
10. Catheters and sensors for implantable active drug-administration devices
11. Implantable active monitoring devices
12. Programmers, software, and transmitters

Source: CEC, Field §2.3

Examples of some devices considered by the European Commission to be AIMDs covered by the AIMDD are shown in Table 16.1.

MISBRANDING

The information provided by the manufacturer with an AIMD is covered by the essential requirements (ERs) of the AIMDD. Annex I of the AIMDD includes specific minimal requirements for information that must be on the label and in the instructions for use (IFU). These requirements include (Directive 90/385/EEC §§14, 15, and 16):

- general labeling provisions,
- particulars to be present on the sterile package label,
- particulars to be present on the sales package label,
- IFU, and
- patient information.

Failure to follow or satisfy these requirements may lead to the device being deemed improperly labeled (i.e., misbranded). Misbranding would be grounds for a notified body to refuse to authorize the manufacturer to affix the CE conformity marking. Discovery of improper labeling of a device that bears the CE conformity marking would be a reportable incident under the vigilance system (CEC, Vigilance §5.5.3). In the extreme case, the failure to properly label the device could cause a competent authority to activate the Safeguard Clause leading to removal of the device from the market. The following sections discuss each of these five areas in detail.

As an aid to cross-referencing to the AIMDD, the number of the ER in Annex 1 (e.g., §14.1) is listed with the corresponding topic. Points or "indents" under a numbered heading are assigned small roman numerals for ease of reference in the remainder of this chapter (e.g., the first indent under ER 14.1 is referenced as 14.1(i)).

GENERAL LABELING PROVISIONS

The AIMDD requires that each device must be accompanied by the information needed to use it safely and to identify the manufacturer. In determining the information required, the manufacturer should take into account the training and knowledge of the potential users.

The AIMDD encourages the use of symbols to convey the required information. Symbols simplify labeling and reduce the need for multiple translations of words into national languages. When a manufacturer wishes to use symbols to convey the required information, it must use the symbols defined in harmonized standards established under the AIMDD. For example, a manufacturer who wishes to use a symbol to advise the user to consult the accompanying documentation for detailed information would use Symbol 11 in Table 15.3.

If a needed symbol is not defined in a harmonized standard, the manufacturer is free to use symbols defined in other recognized sources such as International Electrotechnical Commission (IEC) 60878. If necessary, the manufacturer may develop special symbols, remembering that, for a symbol to be effective, its meaning should be clear when viewed in context of the device. If a symbol is not defined in a harmonized standard, it must be described in the documentation accompanying the device. Unless a symbol that appears in a harmonized standard is very well known through long-term use, it is a good idea to explain it in the accompanying documentation.

PRODUCT IDENTIFICATION (§11)

The AIMDD requires that devices and, if appropriate, their component parts be identified to allow action to be taken in case of any potential risk in connection with the devices and their components. Such identification will enable the manufacturer to recall the device or to take other action (e.g., increased follow-up of patients), if necessary. To satisfy this requirement, the device should bear a batch or serial number (EN 1041 p. 6).

To distinguish the identification code, the batch code can be preceded by the lot symbol (see symbol 3 in Table 15.3). If the identification code is a serial number, this may be identified using the serial number symbol (see symbol 4 in Table 15.3). In either case, the use of the symbol precludes the need for translation.

Although not specifically mentioned in the AIMDD, the European General Standard for AIMDs (EN 45502-1) specifies that the manufacturer's name (or well-known trademark or logo) and the model number should be present on the device (EN 45502-1 p. 14). Incorporating the model designation into a code in the batch or serial number (e.g., YAG-1994001234) has been an acceptable solution for providing this information.

The European General Standard for AIMDs also requires that when an AIMD incorporates a power source, the manufacturer must distinguish between different models of the power source (EN 45502-1 p. 15). One solution is to include a unique code for each model of power source in the serial number. This coding allows users and the manufacturer to track the longevity of a particular model of power source.

NONINVASIVE IDENTIFICATION (§12)

The AIMDD requires that devices bear a code by which they and their manufacturers can be unequivocally identified. In particular, the device code should allow the identification of the type of device and the year of manufacture. This information can be included in the lot or serial number (e.g., YAG-1994001234). For implantable devices with active components (e.g., batteries, electronic circuitry) it should be possible to read this code without the need for a surgical operation. A common method for achieving this objective is to include symbol(s) in the device that are opaque to X-rays. These radio-opaque symbols can be read using an ordinary X-ray.

CONTROLS AND DISPLAYS (§13)

The function of the user-adjustable controls and visual displays on a device or its accessories must be clearly specified in a way that is understandable to the medical practitioner and, when appropriate, the patient. When information must be provided to the patient, consideration should be given to providing a copy that can be retained by the implanting practitioner (EN 1041 p. 6).

PARTICULARS ON THE STERILE PACKAGE LABEL

If a device or a part of a device is provided in a sterile condition, it must be enclosed in a package that allows the contents to be sterilized. The contents of the sterile package must be sterile when placed on the market and must remain sterile until the package is opened or damaged.

Because of the scope of application of the AIMDD, there will be a number of system components (e.g., a pacemaker programmer) that are classified as AIMDs under the directive but that are not provided in a sterile condition. For these devices, the requirements for the sterile package label do not apply.

The label on the sterile package must bear the particulars described in the following sections. If the sterile package is also the package that protects the device during storage and handling by the purchaser,* it must also bear the particulars required on the sales package (see the next section, "Particulars on the Sales Package Label").

MANUFACTURER IDENTIFICATION (§14.1(III))

The sterile package label must bear the name or trade name of the manufacturer (a well-known trademark or logo may be sufficient to specify this.) The label must also bear the address of the manufacturer's registered place of business. The full postal address may not be required if the address is sufficiently detailed to enable the manufacturer to be contacted (e.g., postcode and country) (EN 1041 p. 6).

IDENTITY OF THE DEVICE (§14.1(IV))

The label of the sterile package must provide sufficient detail to allow the intended user to identify the device. For many AIMDs, the identity will be obvious to the intended user from a simple description such as "implantable pacemaker" or "cardiac pulse generator." A transparent package may reduce the need for a detailed description on the label by allowing the intended user to see the device while it remains in the sterile package.

It may be appropriate to list the contents and give the quantity of each item in the package. This would be particularly important if there are accessories that are needed to use the device but that are not included in the package. If it is not practical to list the contents on the sterile package label (i.e., because of space limitations), then the information should be given in the accompanying documentation.

The European General Standard for AIMDs (EN 45502-1) specifies that the device description on the sterile package should include the model number, and, if applicable, the batch or serial number of the device (EN 45502-1 p. 13). The lot symbol (see Symbol 3 in Table 15.3) or the serial number symbol (see Symbol 4 in Table 15.3) can be used to distinguish these codes on the label.

STERILE DEVICE MARKING (§§14.1(I), 14.1(II), AND 14.1(VII))

The label of the sterile package must bear the following information:

---

* The package may be enclosed in another layer of packaging for shipping to the purchaser (i.e., a shipping package).

- An indication that permits the package to be recognized as a sterile package
- A declaration that the contents of the package are in a sterile condition
- The method of sterilization

To avoid the need for translation of this information, symbols have been developed that combine the method of sterilization with the word "STERILE" (see Symbols 7, 8, and 9 in Table 15.3). In this way, the single symbol satisfies all three of these ERs.

A product that is labeled as sterile in Europe must conform to the definition in Subclause 3.4 of EN 556-1:2001. To be considered sterile, a product must be put through a process that achieves a Sterility Assurance Level (SAL) of 10^{-6} (EN 556-1 §4.1).

The label of the sterile package must bear instructions for proper opening of the package so the device can be removed in an aseptic manner (EN 45502-1 p. 14).

Expiration Dating (§14.1(ix))

Virtually all AIMDs that are provided in a sterile package will have an expiration date that reflects a time-related deterioration in safety. The sterile pack must disclose the expiration date beyond which the device may not be used in complete safety. The expiration date may be identified using the "use by" symbol (see Symbol 2 in Table 15.3).

The expiration date should be expressed as a year and month (e.g., in the form yyyy-mm). Some devices require more precision in specifying the expiration date. For these devices, a two-digit day may be added so that the date following the "use by" symbol would take the form yyyy-mm-dd.

Date of Manufacture (§14.1(viii))

The label of the sterile package must disclose the year and month of manufacture in the form yyyy-mm. The year and month of manufacture may be included in the lot or serial number (e.g., 20020701234). For many AIMDs, sterilization is the last manufacturing operation. It is also the operation that determines the expiration date of the product (i.e., a two- or four-year shelf life based on the integrity of the microbial barrier). For these devices, the month and year of manufacture may be combined with one of the methods of sterilization symbols (see Symbols 7, 8, and 9 in Table 15.3). If the month and year of manufacture are included as separate information on the label, they should be identified with the appropriate text or the date of manufacture symbol (see Symbol 5 in Table 15.3).

Connection to Other Devices (§9.(iv))

If the device in the sterile package must be connected to other devices or accessories not in the sterile package in order to operate as required for its intended purpose, the sterile pack label must identify the connector types or configurations required (EN 45502-1 p. 14). The characteristics of the connection must be described in sufficient detail that the user can obtain a safe combination. This description could take the form of a reference to published standards (e.g., IS-1 Connector Standard*) that specifies those characteristics.

Markings for Special-Purpose Devices (§§14.1(v) and 14.1(vi))

For the special marking on investigational and custom-made devices, see the appropriate sections in Chapter 14.

---

* The IS-1 designation refers to the International Connector Standard (ISO 5841-3:2000) whereby pulse generators and leads so designed are assured of a basic mechanical fit.

PARTICULARS ON THE SALES PACKAGE LABEL

The sales package is the packaging layer that protects and identifies the AIMD during the storage and handling by the purchaser. The sales packaging may be enclosed in another layer of packaging (e.g., shipping package) for delivery to the purchaser. ER 14.2 of the AIMDD specifies the minimum information that must be included on the label of an AIMD. If any of the required information cannot be placed on the sales package label, a notice to "see instructions for use" may be placed on the label to inform the user that the accompanying documentation needs to be consulted for important information. This is particularly true if there are warnings, precautions, or other important operating or safety instructions that should be consulted before using the device. To avoid translation, the "see instructions for use" symbol (see Symbol 11 in Table 15.3) may be used to satisfy this requirement.

Manufacturer Identification (§14.2(i))

The label must bear the name or trade name of the manufacturer. A well-known trademark or logo may be sufficient to specify the manufacturer.

The label must also bear the address of the manufacturer's registered place of business. The full postal address may not be required if the address is of sufficient detail to enable the manufacturer to be contacted (e.g., postcode and country) (EN 1041 p. 7).

If the manufacturer's registered place of business in not in the European Union (EU), the label, the outer packaging, or the instructions for use must also bear the name and address of the manufacturer's authorized representative in the EU (EN 1041 p 10).

If neither the manufacturer nor the manufacturer's authorized representative is based in the EU, then the importer is deemed responsible under the terms of the directive for placing the imported product on the EU market. In this case, the label, the outer packaging, or the IFU must also bear the name and address of the importer. The point is to enable the customer to contact someone within the jurisdiction of the EU who can take responsibility for the product.

Identity of the Device (§§14.2(ii), 14.2(iii), and 14.2(iv))

The label on the sales package must provide sufficient detail to allow the intended user to identify the device. For many AIMDs, the identity will be obvious to the intended user from a simple description such as "implantable pacemaker" or "cardiac pulse generator." As part of the device identification, the sales package label should include the model designation of the device (EN 45502-1 p. 11).

If applicable, the sales package label should bear the batch number or serial number of the device. To distinguish the identification code on the label, the batch code may be preceded by the lot symbol (see Symbol 3 in Table 15.3). If the identification code is a serial number, this may be identified using the serial number symbol (see Symbol 4 in Table 15.3).

It may be appropriate to list the contents and give the quantity of each item in the package. This would be particularly important if there are accessories needed in order to use the device that are not included in the package.

The sales pack label should include a description of the purpose for the device if this is not obvious to the intended user from the device description. If necessary for the proper use of the device, the sales package label should bear any additional information and relevant characteristics necessary for identification. This information may be presented in an abbreviated form, provided the full details are given in the accompanying documents (EN 1041 p. 7).

Sterile Device Marking (§14.2(vii))

If the sales package contains an implantable article, the label must contain a declaration that the package contains a sterile device.

Expiration Dating (§14.2(ix))

If the sales package contains an implantable article, the label must bear an expiration date beyond which the article may not be used in complete safety. The expiration date may be communicated with the "use by" symbol (see Symbol 2 in Table 15.3).

The expiration date should be expressed as the year and month (e.g., in the form yyyy-mm). Some devices require more precision in specifying the expiration date. For these devices, a two-digit day may be added so that the date following the "use by" symbol would take the form yyyy-mm-dd.

Date of Manufacture (§14.2(viii))

The label of the sales package must disclose the year and month of manufacture in the form yyyy-mm. The year and month of manufacture may be included in the lot or serial number (e.g., 200207 01234). If the year and month of manufacture are included as a separate piece of information on the label, they should be identified with the appropriate text or the date of manufacture symbol (see Symbol 5 in Table 15.3).

Storage and Handling Conditions (§14.2(x))

If there are exceptional storage, handling, and transport conditions that are critical for the safe and proper functioning of the device, these must be provided on the sales package. If one or more sales packages are enclosed in another container (i.e., a shipping package), the shipping package label must also bear the exceptional storage, handling, and transport conditions. Otherwise, the manufacturer must prepare the device to survive the storage and handling conditions normally expected by the user for devices of the type in question. It would be generally understood that the devices should be protected from extremes of temperature, weather, and electromagnetic radiation. If, however, the device needs to be maintained within a particular range of temperature and relative humidity, or protected from impact, vibration, or pressure, this information should be indicated on the outer label (EN 45502-1 p. 12).

Connection to Other Devices (§9(iv))

If the sales package contains an implantable device that must be connected to other devices or accessories that are not included in the package in order to operate as required for its intended purpose, the sales package label must identify the connector types or configurations required (EN 45502-1 p. 11). The characteristics of the connection must be described in sufficient detail to enable the user to obtain a safe combination. This description could take the form of a reference to published standards that specify those characteristics (EN 1041 p. 13).

Device Containing Radioactive Substances (§8(v))

If the package contains any radioactive substance, the label of the sales package must describe the nature, type, and activity of the radiation.

Markings for Special-Purpose Devices (§§14.2(v) and 14.2(vi))

For the special marking on investigational and custom-made devices, see the appropriate sections in Chapter 14.

INSTRUCTIONS FOR USE (IFU)

The AIMDD requires that when an AIMD is placed on the market, it must be accompanied by IFU. ER 15 of the AIMDD specifies the minimum information that, when appropriate, must be included in the IFU for an AIMD.

PARTICULARS FROM THE LABELS (§15(II))

The AIMDD requires that the IFU must contain all of the applicable information required by the directive to be on the sterile package label (if a sterile package exists) and on the sales package label of the device except for:

- the month and year of manufacture, and
- the expiration date.

There is no requirement in the European General Standard for AIMDs (EN 45502-1) to include the batch or serial number of the device as part of "a description of the device." As a practical point, the manufacturers of AIMDs often provide an adhesive sticker (often several) bearing the device serial number as an aid and convenience to the user, who must record the serial number in different places including the patient's chart and device registration forms.

Specifically, the AIMDD requires that the following information from the labels must appear in the IFU:

- The IFU must include the name and address of the manufacturer. The minimum address information required is the postcode and country, if this is sufficient to allow the customer to contact the manufacturer and, if necessary, the manufacturer's agent within the EU. However, because space is not usually a critical issue in the documentation accompanying the device, it is advisable to include the full postal address along with a telephone number.
- The IFU must provide a description of the device. For many AIMDs, the identity will be obvious to the intended user from a simple description such as "implantable pacemaker" or "cardiac pulse generator." As part of the device identification, the IFU should include the model designation of the device.
- If the IFU accompany a sterile package, they must include a declaration that the implantable parts of the AIMD have been sterilized.
- The IFU must include information about any exceptional environmental or handling constraints (e.g., protection from temperature, humidity, pressure, vibration, or impact) necessary to allow the AIMD to be properly transported and stored.

PERFORMANCES INTENDED BY THE MANUFACTURER (§15(III))

The IFU must describe the intended use for the device and the device specifications, as well as set out the performance claimed by the manufacturer. Care should be taken to describe the performance of the device in clear and objective terms. ER 2 requires that the manufacturer must be able to verify any performance claims.

The manufacturer must warn the user of undesirable side effects reasonably associated with the use of the device. Undesirable side effects should be listed in descending order based on their clinical significance as determined by their severity and frequency. The labeling should provide frequency data from adequately reported clinical studies when the data are not well known to the intended user and/or when this information is needed in deciding between the use of the device and an alternative procedure or approach.

SELECTING A SUITABLE DEVICE (§15(IV))

The IFU must provide the information necessary for the clinician to select the appropriate device for the patient. In addition to the intended uses and side effects discussed in the previous section, this information could include contraindications and functional characteristics important in making a device selection.

In addition, the IFU should describe in sufficient detail the accessories and related devices (e.g., a programmer) needed for the AIMD to achieve its intended purpose.

DEVICE OPERATION (§15(V))

The IFU should provide information that will allow the clinician to use the AIMD safely for the purpose intended. The IFU must alert the clinician to the nature and frequency of maintenance, calibration, and other required activities (e.g., refilling a drug reservoir) to ensure that the device operates properly.

If there are external controls or indicators that must be manipulated or read by the patient, the IFU should provide clear directions so that the patient can properly use the device. An example of such an external control device would be a patient activator for an implanted neurological stimulator, which might enable the patient to activate the implant and set the level of stimulation.

For a device intended to administer medicinal substances, the IFU must advise the clinician of the substances that the device is designed to deliver. Any limitations on the choice of substance should be included (e.g., requiring preservative-free morphine sulfate).

RECIPROCAL INTERFERENCE (§§15(VII) AND 8(IV))

Medical devices may interact with one another or with other medical treatments in ways that harm the patient, damage the devices, or render one or the other of the treatments ineffective. Of particular concern are those treatments and procedures that could permanently damage the AIMD. These include procedures that result in an electrical current through the body from an external source (i.e., electro-cautery), the use of therapeutic ultrasound, therapeutic ionizing radiation, and diathermy (EN 45502-1 p. 28). The IFU must identify the risks of "reciprocal interference" and contain warnings regarding the interactions between the AIMD and other instruments or equipment that are likely to be used in the course of clinical investigations or medical treatments. Examples could include damage to a pacemaker caused by defibrillation of a patient or the risk of ventricular fibrillation induced by currents coupled into the heart from electrocautery being used in the vicinity of an implanted pacemaker/lead system. The IFU should advise the clinician of steps to be taken before the procedure in question (e.g., turning the device off) and/or to carefully monitor the AMID during treatment. The IFU may also need to warn the clinician that damage to the device may not be immediately detectable.

STERILE PACKAGING (§15(VIII))

If the device is provided in a sterile package, the IFU must include instructions to the user in the event of damage to the sterile package. These instructions could range from returning the device to the manufacturer to procedures for resterilizing the device. If resterilization is recommended, then the IFU must provide the details for the appropriate methods. The IFU should also contain warnings about inappropriate sterilization (e.g., "Do not autoclave").

REUSABLE DEVICES (§15(IX))

Typically, implantable parts of an AIMD system are not intended to be reused once they have been implanted in a patient. However, if the manufacturer intends that the device is reusable, then the IFU must include a warning that the device can be reused only if it has been reconditioned, under the responsibility of a manufacturer, to comply with the ERs of the AIMDD.

IMPLANTABLE DEVICES (§15(VI))

If the device is implantable, the IFU must include any appropriate operating instructions, warnings, and precautions to avoid risks uniquely associated with the implantation of the device. These risks could include such things as body rejection phenomena, erosion through the skin, embolism, and infections. This category should include those risks that are recognizable and foreseeable as opposed to those that are unknown and/or improbable (EUCOMED p. 7).

YEAR OF AUTHORIZATION TO AFFIX THE CE MARK (§15(I))

The AIMDD, unlike the other two medical device directives, requires that the IFU bear the year that the manufacturer received authorization to affix the CE conformity marking to the product. This date will appear on the certificate issued by the Notified Body.

DEVICE CONTAINING RADIOACTIVE SUBSTANCES (§8(V))

Devices that contain any radioactive substances should include in the IFU a detailed description of the nature, type, intensity, and distribution of the radiation. In addition to providing information on the nature of the radiation, the IFU should describe means of protecting the patient and the user, ways of avoiding mishandling, and the risks inherent in the use of the device.

MEDICINAL SUBSTANCES INCORPORATED INTO THE DEVICE (§10.)

If a medicinal substance is incorporated as an integral part of a device, the IFU should describe the substance using the International Nonproprietary Name or other commonly used name, the quantity available, and an explanation of the beneficial effect of the medicinal substance.

NONINVASIVE IDENTIFICATION (§12.)

The IFU should include an explanation of the method of interpreting the code used to noninvasively identify the device (EN 45502-1 p. 49).

PATIENT INFORMATION

The information described in the previous sections is directed primarily at the clinician. This information is intended to enable the device to be used safely and for the purpose intended. In contrast, the information in this section is intended both for enabling the clinician to use the device, and to enable qualified medical staff to brief the patient on contraindications or precautions that should be taken into account. As a qualified professional will interpret the information, it need not be written in a manner that can be read and understood by the patient. If the information is not needed to brief the patient, or is obvious, it need not be included (EUCOMED p. 8).

When appropriate, the IFU should contain information on the topics described in the following sections.

LIFETIME OF THE ENERGY SOURCE (§15(x))

If the device has an implantable energy source, the IFU should contain information that enables the clinician to estimate the lifetime of the energy source under the conditions of use.

CHANGES IN PERFORMANCE (§15(xi))

The IFU should explain the precautions that the clinician and the patient should take in the event of changes in the performance of the device. This could include instructions on what signal to look for that would indicate a change in performance and what action should be taken by the patient (i.e., contact a healthcare provider immediately) and the clinician (i.e., replace the device) if any signals are observed. For example, some AIMDs are equipped with an audible alarm intended to alert the patient to conditions such as a low battery, low reservoir volume (i.e., for an implantable drug pump), or other internal errors. The patient must be instructed as to what action to take if such an alarm occurs.

EXPOSURE TO ENVIRONMENTAL CONDITIONS (§15(xii))

Some devices may be affected by exposure to reasonably foreseeable environmental conditions. The AIMDD lists magnetic fields, external electrical influences, electrostatic discharge, and pressure, or variations in pressure, as examples. If there are reasonably foreseeable environmental conditions that the patient should avoid, the IFU should provide the necessary briefing material to the medical staff. For example, the instructions leaflet might tell the medical staff to instruct the patient to avoid work environments where arc welders, induction furnaces, resistance welders, radio, or microwave-frequency transmitters are in use—if the device is susceptible to interference from high-powered electromagnetic fields. Patients might also be instructed to seek medical guidance before entering areas with warning notices advising against entry by pacemaker patients (EN 45502-1 p. 30).

ADMINISTRATION OF MEDICINAL PRODUCTS (§15(xiii))

The IFU must provide adequate information on the medicinal product or products that the device is designed to administer. This includes any restrictions in the choice of substances to be delivered by the device.

THINGS TO REMEMBER

The AIMDD covers the placing on the market and putting into service of any medical device that is, at the same time, both "active" and "implantable." The medical purpose may be achieved by a stand-alone device or as a result of several devices being used as a system. Each part of the system is covered by the AIMDD regardless of whether the part by itself is "active" or "implantable."

"Accessories" to AIMDs are by definition AIMDs in the own right. This does not mean that the definition presupposes that a product described by the manufacturer as an accessory must necessarily meet the attributes "active" and "implantable." Following this logic, a programmer or external activator for controlling an AIMD is covered by the definition of "active implantable medical device."

The information provided by the manufacturer with an AIMD is covered by the ERs in Annex I of the AIMDD. Failure to follow or satisfy these requirements may lead to the device being deemed improperly labeled (i.e., misbranded). The labeling requirements are contained in ERs 14.1 (sterile package), 14.2 (sales package), and 15 (IFU).

17 The *In Vitro* Diagnostic Device Directive (IVDD)

The *In Vitro* Diagnostic Device Directive (IVDD) covers the placing on the market and putting into service of *in vitro* diagnostic (IVD) devices and their accessories. For the purposes of the IVDD, accessories are treated as IVD devices in their own right (Directive 98/79/EC §1(1)). An IVD device is "any medical device which is a reagent, reagent product, calibrator, control material, kit, instrument, apparatus, equipment, or system whether used alone or in combination, intended by the manufacturer to be used *in vitro* for the examination of specimens, including blood and tissue donations, derived from the human body" (Directive 98/79/EC §1(2(b))).

The IVDD entered into force on December 7, 1998. The directive specified that member states must apply the provisions of the directive from June 7, 2000. The directive further specified that for a period of five years following the entry into force of this directive, member states must accept the placing on the market of devices that conformed to the rules in force in their territory on the date this directive entered into force. For an additional two years, these devices might continue to be put into service (Directive 98/79/EC §22).

As with the other directives, the IVDD includes accessories that are intended by the manufacturer to be used together with the IVD device to enable the IVD device to achieve its intended purpose. This directive will consider specimen receptacles, whether vacuum-type or not, specifically intended by their manufacturers for the primary containment and preservation of specimens derived from the human body for the purpose of IVD examination to be IVD devices.

The directive will apply equally to IVD devices intended for professional-use and to IVD devices intended by the manufacturer for normal use in the home environment (i.e., for self-testing).

IMPROPER LABELING

The information provided by the manufacturer with an IVD device is covered by the essential requirements (ERs) of the IVDD. Annex I of the IVDD includes specific minimal requirements for information that must be on the label and in the instructions for use (IFU). These requirements include (Directive 98/79/EC Annex I(8)):

- general labeling provisions,
- devices with a measuring function,
- particulars on the label of reagents,
- particulars on the label of IVD instruments and equipment,
- IFU, and
- devices incorporating dangerous substances.

Failure to follow or satisfy these requirements may lead to the device being deemed improperly labeled. The following sections discuss each of these six areas in detail.

As an aid to cross-referencing to the IVDD, the number of the Essential Requirements (ER) in Annex I (e.g., §8.4) is listed with the corresponding topic.

GENERAL LABELING PROVISIONS

The IVDD requires that each device must be accompanied by the information needed to identify the manufacturer and to use it properly, taking into account the training and knowledge of the potential users. As far as practical and appropriate, the information needed to use the device safely and properly should be placed on the device itself and/or, where appropriate, on the sales packaging. If individual full labeling of each unit is not practical, then the information must be placed on the packaging and/or in the documentation supplied with one or more devices.

As with the other medical device directives, the IVDD encourages the use of symbols to convey the required information. When a manufacturer wishes to use symbols and colors to convey the required information, it must use the symbols and identification colors defined in the harmonized standards.

If a needed symbol is not defined in a harmonized standard, then the manufacturer is free to use symbols defined in other recognized sources such as International Standards Organization (ISO) Publication 15223. If necessary, the manufacturer may develop special symbols, remembering that, to be effective, the meaning of a symbol should be clear when viewed in context of the device. If a symbol or color is not defined in a harmonized standard, it must be described in the documentation accompanying the device. Unless a symbol that appears in a harmonized standard is very well known through long use, it is a good idea to explain the symbol in the accompanying documentation.

Intended Purpose (§8.5)

The IVDD requires that the manufacturer must state the intended purpose of the device clearly in the IFU and, when appropriate, on the label unless the intended purpose is obvious to the user, taking into account the potential user's training and knowledge.

IVDs Intended for "Self-Testing" (§7)

Many IVD devices are intended by their manufacturers to be used normally in the home environment. These devices must be designed and manufactured in such a way that they perform appropriately for their intended purpose, taking into account both skills and means available to the user, and reasonably expected variations in the user's technique and environment. The information and instructions provided by the manufacturer should be easily understood and applied by the user.

Product Identification (§8.6)

The IVDD requires that, whenever reasonable and practical, devices and their detachable components be identified to allow action to be taken in case of any potential risk. Such identification will enable the manufacturer to recall the device and/or its detachable components, if necessary. To satisfy this requirement, the device should bear a batch or serial number. In addition, detachable components that are intended by the manufacturer to be used separately from the original device are to be identified with a batch or serial number (EN 1041 p. 6).

To distinguish the identification code on the label, the batch code is to be preceded by the word "LOT" or the lot symbol (see Symbol 3 in Table 15.3). A serial number may be identified using the serial number symbol (see Symbol 4 in Table 15.3). In either case, the use of the symbol precludes the need for translation.

DEVICES WITH A MEASURING FUNCTION

Devices that are instruments or apparatus having a measuring function must provide adequate stability and accuracy within appropriate limits, taking into account the intended purpose of the device and the availability of appropriate reference measurement standards. The manufacturer must specify the accuracy limits of the device (Directive 98/79/EC Annex I(4.1)).

MEASUREMENT, MONITORING, AND DISPLAY SCALES (§3.6)

The measurement, monitoring, or display scales must be designed and manufactured in line with accepted ergonomic principles, taking into account the intended purpose of the device. This includes color changes and other visual indicators.

UNITS OF MEASURE (§9.2)

Devices that incorporate a measuring function must express the result of the measurement in legal units conforming to the provisions of Council Directive 80/181/EEC.* The basic units are those of the International System of Units (SI), which is described in ISO 1000.

PARTICULARS ON THE LABEL OF REAGENTS

IVDD ER 8.4 specifies the minimum information that must be included on the label of an IVD device. The requirements described in the following sections are constructed around the needs of IVD reagents. Here, the manufacturers need to pay attention to the requirements of key product standards such as EN 375 and EN 376. These standards subdivide the requirements in the directive into the details in the label on the "immediate container," and details in the label on the "outer container." The label requirements in the following sections will follow this format.

The "immediate container" is defined as the packaging layer that protects the content(s) from contamination and/or physical damage (e.g., a sealed vial, an ampule, a bottle, a foiled pouch, or a sealed plastic bag. The "outer container" is the layer of packaging that encloses the immediate container(s), creating a single entity or assembly of different or identical components. These packages are similar in concept to the "sterile package" and "sales package" described in the Active Implantable Medical Device Directive (AIMDD).

IMMEDIATE CONTAINER

The label of the immediate container should provide the information listed in the following sections in legible characters. If the available space is too small to contain this information or if the labeling would interfere with the reading of analytical results, the information may be reduced to the product name, supplier, lot number, expiration date, appropriate warnings, and precautions. The listing of contents, intended use, and storage instructions may be given on the label of the outer container, or in the IFU if this is more appropriate (EN 375 §4.2.1; EN 376 §4.2.1). For small professional-use IVDDs, the indication of the microbiological state (e.g., sterile) may also be given on the label of the outer container or in the IFU if this is more appropriate (EN 375 §4.2.1).

If the outer container encloses components intended to perform a single measurement, the components present in such a container must be identified in the same manner described in the IFU (e.g., name, letter, number, symbol, color, or graphic). If appropriate, the immediate container should carry any needed cautionary statements (EN 376 §4.2.1).

---

* Council Directive 80/181/EEC of 20 December 1979 on the approximation of the laws of the member states relating to units of measurement and on the repeal of Directive 71/354/EEC. 1980. *Official Journal of the European Communities*, 23, no. L 039 (February 15).

Product Name (§13.4(b))

The product name on the label must ensure that the user can properly identify the product. In addition, components of a kit must be identified by name, letter, number, symbol, color, or graphic in the same manner as described in the IFU or on the outer container (EN 375 §4.2.3; EN 376 §4.2.3).

Manufacturer (Supplier) (§8.4(a))

The name and address of the manufacturer must be given on the label of the immediate container. Alternatively, recognized trade name or logo is sufficient (EN 375 §4.2.2; EN 376 §4.2.2). For imported devices, the label of the outer container and the IFU must also contain information about the manufacturer's authorized representative in the community.

Lot Number (§8.4(d))

The label of each immediate container should bear the batch code, preceded by the word "LOT" or the lot symbol (see symbol 3 in Table 15.3). If the contents of the immediate container are serialized, then the immediate container label should bear the serial number preceded by the words "serial number" or the serial number symbol (see Symbol 4 of Table 15.3).

Expiration Date (§8.4(e))

The immediate container must bear an expiration date based on the storage instructions stated on the label (EN 375 §4.2.6; EN 376 §4.2.6). The expiration date may be expressed as the year, month, and day (i.e., yyyy-mm-dd), or as the year and month (i.e., yyyy-mm). If the latter form is used, then the expiration date is the last day of the month indicated (EN 375 p. 5; CEN, EN 376 p. 8). The "use by" symbol may be used to identify the expiration date (see Symbol 2 in Table 15.3).

Contents (§8.4(b))

The label of the immediate container of a professional-use reagent must identify the contents in terms of mass, volume, and/or the number of measurements that can be made with the contents (EN 375 §4.2.7).

For a reagent intended for self-testing, the label should identify the contents by specifying the number of measurements or tests that can be performed (EN 376 §4.2.7).

Intended Use (§§8.4(b), 8.4(g), and 8.4(k))

A reagent intended for self-testing must be clearly labeled as such. The label of the immediate container should contain a brief indication of the intended use of the reagent (e.g., "pregnancy test"). The label must also bear, in lay terms, a clear indication that the device is intended for *in vitro* use (e.g., "not to be swallowed") (EN 376 §4.2.8).

A professional-use reagent may simply be labeled with a general statement (e.g., "for *in vitro* diagnostic use (only)" or "*in vitro* test").

Cautionary Statements (§8.4(j))

If an IVD reagent is considered hazardous, the label of the immediate container must bear the appropriate cautionary symbols and/or statements. These are described later in this chapter in the section, "Devices Incorporating Dangerous Substances."

TABLE 17.1
Examples of Recommended Storage-Temperature Intervals

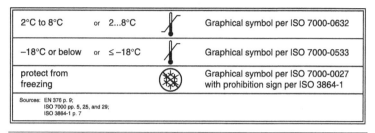

2°C to 8°C	or 2...8°C		Graphical symbol per ISO 7000-0632
−18°C or below	or ≤ −18°C		Graphical symbol per ISO 7000-0533
protect from freezing			Graphical symbol per ISO 7000-0027 with prohibition sign per ISO 3864-1

Sources: EN 376 p. 9;
      ISO 7000 pp. 5, 25, and 29;
      ISO 3864-1 p. 7

Storage Information (§8.4(h))

The label of the immediate container should list the storage conditions necessary to protect the stability of the reagent in an unopened state. Storage conditions for an opened or reconstituted product that are different from those in an unopened state must also be listed on the label of the immediate container. Examples of recommended storage temperature intervals are shown in Table 17.1.

Sterile Device Marking (§8.4(c))

When appropriate, the label of the immediate container should indicate that the contents are sterile, using the word "STERILE." To avoid the need for translation, the sterile symbol may be used (see Symbol 6 in Table 15.3).

If the contents of the immediate container have a special microbiological state, or state of cleanliness, this should be indicated on the label.

Markings for Investigational Use (§8.4(f))

For the special marking on devices intended to be used for gathering performance evaluation data, see the section, "Performance-Evaluation Device Marking" in Chapter 14.

OUTER CONTAINER

The label of the outer container should provide the information listed in the following sections in legible characters.

Product Name (8.4(b))

The product name must appear on the label of the outer container (EN 375 §4.1.3). When appropriate, the catalog reference (product code) should also appear on the label (EN 376 §4.1.3). The catalog number may be identified with the catalog number symbol (see Symbol 10 in Table 15.3).

Manufacturer (Supplier) (§8.4(a))

The name and address of the manufacturer must be given on the label of the outer container. Alternatively, a recognized trade name or logo is sufficient (EN 375 §4.1.2; EN 376 §4.1.2). The full postal address may not be required if the address is of sufficient detail to enable the supplier to be contacted (e.g., postcode and country) (EN 1041 p. 10).

If the supplier is the manufacturer, the label must bear the address of the manufacturer's registered place of business. A manufacturer who does not have a registered place of business in a Member State may place devices on the market under its own name. However, the manufacturer must appoint an authorized representative who is established within the European Union (EU). In this case, the label, the outer packaging, or the instructions for use must also bear the name and address of the manufacturer's authorized representative (EN 1041 p. 10).

If neither the manufacturer nor the manufacturer's authorized representative is based in the EU, then the importer is deemed responsible under the terms of the directive for placing the imported product on the EU market. In this case, the label, the outer packaging, or the IFU must also bear the name and address of the importer.

Lot Number (§8.4(d))

The label of the outer container should bear the batch code, preceded by the word "LOT" or the lot symbol (see Symbol 3 in Table 15.3). If the contents are serialized, then the outer container label should bear the serial number preceded by the words "serial number" or the serial number symbol (see Symbol 4 of Table 15.3).

Expiration date (§8.4(e))

The outer container label must bear the expiration date of the component having the earliest expiration date (see the description of "expiration date" in the section on the immediate container label).

Contents (§8.4(b))

The label of the immediate container of a professional-use IVD device must identify the contents in terms of mass, volume, and/or the number of measurements that can be made with the contents. The components of a kit must be listed and briefly characterized on the label (e.g., "buffer"). Also, the components of a kit must be designated in the same way as on the immediate container label (EN 375 §4.1.7).

Information on additional materials, such as accessories, may be provided on the outer container label or in the IFU.

For an IVD device intended for self-testing, the label should identify the contents by specifying the number of measurements or tests than can be performed (EN 376 §4.1.7).

Identity of the Device and Intended Use (§§8.4(b), 8.4(g), and 8.4(k))

An IVD device intended for self-testing must be clearly labeled as such. The label of the outer container should contain a brief indication of the intended use of the IVD device (e.g., "pregnancy test"). The label must also bear, in lay terms, a clear indication that the device is intended for *in vitro* use (e.g., "not to be swallowed") (EN 376 §4.1.8).

The intended use of a professional-use IVD device may be given by means of the product name or analytical method (e.g., "Glucose, Hexokinase Method" or "ASAT") (EN 375 §4.1.8).

Cautionary Statements (§8.4(j))

If an IVD reagent is considered hazardous, the label of the outer container must bear the appropriate cautionary symbols and/or statements. These are described later in this chapter in the section, "Devices Incorporating Dangerous Substances."

Storage Information (§8.4(h))

The label of the outer container should list the storage conditions necessary to protect the stability of the reagent in an unopened state. Examples of recommended storage temperature intervals are shown in Table 17.1. Other storage conditions that can affect stability (e.g., light or humidity) must be listed on the label of the outer container.

Special Operating Instructions (§ 8.4(i))

The manufacturer must take into account the technical and clinical knowledge and the skill of the intended user when determining the type and extent of the operating instructions that must be included on the label for the safe use of the device. In many cases, operating instructions are required on the label only if the mode of operation of the device is novel or unfamiliar and would not be self-evident to the intended user (EN 1041 p. 12).

Sterile Device Marking (§8.4(c))

If the outer container encloses a reagent that is sterile, or has a special microbiological state, or state of cleanliness, the label must contain a declaration that the package contains such a device.

Markings for Investigational Use (§8.3(f))

For the special marking on devices intended to be used for gathering performance evaluation data, see the section, "Performance-Evaluation Device Marking" in Chapter 14.

PARTICULARS ON THE LABEL OF IVD INSTRUMENTS AND EQUIPMENT

A comparison of the requirements in ER 8.4 of the IVDD and ER 13.3 of the MDD will demonstrate that the basic requirements for the device label are very similar. There are, however, a few important differences in the requirements. The MDD lists several requirements that are not mentioned in the IVDD. These are:

- The IVDD does not require that the label bear an indication that the device is for single use. However, common sense would indicate that a device intended by its manufacturer to be for a single use should be appropriately marked (e.g., by use of the "do not reuse" symbol [see Symbol 1 in Table 15.3]) on the device and/or the package label, and/or the IFU.
- The IVDD does not recognize the "custom-made device" designation.
- The IVDD does not recognize the "cxclusively for clinical investigation" designation. Instead, the IVDD allows devices intended for performance-evaluation studies outside the manufacturer's premises to be marked for performance evaluation only.
- The label of an IVD device is not required by the IVDD to bear the date of manufacture.
- There is no requirement in the IVDD to indicate the method of sterilization on the label.

The IVDD does place the following additional requirements that go beyond the MDD:

- IVD devices are required, when appropriate for the proper use of the device, to bear a statement that the device is for *in vitro* use (§8.4(g)).
- If the device is intended for self-testing, the label must clearly state this fact (§8.4(k)).

The label of an IVD instrument should provide the information listed in the following sections. The markings should be visible from the exterior of the instrument, or be visible after removing a cover or opening a door intended to be removed or opened without the aid of a tool. Marking on the device must remain clear and legible under conditions of normal use and resist the effects of cleaning agents specified by the manufacturer. EN 61010-1 provides a test for durability. Following this test, markings must be clearly legible and adhesive labels must not have come loose or become curled at the edges (EN 61010-1 §5.3).

EN 61010-1 applies to electrical instruments for measurement, control, and laboratory use in general, not just to IVD instruments. To bridge the gap, the Committee for European Standardization (CEN) has developed an additional standard that applies the requirements of EN 61010-1 to the marking of IVD instruments. EN 1658* details the requirements for marking IVD instruments by applying the requirements of EN 61010-1 with some additions. Published in 1996, EN 1658 has been harmonized under the IVDD.

The required markings are not to be placed on the bottom of the instrument except when space is limited on a hand-held instrument (EN 1658 §4.1). In addition, EN 1658 requires that marking must be permanently affixed (EN 1658 §4.10).

Product Name (§13.4(b))

A model number, name, or other means to identify the instrument must be placed on the label. If the instrument bearing the same distinctive designation (model number) is manufactured at more that one location, the equipment from each manufacturing location shall be marked so that location can be identified. The marking of the factory location may be in code and need not be on the instrument's exterior (EN 61010-1 §5.1.2 b)).

Manufacturer (Supplier) (§8.4(a))

The name and address of the manufacturer must be given on the label of the immediate container. Alternatively, a recognized trade name or logo is sufficient (EN 61010-1 §5.1.2(a)). For imported devices, the label of the outer container and the IFU must also contain information about the manufacturer's authorized representative in the community.

Lot Number (§8.4(d))

The instrument label should bear the serial number preceded by the words "serial number" or the serial number symbol (see Symbol 4 of Table 15.3), or the batch code, preceded by the word "LOT" or the lot symbol (see Symbol 3 in Table 15.3). The IVD device does not require the date of manufacture on the labeling of an IVD device. However, EN 1658 requires not only the serial or lot number but also, if appropriate, the date of manufacture in yyyy-mm-dd format (EN 1658 §4.2). Unfortunately, no guidance is provided about the conditions that would make the date of manufacture "appropriate" as a marking on the instrument.

Expiration Date (§8.4(e))

One does not usually think of an instrument as having an expiration date. However, some IVD instruments, particularly ones intended for self testing, may have a date before which they should be used. An example could be an instrument with a battery that cannot be changed by the user. If this is the case, the outer container label and possibly the label of the instrument itself should bear an expiration date. If present, the expiration date may be expressed as the year, month, and day (i.e., yyyy-mm-dd), or as the year and month (i.e., yyyy-mm). If the latter form is used, the expiration

---

* European Committee for Standardization. 1996. *Requirements for marking of* in vitro *diagnostic instruments*. EN 1568. (December). Brussels: Comite Européen de Normalisation.

date is the last day of the month indicated. The "use by" symbol may be used to identify the expiration date (see Symbol 2 in Table 15.3).

Intended Use (§§8.4(b), 8.4(g), and 8.4(k))

An instrument intended for self-testing should be clearly labeled as such. The label on the instrument should contain a brief indication of the intended use of the instrument (e.g., "blood-glucose meter").

Warning Markings (§8.4(j))

Warning markings must be visible when the instrument is ready for normal use. If a warning applies to a particular part of the instrument, the marking should be placed on, or near, this part. If it is necessary for the user to refer to the IFU to preserve the protection afforded by the instrument, the instrument must be marked with Symbol 14 in Table 17.2. Symbol 14 is not required to accompany other symbols that are explained in the IFU (EN 61010-1 §5.2).

TABLE 17.2
Symbols for Marking *In Vitro* Diagnostic Instruments

No.	Symbol[a]	Reference	Description
1		IEC 60417-5031	Direct current
2		IEC 60417-5032	Alternating current
3		IEC 60417-5033	Both direct and alternating current
4		IEC 60417-5032-1	Three-phase alternating current
5		IEC 60417-5017	Earth (ground) terminal
6		IEC 60417-5019	Protective conductor terminal
7		IEC 60417-5020	Frame or chassis terminal
8		IEC 60417-5021	Equipotentiality
9		IEC 60417-5007	"On" (Supply)
10		IEC 60417-5008	"Off" (Supply)
11		IEC 60417-5172	Equipment protected throughout by double or reinforced insulation
12		IEC 61010-1 Table 1 No. 12	Caution, risk of electrical shock
13		IEC 60417-5041	Caution, hot surface
14		ISO 7000-0434	Caution, risk of danger[b]
15		IEC 60417-5268	In position of a bi-stable push control
16		IEC 60417-5269	Out position of a bi-stable push control

[a] The symbols in this table are for convenience of reference only and have no official sanction. The reader is referred to EN 60601-1.
[b] See 5.4.1 of EN 61010-1, which requires manufacturers to state the documentation must be consulted in all cases where this symbol is marked.
Source: EN 61010-1:2001

Warning markings must comply with the following requirements (EN 61010-1 §5.2):

- Symbols shall be at least 2.75 mm high. Text shall be at least 1.5 mm high and contrast in color with the background.
- Symbols or text molded, stamped, or engraved into a material shall be at least 2.0 mm high. If not contrasting in color, they must have a depth or raised height of at least 0.5 mm.

If the IFU state that the user is allowed to gain access, using a tool, to a part of the instrument where the user could come into contact with hazardous voltage during normal use, the instrument must bear a warning that the instrument must be isolated or disconnected from the source of the hazardous voltage before access (EN 61010-1 §5.2).

Other Markings (§8.4(i))

EN 61010-1 describes a number of other marking requirements for instruments. Broadly, these can be considered as part of the particular operation instructions required by ER 8.4(i). They include:

1. Mains supply marking—An instrument that is intended to be connected to a supply main must be marked with the following (EN 61010-1 §5.1.3):
   - Nature of the supply—For a.c. supplies, the label must state the required frequency or range of frequencies (e.g., 50-60 Hz). For information purposes, it may be useful to mark single-phase a.c. equipment with symbol 2 in Table 17.2 or three-phase equipment with Symbol 4 in Table 17.2. For d.c. supplies, the instrument should be marked with Symbol 1 in Table 17.2. For instruments that are suitable to either an a.c. or d.c. supply, Symbol 3 in Table 17.2 can be used.
   - Rated supply voltage—The label on the instrument should state the rated value of the supply voltage(s) or the range of supply voltages (e.g., 120-230 V). If appropriate, the rated voltage fluctuations may also be marked.
   - Power rating—The label on the instrument should state the maximum rated power in watts or volt-amperes, or the maximum rated current, with all accessories or plug-in modules connected. If the instrument can be used with more than one voltage range, separate power ratings should be marked for each voltage range unless the maximum and minimum values do not differ by more than 20 percent of the mean value.
   - Voltage indicator—If the design of the instrument allows the user to set different rated supply voltages, the instrument must provide an indicator to inform the user of the current setting. For portable equipment, this indicator must be visible from the exterior. If the rated voltage can be set without the use of a tool, the action of changing the voltage rating must also change the indicator.
   - Accessory mains socket outlets—If the instrument is equipped with accessory mains socket outlets that can accept standard mains plugs, these outlets must be marked with the voltage if different from mains-supply voltage. If the outlet is intended to be used with specific equipment, it shall be marked to identify the equipment for which it is intended. If not, the maximum rated current or power must be marked, or Symbol 14 in Table 17.2 is to be placed beside the outlet with full details included in the IFU.
2. Fuses—The fuse holder of any fuse that can be replaced by the user must be marked so the user can identify the correct replacement fuse (EN 61010-1 §5.1.4). For example, the codes described in IEC 60127 could be used to satisfy this requirement. For fuses that are not replaceable by the operator, instructions would be provided in the maintenance section of the IFU.

3. Terminals, connections, and operating devices—The purpose of terminals, connectors, control, and indicators should be marked on the instrument (EN 61010-1 §5.1.5). EN 61010-1 qualifies this statement with "if necessary for safety." EN 1658 removes this qualification when applying EN 61010-1 to IVD instruments (EN 1658 §4.6). Special attention is give to any connections for fluids such as gas, water, and drainage. When there is insufficient space on the exterior of the instrument, Symbol 14 in Table 17.2 can be used.

Special requirements exist for terminals for connection to the supply mains. These terminals are to be identified as follows (EN 61010-1 §5.1.5.1):

- Functional earth terminals are to be marked with Symbol 5 in Table 17.2.
- Protective conductor terminals are to be marked with Symbol 6 of Table 17.2, except when the protective conductor terminal is a part of an approved mains-appliance inlet. In this case, the symbol can be placed close to or on the terminal.
- Terminals of measuring and control circuits that are permitted by EN 61010-1 to be connected to accessible conductive parts must be marked with Symbol 7 of Table 17.2 if this connection is not self-evident. This symbol may be considered as a warning in that it indicates that a hazardous voltage must not be connected to the terminal. The symbol should also be used if the user could make such a connection inadvertently.
- Interior terminals that have a hazardous voltage present are to be marked with the voltage, current, charge, or energy value or range of values, or with Symbol 14 of Table 17.2. This requirement does not apply to mains-supply outlets when a standard mains socket outlet is used.
- Accessible functional earth terminals connected to accessible conductive parts are to be marked with an indication that this is the case, unless it is self-evident. Symbol 8 of Table 17.2 is acceptable for this marking.

Unless a clear indication is provided on the instrument that voltage and current-measuring circuit terminals are not intended to be connected to voltage-to-earth of 50 V a.c. or 120 V d.c., the terminals are to be marked as follows (EN 61010-1 §5.1.5.2):

- Measuring-circuit terminals for measurements within Category I as described in Sub-clause 6.7.4 of EN 61010-1 are marked with the rated voltage or current as applicable and with Symbol 14 of Table 17.2.
- Measuring-circuit terminals for measurements within Category II, III, and IV as described in Subclause 6.7.4 of EN 61010-1 are marked with the rated voltage or current as applicable and with the relevant measurement category. The measurement categories marking are "CAT II," "CAT III," or "CAT IV."

Examples of acceptable indications that the input is intended to be less than 50 V a.c. or 120 V d.c. to earth include:

- the full-scale deflection marking on a single-range indicating voltmeter or maximum marking of a multirange voltmeter,
- the maximum range marking of a voltage-selector switch, and
- the marked intended function of the instrument (e.g., millivolt meter).

Voltage and current-measuring circuits that are permanently connected and not accessible need not be marked. The measurement category and the rate-maximum voltage and current for such terminals can be included as part of the installation instructions. An exception is also permitted for circuit terminals that are dedicated for connection to specific terminals on other equipment, provided that there is a means for identifying these terminals. Markings should be placed adjacent to the terminals unless there is insufficient space. In this case, the markings may be placed on the rating plate or scale plate, or the terminals may be marked with Symbol 14 in Table 17.2 and the required information appearing in the IFU.

4. Switches and circuit breakers—If the power-supply switch or circuit breaker is used to disconnect the instrument from the supply mains, the on and off positions must be clearly marked. Symbols 9 and 10 in Table 17.2 can, in some cases, also be suitable as the switch identification. A lamp alone is not considered to be a satisfactory marking. Symbols 9 and 10 are not to be used for switches other than power-supply switches. If a push-button switch is used as the power-supply switch, Symbols 9 and 15 in Table 17.2 may be used to indicate the on position or Symbols 10 and 16 to indicate the off position. The pairs of symbols must appear close together (EN 61010-1 §6.1.6).

5. Instruments protected by double or reinforced insulation—Instruments that are protected throughout by double or reinforced insulation are to be marked with Symbol 11 of Table 17.2 unless the instrument is provided with a protective conductor terminal. Instruments that are only partially protected by double or reinforced insulation should never bear Symbol 11 of Table 17.2 (EN 61010-1 §6.1.7). These markings must be on those parts of the instrument that cannot be removed by the user (EN 1658 §4.7).

6. Field-wired terminal boxes—If the temperature of the terminals or the enclosure of a field-wired terminal box or compartment exceeds 60 °C in normal conditions when measured at an ambient temperature of 40 °C, or the maximum rated ambient temperature is higher, there must be a marking of the maximum temperature rating of the cable to be connected to the terminals. This marking should be visible before and during connection, or be beside the terminals (EN 61010-1 §6.1.8).

7. Batteries and battery charging—If the instrument is equipped with a means of recharging batteries, and if nonrechargeable batteries would fit in the battery compartment, there must be a warning in or near the battery compartment. The warning marking must warn against charging nonrechargeable batteries and indicate the type of rechargeable battery that can be used with the recharging circuit. If an explosion or fire hazard could occur if the wrong battery is used in the instrument, a warning must also be placed on or near the battery compartment. Symbol 14 in Table 17.2 is an acceptable marking (EN 61010-1 §13.2.2). These markings must be on those parts of the instrument that cannot be removed by the user (EN 1658 §4.8.1). For battery-powered IVD instruments and equipment, the specific battery type and polarity of the battery connections must be marked in or near the battery compartment (EN 1658, § 4.8.2).

Markings for Investigational Use (§8.4(f))

For the special marking on devices intended to be used for gathering performance evaluation data, see the section, "Performance-Evaluation Device Marking" in Chapter 14.

INSTRUCTIONS FOR USE (IFU)

The IVDD requires that each device must be accompanied by the information needed to use it properly. IFU must be on the device itself and/or on the packaging for each unit, or, where appropriate, on the sales packaging. When full labeling of each unit is not practical, the IFU must be set out on the packaging and/or the instruction leaflet supplied with one or more devices.

An exception to this requirement is allowed if these devices can be used safely without such IFU. These would be devices where the users, because of their training and experience, would be familiar with their proper use.

ER 8.7 of the IVDD specifies the minimum information that, when appropriate, must be included in the instructions for use for an IVD device.

Instructions for Use for Reagents

IVD reagents must be supplied with IFU. It is common practice to supply the IFU as a package insert. In special cases, the IFU may be given on the outer container or in an operation manual that, in conjunction with the IFU of an instrument or other part of the analytical system, allows the user to safely and properly carry out the test procedure (EN 375 §5.1). The information provided must be sufficiently detailed to ensure proper performance of the procedure and safe use of the reagent. For reagents intended for self-testing use, an explanation of the measurement results and necessary follow-up action should be included. Any symbols and identification colors used on any labels that are not in European harmonized standards must also be described (Directive 98/79/EC Annex 1(8.2)).

Particulars from the Labels (§8.7(a))

The IFU must contain all of the applicable information required on the outer container label of the IVD, except:

- the lot number of the device, and
- the expiration date.

Application and Intended Use (§8.7(d))

The intended use is to be described in terms of the field of application. For professional-use products, this may be sufficiently described by the product name or analytical method (e.g., "Glucose, Hexokinase Method" or "ASAT") (EN 375 §5.5). For self-test products, the intended use must be described in terms that are readily understood by the layperson (e.g., "pregnancy test").

All products should bear a clear statement that they are for *in vitro* use (e.g., "for *in vitro* diagnostic use [only]"). For self-test products, this should be aimed at the layperson (e.g., "not to be swallowed") (EN 376 §5.6).

Composition of Reagents (§8.7(b))

In a kit, each component must be identified using the same designations (i.e., name, letter, number, symbol, color, or graphic) as on the label of the immediate container (EN 375 §5.7).

In addition, the instructions for use for professional-use products should contain a description of the nature and amount of the active ingredient(s) of each reagent(s) expressed in SI units. An explanation of other ingredients (e.g., stabilizers, host media) that influence the reaction should be described.

Additional Materials and Devices (§8.7(e))

A list of any materials or equipment required but not provided with the reagent (e.g., stopwatch, cotton wool) must be included so the user can identify and obtain the proper supplies and instruments.

Methodology (§8.7(h)(1))

The IFU must include information on the principle behind the method of the test. For self-test products, a brief description of how it works in basic terms, understandable to the layperson should be included (EN 376 §5.12.1).

For professional-use products, a more extensive explanation of the method is appropriate. This explanation should indicate the type of reaction (e.g., chemical, microbiological, or immunochemical) and provide a description of the indicator or detection system (EN 375 §5.12.1).

Performance Criteria, Limitations, and Possible Errors (§§8.7(d), 8.7(h)(2), and 13.7(h)(4))

The factors and circumstances that can affect the results of the test (e.g., fasting or medication) together with precautions to avoid possible errors must be included in the IFU.

For professional-use products, the analytical performance characteristics such as sensitivity, specificity, precision, repeatability, reproducibility, and accuracy must be explained. Any limitations of the method, and, when appropriate, information on the diagnostic sensitivity and specificity should be provided (EN 375 §5.12.2). The professional user also needs to be informed about the use of available reference measurement standards and materials. Any special training that is required should be described.

Reagent Preparation (§8.7(h)(3))

If required, the IFU must describe the steps that are needed to prepare the reagent for use (e.g., reconstitution, incubation, dilution, instrument checks).

Storage and Shelf Life after Opening (§§8.7(c) and 8.7(j))

The IFU must give the storage conditions and shelf life following the first opening of the immediate container. In addition, the storage conditions and stability of working reagents are to be provided. This information is required if the storage conditions and shelf life are different than those specified on the container.

When appropriate, the IFU should tell the user how to identify when reagents have deteriorated to a point that the analytical performance of the IVD is affected (e.g., by change in color).

Specimens (§8.7(f))

The IFU must explain the type of sample, the method for obtaining the specimen, any pretreatment required, and, if necessary, the storage conditions for maintaining the sample. For a professional-use test, the IFU should include instructions for the preparation of the patient (EN 375 §5.10).

Test Procedure (§8.7(g))

A detailed test procedure that can be followed by the operator must be provided in the IFU of all IVD devices. For a self-testing device, the detailed procedure ("how to carry out the test") should explain the steps for preparing the working reagents. When applicable, procedures for carrying out a control should be given. These procedures must be clear to a layperson. Illustrations or line drawings may be useful in simplifying the instructions.

If any significant changes from previous editions of the IFU have been made, the changes should be highlighted and explained (EN 375 §5.14; EN 376 §5.14).

Reading and Explanation of Results (§§8.7(i), 8.7(k)(1), 8.7(k)(2), and 8.7 (l))

For a self-test device, the IFU should include a section on the procedures for reading the test results. The results need to be expressed and presented in a way that is readily understandable to a layperson. This section of the IFU should describe the meaning of the test results in the light of the intended use for the test. When applicable, measuring intervals should be included. Proposals for action to be taken by the user in the event of unexpected results should be included (EN 376 §5.13).

A professional-use device, on the other hand, should provide the mathematical formula used to calculate the results. This could be accomplished by providing the name of a computer program used for that purpose. Suitable quality-control material should be provided together with the imprecision and inaccuracy to be expected, as well as procedures to validate the results. Where appropriate and available, reference intervals should be given for the analyte(s) being determined. including the appropriate reference population.

Follow-up Action (§§8.7(t)(1), 8.7(t)(3), and 8.7(t)(4))

The user of a self-test product must be informed of the appropriate follow-up action to be taken in case of a positive, negative, or indeterminate result (e.g., "consult your physician") (EN 376 §5.14). The user should be informed of the possibility of false positive or false negative results. The IFU should contain a statement clearly directing the user not to take any decision of medical relevance without consulting a medical practitioner. If the IVD is being used to monitor a preexisting condition (e.g., home blood-glucose testing), the IFU should tell the patient to refrain from adapting the treatment unless he or she has received appropriate training.

Precautions and Warnings (§8.7(s))

If there are hazards (e.g., chemical, radioactive, or biological) associated with the product or its use, appropriate warnings and precautions should be included in the IFU. These should include the risks reasonably associated with misuse.

If there are unusual risks or special precautions that should be taken when disposing of the material, these need to be covered in the IFU. Particular attention should be drawn to proper handling of substances of human or animal origin to prevent the spread of infection.

Sterile Package (§§8.7(o) and 8.7(p))

If the product is provided in a sterile condition or in some other special microbiological state or state of cleanliness, the IFU must include instructions to the user in the event of damage to the protective package. These instructions could range from returning the device to the manufacturer to procedures for resterilization or decontamination. If resterilization is recommended, then the IFU must provide the details of the appropriate methods for sterilization. The IFU should also contain warnings about inappropriate sterilization (e.g., "Do not autoclave").

If the user is required to perform any special processes (e.g., sterilization) or other special handling before the IVD device can be used, these must be detailed in the IFU.

Radiation-Emitting Products (§5.3)

IVD devices that emit radiation must include in the IFU a detailed description of the nature, type, intensity, and distribution of the radiation. In addition to providing information on the nature of the radiation, the IFU must describe means for protecting the user and ways to avoid mishandling.

Literature References (§§8.7(h), 8.7(i), 8.7(k), and 8.7(l))

The instructions for use of a professional-use IVD device should provide applicable literature references in a bibliography (EN 375 §5.18).

Particular Information that May Be Omitted (§8.7(t)(2))

For a self-test IVD device, specific particulars required in the IFU (such as composition of the reagents) may be omitted if the information provided is sufficient to enable the layperson to know how to use the IVD device and to understand the results of the test. The supplier should be able to provide adequate information on the product (e.g., expected accuracy, precision, analytical sensitivity, and specificity) in response to a request from the physician, pharmacist, and/or consumer.

Date of Issue for the Instructions for Use (IFU) (§8.7(u))

All material containing the IFU must bear the date on which the instructions were issued or the date on which they were last revised.

INSTRUCTIONS FOR USE (IFU) FOR INSTRUMENTS AND EQUIPMENT

IVD instruments must also be supplied with IFU. It is common practice to supply the IFU in a separate user manual, which provides the necessary information for the safe and correct operation, maintenance, and basic troubleshooting of the instrument. For instruments intended for self-testing use, the user manual needs to be expressed in simple, easy-to-understand terms.

A comparison of the requirements in ER 8.7 of the IVDD and ER 13.6 of the MDD will demonstrate that there are a number of similarities between the basic requirements. For instruments and equipment, the basic requirements described in Chapter 15 can be used as a starting point, paying special attention to the requirements of the standards used to demonstrate conformity (i.e., EN 61010-1).

There are, however, several important differences in the requirements. The MDD lists a few requirements that are not mentioned in the IVDD. These requirements relate to the use of the medical device with a patient (i.e., risk associated with implantation [ER 16.6(e)], risk of reciprocal interference [ER 13.6(f)], and medicinal products delivered by or incorporated into the device [ERs 13.6(m) and 13.6(o)]). There are also several requirements (ERs 13.6(k) through 13.6(p)) in the MDD that discuss material that must be available to allow the medical staff to brief the patient on contraindications or precautions to be taken. In Chapter 15, these subjects are described under the heading of "Patient Information." However, when applicable to the IVD instrument in question, this information must be present in a form that is understandable by the intended user, be that user a trained professional or a layperson.

The IVDD contains a number of additional requirements that are unique to IVD instruments and equipment. These special requirements are described in the following sections.

Particulars from the Labels (§8.7(a))

The IFU must contain all of the applicable information required on the outer container label of the IVD, except:

- the lot number of the device; and
- the expiration date, in those cases where there is one associated with the instrument.

Application and Intended Use (§8.7(d))

The intended use is to be described in terms of the field of application. For professional-use products, this may be sufficiently described by the product name or analytical method. For self-test products, the intended use must be described in terms that are readily understood by the layperson (e.g., "blood-glucose monitoring system").

All products should bear a clear statement that they are for *in vitro* use (e.g., "intended for *in vitro* diagnostic use").

Additional Materials (§8.7(e))

This section of the user manual should describe any special materials needed for the effective use of the device, if these materials are not supplied with the instrument. This would include a list of needed reagents, in sufficient detail (i.e., product name, manufacturer, concentration, etc.) to allow the appropriate materials to be obtained.

Methodology (§8.7(h)(1))

The user manual should include information on the principle behind the method of the test. For self-test products, a brief description of how it works in basic terms, understandable to the layperson, should be included.

For professional-use products, this should indicate the type of reaction (e.g., chemical, microbiological, or immunochemical) and a description of the indicator or detection system. A bibliography with references that support the material presented should be included.

Performance Criteria, Limitations, and Possible Errors (§§8.7(h)(2) and 8.7(h)(4))

The factors and circumstances that can affect the results of the test (e.g., fasting or medication) together with precautions to avoid possible errors must be included in the user manual. The user manual should provide information about the accuracy of the instrument claimed by the manufacturer.

For professional-use products, the analytical-performance characteristics such as sensitivity, specificity, precision, and accuracy must be explained. Also, the professional user needs to be informed about the use of available measurement standards. Any special training that is required should be described.

Specimens (§8.7(f))

The manual should explain the type of sample, the method of obtaining the sample, any pretreatment required, and, if necessary, the storage conditions for maintaining the sample.

Instrument Operation and Test Procedure (§8.7(g))

Detailed operating instructions that can be followed by the operator to successfully carry out the intended function of the instrument must be provided. If there are any significant changes from previous editions of the user manual, the changes should be highlighted and explained.

The IFU should describe, in a level of detail appropriate to the intended user, the following information (EN 61010-1 §5.4.4):

- Identification of the operating controls and their use in all operating modes
- An explanation of all safety-related symbols used on the equipment. Other symbols and identification colors used on any labels that are not in the European harmonized standards must also be described (Directive 98/79/EC Annex 1(8.2))
- An instruction on how to position the instrument for proper operation and a warning, if appropriate, not to position the instrument so that it is difficult to operate the device or to disconnect the instrument from the mains
- Instructions for connection to accessories and other equipment, including an indication of suitable accessories, detachable parts, and any special materials needed
- Instruction for replacement of consumable materials

- Cleaning and decontamination instructions (see the description of maintenance and service requirements below)
- Limits on continuous operation (i.e., the duty cycle of the instrument)
- Warnings about potentially poisonous gases that can be liberated from the instrument and possible quantities
- Detailed instruction about risk-reduction procedures when the instrument contains or uses flammable liquids

For a self-test device, the detailed procedure ("how to carry out the test") should explain the steps for preparing the instrument and for the test itself. When applicable, procedures for carrying out a control should be given. These procedures must be clear to a layperson. Photographs, illustrations, or line drawings may be useful in simplifying the instructions.

Reading and Explaning of Results (§§8.7(i) and 8.7(l))

For a self-test device, the IFU should include a section on the procedures for reading the test results. This section should describe the meaning of the results in the light of the intended use of the test. When applicable, measuring intervals should be included. Proposals for action to be taken by the user in the event of unexpected results should be included (EN 376 §4.4.6).

A professional-use device, on the other hand, should provide the mathematical formula used to calculate the results. This could be accomplished by providing the name of a computer program that can be used to calculate the results.

Follow-Up Action (§§8.7(t)(1), 8.7(t)(3) and 8.7(t)(4))

The user manual for a self-test product must inform the intended user of the appropriate follow-up action to be taken in case of a positive, negative, or indeterminate result (e.g., "consult your physician") (EN 376 §4.6.7). The user should be informed of the possibility of false positive or false negative results. A statement clearly directing the user not to take any decision of medical relevance without consulting a medical practitioner should be prominently placed in the user manual. If the IVD device is being used for monitoring a preexisting condition (e.g., home blood-glucose testing), the IFU should tell the patient to refrain from adapting the treatment unless he or she has received appropriate training.

Internal Quality Control (§8.7(k))

Suitable quality-control material should be provided. The imprecision and inaccuracy to be expected in the measurement should be explained, and procedures to validate the results included. Where appropriate and available, reference intervals should be given for the analyte(s) being determined. If calibration or adjustment of the instrument is required, this should be explained. Procedures should be provided that are keyed to the knowledge and skills of the expected user.

Literature References (§§8.7(h), 8.7(i), 8.7(k), and 8.7(l))

The user manual for a professional-use IVD instrument should provide applicable literature references in a bibliography.

Installation, Calibration, and Changes in Performance (§§8.7(n), 8.7(q), and 8.7(j))

If the IVD device is to be used in conjunction with, installed with, or connected to other medical devices or equipment, the IFU must provide sufficient details to enable the user to identify the correct devices. The user needs to know the characteristics of the interfaces and any limitations in order to obtain a safe and proper combination that will allow the device to perform its intended purpose.

When necessary, the IFU must provide directions for proper installation of the instrument. This should include the following, as appropriate (EN 591 §5.8):

- The actions that the user should take on delivery of the instrument including instructions for unpacking, checking that all materials are present, and inspection for obvious damage during transport
- Any site-preparation work that is necessary for proper installation, including the necessary technical specification (e.g., power-supply wiring requirements, instructions for protective earthing, requirements for circuit breakers, and other protective devices, load-bearing capacity, ventilation requirements, and other special services such as air or cooling liquid)
- If the instrument emits excessive noise, instructions for dealing with sound-pressure levels
- The steps to commission the instrument including set-up procedures and checks for proper installation

Periodic maintenance and calibration may be necessary in order to ensure that the instrument functions properly. The IFU must describe these procedures in sufficient detail to allow the user (professional or layperson) to perform the required tasks. These include (EN 591 §§5.20 and 5.21):

1. Maintenance and service—The IFU need to describe the nature and frequency of required preventive maintenance. This should include instruction for cleaning, and, if required, decontamination, disinfection, and sterilization. The manufacturer must state if there are any restrictions on the number of reuses of the instrument.

   The manufacturer should also provide a component list including relevant working materials and tools, and a list of consumables (e.g., fuses) necessary to perform the maintenance procedures. Special attention should be given to the inspection and replacement, if necessary, of hoses and other parts containing liquids, if their failure could cause a hazard.

   For instruments using replaceable batteries, the specific battery type must be stated in the IFU. If an explosion or fire hazard could occur if the wrong battery is used in the instrument, a warning must appear in the IFU (EN 61010-1 §13.2.2).

   If the user is intended to service the instrument, additional instructions for the service personnel should be provided. These may be in the IFU or they may be in a separate service manual depending on the nature and complexity of the service procedures and whether or not the service is to be performed by specially trained personnel. At a minimum, the service instructions should describe the procedures to be followed and provide a list of recommended spare parts. Any parts that are required to be inspected by or supplied by the manufacturer or an authorized agent must be clearly identified. If there are special requirements for the service personnel, such as factory-authorized training, these must be described. Depending on the nature of the service to be performed, the service instructions may need to provide detailed technical descriptions including circuit diagrams and testing and troubleshooting procedures. The maintenance procedures should include any tests necessary to check that the instrument is operating properly and is still in a safe condition after the service is performed.

2. Troubleshooting—A troubleshooting guide should be provided that helps the user understand the meaning of error messages and other alarm signals such as warning lights and

alarm sounds. This information should be organized to help the user quickly locate the needed information and should provide a concise explanation of the cause and the actions required to correct or eliminate the problem. Those situations that require servicing the instrument must be clearly identified.

The troubleshooting guide should explain to the user the actions to be taken, including recalibration or other service procedures, if a change in the analytical performance of the instrument is observed.

3. Calibration—Some instruments, especially those for professional use, may require periodic calibration in order to maintain their performance within the manufacturer's specified limits. The IFU must specify the frequency of routine calibration, the calibration procedures, and any required reference. The user should be advised to maintain records for the calibration of the instrument for quality-control purposes. The manufacturer may provide service calibration-log forms or other record-keeping materials as a customer service.

Technical Specification (§§8.7(h)(2) and 8.7(r))

The IFU should contain the technical specifications for the instrument, including (IEC 61010-1 §5.4.2):

- the dimensions and mass of the instrument;
- the supply voltage or voltage range, frequency or frequency range, and the power or current rating;
- requirements for other inputs such as gas or water pressure and the consumption rating;
- a description of all input and output connections and any restriction on those connections;
- the range of environmental conditions in which the instrument is designed to operate (e.g., temperature, humidity, pressure) and precautions to take regarding exposure to reasonably foreseeable environmental conditions, to magnetic fields, external electrical influences, electrostatic discharge, pressure or variation in pressure, acceleration, thermal ignition sources, and so on;
- the requirements for ventilation, cooling water, and so on;
- the sensitivity of the instrument to electromagnetic disturbances in the reasonably foreseeable environment of use and the level of electromagnetic emissions from the instrument;
- power-on default settings provided by the manufacturer;
- warning if an explosion or fire hazard exists if the wrong batteries are used in the instrument (EN 61010-1 §13.2.2); and
- if appropriate, an indication of the degree of protection against the entry of liquids or particulate matter into the instrument enclosure, for example, by using the rating system in IEC 60529.

Particular Information that May Be Omitted (§8.7(t)(2))

For a self-test instrument, specific particulars required in the user manual may be omitted if the information provided is sufficient to enable the layperson to know how to use the instrument, and to understand the results of the test. The supplier should be able to provide adequate information on the product (e.g., precision, analytical sensitivity, and specificity) in response to a request from the physician, pharmacist, and/or consumer (EN 376 §5).

Date of Issue for the Instructions for Use (IFU) (§8.7(u))

All material containing the IFU must bear the date on which the instructions were issued, or the date on which they were last revised.

DEVICES INCORPORATING DANGEROUS SUBSTANCES

IVDs may incorporate hazardous substances that are subject to regulation under the Classification, Packaging, and Labeling of Dangerous Substances Directive (DSD).* Such devices must be labeled according to DSD Articles 23, 24, and 25 (Directive 92/32/EEC pp. 154/12–154/13).

As an aid to cross-referencing to the DSD, the number of the article in the directive (e.g., Article 23) is listed with the corresponding topic.

HAZARDOUS SUBSTANCE LABELING (ARTICLE 23)

Under the provisions of Article 23 of the DSD, every package must be clearly and indelibly labeled with the following information.

- The package label must display the name of the hazardous substance under one of the designations given in Annex I of the latest amendment to the DSD. Annex I is published periodically in the *Official Journal of the European Communities*. If the substance is not yet listed in Annex I, the name must be given using an internationally recognized designation.
- The package label must include the name, full address, and telephone number of the person established in the EU who is responsible for placing the substance on the market. That person could be the manufacturer, the importer, or the distributor.
- The label of the inner and outer container must contain the appropriate danger symbol and signal word shown in Table 17.3. The symbol is printed in black on an orange-yellow background. The particular danger symbol and signal word to be used for each substance is indicated in Annex I of the DSD. For dangerous substances not yet appearing in Annex I of the DSD, the danger symbol and signal word are assigned according to the rules in Annex VI of the DSD.

TABLE 17.3
Hazard Symbols and Indications of Danger

No.	Symbol [a]	Code	Signal Word	Code	Signal Word
1		E	Explosive		
2			Oxidizing		
3		F	Highly flammable	F+	Extremely flammable
4		T	Toxic	T+	Very toxic
5		C	Corrosive		
6		Xn	Harmful	Xi	Irritant
7		N	Dangerous for the environment		

[a] The symbols in this table are for convenience of reference only and have no official sanction. The reader is referred to Directive 2001/59/EC. For official translation of the signal words, see Annex II of 2001/59/EC.

Source: Directive 2001/59/EC p. 225

---

* Council Directive 67/548/EEC of 27 June 1967 on the approximation of laws, regulations and administrative provisions relating to the classification, packaging and labelling of dangerous substances. 1976. *Official Journal of the European Communities*, 10, no. P 196, as amended (August 6).

- When more than one danger symbol is assigned to a substance:
  - The requirement to display the toxic symbol (see Table 17.3, Symbol 4) makes the harmful symbol (see Table 17.3, Symbol 6) and the corrosive symbol (see Table 17.3, Symbol 5) optional, unless Annex I of the DSD provides otherwise;
  - The requirement to display the corrosive symbol (see Table 17.3, Symbol 5) makes the display of the harmful symbol (see Table 17.3, Symbol 6) optional; and
  - The requirement to display the explosive symbol (see Table 17.3, Symbol 1) makes the display of the flammable symbol (see Table 17.3, Symbol 3) optional.
- Standard risk phrases indicate the special risks arising from the danger involved in using the substance. The wording of the risk phrases is established in Annex III of the DSD. The particular risk phrase to be used for each substance is indicated in Annex I of the DSD. For dangerous substances not yet appearing in Annex I of the DSD, the risk phrase is assigned according to the rules in Annex VI of the DSD.
- Standard safety phrases relating to the safe uses of the substance are to be included. The wording of the safety phrases is established in Annex III of the DSD. The particular safety phrase to be used for each substance is indicated in Annex I of the DSD. For dangerous substances not yet appearing in Annex I of the DSD, the safety phrase is assigned according to the rules in Annex VI of the DSD.
- The label must display the European Communities (EC) number assigned to the substance (Directive 97/69/EC §1(2)(a)). The European Commission assigns the EC number to each substance covered by the DSD. In addition, the label of a product containing a substance appearing in Annex I of the DSD must include the words "EC label" (Directive 97/69/EC §1(2)(b)).

For substances that are classified as irritant, highly flammable, flammable, or oxidizing, the applicable risk phrase and safety phrase need not be provided on the label if the package does not contain more than 125 milliliters of hazardous substance. A harmful substance that is not retailed to the general public need not bear the risk phrase and safety phrase on its label if its package does not contain more than 125 milliliters of the harmful substance.

The label of a product containing a substance subject to the DSD must not contain indications such as "nontoxic," "nonharmful," or any other similar indications.

Where it is not yet possible to label them in accordance with the principles set out in Article 23, the label should bear, in addition to the label deriving from the tests already carried out, the following warning (Directive 92/32/EEC §8(5)):

CAUTION - SUBSTANCE NOT YET FULLY TESTED.

IMPLEMENTATION OF LABELING REQUIREMENTS (ARTICLE 24)

A label bearing the particulars required by DSD Article 23 must be affixed to one or more surfaces of the package so that the information can be read horizontally when the package is resting in its normal position. The minimum dimensions of the label based on the size of the container are shown in Table 17.4.

Each symbol must cover at least one-tenth of the surface area of the label. However, the symbol cannot be less than 1 cm^2. The label must be affixed to the package immediately containing the substance. These dimensions are intended solely for provision of the information required by the DSD and any necessary supplementary health and safety indications. The required information may be printed on a label affixed to the package or may be printed directly on the container.

TABLE 17.4
Minimum Hazard-Label Sizes

Capacity of the Package (liters)	Minimum Label Dimension (in millimeters)
≤3	52 x 74
3 < Capacity ≤ 50	74 x 105
50 < Capacity ≤ 500	105 x 148
>500	148 x 210

Source: Directive 92/32/EEC Article 24(1)

The danger symbol and its background must stand out clearly on the package. The printed information required by the DSD must stand out clearly from the background and be of a size and spacing that can be easily read.

Member states may require that the labeling specified in the DSD be printed in their official language or languages before the product is allowed on the market.

For the purpose of the DSD, the labeling requirements will be considered satisfied if the following conditions are met:

- In the case of an outer package containing one or more inner packages, the requirements are satisfied if the outer package is labeled in accordance with international rules on the transport of dangerous substances and the inner packaging is labeled according to the DSD.
- In the case of a single package, the requirements are satisfied if the package is labeled in accordance with international rules on the transport of dangerous substances. In addition, the label must include all of the information required by DSD Article 23 except the "danger symbol" in Table 17.3.

Exemptions from Labeling and Packaging Requirements (Article 25)

Member states may permit deviations from the requirements of DSD Articles 23 and 24 under the following conditions:

- A member state may permit the labeling required by Article 23 to be applied in some other appropriate manner on packages that are either too small or otherwise unsuitable for labeling according to DSD Article 24(1) and (2).
- A member state may permit a package containing dangerous substances that are explosive, very toxic, or toxic to be labeled in some other appropriate way if it contains such small quantities that there is no reason to fear any danger to people.
- A member state may permit a package containing dangerous substances that are not explosive, very toxic, or toxic to either be unlabeled or to be labeled in some other way if it contains such small quantities that there is no reason to fear any danger to people.

These exceptions do not permit the use of symbols, signal words, risk phrases, or safety phrases that are different from those specified in the DSD.

THINGS TO REMEMBER

The IVDD covers the placing on the market and putting into service of IVD devices. The information provided by the supplier (the manufacturer, the manufacturer's agent registered in the EU, or the importer) with an IVD device is covered by the ERs in Annex I of the IVDD. Failure to follow or satisfy these requirements may lead to the device being deemed improperly labeled.

The labeling requirements are contained in ER 13 in Annex I. The directive specifies the contents of the label (ER 13.4) and the IFU (ER 13.7). Practically, the requirements can be separated into those applicable to reagent products and those applicable to instruments and equipment. The labeling for regent products can be divided into the requirements for the immediate container label, the outer container label, and the IFU. For instruments and equipment, the requirements are divided into those applicable to the labels on the device and/or the container, and the user manual. The requirements for instruments and equipment are similar to those described in Chapter 15 for medical devices. The unique requirements for IVD devices are described in this chapter.

IVD devices may incorporate hazardous substances that are subject to regulation under the DSD.

Part VII

Japan

18 The Pharmaceutical Affairs Law of Japan

The government of Japan has a long history of regulating the quality, efficacy, and safety of medical products. Enacted in 1948, the Pharmaceutical Affairs Law (PAL) of Japan places the responsibility for examining the safety, effectiveness, and quality of medical products under the Pharmaceutical and Medical Safety Bureau (PMSB) of the Ministry of Health, Labor, and Welfare (MHLW). The PMSB is one of two bureaus within the MHLW responsible for assuring the quality, efficacy, and safety of drugs, quasi drugs, cosmetics, and medical devices. The Health Policy Bureau is in charge of the promotion of research and development, the production and distribution of drugs, and so on. The PMSB is the successor to the Pharmaceutical Affairs Bureau (PAB) (Pharmaceutical p. 11). The PAL was revised in 1960 to control and regulate matters related to drugs, quasi drugs, cosmetics, and medical devices to assure their quality, efficacy, and safety (PAL §1). The 1960 revision added coverage of a variety of instruments and apparatus developed for the diagnosis, prevention, and treatment of disease. Since 1960, the PAL has been amended numerous times to keep pace with social needs, technological evolution, and changing market requirements. In 1983, the PAL was modified to allow foreign medical device manufacturers to apply directly to the Japanese government for approval of their products intended for importation into Japan. The PAL was dramatically updated in 1994 to restructure the regulations to take greater account of the range and complexity of modern medical devices.

Medical devices are defined in Article 2.4 of the PAL as "equipment or instruments intended for use in the diagnosis, cure, or prevention of disease in humans or animals, or intended to affect the structure or functions of the body of humans or animals, and which are designated by cabinet order." Not all equipment or instruments intended for use in the diagnosis, cure, or prevention of disease are regulated by the PAL. Only those devices listed in a cabinet order are subject to regulation under the PAL.

Under the 1960 law, medical devices are categorized based on their purpose. Devices are divided into five types: (1) instruments and apparatus, (2) medical products, (3) dental materials, (4) sanitary products, and (5) medical devices exclusively for animals (Guide p. 2). Article 83 of the PAL places responsibility for devices that are intended only for animal use under the jurisdiction of the Ministry of Agriculture, Forestry, and Fisheries. Veterinary devices regulated by the Ministry of Agriculture, Forestry, and Fisheries are not covered in this book.

The products included in the four types of human-use medical devices are listed in a table that is attached to the Enforcement Ordinance of the Pharmaceutical Affairs Law (EOPAL). This table "Attached Table 1" is further subdivided into 103 categories. In 1983, the EOPAL was amended by cabinet order to include the generic names of medical devices (PAB No. 752). The 103 categories of human-use medical devices, along with representative devices within many of the categories, are listed in Table E.1 in Appendix E.

1994 REVISION TO THE PAL

In 1994, the Japanese Diet adopted a major overhaul of the PAL related to medical devices. The MHLW began enforcing the provisions of the amendments on July 1, 1995, with various provisions of the amendments coming into force within two years of its adoption.

The 1994 amendments are intended to deal with a number of issues with medical devices that have emerged in the years since the 1960 law was enacted. These include advancements in technology, the perceived need for postmarket surveillance of novel devices, the tracking of certain life-sustaining devices, and the increase in the home use of medical devices.

As home-use medical devices are intended to be used by person with no professional training, emphasis is placed on safety without the need for special training. Information for lay users, such as the instructions for use (IFU), should be described in plain language (Guide pp. 361–362).

For medical devices used exclusively by professionals, the provisions of the 1994 amendment are based on a four-tiered risk-classification scheme. The classification scheme was originally described in a proposal for reclassification of medical devices written in October 1993 by a Study Group on Policies for Medical Devices chaired by Dr. Takemochi Ishii. The four classes, along with a summary of the regulatory requirements for each class, are described in Table 18.1.

Class IV includes devices are life supporting or life sustaining and have an immediate effect in case of a failure. These are the devices designated by the MHLW as those whose location must be known in order to not jeopardize public health and hygiene. Examples include devices implanted in the human body or other devices that may be used outside facilities providing medical treatment (PAL §77-5). Collectively, these devices are known as "designated medical devices." In addition to approvals and licenses, the manufacturer or importer of designated medical devices are required to register patients and provide postmarket tracking. The devices currently designated by the MHLW are listed in Article 64-6 of the Enforcement Regulations of the Pharmaceutical Affairs Law (ERPALs). This list is reproduced in Table 18.2.

The 1994 amendments introduce the concept of a "new medical device" into the Japanese regulations. New medical devices are defined as "medical devices for which the structure, method of use, indications or performance are different from those medical devices which have already been approved for manufacture or import" (PAL §14-4). The MHLW will establish the reevaluation period at the time the initial license is granted. The reevaluation period will typically be three to four years but may be extended by the MHLW for up to seven years. Labeling can effect this classification through the claims made with respect to the method of use, indications, or performance of the device if they appear to the be significantly different from devices already approved.

When developing the 1994 amendments, the MHLW was concerned that the approval process for medical devices could become an unreasonable barrier to making medical devices of better quality and performance available promptly. The 1994 amendments also require the MHLW to allocate resources to ensure effectiveness and safety through the expansion of postmarket surveillance programs and other measures. To allow the MHLW to focus its resources, the amendment allows the government to entrust a part of the review process for the manufacturing or import approval of medical devices to third-party organizations, known as "designated review organizations" (PAL §14.3). The MHLW would approve or reject the manufacturer's or importer's application based on the results of the third-party investigation.

The 1994 amendments place greater emphasis on the manufacturer providing procedures in the package insert for proper maintenance of specific medical devices. This amendment also clarifies that manufacturers and importers must provide information to medical and pharmaceutical professionals on the proper maintenance of devices.

THE REGULATIONS

The MHLW implements the specific requirements of the PAL through cabinet ordinances and enforcement regulations. The ordinances and regulations relating to medical devices are contained

TABLE 18.1
Summary of the Regulatory Requirements for the Four Classes of Medical Devices in Japan

Class	Description	License Required	Premarket Approval Required	Clinical Trial Required
Class I	Devices to which any of the following applies: (1) It does not come into contact with the body, or comes into contact only with normal, healthy skin, and does not require power, does not supply any matter or energy, and furthermore does not give off radiation or electromagnetic waves. (2) It does not come into contact with the body, or comes into contact only with normal, healthy skin and (1) does not apply; even if it malfunctions, the probability of serious danger to the body is considered to be relatively small. (3) It comes into contact only with the teeth and/or the oral cavity mucous membranes; it is not used to fill the teeth and does not come in prolonged contact with the oral cavity mucous membranes; even if it malfunctions, the probability of serious danger to the body is considered to be relatively small. (4) It comes into contact with tissues, wound sites, etc.;[a] it is not implanted or placed; even if it malfunctions, the probability of serious danger to the body is considered to be relatively small.	Yes	No	No
Class II	Devices to which any of the following applies: (1) It does not come into contact with the body, or comes into contact only with normal, healthy skin; Class I (1) does not apply; if it malfunctions, although the probability of a direct link to danger to human life is relatively small, the probability of serious danger to the body is considered to be relatively large. (2) It comes into contact only with the teeth and/or the oral-cavity mucous membranes; it is not used to fill the teeth and does not come into prolonged contact with the oral cavity-mucous membranes; if it malfunctions, the probability of serious danger to the body is considered to be relatively large. (3) It comes into contact with tissues, wound sites, etc.,[a] it is not implanted or placed; even if it malfunctions, the probability of a direct link to danger to life or serious functional disorder is considered to be relatively small.	Yes	Yes[b]	No
Class III	Devices to which any of the following applies: (1) It does not come into contact with the body, or comes in contact only with normal, healthy skin; if it malfunctions, the probability of a direct link to danger to life is considered to be relatively large. (2) It is used to fill the teeth or comes into prolonged contact with the oral-cavity mucous membranes; if it malfunctions, although it is not considered to be directly linked to danger to human life, the probability of serious danger is considered to be relatively large. (3) It comes into contact with tissues, wound sites, etc.;[a] it is not implanted or placed; if it malfunctions, the probability of a direct link to danger to human life or serious functional disorder is considered to be relatively large. (4) It is implanted or placed; if it malfunctions, the probability of a direct link to danger to life is considered to be relatively small.	Yes	Yes	Yes (on a case-by-case basis)
Class IV	Devices to which any of the following applies: (1) It utilizes human or animal tissue or cells. (2) It is implanted or placed; if it malfunctions, the probability of a direct link to danger to life is considered to be relatively large.	Yes	Yes	Yes (generally)

[a] Includes contact with mucous membranes and the cornea, and contact by means of blood and medical fluids; does not include contact with only the teeth and/or oral cavity mucous membranes.
[b] For Class II devices, submission of a registration is not required if the manufacturer can demonstrate conformity to relevant standards.

Source: Hasegawa p. 196

TABLE 18.2
Designated Medical Devices in Japan

1. Implantable-type cardiac pacemakers
2. Implantable-type cardiac pacemaker leads
3. Implantable defibrillators
4. Implantable defibrillator leads
5. Artificial-heart valves
6. Artificial-valve rings
7. Artificial blood vessels (limited to those used in the coronary arteries, thoracic aorta, and abnormal aorta)

Source: ERPALs Article 64-6.

in the EOPAL and the ERPALs. Other regulations and guidelines are promulgated through official notifications from the Ministry to the Prefectural authorities (e.g., PAB Notification No. 127 of February 13, 1989, The Guidelines for Exhibition of Not Yet Approved Medical Devices).

The PAL is the primary law under which the MHLW derives authority to take action with regards to the labeling of regulated medical devices. Specifically:

- PAL Article 52 defines the information required on the package insert enclosed with the device, or on the container or wrapper;
- PAL Article 53 sets the standard for prominence of required information;
- PAL Article 54 establishes impermissible claims; and
- PAL Article 63 establishes requirements for the information that must appear on the immediate container or wrapper of the medical device or on the device itself.

Articles 52, 53, and 54 are written to apply to drugs. However, PAL Article 64 makes them applicable to medical devices *mutatis mutandis.**

ENFORCEMENT ORDINANCE OF THE PHARMACEUTICAL AFFAIRS LAW (EOPAL)

The EOPAL was enacted in 1961 by the Japanese Cabinet under articles of the PAL that require that the regulations necessary to implement the provision of the PAL be laid down by ministerial order. The EOPAL deals primarily with administrative matters such as procedures for registration of manufacturers. The EOPAL contains the table that establishes the range of medical devices subject to regulation under the PAL (see Table E.1 in Appendix E). The EOPAL contains a few labeling requirements for drugs granted special license before approval. However, there are no direct labeling requirements specified in the EOPAL for medical devices.

ENFORCEMENT REGULATIONS OF THE PHARMACEUTICAL AFFAIRS LAW (ERPALs)

The MHLW established the ERPALs in 1961 under authority of the PAL and Article 16 of the EOPAL. The ERPALs contain specific requirements dealing with product labeling as well as many other aspects of the product- and manufacturer/importer-registration process. These include permissible exceptions or variations in the information required on the immediate container or wrapper of the medical device or on the device itself. These requirements are discussed in Chapter 19.

---

* The Latin phrase, *mutatis mutandis*, meaning "the necessary changes having been made," is used extensively in the PAL to apply provisions that were written to cover drugs and other regulated items including medical devices.

ADULTERATION AND MISBRANDING

Although the PAL does not use the terms "adulteration" or "misbranding," these concepts do appear in the Japanese law. Storing, exhibiting, or conveying, whether free of charge or otherwise, an adulterated or misbranded medical device is a criminal offense. A person who is found guilty of a violation of the PAL is subject to penal servitude or a substantial fine, or both.

Article 65 of the PAL defines those conditions under which a medical device could be considered adulterated. These conditions include:

- a medical device whose properties, quality, or performance does not conform to the properties, quality, or performance approved by the MHLW under Articles 14 and 19.2 of the PAL;
- a medical device that does not conform to standards established for the device by the MHLW;
- a medical device that consists of any impure, putrid, or decomposing substance;
- a medical device in or on which any foreign matter is found;
- a medical device that is contaminated, or is likely to be contaminated, by pathogenic microorganisms; or
- a medical device that, when used, might jeopardize public health or hygiene.

Article 55 of the PAL defines the conditions under which a device could be considered misbranded. A device could be deemed misbranded if the package insert, container, or wrapper (including the inner package) of the medical device, or the labeling on the medical device itself:

- contains false or misleading claims about the medical device;
- contains a statement of indications or effects not approved by the MHLW pursuant to provisions of Articles 14 and 19.2 of the PAL;
- contains directions, dosage, or durations of use that might jeopardize public health or hygiene;
- fails to provide the information required by Article 63, or the appropriate material required by Article 53, and additional material laid down in MHLW ordinances or standards established under Article 42 of the PAL; or
- fails to display the material required by the law and regulations more prominently than other material in the labeling, and fails to communicate the required material in terms that renders it easily read and understood by the ordinary purchaser or user of the product.

A medical device that is misbranded may not be sold, leased, given, stored, or exhibited for the purpose of sale, leasing, or giving.

FALSE OR MISLEADING LABELING

The PAL declares that a device is misbranded if its labeling makes false or misleading claims. The PAL imposes a particularly stiff penalty on a person found guilty of offering for sale or otherwise conveying a device whose labeling contains false or misleading claims. A person who violates this provision of the PAL is subject to a term of penal servitude not exceeding two years or a substantial fine, or both.

ADVERTISING AND PROMOTION

Article 66 of the PAL makes it a crime to explicitly or implicitly advertise, describe, or circulate false or exaggerated statements regarding the name, manufacturing process, indications, effects, or properties of a medical device. Advertising, describing, or circulating statements that would lead to the false impression that a physician or other person has certified the indications, effects, or properties of a medical device is also construed as a prohibited practice. Further, statements or diagrams suggesting criminal abortion, as well as any obscene statements or diagrams, may not be used in conjunction with a medical device. A person found guilty of violating this provision of the PAL is subject to penal servitude or a fine or both.

Article 67 of the PAL makes it a crime to advertise the name, manufacturing process, indications, effects, or properties of a medical device that requires approval of the MHLW, prior to receiving the approval. A person found guilty of a violation of this provision is also subject to penal servitude or a fine, or both.

The provision in Article 67 not withstanding, the MHLW recognizes that it is an internationally accepted practice to display medical devices at exhibitions before they are approved for market release. To facilitate this process, and to see that it is properly controlled, the Ministry has published comprehensive guidelines for exhibition of medical devices that have not as yet been approved (Japanese MHLW, PAB No. 127). Directed to the prefectural authorities, these guidelines lay out the conditions under which an unapproved device can be exhibited. The prefectural authorities are also instructed to follow the spirit of the guideline when dealing with those areas not specifically covered by the guideline.

The guideline recognizes three types of exhibits where an unapproved device may be shown. They are:

- exhibitions for specialists within a field that are aimed at promotion of academic research;
- exhibitions for the general public that are aimed at promotion of scientific or technical issues and/or related industry; and
- exhibitions for the general public that are aimed at giving general information, including the design of medical devices, other than product name, manufacturing method, efficacy, or performance.

The permissible practices at each of the three types of exhibition are discussed in the following sections.

EXHIBITIONS FOR SPECIALISTS PROMOTING ACADEMIC RESEARCH

Such exhibits must be sponsored by official academic circles composed of academic researchers who are involved with the promotion and development of academic research. Any research group that is deeply connected with any private company is excluded. Exhibitions sponsored by academic research groups registered in the Japan Academic Conference, for example, would be eligible under this provision.

The information on the unapproved medical device must be presented by a researcher or by the academic society. The presentation must take place within the hall of the meeting of the scientific society, or within the hall designated by the scientific society. The exhibit must clearly indicate that the device is not yet approved for sale, and the device cannot be sold or given to anyone. Any information presented in the exhibit about manufacturing methods, efficacy, or performance must be based only on objective and scientific data or fact. In principle, any documents related to the unapproved device should not be given out in conjunction with the exhibit. The only exception is to supply, at the request of a physician or other specialists, a scientifically evaluated paper such as those reported in the scientific society.

At the conclusion of the exhibit, the device must be properly disposed of or returned to the manufacturer. It cannot be sold or given away. However, the following exceptions are allowed:

- The device may be used in support of an application for approval after going through the necessary formalities (e.g., for the purpose of clinical trials).
- The device may be stored in a warehouse if approval is expected in the near future.
- The device may be used for other purposes that are allowed under the law.

EXHIBITIONS FOR THE GENERAL PUBLIC PROMOTION OF SCIENTIFIC/TECHNICAL ISSUES AND/OR RELATED INDUSTRY

These events must be sponsored by official organizations. Examples would include exhibitions sponsored by government organizations, local governments, foreign governments, embassies, or other persons having a special status.

The information on the unapproved medical device must be presented by the sponsor of the exhibit. The presentation must take place within the hall designated by the sponsor. The exhibit itself must clearly indicate that the device is not yet approved for sale; it cannot be sold or given to anyone. The product name may not be mentioned in the exhibit, with the exception of an imported product that is allowed to display the product name in a language other than Japanese. Any information mentioned about manufacturing method, efficacy, or performance must be based only on objective and scientific data or fact. Documents relating to the unapproved device should not be given out. The only exception is to supply general introductory scientific documents made by the sponsor of the exhibition and having nothing to do with any specific company or commodity.

At the conclusion of the exhibit, the device must be properly disposed of or returned to the maker. It cannot be sold or given away. However, the following exceptions are allowed:

- The device may be used for application for approval after going through the necessary formalities (e.g., for the purpose of clinical trials).
- If approval is expected in the near future, the device may be stored in a warehouse.
- The device may be used for other purposes as allowed under the law.

EXHIBITIONS FOR THE GENERAL PUBLIC PROVIDING GENERAL INFORMATION

These exhibitions must be sponsored by one of the official organizations mentioned in the previous section or by a public corporation.

The information on the unapproved medical device must be presented by the sponsor of the exhibit. The presentation must take place within the hall designated by the sponsor. The exhibit itself cannot mention the product name, manufacturing method, efficacy, or performance of the device. However, imported products can display the product name printed on the device at the time of assembly in the exporting country if it is not in Japanese. Any documents relating to the unapproved device should not be given out. The only exception is to supply general introductory scientific documents made by the sponsor of the exhibition and having nothing to do with any specific company or commodity.

At the conclusion of the exhibit, the device must be properly disposed of or returned to the manufacturer. It cannot be sold or given away. However, the following exceptions are allowed:

- The device may be used for application for approval after going through the necessary formalities (e.g., for the purpose of clinical trials).
- If approval is expected in the near future, the device may be stored in a warehouse.
- The device may be used for other purposes as allowed under the law.

IN VITRO DIAGNOSTIC (IVD) PRODUCTS

In Japan, *in vitro* diagnostic (IVD) reagents are classified as drugs, while the apparatus used to analyze the results of a test is classified as a medical device. Examples of *in vitro* test apparatus include analyzers for clinical chemistry, immunological analyzers, and blood-type analyzers.* However, since diagnostic reagents are used outside the human body, they have been placed in a special category with respect to public health and hygiene (PAB No. 662). These "*in vitro* test drugs" are placed in one of three classes depending on the novelty of the measured items or measuring methods. The licensing procedure has been simplified based on the class in which the *in vitro* test drug has been placed (Standards p. 27). In addition, Article 1-2-2 of the EOPAL exempts *in vitro* test drugs from the provisions of drug Good Manufacturing Practice (GMP) regulation. Even though IVD reagents are regulated as drugs, the labeling requirements are discussed in Chapter 19.

BRINGING DEVICES TO MARKET IN JAPAN

Any firm that intends to manufacture or import medical devices that fall into one of the categories in Table E.1 in Appendix E must obtain a license, called a *kyoka*, from the MHLW. The manufacturer, or importer, must also obtain a product approval, called a *shonin*, for each medical device that has not been exempted from the approval process by Article 18 of the ERPALs. The requirement for a *shonin* apply equally to devices manufactured in, or imported into, Japan. The basic procedures and requirements for obtaining licenses and product approvals are outlined in the following sections.

MEDICAL DEVICE MANUFACTURER'S (IMPORTER'S) LICENSE

A firm that intends to manufacture or import a medical device that fall into one of the categories in Table E.1 in Appendix E must obtain a license, called a *kyoka*, for each manufacturing plant or importing office in Japan. A license is granted after the examination of an application submitted by the manufacturer or importer. For a manufacturing facility, the examination process stresses:

- the material conditions, such as building and facilities, under which the devices are manufactured;
- that directors in charge of the business are qualified to hold a license under the PAL;
- that the person designated as technical director** in charge of manufacturing is qualified; and
- if applicable, that product approval has been obtained for the product to be manufactured.

Since the importer is held responsible for the quality, efficacy, and safety of the device, it must meet the same basic requirements as those of a manufacturer (Guide pp. 29-30). The importer must have adequate facilities for storing imported devices under sanitary and protective conditions. The importer must also provide adequate facilities, equipment, and utensils for testing the imported devices. In lieu of maintaining the testing facilities on the premises, the importer may, on its own responsibility, use testing facilities located at other institutions.

The 1994 revision to the PAL added requirements for manufacturing control and quality control at manufacturing plants. Referred to as "medical device good manufacturing practices," or simply

---

* These devices are classified as hematological testing apparatus in Category 17, Hematological Testing Apparatus, in the table attached to the EOPAL. This is reproduced in Table E.1 – JAPANESE MHWL CATEGORIES OF MEDICAL DEVICES in Appendix E – JAPANESE MEDICAL DEVICE CLASSIFICATION.

** The term "technical director" is the "responsible manager" in the standards for good manufacturing practices (GMP) for medical devices contained in MHLW Ordinance No. 40.

as "medical device GMPs," this ordinance* establishes the requirement for a quality-assurance (QA) system for manufacturers of certain medical devices. Some medical devices are exempted from the requirements of medical device GMPs. These are listed in Table E.5 of Appendix E.

The medical device GMPs is established by ordinance and are a requirement for obtaining a *kyoka* for manufacturing medical devices not specifically excluded in the regulation. The MHLW has published complementary guidelines under the title, "Quality Assurance Standards of Medical Devices (standards of medical device QA system) and Quality Assurance Standards for Manufacturing Plants of Medical Devices Such as Medical Illuminators (GMP for medical illuminators etc)."** The standards for medical device QA systems are based on ISO Standard 9001:1994. These standards are more detailed than ISO 9001 and include provisions for control of labeling. If followed, the GMPs specified by MHLW ordinance will be satisfied with the exception of a few additional matters, such as standards for manufacturing at two or more plants (Guide p. 269).

The license to manufacture or import a medical device must be obtained for each manufacturing plant or importing office in Japan. The license becomes invalid unless it is renewed every third year. A license is granted to manufacture or import each product. Therefore, a firm that wishes to begin manufacturing or importing a new product must obtain an additional or supplementary license for the new product.

Article 12 of the PAL presumes that any given device is completely manufactured within a single factory. A manufacturer may have multiple factories, each producing a particular device, but all work on a given device takes place with the single factory. The PAL was amended in 1995 to create a partial licensing process, called a *kubun-kyoka*, that covers those situations where a manufacturer subcontracts part of the manufacturing process to a third party. For the purposes of licensing, the third party could be another company or another plant belonging to the same manufacturer. A manufacturer cannot subcontract the final inspection of the product. A subcontractor cannot further subcontract any part of the work it has undertaken on behalf of the original device manufacturer.

The manufacturing processes that are licensable under a *kubun-kyoka* are (Guide p. 251):

- radiation sterilization (gamma ray, electron beam);
- ethylene oxide (EtO) sterilization; and
- coating of lenses for glasses

The medical device classifications to which a *kubun-kyoka* can be applied are listed in Table 1-3 attached to the ERPALs. That table is reproduced in Table E.4 of Appendix E.

The subcontractor may obtain a *kubun-kyoka* for a class of devices. The subcontractor is not required to obtain a license for each product (Guide p. 253).

MEDICAL DEVICE APPROVAL

In principle, a manufacturer or importer must also obtain a product approval, called a *shonin*, for each medical device that falls into one of the categories listed in Table E.1 in Appendix E. The manufacturer or importer must obtain the product approval on a product-by-product basis. As with the *kyoka*, a *shonin* is granted after the examination of a written application submitted by the manufacturer or importer. The examination emphasizes the assurance of the safety, effectiveness, and quality of the device.

---

* *Standard for Manufacturing Control and Quality Assurance of Medical Devices.* 1995. MHLW Ordinance No. 40. (June 26).
** Japanese Ministry of Health and Welfare, Pharmaceutical Affairs Bureau. 1994. PAB Notification No. 1128. (December 28). Tokyo: Printing Bureau, Ministry of Finance.

The information required in the *shonin* application is fixed in Article 17 of the ERPALs. The manufacturer or importer must submit the required data on MHLW Form 10(2). This includes appropriate warnings, contraindications, and the directions for the proper use of the device. In addition to the data listed on Form 10(2), the manufacturer or importer must attach additional data unless it is confirmed that the device is well known in medical field or these is some other justifiable reason. The scope of this data is specified in Paragraph 1, Item 4, of Article 18.3 of the ERPALs and in Medical Device Division Notification No. 100 of June 27, 1995 (Guide p. 72). The written application and the accompanying data must be prepared in Japanese. For documents that are difficult to prepare in Japanese, such as certificates issued by foreign governments, a Japanese translation must be provided (Guide p. 61).

Not all of the devices listed in Table E.1 in Appendix E require a product-by-product approval. Before the 1994 revision, there were 33 categories of devices that were exempt from approval on a product-by-product basis because of their exclusive use by specialists, established effectiveness and safety, and reliable operating technique. The 1994 revision of the PAL significantly expanded the number of devices exempted from the requirement to obtain a *shonin*. Table 1 attached to Article 18 of the ERPALs lists those devices that are exempt from approval on a product-by-product basis. This table is reproduced in Table E.2 in Appendix E. The list of devices exempted from the requirement for individual approval was expanded based on the idea that such approval is only needed for devices that pose a significant risk to the human body (Guide p. 31).

Medical devices that conform to Japanese Industrial Standards (JIS) are included in those that are exempt from approval on a product-by-product basis. Such an exemption is granted on the basis that these devices are widely used, and their quality and description have been established "in accordance with comprehensive knowledge of current medical science and engineering practice" (Guide p. 31).

The manufacturer or importer of a device that is not included in Table E.2 in Appendix E is required to obtain a *shonin* for the device before it is legal to sell it in Japan. In addition, the manufacturer or importer of a device that is listed in Table E.2 will be required to obtain a *shonin* if a new device is not substantially equivalent to a device already approved in Japan. Under the 1994 revision to the PAL, a device listed in Table E.2 is considered a "new medical device" when the structure, usage, indications, effects, or performance differs significantly from a device that has already been approved for manufacturer or import (PAL §14-1, 6(2)).

The conditions for approval are set out in Article 14 of the PAL. Approval of a medical device is to be based on an examination of the name, ingredients and quantities, directions and dosage, indications and effects, properties, side effects, and so on. Approval will not be granted if any of the following conditions exist:

- The device does not possess the indications, effects, or properties indicated on the application.
- The device has no value as a medical device because it has harmful actions that outweigh its indications, effects, and properties.
- In addition to the conditions listed above, the device is designated by MHLW ordinance as inappropriate as a medical device.

DETERMINE SUBSTANTIAL EQUIVALENCE AND REEXAMINATION

The 1994 revision of the PAL introduced the concept of a "new medical device" to the Japanese regulatory system. A new medical device is defined in Article 14 of the PAL as one that has a "distinctly different structure, quantities, method of use, indications, performance, etc. from those medical devices, which have already been approved for manufacture or import." As noted before, this would include devices described in Table 1 attached to the ERPALs (Table E.2 in Appendix E) if they are not substantially equivalent to already approved devices. Under the new system, a device is classified as either a "new medical device," an "improved or modified medical device," or a "me-too medical device."

During the application process, the MHLW will determine if the submitted device is substantially equivalent to an already-approved device. If substantial equivalence is established, the device is exempted from the reexamination provision of Article 14-4 of the PAL. Because the MHLW may not have the necessary resources to make the substantial equivalence determination, the 1994 revision of the PAL allows the Minister to designate external review organizations who will perform this task. The "designated review organizations" are responsible to the MHLW and operate under rules established in the EOPAL. At the time the regulations were implemented, the MHLW granted the Japan Association for the Advancement of Medical Equipment (JAAME) the authority to make the substantial-equivalence determination.

A separate organization, the Pharmaceutical and Medical Device Examination Center (PMEDC) was set up by the MHLW to handle the review of "new medical devices" or "improved or modified medical devices." Applications for these devices are submitted directly to the PMEDC, without need for JAAME involvement. The manufacturer is responsible for determining which of the three categories apply to a device. However, if an application for a "me-too medical device" is judged to be an "improved or modified medical device," the JAAME will return the application to the originator, who must begin the process anew with the PMEDC.

If a device is determined to be a "new medical device," its safety and effectiveness must be reexamined after it has been in the market for a time. The time for reexamination is established by the MHLW at the time of initial approval and may be from three to seven years. The reexamination of the safety and effectiveness of a new medical device is intended to take advantage of any new scientific and technological advances after the time of original approval.

CHANGES IN APPROVED DEVICES

A person who intends to make a change to an approved device may be required to seek prior approval from the MHLW for the change under Article 14, Paragraph 6, of the PAL. A moderate change in the structure, raw material, or performance of the device would require approval. A change in the name, form, dimensions, indications and effects, method of operation or use, or standards and test method may also be subject to approval by the MHLW. A more drastic change in the structure, raw materials, or performance would create a "new" device, requiring a new application.

The determination of when to file for an approval of a partial change should be based on a comprehensive evaluation of the specifics of the change. If the alteration is a slight improvement, minor change, or the addition or deletion of components or accessories such that the identity of the existing device is maintained, then an application for approval will not be required. Overall, the change must have no effect on the safety and effectiveness of the device and must preserve the original identity of the device.

The person (manufacturer or importer) responsible for the device must make an informed judgment on a case-by-case basis about when an application is required. The following are examples of when an application is not required (Guide p. 168):

- Changes in serial numbers and catalog numbers
- Changes in color, coating, and plating methods for cabinets, assemblies, and other parts (Changes in materials where biocompatibility is an issue are not exempt.)

- Changes in the location of keyboards, switches, displays, and other components on a control panel
- Changes in the external dimensions within ±10 percent of the approved dimensions unless the changes would affect the performance or use of the device
- Changes in software that do not involve changes in specification, functions, and so on
- Changes in the packaging units (e.g., from a 10-g unit to a 20-g unit or from a 5-piece unit to a 10-piece unit)
- Changes in packaging materials (excluding packaging materials for sterilized medical devices)
- Minor changes in the method or operation or use associated with other changes that do not require partial-change approval

Moderate changes in an approved device require a partial-change approval from the MHLW. A moderate change could include a change in name, form, dimensions, indications and effects, method of operation or use, or standards and test methods. A change to the structure, raw materials, ingredients and quantity, or performance could be considered moderate depending on the individual circumstances. However, a change structure, raw materials, ingredients and quantity, or performance could be viewed as drastic, in which case the change creates a new device that requires a new device application (Guide p. 165). A change in labeling that effects the indications, or method of operation or use could constitute a change requiring partial approval.

Medical Device Approval for Imported Products

As can be seen in the previous discussion, the PAL makes little distinction between a manufacturer and importer as far as the device approval process is concerned. The importer is responsible for the assurance of the safety, effectiveness, and quality of the approved products produced by foreign manufacturers. An importer does not have to obtain an approval when the foreign manufacturer has obtained the approval directly from the MHLW as described in the following section. The importer of the medical devices into Japan must obtain the facility license (*kyoka*) for each business office in Japan.

Role of the Prefecture Governments

In the Japanese system, the prefectural governments act as intermediaries between the manufacturer or importer and the MHLW. Applications for both *kyoka* licenses and *shonin* product approvals are submitted to the governor of the prefecture where the manufacturing facility or business office is located. In some cases, prefectural government officials will carry out on-the-spot inspections of buildings and facilities and other items described in the application. If the application remains viable after the inspection, the prefectural governor forwards the application with a rider to the MHLW. In other cases, the prefectural governor simply forwards the application to the Ministry. Once the examination is complete, the letter of license and/or letter of product approval is delivered to the applicant through the prefectural government.

Some medical devices are approved by the prefectural government without the intervention of the MHLW. These products include some groups of injection needles and puncture needles, glass syringes, dental scalers, dental-filling instruments, impression or articulation instruments, and sight-corrective ophthalmic lenses. For these products, the prefectural governor is completely responsible for the examination of both the *kyoka* licenses and *shonin* product-approval applications. However, under the 1994 revision of the PAL, most of these devices no longer require a product-by-product approval. Only sight-corrective ophthalmic lenses that do not conform to JIS standards require a product approval.

MEDICAL DEVICE APPROVAL FOR PRODUCTS MANUFACTURED BY FOREIGNERS

Since 1983, foreign manufacturers may apply to the Japanese government for approval of their products intended for importation into Japan. To take advantage of this program, a foreign manufacturer must designate an in-country caretaker (agent). The caretaker manages the application procedures for the manufacturer (Guide pp. 201–202). The in-country caretaker plays an important role in the approval process. Hence, the in-country caretaker must meet certain qualifications and fulfill the duties specified in the PAL and the ERPALs. The in-country caretaker submits applications through the prefectural governor of the prefecture where the in-country caretaker resides. In this case, the prefectural governor simply forwards all applications to the MHLW. The letter of approval is returned through the prefectural governor to the in-country caretaker.

The foreign manufacturer receives the letter of approval through the in-country caretaker. Once the letter of approval has been issued to the foreign manufacturer, a domestic firm that intends to import that product need only obtain a *kyoka* license to import the device (Guide p. 36).

DESIGNATED MEDICAL DEVICES

Certain life-supporting or life-sustaining devices that can have an immediate effect in case of a failure fall into a special category under Article 77-5 of the PAL. These "designated devices" are listed in Article 64-6 of the ERPALs, which is reproduced in Table 18.2. If a serious defect is found in the product after its introduction, it may be necessary to track users of the device so that prompt medical care can be provided. In this context, the person who has a designated device such as animplanted pacemaker would be considered the user. To comply with this provision of the law, the manufacturer or importer who holds the *shonin* for the product must create and retain specific records about each device in a tracking system. If a foreign manufacturer holds the *shonin*, the in-country caretaker must maintain the tracking system.

Under the PAL, physicians or other healthcare professionals handling designated devices are to provide the required information required by the ERPALs Article 64-7 to the holder of the *shonin* for the product or his or her in-country caretaker (PAL §77-4(2)). Under the Japanese system, the user of the designated device is presumed to consent to the collection of this information. However, the user may object, in which case the data will not be collected or maintained.

The personnel involved in assembling and maintaining the tracking system are legally responsible to maintain the confidentiality of the data collected and may only release the information if there is a valid reason. A failure to protect the confidentiality of the tracking data may subject the perpetrator to a stiff fine under Article 87 of the PAL.

The holder of the *shonin* or his or her in-country caretaker must maintain the tracking records until (ERPALs §4-10):

- the death of the user of the designated device;
- the designated device is no longer in use; or
- when, for another valid reason, the data is no longer needed.

Information on the data elements required in the tracking record is found in the next section of this chapter.

MEDICAL DEVICE VIGILANCE

In 1993, the MHLW amended the regulations to strengthen the requirements for tracking of certain medical devices and report of adverse events. The specific types of data required to be retained in the tracking system are outlined in the ERPALs. At a minimum, the tracking system systems should contain data about the device, the user, and adverse events.

ADVERSE EVENT REPORTING

In March 1993, the MHLW amended the ERPALs to provide for increased reporting of adverse incidents involving medical devices approved in Japan, regardless of where the adverse incident occurred. The amendments are intended to broaden the scope of reporting and to collect information on "side effects" as early as possible so that appropriate action can be taken (PAB No. 333 pp. 2-4). The MHLW began enforcement of the new system on April 1, 1994 (PAB No. 333 p. 1).

An incident "caused by a malfunction of a medical device" that results in a death or serious injury is reportable unless the incident is clearly attributable to the user's lack of knowledge or skill (PAB No. 333 pp. 6-7). From the point of view of device labeling, the occurrence of consequences that are included in the warnings and precautions for use is reportable if there is a definitive change in the number, frequency, or conditions of occurrence. The occurrence of such consequences is also reportable if the severity of the outcome cannot be anticipated from the descriptions in the precautions for use. Also, any case that is suspected of being caused by a device malfunction that is not mentioned in the warnings, precautions, or accompanying documentation is potentially reportable. This type of case becomes reportable if the practitioner in charge of the case believes the incident to be significant (PAB No. 333 p. 7).

DEVICE TRACKING

The Japanese law requires the manufacturer or importer of a medical device listed in Table 18.2 to create and maintain a record of the patient who has received the device. These tracking records are intended to expedite and facilitate prompt medical care in the event of a device-related emergency. The content of the tracking record is specified in ERPALs Article 64–7 as amplified by PAB Notification No. 600 of June 26, 1995, and must include (Guide p. 294):

- the name, model number, and serial or lot number of the device;
- the name, address, date of birth, sex, and telephone number of the user of the device (i.e., the person who has the device implanted);
- the date when the device was implanted and the region of implant;
- the name, address, and telephone number of the medical institution where the device was implanted;
- the name, address, and telephone number of a medical facility where the user may consult someone about the device; and
- the date of registration.

THINGS TO REMEMBER

The PAL of Japan is the primary law under which the PMSB of the MHLW derives authority to take action with regard to medical devices. The PAL was revised in 1960 to control and regulate matters related to drugs, quasi drugs, cosmetics, and medical devices to assure their quality, efficacy, and safety. For manufacturers located outside Japan, the most significant amendment to the PAL occurred in 1983, when foreign manufacturers were allowed to apply directly to the Japanese government for approval of their products intended for importation into Japan.

In 1994, the MHLW overhauled the PAL by introducing a four-tiered, risk-based classification for medical devices. The new system, which is similar to the system used in the European Union and Canada, was adopted by the Japanese Diet in June 1994. The revision includes a mandate that the name of the device, and other required information, appear on the outer container.

The PAL requires that any firm that intends to manufacture or import medical devices that are subject to regulation must obtain a license, called a *kyoka*, from the MHLW. The manufacturer, or importer, must also obtain a product approval, called a *shonin*, on a product-by-product basis for certain medical devices manufactured in or imported into Japan.

Although the PAL does not use the terms "adulteration" or "misbranding," these concepts do appear in the Japanese law. A person who is found guilty of storing, exhibiting, or conveying an adulterated or misbranded medical device is subject to penal servitude or a substantial fine or both. The PAL declares that a device is misbranded if its labeling makes false or misleading claims. The PAL imposes a stiff penalty on a person found guilty of offering for sale or giving a device whose labeling contains false or misleading claims.

The PAL also regulates advertising or promotion of medical devices. It a crime to explicitly or implicitly advertise, describe, or circulate false or exaggerated statements regarding the name, manufacturing process, indications, effects, or properties of a medical device. Advertising the name, manufacturing process, indications, effects, or properties of a medical device that requires approval by the MHLW prior to receiving that approval is also a crime.

The MHLW has recognized that it is an internationally accepted practice to display medical devices at exhibitions before they are approved for market release. To facilitate this process, and to see that it is properly controlled, the Inspection and Guidance Division of the PAB has published comprehensive guidelines for exhibition of medical devices prior to their approval.

In Japan, IVD reagents are classified as drugs, while the apparatus used to analyze the results of the tests are classified as medical devices. However, since diagnostic reagents are used outside the human body, they have been placed in a special category with respect to public health and hygiene. These *in vitro* test drugs are placed in one of three classes depending on the novelty of the measured items or measuring methods. The licensing procedure has been simplified based on the class.

19 General Medical Device Labeling in Japan

The Pharmaceutical Affairs Law (PAL) establishes the requirements for medical device labeling in Japan. The PAL sets the minimum required information that must be included in the labeling on the device, on its immediate container, and in the package insert. The requirements in the PAL are amplified in the Enforcement Regulations of the Pharmaceutical Affairs Law (ERPALs), and in performance and approval standards adopted by the Ministry of Health, Labor, and Welfare (MHLW).

MISBRANDING

Although the PAL does not use the term "misbranding," the law does define a set of conditions under which a medical device could be considered misbranded. A person who is found guilty of selling, storing, or exhibiting for sale a misbranded medical device is subject to penal servitude or a substantial fine, or both. A device can be deemed misbranded if the package insert, container, or wrapper (including the inner package) of a medical device, or the labeling on the medical device itself:

- contains false or misleading claims concerning the medical device;
- contains a statement of indications or effects not approved by the MHLW pursuant to provisions of Article 14 of the PAL;
- contains directions, dosage, or durations of use that are dangerous to health;
- fails to provide the information required by Article 63 or the appropriate material required by Articles 52 through 54, and additional material laid down in MHLW ordinances or standards established under Article 42 of the PAL; or
- fails to display the material required by the law and regulations more prominently than other material, or in terms so as to render them easily read and understood by the ordinary purchaser or user of the product.

A medical device that is misbranded may not be sold or given, or stored or exhibited for purpose of sale or giving.

GENERAL LABELING PROVISIONS

General labeling requirements are defined in Articles 52, 53, and 63 of the PAL. Articles 52 and 53 are written to apply to drugs. However, PAL Article 64 makes them applicable to medical devices *mutatis mutandis*. These articles cover the basic requirements for material to be included on the immediate container (or the device itself) and on the package insert, as well as the prominence of required information relative to other text, descriptions, diagrams, or designs. These requirements are amplified in Chapter V of the ERPALs.

The drug-labeling provisions of the PAL specify that the information required on the immediate container or wrapper must be easily legible through the outer container or wrapper; or the same material must be indicated on the outer container or wrapper. For medical devices, the required information must appear on the immediate container or wrapper, or on the device itself. In 1994, the MHLW revised the regulations so that the name of the device and other required information should appear on the outer container.

The following sections discuss each of these requirements in detail. The number of the corresponding article in the PAL and/or ERPALs (e.g., PAL Article 52 or ERPALs Article 61) is listed with each topic for reference.

Immediate Container (PAL Article 63)

The items that must appear on the immediate container or wrapper of a medical device, or on the medical device itself, are listed in the PAL. Exemptions to the following requirements may be established by regulations prescribed by MHLW ordinance. The required information includes:

- the name and address of the manufacturer or importer;
- for a medical device designated by the Minister,
  - the serial number or lot number;
  - the quantity of the contents in terms of weight, volume, number, and so on; and
  - the expiration date; and
- any matters laid down by MHLW ordinances supplementary to those specified in the preceding items.

The variations and additional requirements established in the ERPALs are discussed in the following sections.

Name and Address of the Manufacturer or Importer (ERPALs Article 61)

For those medical devices listed in Table 4 attached to the ERPALs, the full name and address of the manufacturer (or importer) can be replaced by either of the following:

- The name and address of the manufacturer (or importer) can be replaced by the abbreviated name of the manufacturer (or importer) and the name of the prefecture or city where the manufacturer (or importer) is located. For example, the abbreviated name of a manufacturing company called "Ophthalmic Lenses Co., Ltd." would be "Ophthalmic."
- Alternatively, the name and address of the manufacturer (or importer) can be replaced by the trademark, registered under the Trademark Law of Japan,* of the manufacturer (or importer) and the name of the prefecture or city where the manufacturer (or importer) is located.

The categories of products for which the substitution of an abbreviated name/trademark of the manufacturer (or importer) is allowed are listed in Table E.8 in Appendix E.

Performance Standards Established by Ordinance (ERPALs Article 60–2, Item 1)

The MHLW has established, under provisions of Article 42, Paragraph 2 of the PAL, a set of performance standards for medical devices. The devices covered by standards in place at the time of publication are listed in Table 19.1. The labeling on the medical device or its immediate container or wrapper must include material required in the applicable standards.

---

* *Trademark Law of Japan.* 1959. Law No. 127.

TABLE 19.1
Medical Devices Subject to Japanese Mandatory Performance Standards

1. Disposable syringe needles
2. Syringe needles
3. Syringes
4. Disposable syringes
5. Disposable transfusion sets and infusion sets
6. Disposable blood donor sets
7. Blood donor sets and blood transfusion sets
8. Polyvinyl chloride (PVC) resin blood sets
9. Disposable sets for artificial heart-lung machine
10. Artificial blood vessels
11. Adhesives for medical use
12. Contact lenses for visual correction
13. Plastic sutures
14. Latex condoms
15. Dutch pessaries
16. Sanitary tampons
17. Artificial heart valves
18. Cardiac pacemakers
19 Medical X-ray apparatus
20. Monofocal plastic lenses for visual correction

Source: Guide p. 69

TABLE 19.2
Medical Devices Subject to Japanese Approval Standards

1. Hemodialysis apparatus
2. Dental casting nickel-chromium alloy (for Crown)
3. Intraocular lenses
4. X-ray apparatus

Source: Guide p. 70

Approval Standards Established by Ordinance (ERPALs Article 60–2, Item 1)

The Pharmaceutical and Medical Safety Bureau (PMSB) of the MHLW may establish approval standards for medical devices under provisions of Article 42, Paragraph 2, of the PAL. The devices covered by standards in place at the time of publication are listed in Table 19.2.

Medical Devices Manufactured in Foreign Countries (ERPALs Article 60–2, Item 2)

For medical devices approved for import into Japan under the provisions of Article 19–2 of the PAL, the ERPALs requires the following information to appear on the immediate container or wrapper of a medical device, or on the medical device itself:

- The name of the manufacturer and the name of the country where the manufacturer is located must be present. For those medical devices listed in Table 4 of the ERPALs, the name of the manufacturer can be replaced by either:
  - The abbreviated name of the manufacturer or
  - The trademark of the manufacturer, registered under the Trademark Law of Japan.
- The name and address of the in-country caretaker must also be present. For those medical devices listed in Table 4 of the ERPALs, the name of the in-country caretaker can be replaced by the abbreviated name of the in-country caretaker and the name of the prefecture or city where the in-country caretaker is located.

The categories of products for which the substitution of the abbreviated name and/or trademark of the manufacturer (or importer) and in-country caretaker is allowed are listed in Table E.8 in Appendix E.

PACKAGE INSERT (PAL ARTICLE 52)

The insert (instruction manual or similar material) enclosed or provided with a medical device must contain the information prescribed in Article 52 of the PAL. If there is no insert provided with the device, then the container or wrapper of a medical device must contain the required information, as follows:

- The directions, dosage, or other necessary precautions for use and handling of the device must appear. The insert should contain a statement of the purposes for which the device is intended by the manufacturer and any contraindications to its use. It must also include clear instructions for the safe and correct operation of the device, as well as appropriate instructions for the installation, assembly, storage, maintenance, and repair of the device.
- For a medical device recognized by the *Japanese Pharmacopoeia*, the insert, or the container or wrapper, must include the material specified by the *Japanese Pharmacopoeia*.
- For a medical device for which the standards have been developed under the provisions in Article 42, Paragraph 2, the insert, container, or wrapper must contain the material required by the standard.
- In addition to information specified in the preceding items, the insert, container, or wrapper must include any information required by MHLW ordinance. Those requirements are discussed later in this chapter.

The PAL provides that exemptions to the preceding requirements may be established by regulations prescribed by MHLW ordinance. The presently allowed variations are discussed later in this section.

The package insert should have clear and legible printing that can be read by the intended user in the environment where the device is likely to be used. For example, a device intended for use by the elderly may require larger printing on the package insert. As a rule, when the package insert is an instruction manual, it should be printed on A4- or B5-size paper.* When a number of sheets are bound together into a volume, they must all be of the same dimension. If the instruction manual is bound into multiple volumes to facilitate its use, each volume must be numbered or otherwise marked to make clear the relationship between volumes (JIS T 1005 pp. 2–3).

The package insert should also contain the complete address and telephone number of the manufacturer and distributor of the device, as well as the addresses and telephone numbers of branch and business offices, and service facilities. These addresses and telephone numbers should be listed on the cover (including the inside and back covers) of the insert (JIS T 1005 p. 3).

Warnings and Directions for Use (PAL Article 52, Item 1)

The manufacturer of a medical device is required to include in the package insert the directions, dosage, or other necessary precautions for proper use and handling of the device. These required elements are sometimes referred to as the "warnings and directions for use" (Guide p. 71). The warnings and directions for use should contain contraindications in addition to the directions for use. The rules to follow when creating these elements are listed in Pharmaceutical Affairs Bureau (PAB) Notification No. 1330.** They are:

---

* Paper sizes are specified in JIS T 0138.
** Japanese Ministry of Health and Welfare, Pharmaceutical Affairs Bureau. 1980. *On Enforcement of the Law for Amending the Pharmaceutical Affairs Law.* PAB Notification No. 1330. (October 9).

- Mention the most important information first.
- Make headlines and important information conspicuous. For example, use a boldface type.
- Explain warnings and directions in simple language.
- Give precise and scientific information, and describe the reasons for, and background of, the information.
- A deletion or change in information, if any, shall be made on adequate grounds.

These rules are described in PAB Notification 1330 as the "minimum" requirements for the warnings and directions for use.

The warnings and directions for use may provide details on some or all of the following topics as appropriate to the individual device (JIS T 1005 pp. 4–8):

- The purposes and conditions for use.
- An explanation of the device performance, giving special attention to the physiological effect and outputs that may be hazardous to the patient or user.
- The conditions for safe and correct use of the device. This should include the:
  - preparations for use;
  - steps to confirm the safe operation of the device, including a description of alarms and indicators;
  - operation immediately before use, paying particular attention to abnormal displays, the underlying trouble, and the interlocks to prevent hazardous output;
  - operation during use, including the sequence of operations, the methods for reading displays, and necessary adjustments for maintaining normal function;
  - measures following use to prepare for the next use of the device; and
  - measures for both short- and long-term storage of the device.
- The appropriate environmental conditions required for the accurate, safe, and correct use of the device (i.e., ambient temperature, relative humidity, and atmospheric pressure).
- If applicable, a description of the required power source (i.e., voltage, frequency, power rating, and allowable ranges).
- The required site-preparation and installation procedures, including such items as the space required for use, environmental conditions, protective barriers (radiation, ionizing, and electromagnetic), and so on.
- The construction of the equipment (e.g., using photographs, block diagrams, wiring diagrams), with special attention to the interconnection of system elements.
- A description of the mechanical structure required for the safe use of the device, which includes such items as precautions to be taken with movable parts, motor circuits, radiation shields (like those on X-ray and laser equipment), and mobile equipment.
- An explanation of the activation principle for the safe and correct use of the device.
- Clear descriptions of the methods for cleaning, disinfection, and sterilization of the device; any limits or precautions to be taken to prevent exfoliation and/or damage to name plates and other labels; and any items that can damage the safety and performance of the device.
- For portable devices, the preparation of the device for safe transport.

The warnings and directions for use should describe the meaning of any symbols (letter symbols and graphical symbols) used in the package insert and on the device labels.

Warnings and Directions for Use by Device Category (PAL Article 52, Item 3)

There are many medical devices for which the Japanese Standards Association (JSA) has published Japanese Industrial Standards (JIS). Many of these standards are listed in Table E.3 in Appendix E. They describe performance and safety requirements for individual devices (i.e., mechanical safety, electrical safety, biological safety, radiation protection, etc.). They also may specify the form and substance of information that must be included in the labeling of the device.

Warnings and Directions for Use Specified by Ordinance (PAL Article 52, Item 4)

The MHLW has established specific requirements for warnings and directions for use for certain device categories. These minimum requirements are given in the MHLW notifications that are listed in Table 19.3.

Prominence of Required Statements (PAL Article 53)

The words, statements, or other matters specified in Article 52* and Article 63 must be exhibited in the labeling more prominently than other texts, descriptions, diagrams, or designs. These matters must be accurately indicated, pursuant to the provisions prescribed by MHLW Ordinance, in such terms as to render them easily read and understood by the ordinary purchaser or user of the medical device concerned. Article 57, Item 1, of the ERPALs goes on to state that the information that must appear on inserts enclosed with a medical device or on the container or wrapper of a medical device, in accordance with provisions of the PAL, must be especially clearly indicated.**

LANGUAGE REQUIREMENT (ERPALs ARTICLES 58 AND 62)

The information required on package inserts (PAL Article 52) and on the immediate container or wrapper of a medical device, or on the device itself (PAL Article 63), must be written in the Japanese language.

TABLE 19.3
Warnings and Directions by Device Category

Category	Notification Title	Notification No.
Hyperbaric oxygen chamber	"Warnings and directions for use are to be printed in the package insert for hyperbaric oxygen chambers"	PAB Notification No. 704 (September 12, 1969)
Sanitary tampons	"Establishment of standards for artificial blood vessels and amendments to the standards for blood preservatives in blood collection bottles"	PAB Notification No. 863 (October 6, 1970)
Surgical laser apparatus	"Surgical laser apparatus"[a]	ERD Notification No. 524 (April 22, 1979)
Hemodialysis apparatus	"Approval standards for hemodialysis apparatus"[b]	PAB Notification No. 494 (June 20, 1983)
Electric medical instruments and apparatus	"Warnings and directions for use are to be printed in the package insert for electric medical instruments and apparatus"	PAB Notification No. 495 (June 1, 1972)

[a] There are two separate notices that should be consulted. They are: "Warnings and directions for use of surgical laser apparatus," and the notice of August 6, 1991, on "Exempt from submission of clinical data on surgical laser apparatus."

[b] There are two notices attached to this PAB Notification that should be consulted. They are: "Warnings and directions for use of dialyzers," and "Warnings and directions for use of the dialysis solution supplying part and the monitoring equipment."

Source: Guide p. 72

---

* This includes the material that is applicable to medical devices *mutatis mutandis* under Article 64 of the PAL.
** Article 62, Item 3, of the ERPALs applies Article 57, Item 1, to medical devices *mutatis mutandis*.

When preparing Japanese text, sentences are to be written laterally* with kanji and kana mixed.** For kanji, Chinese characters in common use are to be used. If writing in kana is difficult to understand, then kanji consisting of Chinese characters not in common use may be written followed by kana in parentheses.

Katakana is used only for words of foreign origin and foreign words. If necessary, the foreign words may be inserted using parentheses following the katakana. For other words, hiragana shall be used.

For technical terms, terms specified in JIS and Japanese Scientific Terms established by the Ministry of Education are generally used. For terms not specified in either of these sources, the terms used conventionally in institutions, societies, and industrial associations are to be used.

TRADE NAMES

The manufacturer of a medical device is free to assign any trade name that is desired. However, because it is important to maintain the dignity of the product and to avoid confusion, it is the policy of the MHLW to not approve applications for products where the trade name meets any of the following conditions (Guide p. 65):

- False or exaggerated names
- Names that lower the dignity of products as medical devices
- Names that are the same as the name of an already-approved medical device
- Names that obviously infringe on the trademark right of another manufacturer
- Trade names composed merely of roman characters and/or arabic or roman numerals
- Names that may cause confusion with nonmedical devices
- Trade names that are the same as generic names in the list attached to MHLW Notification No. 1008 that represent a different category from the product to which the brand name is applied (The generic names are listed in Table E.1 of Appendix E.)

As a matter of principle, the MHLW gives a single trade name to a single device.

Trade names, particularly those of household-use devices, should not suggest unapproved indications or effects. Neither should the trade name suggest undesirable methods of use. Such trade names are considered unsuitable (Guide p. 155).

Trade names must be written in Japanese. However, the Japanese characters may be combined with alphanumeric characters as illustrated in Figure 19.1 (MDR Japan, p. 40).

EXPORT OF MEDICAL DEVICES (ERPALS ARTICLE 66)

Article 80 of the PAL specifies that a cabinet order may exempt medical devices that are intended for export from the provisions of the PAL.

レーザー手術装置YAG-1

FIGURE 19.1 Example of a trade name containing Japanese and alphanumeric characters.

---

* Traditional-style Japanese is written in columns from right to left.
** After the Japanese learned to write Chinese, they began adapting the Chinese characters (called kanji) to their own tongue. Certain Chinese symbols are used for their sound value, disregarding their Chinese meaning. The result was two syllabaries, each containing about 50 symbols called kana. The ordinary rounded symbols known as hiragana developed from the "grass-hand" forms of the characters, and the squarish symbols called katakana, used as a kind of italics to write foreign or unusual terms, were taken from abbreviations of the "square-hand" (printed) forms. After World War II, the Minister of Education simplified the shapes of many kanji and restricted the number for general use to less than 2,000.

Article 15 of the EOPAL describes the conditions under which the manufacture (or import) of medical devices intended for export can be carried out. The manufacturer (or importer) must notify the MHLW, in advance, of the medical devices covered by MHLW regulations that are going to be manufactured or imported for purposes of export. The notification, which contains the details of the product and other information required by MHLW ordinances, is made via the governor of the prefecture where each factory or business office is located. The licensing standards for a manufacturer (or importer) in PAL Article 13, Paragraph 1, and Article 18 do not apply as long as the manufacturing or import is performed in accordance with the contents of the notification. Article 66 specifies the information that must be reported by the manufacturer (or importer) of a medical device intended for export.

The provisions of Article 43* and Chapter 7** of the PAL, which includes the requirements for labeling of medical devices, do not apply to the manufacture, import, marketing, giving, storage, or display of medical devices for export. However, this exemption applies only in the cases where the medical devices are manufactured (or imported) based on the specifications in the notification described above.

IMPORT OF MEDICAL DEVICES (ERPALS ARTICLES 53–2, ITEM 3, AND 60–2, ITEM 2)

Beginning in 1983, foreign manufacturers could apply to the Japanese government for approval of their products intended for importation into Japan. To take advantage of this program, a foreign manufacturer must designate an in-country caretaker (agent). The caretaker manages the application procedures for the manufacturer (Guide p. 201). The products that are imported into Japan must meet the same labeling requirements as products produced in Japan. In addition, the ERPALs specify that the labeling must bear the name of the recipient of the approval for manufacturing abroad, the name of the country where the manufacturer resides, and the name and address of the in-country caretaker.

TESTING OF MEDICAL DEVICES (ERPALS ARTICLE 43)

Article 43 of the PAL prohibits the sale of a medical device regulated by the MHLW unless the device has passed the test performed by a person designated by the MHLW. Articles 8 through 11 of the EOPAL set out the procedure to be followed to obtain a test certificate. The manufacturer (or importer) applies to the testing institution via the prefectural governor, with the fee indicated by the MHLW. The item to be tested is collected by a pharmaceutical inspector and sent to the testing institution. The testing institution performs the test and notifies the prefectural governor of the results. If the medical device passes the test, the prefectural governor issues a certificate.

When an applicant wishes to have a test performed, the manufacturer (or importer) places the device in the container or packaging that is used when the device is sold. That container is placed in a box or further container suitable to be sealed. The following information must be indicated on the outer box or container:

- The name of the applicant
- The name of the medical device

---

* Article 43 prohibits the sale, storage, or exhibition for purpose of sale of medical devices until they have passed the tests prescribed by the MHLW.
** Devices intended for export are subject to the adulteration provisions of the PAL if (a) they contain impure, putrid, or decomposing substances; (b) they contain foreign matter; (c) they are contaminated, or are likely to be contaminated, by pathogenic microorganisms; or (d) their use is likely to be injurious to health.

- The manufacturing number (in the case of imports, the manufacturing number and the name of manufacturer from whom the device was imported)
- The date of manufacture (in the case of imports, date of manufacture and date of import)
- The quantity

Once the carton is sealed, it may be transported by the pharmaceutical inspector to the testing institution.

CLINICAL TRIALS (ERPALS ARTICLES 67, 68, 69–2, AND 70)

The development of safe and effective medical devices is a responsibility imposed on the manufacturer. A clinical trial of a new medical device in humans is often an important element of the testing necessary to demonstrate the safety and effectiveness of a device. In order to ensure that clinical trials are conducted in a way that leads to satisfactory results, the PMSB has published a set of good practices for conducting clinical trials involving human subjects. The requirements in this notification apply to clinical trials conducted to collect data to be submitted for approval to manufacture (import) the medical devices covered under Article 14 of the PAL. These requirements do not apply to clinical trails to collect data on medical devices using *in vitro* reagents (PAB No. 615 p. 2).

The sponsor of a clinical trial is responsible under Article 80–2 of the PAL for carrying out the duties listed in Article 15 of the Good Clinical Practice (GCP) rules. (PAB No. 615 p. 6) These include proper labeling of the investigational device. The label on the device or its container or wrapper must bear the following information in Japanese (ERPAL §67(7)):

- The fact that the medical device is for trial use (e.g., "Investigational Device" or "Clinical Trials")
- The name and address of the sponsor of the clinical trial (If the sponsor is not a resident in Japan, the name and country of the sponsor and the name and address of the in-country clinical trial administrator must be used.)
- The identification number of the device
- The serial number or batch code, as applicable
- Storage instructions and "use before" date, as applicable

The following information may not appear on any documentation accompanying the device or on any label attached to the device, its container, or wrapper (including the inner wrapper):

- The anticipated brand name of the device
- The anticipated indications, effects, or properties of the device
- The expected usage and dosage

WARNINGS AND DIRECTIONS FOR USE BY DEVICE CATEGORY (PAL ARTICLE 52, ITEM 3)

There are many medical devices for which JIS have been published by the JSA. Many of these standards are listed in Table E.3 in Appendix E. They describe performance and safety requirements for individual devices (i.e., mechanical safety, electrical safety, biological safety, radiation protection, etc.). They also may specify the form and substance of information that must be included in the labeling of the device. The following sections describe requirements for representative groups of devices.

ELECTRICAL MEDICAL EQUIPMENT

This device category, in general, covers most electrical instruments and apparatus used in the patient environment and those electrically-powered devices having a direct effect on the safety of the patient. This equipment is intended to diagnose, treat, or monitor patients under medical supervision; it is equipment that (a) makes physical or electrical contact with the patient, (b) transfers energy to or from the patient, and/or (c) detects energy transfer to or from the patient. These devices should conform to the requirements in JIS T 1001,* unless the manufacturer can justify the use of substitute standards (Guide p. 93). The requirements in JIS T 1001 are essentially identical to those prescribed in the International Electrotechnical Commission (IEC) 60601-1.

All required markings must be clearly legible and removable only with a tool or by application of considerable force. If the equipment is permanently installed, then all required markings must be visible when the equipment is mounted in its normal use position. For transportable equipment, it is preferred that required markings are visible during normal use. However, when necessary, the required marking must become visible when the equipment is moved away from a wall against which it has been positioned or has been removed from the rack for dismountable rack units. Warning statements and operating instructions must be affixed in a prominent location so they are legible with normal vision from the operator's position (JIS T 1001 p. 49).

Markings required on the outside of the equipment depend on the type of power source. JIS T 1001 defines three configurations. They are:

- Equipment that receives its energy from the supply mains in the facility where it is used (mains-powered).
- Equipment that is able to operate from an internal electrical-power source (internally powered).
- Equipment that is supplied by another specified power source (other than supply mains). Such equipment must be isolated from the supply mains.

The markings required on the outside of the equipment are described in Table 19.4. For permanently installed equipment, the nominal supply voltage or voltage range may be marked either on the outside or inside of the equipment enclosure. The preferred location of the markings is adjacent to the supply-connection terminals.

When the size of the equipment or the configuration of the enclosure makes it impractical for all of the required information in Table 19.4 to be affixed to the outside of the equipment, then at least the information in items 1, 2, and 3, and, if applicable, 7 and 11 should be included on the outside of the enclosure. The remaining items may then be included in the accompanying documentation. If it is impractical to include any markings on the outside of the enclosure, then all of the required information may be placed in the accompanying documentation (JIS T 1001 p. 49).

JIS T 1001 specifies requirements for marking controls and instruments on medical electrical equipment, as well as marking on the inside of the equipment. These markings on the controls must include the following information, as applicable (JIS T 1001 pp. 53–54):

---

*In December 1999, JSA adopted JIS T 0601, which is identical to IEC 60601 and its amendments. This standard is scheduled to replace JIS T 1001 in 2005. A transition period of 5 years is expected.

TABLE 19.4
Markings on the Outside of Medical Electrical Equipment

Items to be Marked	Mains Operated Equipment	Internally Powered Equipment[a]	Equipment Supplied from a Specified Power Source[b]
1. Name and/or trademark and address of the manufacturer or supplier	*	*	*
2. Model identification	*	*	*
3. Mains-supply characteristics (rated voltage/voltage range; current waveform—a.c., d.c., or dual supply; number of phases; etc.)	***		
4. Supply frequency	***		
5. Rated power input	***		
6. Auxiliary-mains socket output(s)	**		
7. Equipment classification[c]	**	**	**
8. Mode of operation[d]	**	**	**
9. Fuses	**	**	**
10. Output	**	**	**
11. Physiological effects[e]	**	**	**
12. Category AP/APG equipment (see Appendix C, Symbols 23 and 24)	**	**	**
13. High voltage terminals accessible without a tool (see Appendix C, Symbol 25)	**	**	**
14 Cooling requirements	**	**	**
15. Mechanical-stability marking if special precautions are required	**	**	**
16. Protective packaging, if required during transport and storage	**	**	**
17. Earth-terminals markings[f]	**	**	**
18. Removable protective means	**	**	**
19. Other legally required markings	**	**	**

* Marking is required on all equipment.

** Marking is required if applicable to the equipment.

***Marking is not required on permanently installed equipment that is marked appropriately on the inside.

[a] Internally powered equipment is capable of operating from an internal electrical-power source.

[b] The equipment is intended to be supplied from a power source other than supply mains.

[c] Classification includes the degree of protection against electrical shock (see Appendix C, Symbol 1), the protection against leakage current (see Appendix C, Symbols 17, 18, and 19), and the harmful ingress of liquids using the IP Code defined in IEC 60529.

[d] If equipment is unmarked, continuous operation is assumed.

[e] Equipment producing physiological effects that cause danger to the patient and/or operator must prominently display a symbol instructing the patient/operator to consult the accompanying documentations for warnings (see Appendix C, Symbol 14). If the device produces non-ionizing radiation such as high-power microwaves, then Symbol 43 in Appendix C must be displayed.

[f] Earth terminal markings include the markings for connection to a potential equalized conductor (see Appendix C, Symbol 8), a functional earth terminal (see Appendix C, Symbol 5), and a protective earth conductor (see Appendix C, Symbol 6).

Source: JIS T 1001 §15.2

- The main power switch must be clearly identified with its "ON" and "OFF" positions indicated by a symbol, adjacent indicator light, or other unambiguous means (see Appendix C, Symbols 9 and 10). Indicator lights and push buttons should follow the conventions listed in Table 19.5.
- Controls and indicators with a safety function must be clearly identified.
- The change-over positions of control devices and switches must be indicated by figures, letters, or other visual means. If the control is intended to adjust a device setting during operation, the control must be marked to indicate the direction in which the magnitude of

TABLE 19.5
Colors for Indicator Lights[a] and Push Buttons[b] and their Recommended Meanings

Color	Meaning	Required Usage	Recommended Usage
Red	Warning of danger and/or the need for immediate action	X	
Yellow	Caution or attention required		X
Green	Ready for use		X
Any other color	A meaning other than that associated with red or yellow		X

[a] JIS T 1001 does not consider dot matrix or other alphanumeric displays to be indicator lights.

[b] An unilluminated push button may be colored red only if it is used to interrupt the function of the equipment in the event of an emergency.

Source: JIS T 1001 §15.7

the equipment function will change. The marking on the control may be omitted if there is another indicator that will clearly show the magnitude of the change, and if there is no safety hazard to the patient created by moving the control in an unfavorable direction.

A label near the point where the power-supply conductors connect to the equipment must show the correct method of connecting the supply connector, unless no hazard would occur if the connections were interchanged. If there is insufficient room to appropriately label the connections on the equipment, the instructions can be included in the accompanying documentation.

In permanently mounted equipment, the terminal for connection to the neutral supply conductor must be indicated by a capital letter "N."

The protective earth terminal must be marked (see Appendix C, Symbol 6) adjacent to this terminal, unless the protective earth terminal is part of a detachable power cord.

Permanently connected equipment that has a terminal box that reaches a temperature of more than 75 °C when tested according to the method specified in JIS T 1002 must be marked near the point of connection with the following statement (JIS T 1001 p. 53):

FOR SUPPLY CONNECTION, USE WIRING MATERIALS SUITABLE FOR AT LEAST __ °C.

The labels identifying the correct method of connecting the supply, the supply-neutral terminals, the protective earth terminal, and instructions about high-temperature materials may not be affixed to any part that must be removed in order to make the supply connections. In addition, the labels must remain visible after the connections are made.

The presence of parts that carry more than 1,000 V a.c., 1,500 V d.c, or 1,500 V peak value must be clearly marked (see Appendix C, Symbol 25).

The functional earth terminal must be marked (see Appendix C, Symbol 5).

For those devices that contain heating elements or heating lamps, the maximum power loading must be permanently and indelibly marked near the heating element or lamp holder.

For devices with internal fuses, the fuse ratings and characteristics must be marked near the fuse. Information on fuses may also appear in the instruction manual accompanying the equipment.

Equipment that incorporates batteries must have the type of battery and mode of insertion marked adjacent to the battery holder.

The requirements for the instruction manual for medical electrical equipment are set out in JIS T 1005. The general information content of the instruction manual is discussed in the previous section on general warnings and directions for use.

ULTRASONIC DIAGNOSTIC EQUIPMENT

As a type of medical electrical equipment, ultrasonic diagnostic equipment should follow the general labeling and documentation requirements for that category of equipment. Requirements for specific types of diagnostic ultrasound devices can be found in JIS standards T 1503 through T 1507. These standards are listed in the medical electric equipment section of Table E.3 in Appendix E. In addition, the instruction manual should describe in detail (Guide pp. 104–105):

- the working modes of the equipment;
- the scanning method employed;
- the focusing method for the ultrasonic beam;
- the method used for processing the resulting image;
- an explanation of any diagnostic data processing functions;
- any auxiliary functions of the equipment;
- the type of sound energy employed;
- the area to which the sound energy is applied;
- the performance parameters of the equipment, including the (a) total sensitivity, (b) resolution, (c) ultrasonic frequency, (d) ultrasonic output, and (e) indicated precision; and
- the details of any phantoms used in evaluating the performance of the device.

Because electromagnetic interference (EMI) can effect the quality of the image produced by the device, the instruction manual should include appropriate warnings about EMI.

RADIATION-RELATED APPARATUS

Radiation-related apparatus includes medical X-ray equipment, sealed radiation sources for medical use, personal and area radiation-protective devices, medical X-ray-film viewers, and other medical radiation devices such as medical electric accelerators. This equipment should follow the general labeling and documentation requirements for electrical medical equipment. Requirements for specific types of radiation-related devices can be found in the JIS standards listed in the radiation-related apparatus section of Table E.3 in Appendix E. In addition to technical requirements, these standards specify the information that is to be included in the instructions for use (IFU). This includes information such as the generator rating and duty cycle, as well as the X-ray-tube rating.

The directions for use should describe concisely the complete operating procedure for the device, beginning with equipment preparation and continuing through safety assurance, preoperating procedure, operation, postoperation procedure, and maintenance of the device (Guide p. 110). The directions for use should include adequate instructions concerning any radiological safety procedures and precautions that may be necessary. Because these devices can generate large electric fields, a warning about EMI should be included (Guide p. 111).

Some of the devices in this category, such as X-ray films, deteriorate with time. These devices require an expiration date that must be printed on the immediate container or wrapper (Guide p. 110).

SURGICAL LASER APPARATUS

Surgical laser apparatus includes carbon dioxide and yttrium aluminum garnet (YAG) surgical lasers. This equipment should follow the general labeling and documentation requirements for electrical medical equipment. The provisions in JIS C 6802 "Radiation safety standards for laser products" should be consulted.

The directions for use should describe the intended purpose, using terms such as "incision, hemostasis, coagulation, and transpiration of human tissue" (Guide p. 112). The performance specification should include:

- a statement of the laser medium, irradiation wavelength, laser-oscillation system, laser modulation system (if available), and the working mode of the laser;
- a statement of the magnitude, in appropriate units, of the pulse duration(s), maximum radiant power, and, where applicable, the maximum radiant energy per pulse under the most severe conditions of the accessible laser radiation; and
- the irradiation area.

The labeling on the laser apparatus or its immediate container should include the following information (Guide p. 114):

- The type of laser
- The rated power output of the equipment
- The power-supply characteristics (rated voltage/voltage range, current waveform [a.c., d.c., or dual-supply], number of phases, etc.)
- The rated power input
- The production number or production symbol
- Other information required for the proper and safe operation of the laser

Procedures for cleaning the equipment, sterilizing, and disinfecting the parts that enter the sterile surgical field must be described in the directions for use.

The MHLW has established specific requirements for the warnings and directions for use for surgical lasers. These were discussed in the previous section, "Warnings and directions for use specified by ordinance." In addition to these requirements, the warnings and directions for use should include instructions for safe irradiation to prevent embolism due to cooling gas ejected from the fiber tip. Since the equipment may be operated in environments with anesthetic gas, precautions for avoiding fire and explosion should also be included. Steps should be taken to draw the user's attention to these warnings. For example, printing in red ink, enclosing the information in a red box, or other techniques to make the warnings conspicuous could be employed.

Dental Materials

Dental materials should follow the general labeling and documentation requirements described earlier in this chapter. Requirements for specific types of dental materials can be found in JIS standards. These standards are listed in the dental materials section of Table E.3 in Appendix E (Guide p. 148). Labeling requirements for metals for dental use are set out in ERPALs Article 60–3, which are discussed later in this chapter.

Disposable Products

This classification covers a wide variety of products intended for single use. They include products ranging from filters and dialyzers for artificial kidney machines to lancets used with home blood-glucose test kits. They may be provided in a sterile or nonsterile form. However, in each case, the device is intended for a single use and must bear prominently the warning (Guide p. 124):

THIS IS A DISPOSABLE PRODUCT AND SHOULD BE DISCARDED AFTER USING.

The labeling must provide complete information on the intended use, performance, indications, or effects of the device. The method of operation or use must be clearly and concisely described. Diagrams and charts may be useful vehicles for communicating the method of operation. The labeling for devices intended for use by specialists such as physicians, dentists, nurses, and midwives—for whom the device is well known and understood—may omit this information, except for the statement of intended use (Guide p. 124).

If the device will deteriorate unless stored under special conditions (e.g., cool temperatures, low light), these conditions must be described on the label. If the device deteriorates over time, an expiration date must be included on the inner container or wrapper (Guide pp. 125–126).

SMALL STEEL DEVICES

This category includes a myriad of devices, including surgical instruments, dental scalers, and so on. The method of operation should be described. However, this information can be omitted if specialists such as physicians, dentists, nurses, and midwives, who are familiar with such products, primarily use the device. There is no need to describe the purpose, performance, and indications or effects unless the device is new in form or structure. In this case, the purpose may not be readily apparent and should be briefly described.

CONTACT LENSES

Warnings and directions for use should be prepared for both the ophthalmologist and the patient. The ophthalmologist's directions should include contraindications, tests before use, warnings during wear, duration of use, test schedule, and sterilization of the lenses. The patient's version of the directions should include instructions for wear, duration of use, regular checkups, and sterilization of the lenses.

The warnings and directions for use should also meet the "Self-imposed Standards for Writing Instructions for Proper Care of Contact Lenses" attached to MHLW Notification No. 123 of November 2, 1988,* Revision of the "Self-imposed Standards for Written Instructions for Proper Use of Contact Lenses" (Guide p. 137).

INTRAOCULAR LENSES (IOLs)

The warnings and directions for use of IOLs should include the following information:

- The indications for use of the device (e.g., aphakia after cataract surgery [senile cataract and traumatic cataract]).
- The contraindications: infants, uncontrolled glaucoma, progressive diabetic retinopathy, uveitis, iridovascularization, retinal ablation, serious complications during surgery, and other cases of serious ophthalmic diseases for which IOLs are contraindicated in the judgment of an ophthalmologist.

---

* At the time this notice was published, the responsible organization was called the Pharmaceuticals and Chemicals Safety Division.

- An indication that the product should be used with caution in young patients and patients with corneal endothelial disorders, glaucoma, diabetic retinopathy, advanced myopia, congenital ophthalmic anomaly, and patients with a history of retinal ablation.
- Other warnings and directions, which might include:
  - Follow the directions for use.
  - Do not resterilize after use.
  - Do not use if the sterility of the product is suspect due to damage of the packaging.

ELECTRICAL THERAPY APPARATUS FOR HOUSEHOLD USE

Electrical therapy apparatus for household use are electrical devices intended to be used in the home, not in the hospital or medical practitioner's office. The device should have a power input of not more that 300 W and those with a battery of not more than 30 V. Examples include the following devices intended for household use: low-frequency therapy apparatus, static electricity therapy apparatus, infrared lamps, ultraviolet lamps, and very-high frequency (VHF) therapy apparatus. See the section on "Labeling requirements for household use devices" later in this chapter.

LABELING REQUIREMENTS FOR *IN VITRO* DIAGNOSTIC (IVD) PRODUCTS

In Japan, *in vitro* diagnostic (IVD) reagents are classified as drugs, while the apparatus used to analyze the results of the tests are classified as medical devices. For example, a test kit to perform a radioimmunoassay for human chorionic gonadotropin (hCG) in human serum or plasma is a medical device. The standards used with the test kit are classified as drugs.

As a medical device, the *in vitro* test apparatus must meet the general labeling requirements listed in this chapter. In addition, the labeling of the *in vitro* test apparatus should include the operating principles of the device (e.g., discrete method), the number of simultaneous testing items, the testing capacity, the method of analysis, and the testing subjects (e.g., blood, urine).

The documentation should also include a tabulation of the reagents (*in vitro* test drugs) that are intended to be used with the apparatus. Figure 19.2 contains an example of a tabulation of reagents for the A² Medical hCG RIA Test Kit. The tabulation of reagents should include the following information in the sequence shown:

- The name of the test for which the reagent is used (e.g., hCG)
- The MHLW approval number for the reagent (e.g., 63AM No. 1234)
- The name of the reagent manufacturer (e.g., A² Medical Japan)
- The name of the reagent (e.g., human chorionic gonadotropin [hCG] standard)

If the *in vitro* test apparatus is electrically operated, it must meet the electrical safety and labeling requirements for medical electrical equipment in JIS T 1001, T 1002, and T 1005.

A² Medical RIA Test Kit			
Test Name	MHLW Approval No.	Name of Manufacturer	Name of Reagent
hCG	63AM No. 1234	A² Medical Japan	Human chorionic gonadotropin (hCG) standard

FIGURE 19.2 Table of reagents.

Inner Label or Wrapper of an *In Vitro* Diagnostic Reagent (ERPALs Article 56–2)

Because diagnostic reagents are used outside the human body, they have been placed in a special category with respect to public health and hygiene (PAB No. 662). As a drug, however, the items that must appear on the immediate container or wrapper are listed in PAL Article 50. For *in vitro* test drugs, the requirements for labeling on the immediate container or wrapper have been modified by ERPALs Article 56–2. If the outermost container or wrapper is clearly labeled "*in vitro* diagnostic agent*," the required information on the immediate container or wrapper of an *in vitro* test drug includes:

1. The name and address of the manufacturer or importer must be displayed. These may be replaced on the immediate container or wrapper by any of the following:
   - The abbreviated name of the manufacturer or importer. (For example, if the name on the outermost container or wrapper is "Diagnostic Reagent Co., Ltd.", then the abbreviated name "Diagnostic" can appear on the immediate container label.)
   - The trademark of the manufacturer or importer registered under the Trademark Law of Japan.
   - The abbreviation of the manufacturer or importer. (This is limited to cases where the name and address of the manufacturer or importer can be easily confirmed by verification with the entries on the outermost container or wrapper of the drug concerned. For example, if the name on the outermost container or wrapper is "Diagnostic Reagent Co., Ltd. (DRC)," then the abbreviation "DRC" can appear on the immediate container label.)
   - The abbreviated name, trademark registered pursuant to the trademark legislation, or the abbreviation of the manufacturer in the country of origin (This is limited to cases where the name and address of the manufacturer or importer can be easily confirmed by verification with the entries on the outermost container or wrapper of the drug concerned.)
2. The name of the drug must appear. (For a drug recognized in the *Japanese Pharmacopoeia*, the name given in the pharmacopoeia must be used. For any other drug, the generic name must be used, if applicable). The entry may be replaced on the immediate container or wrapper by an abbreviated name or an abbreviation in cases where the name can easily be confirmed by verification with the entries on the outermost container or wrapper of the drug concerned.
3. The serial number or lot number must be present.
4. The quantity of contents in terms of weight, volume, number, and so on, must appear. This entry may be omitted on the immediate container or wrapper.
5. For a drug recognized in the *Japanese Pharmacopoeia*, the words "Japanese Pharmacopoeia" and the matters specified in the *Japanese Pharmacopoeia* to be indicated on the immediate container or wrapper must appear. The required information, except for the expiration date when it is required, may be replaced on the immediate container or wrapper by "JP" or the equivalent abbreviation in Japanese.
6. For a drug for which the standards have been laid down under the provisions of PAL Article 42, Paragraph 1, the method of storage, effective period, or any other matters specified by the standards to be indicated on the immediate container or wrapper are to appear. The entry may be omitted on the immediate container or wrapper except for the expiration date, when it is required.
7. For a drug not recognized by the *Japanese Pharmacopoeia*, the name (generic name, if applicable) and the quantity of each ingredient (for a drug with unknown active ingredients,

the drug's nature, and an outline of its manufacturing process) are to appear. This entry may be omitted on the immediate container or wrapper.

8. For a drug designated as habit-forming by the Minister, the words "Caution—Habit-forming" must be used. The entry may be replaced on the immediate container or wrapper with the words "Habit-forming."

9. For a drug designated as a prescription drug by the Minister under the provision of PAL Article 49, Paragraph 1, the words "Caution—Use only pursuant to the prescription or directions of a physician, etc." must be used. The entry may be replaced on the immediate container or wrapper by "Prescription required."

10. For a drug designated by the Minister as requiring expiration dating, the expiration date must be shown.

11. Any matters laid down by MHLW ordinances supplementary to those specified in the preceding items must appear. For *in vitro* test drugs that have been approved in accordance with the provisions in Article 19–2 of the PAL, ERPALs Article 53–2 specifies that the label must bear "the name of the recipient of the approval for manufacturing abroad and the name of the country where he is domiciled and the name and address of the local administrator." This entry may be replaced on the immediate container or wrapper by any of the following:

    • The abbreviated name of the person who obtained the foreign manufacturing approval.
    • The trademark of the person who obtained the foreign manufacturing approval registered pursuant to the Trademark Law of Japan.
    • The abbreviation of the person who obtained the foreign manufacturing approval. (This is limited to cases where the name and address of the manufacturer or importer can be easily confirmed by verification with the entries on the outermost container or wrapper of the drug concerned.)

The information required by PAL Article 50 to appear on the inner label or wrapper must be written in the Japanese language.

In the event the minimum information required cannot be clearly indicated on the immediate container or wrapper because of space limitations, the information may be omitted from the immediate container or wrapper provided that (a) the information required by PAL Article 50 is indicated on the outer container or wrapper and (b) a license permitting such labeling is granted by the MHLW.

If the IVD reagent has been designated as a poisonous drug by the Minister of Health, Labor, and Welfare, then the name and the word "Poison" must be exhibited on the immediate container or wrapper in white on a black background framed in white (PAL §44(1)). If the reagent has been designated as a powerful drug by the Minister, then the name and the word "Powerful" must be exhibited on the immediate container or wrapper in red on a white background framed in red (PAL §44(2)).

PACKAGE INSERT FOR *IN VITRO* DIAGNOSTIC (IVD) REAGENTS (ERPALS ARTICLE 57)

The package insert enclosed with an IVD drug must contain information as follows:

• The directions for use, dosage, and other necessary precautions for use and handling of the reagent must be included.
• For a drug recognized in the *Japanese Pharmacopoeia*, the matters specified in the Japanese Pharmacopoeia to be indicated on the insert, container, or wrapper must appear.

- For a drug for which the standards have been laid down under the provisions of PAL Article 42, Paragraph 1, the matters specified by the standards to be indicated on the insert, container, or wrapper must appear.
- In addition, the insert must contain any information required by MHLW ordinance.

ERPALs Article 57 mandates that the required information listed above must be clearly indicated in the package insert. It must be written in the Japanese language. Finally, when a drug is recognized in the *Japanese Pharmacopoeia*, the name specified in the pharmacopoeia must appear on the package insert and the on the container or wrapper. If another name also appears, the name specified in the pharmacopoeia must be printed in such a way as to be as obvious as the other name.

Outer Container Label or Wrapper of *In Vitro* Diagnostic (IVD) Reagents (ERPALs Article 56–2)

The outermost container or wrapper of an IVD reagent must be clearly labeled, "*In vitro* diagnostic agent.*"

The following information is required to be legible through the outer container or wrapper:

- The name and address of the manufacturer or importer must be visible.
- The name of the drug must appear. (For a drug recognized in the *Japanese Pharmacopoeia*, the name given in the pharmacopoeia must be used. For any other drug, the generic name must be used, if applicable.)
- If the IVD reagent has been designated as a poisonous drug by the Minister of Health, Labor, and Welfare, then the name and the word "Poison" must be visible through the outer container or wrapper. The lettering must be in white on a black background framed in white.
- If the reagent has been designated as a powerful drug by the Minister of Health, Labor, and Welfare, then the name and the word "Powerful" must be visible through the outer container or wrapper. The lettering must be in red on a white background framed in red.

If the required information it is not legible through the outer container or wrapper, it must be reproduced on the outer container or wrapper.

LABELING REQUIREMENTS FOR METALS FOR DENTAL USE (ERPALS ARTICLE 60–3)

Metals for dental use are considered medical devices. They must meet the general labeling requirements described in this chapter. In addition, the labeling must include the names of the ingredients (chemical name or generic name, if any). When a substance described in an official standard such as the *Japanese Pharmacopoeia* is used, the name given in the standard should be used in the labeling (Guide p. 148).

The quantity of each ingredient must be listed. However, the quantity of ingredients other than gold, silver, platinum, ruthenium, rhodium, palladium, osmium, iridium, or iridosmine need not be listed if they are present in amounts of not more than 5 percent by weight. The quantities of ingredients are to be expressed as percentages by weight. These values are to have one significant figure after the decimal point in the cases of gold and mercury, and are to be whole numbers in the cases of alloys. The quantities of ingredients should never be given in a range (Guide p. 149).

LABELING REQUIREMENTS FOR HOUSEHOLD-USE DEVICES

Household use devices are those intended to be used by laypersons having little knowledge of medicine or the technology employed in the devices. The directions for use and the method of operation must be described in detail and in plain language that can be easily understood by the intended user. Ample use of charts and diagrams may be helpful in providing clear and simple directions for use. It is essential that the directions for use include measures to be taken when there is no beneficial effect from use of the device. See Chapters 27 and 28 for recommendations on preparing the directions for use.

The device labeling should contain an accurate statement of the principal intended effects of the device. It should also contain a clear warning statement if the device cannot be used in conjunction with other therapeutic devices.

The warnings and directions-for-use section of the device labeling should contain the following information in the order listed (Guide p. 157):

- Contraindications. Clearly describe the situations in which the device should not be used because the risk of use clearly outweighs the benefit. Examples would include use of VHF-therapy apparatus or magnetic-induction therapy apparatus with patients that have implanted metal devices (e.g., cardiac pacemakers, artificial metal bones, bone screws).
- The environment and conditions for use. For example, this section should indicate whether the device is safe to be used in or near water. The duration (time) and frequency of use should be included without fail.
- Warnings and directions during operation (e.g., Do not use on sensitive or irritated skin. If symptoms such as eruption, redness, and itching occur, discontinue use and consult a physician.)
- Warnings and directions after use and storage. This section should include special storage conditions (e.g., store in a cool, dark place) that are necessary to prevent serious qualitative changes or deterioration of the device. If the device is likely to deteriorate with time, an expiration date must appear on the immediate container or wrapper of a medical device, or on the medical device itself.
- Warnings and directions in case of a failure or trouble with the device, and a warning against remodeling.

In the case of a device that uses an adhesive agent, such as a polyacrylic acid gel, the directions for use should include the statement: "If such symptoms as eruption, redness, and itching occur after use of this device, discontinue use and consult a physician" (Guide p. 157).

The warnings and directions for use should include any device-specific warnings. Examples include (a) use of VHF-therapy apparatus with clothing combining metal materials such as metal threads or metal buttons and (b) use of magnet-induction-therapy devices in the vicinity of magnetic-sensitive objects such as watches or magnetic cards.

THINGS TO REMEMBER

The PAL sets the minimum required information that must be included in the labeling on the device, on its immediate container, and in the package insert. The requirements in the PAL are amplified in the ERPALs and in performance and approval standards adopted by the MHLW. Although the PAL does not use the term "misbranding," the law does define the conditions under which a medical device could be considered misbranded. A person who is found guilty of selling or giving a misbranded device, or storing or exhibiting a misbranded device for purpose of sale, is subject to penal servitude or a substantial fine, or both.

General labeling requirements are defined in Articles 52, 53, and 63 of the PAL. Articles 52 and 53 are written to apply to drugs. However, PAL Article 64 makes them applicable to medical devices *mutatis mutandis*. These articles cover the basic requirements for material to be included (a) on the immediate container (or the device itself) and (b) the package insert. They also state the prominence of required information relative to other text, descriptions, diagrams, or designs. These requirements are amplified in Chapter V of the ERPALs. The articles in Chapter V of the ERPALs deal with the labeling requirements for (a) the immediate container or wrapper or the device itself and (b) on the package insert.

In Japan, IVD reagents are classified as drugs, while the apparatus used to analyze the results of the tests are classified as medical devices. The *in vitro* test apparatus must meet the general labeling requirements for medical devices. Because diagnostic reagents are used outside the human body, they have been placed in a special category with respect to public health and hygiene; however, *in vitro* reagents must meet the labeling requirements of the PAL and the ERPALs for drugs rather than the requirements for medical devices.

Metals for dental use are considered medical devices in Japan. As such, they must meet the general labeling requirements for medical devices in addition to specific requirements covered in the ERPALs.

Also covered in this chapter are the labeling requirements for unapproved devices being submitted to testing as part of the device-approval process, and for devices undergoing clinical investigation.

Part VIII

United States

20 The Federal Food, Drug, and Cosmetic Act

At the turn of the twentieth century, the unsanitary conditions in the United States (US) food and drug industries were appalling. Upton Sinclair's novel, *The Jungle*, described the filth of the meat packing industry and made Americans leery of eating meat. The Food and Drug Act of 1906 (also known as the Wiley Act) was enacted to give the government a role in protecting the public against unsafe or adulterated foods and drugs moving in interstate commerce. Although the Food and Drug Act of 1906 defined adulterated and misbranded drugs, it was the responsibility of the government to prove that the seller intended to defraud the buyer. Medical devices were not addressed at all in the 1906 act.

Unfortunately, major human catastrophes are often the precursors of legislation that protects society. The tragic deaths caused by the misformulated elixir of sulfanilamide* prompted the enactment of the federal Food, Drug, and Cosmetic Act of 1938 (FD&CA). The FD&CA established the statutory framework under which the Food and Drug Administration (FDA) regulates new foods, drugs, devices, and cosmetic products in order to protect public health. For the first time, mechanical devices intended for curative purposes and devices to bring about changes in the structure of the body were covered. Manufacturers were required to prove that their devices were not adulterated or misbranded.

For more than sixty years, the FDA has been at work creating and enforcing rules that set the requirements that the food, drug, device, and cosmetic industries are obligated to meet for products administered to humans and animals. Fundamentally, most legal experts from the FDA, regulated industries, consumer groups, and Congress have concluded that the FD&CA has functioned well, as written. Even with expected socioeconomic and technological challenges, most legal experts think the basic structure provided by the FD&CA will endure well into the twenty-first century (Peck p. 1).

Congress has amended the FD&CA a number of times to enable the FDA to respond to changing circumstances. Even with the challenges of rapidly evolving technology, pressure to bring to the market new treatments for diseases such as AIDS, and the fundamental restructuring of the US healthcare system (which is almost certain to occur) the overall mission of the FDA will likely remain the same.

The five amendments to the FD&CA that bear most directly on the medical device manufacturers doing business in the United States are the Medical Device Amendments of 1976, the Safe Medical Devices Act of 1990 (SMDA), the FDA Export Reform and Enhancement Act of 1996 (ER&EA), the Food and Drug Administration Modernization Act of 1997 (FDAMA) and the Medical Device User Fee and Modernization Act of 2002 (MDUFMA).

The Medical Device Amendments of 1976 was a sweeping revision to the FD&CA that brought all medical devices intended for human use under the jurisdiction of the FDA. Generally, the law amplified the definition of medical devices and created a three-tiered classification system. Devices are classified based on the degree of control required to provide reasonable assurance of safety and effectiveness. Class I devices are those for which general controls are sufficient. Class II devices require additional controls contained in FDA-defined performance standards. Class III devices, because of their life-sustaining nature, or because of some other risk of illness or injury, require

---

* Congress acted in response to the tragic deaths in 1937 of at least 73 persons who had taken the misformulated drug.

premarket approval. The FDA, with the assistance of expert panels, classifies devices under criteria defined in the statute (O'Keefe p. 4).

Under the FD&CA, the FDA has the authority to restrict the sale of or to ban a device under specific conditions. The law establishes specific requirements for record keeping and good manufacturing practices, imposes import and export controls, regulates custom devices and devices for investigational use, provides for notification to the public of hazardous devices, and specifies the essential content of device labeling.

The SMDA* expanded the FDA's authority to regulate medical devices. The Act provides the FDA with greatly expanded enforcement powers, extends adverse device reporting to user facilities, requires compliance with device tracking regulations for designated types of devices, and significantly affects the premarket approval and premarket notification submissions (Holstein p. 91).

The ER&EA** provided a new option for exporting unapproved devices, and added a new provision that permits the importation of certain components, parts, and accessories of devices for further processing or incorporation into products intended for export.

Almost ten years passed between the adoption of the SMDA and the next major reform for the FD&CA. Signed into law on November 21, 1997, FDAMA*** contains provisions related to all products under the FDA's jurisdiction. With respect to medical devices, the FDA is directed to focus its resources on the regulation of those devices that pose the greatest risk to the public and those that offer the most significant benefits. The FDA must base its decisions on clearly defined criteria and provide for appropriate interaction with the regulated industry. The new legislation assumes that enhanced collaboration between the FDA and regulated industry will accelerate the introduction of safe and effective devices to the United States.

The most recent amendment to the FD&CA is the MDUFMA.**** The MDUFMA contains three particularly significant provisions relating to medical devices. For the first time, FDA is authorized to charge a user fee for reviewing a premarket approval application (PMA) or a premarket notification (510(k)). Fees will add at least $25 million to FDA's medical device budget. The FDA will use this revenue to pursue a set of ambitious performance-improvement goals. The MDUFMA allows facility inspection to be conducted by accredited third-parties under carefully prescribed conditions. Finally, MDUFMA sets new regulatory requirements for reprocessing single-use devices. Included in the reform is a new category of premarket submission, the premarket report.

BACKGROUND AND GENERAL INTENT OF THE LAW

When the FD&CA became law in 1938, most of the legitimate medical devices on the market were relatively simple. They employed basic scientific principles so that an expert using the device could recognize that it was functioning properly. The major concern was that the claims made for the device in the labeling were truthful. Much of the FDA's activity with respect to the regulation of medical devices during the first 20 years following enactment of the FD&CA involved protecting Americans from fraudulent devices. In fact, a major objective of Congress in passing the 1938 Act was to give the FDA authority to regulate fraudulent devices (House, Interstate and Foreign Commerce p. 7).

By 1960, the FDA's attention began to shift to the hazards that might result from the use of legitimate medical devices. Developments in the electronics, plastics, metallurgy, and ceramics industries contributed to inventions such as the cardiac pacemaker, kidney dialysis machine, defibril-

---

*Safe Medical Device Act of 1990. US Public Law 101-629 November 28, 1990.

** Export Reform and Enhancement Act of 1996, as amended. Public Law 104-134. April 26, 1996.

*** Food and Drug Administration Modernization Act of 1997. US Public Law 105–115. November 21, 1997.

**** Medical Device User Fee and Modernization Act of 2002. US Public Law 107–250. October 26, 2002.

lator, cardiac and renal catheters, orthopedic implants, artificial vessels and heart valves, intensive-care monitoring units, and a wide spectrum of other diagnostic and therapeutic devices. While the new technology had saved or improved many lives, the potential for harm to the consumer had also increased. The House Committee on Interstate and Foreign Commerce concluded in its report that: "In the search to expand medical knowledge, new experimental approaches have sometimes been tried without adequate premarket clinical testing, quality control in materials selected, or patient consent" (House, Interstate and Foreign Commerce p. 8).

In the absence of specific language in the FD&CA, the courts expanded the FDA's authority by ruling that some medical devices were, in fact, drugs and could be regulated as such. Examples of products that were regulated as drugs included intrauterine contraceptive devices (IUCDs), contact lenses, weight-reducing kits, and some *in vitro* diagnostic (IVD) products (Parr, Barcome, Regulatory Requirements p. 1).

Presidents Kennedy, Johnson, and Nixon recognized the need for more comprehensive authority to regulate medical devices. In 1969, the Secretary of Health, Education, and Welfare convened a medical device study group composed of experts in medicine and technology. Dr. Theodore Cooper, who was the Director of the National Heart and Lung Institute, chaired the committee. The Cooper Committee, as it became known, was charged with evaluating the alternative approaches to regulating medical devices and recommending new comprehensive device legislation.

The Cooper Committee published its report in September 1970. The committee concluded "...that problems do in fact exist and that a predictable increase in the complexity and sophistication of medical devices requires action now to prevent the emergence of even more serious and complex problems in the foreseeable future....the study group believes that present and potential hazards, and the need for reliability and effectiveness of devices necessitates explicit legislation" (House, Interstate and Foreign Commerce p. 9). The recommendations of the Cooper Committee eventually became the outline of the Medical Device Amendments of 1976.

Among other findings, the Cooper Committee concluded: "Many hazards associated with medical devices do not arise from their design or manufacture, but rather from the manner in which the devices are used. Devices are often not properly labeled as to specific operating instructions. Accordingly, electrical and other hazards may result from improper installation and interconnection of devices with one another and with the patient" (House, Interstate and Foreign Commerce p. 10).

As a result of the Medical Device Amendments of 1976, the definition of a misbranded device was broadened considerably. A restricted device is deemed to be misbranded if its labeling fails to include a true statement of the device's established name and a statement of the intended uses of the device and relevant warnings, precautions, side effects, and contraindications (Public Law 94-295 §3(e)).

The Medical Device Amendments of 1976 were described at the time as "complex legislation negotiated by lawyers for lawyers" (Munsey pp. 350–354). While it tightened controls on legitimate devices being marketed as safe and effective without adequate premarket testing, the Medical Device Amendments of 1976 did not adequately deal with a number of fundamental issues, of which three are often cited. The first is the use by the FDA of a "modest statutory provision" contained in Section 510(k) of the FD&CA. This "provision requiring agency notification of a manufacturer's intent to market a product" has been the main path for FDA approval of new devices reaching the market (Munsey p. 366). Second is a concern that there has been consistent underreporting by the medical device industry of field problems under Section 519 of the FD&CA. Third is the need to establish a formal system for collecting useful information on the safety and effectiveness of a device after it has been introduced into interstate commerce. Legislative pressure for reform of the FD&CA to address these concerns began building in the early 1980s.

On October 27, 1990, Congress passed the SMDA. President George H. W. Bush signed the Act into law on November 28, 1990. The impact of the SMDA on the medical device industry and the user may not be fully understood for years. Many of the provisions of the law are ambiguous and require the FDA to develop a large number of new regulations. However, the two areas in which the medical device industry will be directly affected are compliance and product approvals (Gibbs, New Medical Device p. 505).

The SMDA strengthened the FDA's enforcement powers by giving the FDA the power to impose fines on manufacturers for virtually any violation of the FD&CA. The FDA was given power to "stop shipments," and its recall authority was expanded (Public Law 101-629 §8). The power to order a recall was in the Medical Device Amendments of 1976. However, the SMDA streamlines the process, making mandatory recalls far more likely (Gibbs, New Medical Device p. 506). The FDA can temporarily suspend the approval of a premarket approval application (PMA) if there is reasonable probability that continued distribution of the device would lead to "serious, adverse health consequences or death" (Public Law 101-629 §9). Finally, Congress required that companies promptly report any device removal or correction undertaken to reduce a risk to health. Companies must also report actions taken to remedy a violation of the FD&CA caused by a device that may present a health risk (Public Law 101-629 §7).

The SMDA also made substantial changes to the 510(k) premarket notification process. A company claiming substantial equivalence to a Class III device must conduct "a reasonable search of all information known" about the predicate device (Public Law 101-629 §4). The manufacturer must include a summary of, and a citation to, all adverse safety and effectiveness data. The FDA will make the 510(k) summary of safety and effectiveness data available to the public within 30 days of the determination of substantial equivalence (Medical Devices, Substantial Equivalence p. 64296).

In Section 515(b) of the FD&CA, the FDA was directed to publish regulations requiring the submission of PMAs for preamendment Class III devices. In the years since the enactment of the Medical Device Amendments of 1976, the FDA has issued rules requiring a PMA for some preamendment devices such as the replacement heart valve.* One of the objectives of the SMDA was to motivate the FDA to move forward with the rulemaking process for pre-amendment Class III devices. As a result, the FDA published, in April 1994, a strategy for rulemaking involving preamendment Class III devices not already subject to PMA. The FDA grouped the 117 preamendment Class III devices on which no action has been taken into the following three categories (Alpert p. 2):

Group 3 This group includes 42 devices for which the FDA plans to issue rules requiring PMAs.

Group 2 This group includes 31 devices that the FDA believes have a high potential for reclassification into Class II.

Group 1 Included in this group are 44 devices that have fallen into disuse or limited use. No call for a PMA for a device in this group is anticipated.

Beginning in 1994 and extending over a three-year period, the FDA planned to initiate PMA rulemaking for 15 high priority devices in Group 1. For the remaining devices in Group 1, the FDA planned to follow the same program of evaluation and prioritization that it will pursue for Group 2 devices. Although the remaining Group 3 devices were not considered candidates for reclassification, the FDA's evaluation of the data collected may lead to moving one or more of these devices to Group 2 (Alpert p. 2).

---

* See 21 CFR 870.3925.

In the SMDA, Congress encouraged the FDA to reclassify Class III devices into Class II. The FDA was given authority to impose new controls over Class II devices (Public Law 101-629 §5). In keeping with this provision, the FDA planned to issue orders requiring manufacturers to submit all available safety and effectiveness information, including adverse experience data, to them for Group 2 devices. Based on the evaluation of the submitted data, the FDA might reclassify these devices into Class II (Alpert, p. 3).

Group 1 devices have fallen into disuse. While they do raise significant safety and effectiveness questions, their infrequent use is unlikely to result in viable PMAs or reclassification petitions. These devices will be dealt with in a common rulemaking process (Alpert p. 3).

In an effort to make better use of its resources, the FDA moved to exempt additional Class I devices from the requirement that the manufacturer submit a 510(k) premarket notification. In one of the largest reclassification actions in its history, the FDA has added 148 generic types of Class I devices to the list of devices that are exempt from the premarket notification requirements of the FD&CA (Medical Devices, Exemptions pp. 63005–63006).

The procedure for establishing performance standards for Class II devices was streamlined (Public Law 101-629 §6). This provision was intended to make it easier for the FDA to establish performance standards (Gibbs, New Medical Device, p. 508).

The Medical Device Amendments of 1976 authorized the FDA to require manufacturers to report information that might reasonably suggest that a device has caused or contributed to a death or serious injury to a patient (Public Law 94-295 §519). People, both inside and outside the FDA, have asserted that relying solely on the medical device manufacturers has resulted in consistent underreporting of device problems. Congress addressed this concern by adding a user-reporting provision to the FD&CA. Under this provision, user facilities are required to submit Medical Device Reports (MDRs) to the FDA within 10 working days if there is reason to believe that a device caused or contributed to a death. If a user facility has reason to believe that a device caused or contributed to a serious injury to a patient of the facility, the facility has 10 days to submit an MDR to the manufacturer (Public Law 101-629 §2).

To establish a formal system for collecting useful information on the safety and effectiveness of a device after it has been introduced into interstate commerce, Congress gave the FDA the power to require a manufacturer to conduct postmarket surveillance on certain devices introduced after January 1, 1991. A device qualifies for postmarket surveillance if it "(a) is a permanent implant where failure may cause serious, adverse health consequences or death; (b) is intended for use in supporting or sustaining human life; or (c) potentially presents a serious risk to human health" (Public Law 101-629 §10). Devices that do not meet these conditions may also be subject to postmarket surveillance if the FDA is convinced that such surveillance is necessary to protect the public health or to provide safety and effectiveness data.

The FD&CA stated that a device intended for export would not be considered adulterated or misbranded if the product (1) met the foreign purchaser's specifications, (2) was not in conflict with the laws of the country to which it was being exported, (3) was labeled on the outside of the shipping package that the product was intended for export, and (4) was not sold or offered for sale in domestic commerce. This authority remained unchanged until the Medical Device Amendments Act of 1976 amended the provision of the FD&CA dealing with device export. Under the revised statute, the four criteria did not apply to an unapproved device unless, in addition to requiring compliance with Section 801(e)(1) of the act, the FDA determined that exportation of the device would not be contrary to the public health and safety and the device had the approval of the foreign country that would receive it.

The statutory requirement that the FDA approve device exports began to generate criticism from the device industry about the amount of time the FDA took to determine whether an export request met the statutory criteria. Despite reductions in processing time, the statute's export approval requirements were seen as adversely affecting the ability of US firms to enter or to compete in foreign markets. In response to industry pressure, Congress again amended the FD&CA by passing the ER&EA. President Clinton signed the ER&EA into law on April 26, 1996.

The ER&EA permits the import of component parts, accessories, or other articles of a device that do not comply with other provisions in the FD&CA if those component parts, accessories, or other articles are intended for incorporation or further processing by the initial owner or consignee into a device that will be exported.

The ER&EA eliminated the requirement for prior FDA approval for export of devices approved in Australia, Canada, Israel, Japan, New Zealand, Switzerland, South Africa, or any member nation of the European Union (EU) or the European Economic Area (EEA), or for export of devices destined for clinical investigations in one of these countries. The law created an administrative mechanism for the Secretary of the Department of Health and Human Services (DHHS) to add countries to the list and for the FDA to approve exports of specific products to unlisted countries. The ER&EA also authorized the export of unapproved devices to a listed country in anticipation of marketing approval in that country. A simple notification process was set up for exported devices (as opposed to the application process under Section 801(e)(2) of the act). Notification is not required for devices exported for investigational use to a listed country or for devices exported in anticipation of marketing authorization in the listed country. Finally, the ER&EA authorized the FDA to permit the export of unapproved devices intended to treat tropical diseases or other diseases that are "not of significant prevalence in the United States."

FDAMA was a major legislation initiative focused on reforming the regulation of food, medical products, and cosmetics. The law enacts many of the initiatives that the FDA undertook under Vice President Al Gore's Reinventing Government program. Congress also attempted to address some of the medical device industry's most significant complaints about the FDA's regulatory process and to make that process less confrontational. Initiatives include programs such as the early collaboration on data requirements for clinical studies and the PMA collaborative review process. Under the early data review program, sponsors that intend to perform a clinical study of any Class III device or any implantable devices in any class can have their investigational plan, including the clinical protocol, discussed with the FDA for the purpose of reaching an agreement on the investigational plan before they apply for an investigational device exemption (IDE). The PMA collaborative review process is intended to promote earlier and ongoing communications between the FDA and the applicant regarding deficiencies in the application.

The device tracking requirement has been changed to allow the FDA to order that certain devices must be tracked but to delete any automatic requirements to track devices unless there is such an order. Manufacturers will no longer be automatically required to conduct postmarket surveillance studies for particular devices. Rather, the FDA may order such studies to be conducted for certain Class II and Class III devices. The FDA can order postmarket surveillance for any Class II and Class III device (FDAMA Guidance §11):

- if it is reasonably likely that a failure would have serious adverse health consequences,
- if it is intended to be implanted in the human body for more than one year, or
- if it is intended to be a life sustaining or life supporting device used outside a healthcare facility.

FDAMA adds a system for recognizing national and international standards in product reviews. The FDA may, through publication in the *Federal Register*, recognize all or part of an appropriate standard established by a nationally or internationally recognized standards development organization.

An applicant may reference the recognized standard in a Declaration of Conformity, which can be used to satisfy a premarket submission requirement (PMA or 510(k)) or other requirement under the FD&CA to which such a standard applies.

The user fee provision of the MDUFMA is a major refom of the way medical device regulatory process is funded in the US. The medical device manufacturers will directly fund the process through payments made to the FDA when submitting various application. This process mirrors a similar funding mechamisim in place for drug regulatory reviews for several years.

The MDUFMA authorizes FDA-accredited third-parties to inspect qualified manufacturers of Class II and Class III devices. Once a third-party is accredited, an eligible establishment would then be permitted to select any accredited third-party to conduct an inspection in lieu of an FDA inspection. To qualify, an accredited third-party may not be owned by or be affiliated with a device manufacturer, supplier, or vendor, and cannot be engaged in the design, manufacture, promotion, or sale of FDA-regulated products. Not all manufacturers are eligible to take advantage of third-party inspections. Among the conditions, the manufacturer must have a clear inspection record and must market a device in the US and in at least one other country. Third-party inspections are intended to apply to manufacturers that operate in a global market and are subject to multiple inspection requirements (MDUFMA Summary).

The MDUFMA significantly modifies the regulatory requirements for reprocessing of single-use devices. This section of the law requires the FDA to reassess all critical or semi-critical reprocessed devices that have been expected from the requirement to submit a 510(k) and to identity those whose exemption from 510(k) should be ended. By April 2003, the FDA must review the types of repro-cessed single-use devices that currently are the subject to 510(k) clearance and identify those for which FDA will require "validation data . . . regarding cleaning and sterilization, and functional performance" to demonstrate that the reprocessed device "will remain substantially equivalent . . . after the maximum number of times the device is reprocessed as intended" by the reprocessor who submits the 510(k) (FD&CA §510(o)(1)(A)). New labeling is required on all reprocessed devices place on the market after January 2004. These requirements are described in detail in the next chapter.

RELATED LAWS

In addition to the FD&CA, the FDA also administers 15 other acts passed by Congress that apply to food, drugs, cosmetics, biologics, and radiation-emitting electronic products, as well as medical devices. In addition to the FD&CA, the FDA currently administers three other laws that directly address the labeling of medical devices:

- The Fair Packaging and Labeling Act (FPLA)
- The Radiation Control for Health and Safety Act of 1968 (RCHSA)
- The Public Health Service Act (PHSA)

Through the FPLA, Congress intended to ensure that packaging and labeling would enable consumers to obtain accurate information on the quantity of the contents of a package and to facilitate value comparison (FPLA §1451). Section 1454(a) of this act vests in the Secretary of DHHS the authority to issue regulations under this act with respect to any consumer commodity that is a food, drug, device, or cosmetic. Requirements of the FPLA apply to over-the-counter medical devices distributed by retail outlets.

Congress has declared that the public health and safety must be protected from the dangers of electronic product radiation (RCHSA §263b). To accomplish this, the Secretary of the DHHS is authorized to establish an electronic radiation control program that includes the development and administration of performance standards to control the emission of electronic product radiation. Section 358(h) of the RCHSA requires manufacturers or distributors of radiation-emitting electronic products, including medical devices, to place certification labeling on their devices.*

The PHSA, among other provisions, deals with the import, export, and sale of products of biological origin (e.g., virus, therapeutic serum, blood, blood compounds or derivatives) applicable to the prevention, treatment, or cure of diseases or injuries to humans. In particular, some IVD devices incorporate reagents of biological origin that fall under the provisions of this act.

SCOPE OF THE FDA REGULATIONS

In general, the FDA's authority to enforce the FD&CA applies to products in interstate commerce. Congress and the courts have given the term "interstate commerce" a very broad definition. The FDA may exercise its enforcement powers over a finished product that moves in interstate commerce, as well as over products made from components that have moved in interstate commerce. While not considering every possible circumstance, the Ninth Circuit Court held in *Baker v. United States*** that if an active ingredient or significant component of a product has moved in interstate commerce, the FDA may regulate the finished product, whether or not the finished product moves in interstate commerce. In addition, Congress has defined the terms "food," "drug," "device," and "cosmetic" to include not only the finished product, but components of these products as well. Using this interpretation, the FDA has authority over components moving in interstate commerce before they are incorporated into finished devices.

A food, drug, or cosmetic made in whole and in all parts within one state and with sales confined to that state alone are not regulated by the FDA. However, the FDA may seize misbranded or adulterated medical devices and counterfeit drugs regardless of whether or not these products or their components have moved in interstate commerce. The FDA may take advantage of the provision found in Section 709 of the FD&CA, which states that: "In any action to enforce the requirements of this Act respecting a device, the connection with interstate commerce required for jurisdiction in such action shall be presumed to exist." FDA seizures are not common, but they do happen. For example, on July 11, 2002, the FDA seized and destroyed 40,000 OB/GYN and surgical devices manufactured by a small Georgia firm because of serious sterilization problems that *could* have resulted in serious and possibly life-threatening infections. The FDA took regulatory action even though no complaints of injury were reported (Unsafe p. 1).

THE REGULATIONS

The specific requirements of the laws administered by the FDA are implemented by the Secretary of the DHHS through regulations. Proposed regulations are prepared by the Commissioner of Food and Drugs. The regulations are reviewed and approved following administrative procedures established by Congress. Once a final regulation is published in the *Federal Register,* it has the force of law.

Labeling regulations for medical devices promulgated by the Commissioner of Food and Drug under authority of the FD&CA, FPLA, RCHSA, and PHSA are found in the Code of Federal Regulations (CFR) under Title 21—*Food and Drugs.* The parts of the 21 CFR that currently deal with medical device labeling requirements are:

---

* This statute, which was originally part of the *Public Health Service Act* (PHSA) at US Code, vol 42, §263b *et seq.*, was recodified by Section 19(3) of Public Law 101–629, which repealed Section 354 and redesignated Sections 355 through 360F of the PHSA as Section 531 through 542 of the *Federal Food, Drug, and Cosmetic Act* (FD&CA), 21 US Code Vol 21 §360hh *et seq., supra.*
** *Baker v. United States*, 932 F. 2d 813 (9th Cir. 1991).

- General Device Labeling—§801
- Premarket Notification Procedure—§807
- *In Vitro* Diagnostic Products—§809
- Investigational Device Exemption—§812
- Investigational Exemptions for Intraocular Lenses—§813
- Premarket Approval of Medical Devices—§814
- Good Manufacturing Practices—§820
- General Electronic Products—§1010

The primary law under which the FDA derives authority to take action against the labeling of regulated medical devices is the FD&CA. Specifically:

- Sections 201(k) and 201(m) define "labels" and "labeling."
- Sections within Chapter III address prohibited acts with respect to medical devices. These prohibitions are divided into two major offenses: "adulteration" and "misbranding."
- Sections of Chapter V describe the specific instances under which the FDA will consider a device adulterated or misbranded.

The FPLA applies only to medical devices intended for over-the-counter sales through retail outlets. Regulations based on the labeling requirement in the FPLA are found under 21 CFR 801, in particular, Subpart C—*Labeling Requirements for Over-the-Counter Devices.*

The RCHSA applies to a range of electronic devices that emit radiation that could be hazardous to public health and safety, including a number of medical devices. Regulations based on the labeling requirement in the RCHSA are found in 21 CFR 1010, in particular, Subpart A—*General Provisions.*

The PHSA deals with the labeling of products that contain virus, therapeutic serum, toxin, antitoxin, vaccine, blood, and blood compounds or derivatives (or any other trivalent organic arsenic compound) used for the prevention, treatment, or cure of diseases or injuries to humans. For medical devices, this applies most directly to IVD products. Regulations that incorporate the requirements of the PHSA are found in 21 CFR 809, in particular, Subpart B—*Labeling.*

LABELS AND LABELING

Section 201 of the FD&CA draws a distinction between label and labeling. These terms are related, but they are not interchangeable. Section 201(k) defines a "label" as "a display of written, printed, or graphic matter upon the immediate container of any article." Section 201(l) further stipulates that "immediate container" does not include package liners. For a device to be in compliance, any information required by the law must appear "on the outside container or wrapper, if any there be, of the retail package of such article, or is easily legible through the outside container or wrapper." For practical purposes, the term "label" covers all written, printed, or graphic information applied to a medical device or any of its containers or wrappers.

"Labeling," on the other hand, is defined in Section 201(m) as "all labels and other written, printed, or graphic matter (1) upon any article or any of its containers or wrappers, or (2) accompanying such article." It is the term "accompanying" that gives labeling its increased scope. FDA interprets "accompanying" liberally to include more than physical association with the medical device (Cardamone p. 2). The FDA includes posters, tags, pamphlets, circulars, booklets, brochures, instruction books, direction sheets, fillers, and so on, regardless of whether they are shipped with the device, or brought together with the device after shipment.

LABELING AND ADVERTISING

The FDA takes the view that there is little distinction between labeling and advertising. Both are used to draw attention to a product for sale. In its guidance document on labeling, the FDA supports this contention by citing an appellate court decision: "Most, if not all advertising, is labeling. The term `labeling' is defined in the FD&CA Act as including all printed matter accompanying any article. Congress did not, and we cannot, exclude from the definition printed matter which constitutes advertising" (Cardamone p. 3).

In addition to the requirements of the FD&CA, advertising of medical devices is covered by the Federal Trade Commission Act (FTCA). The FTCA makes it unlawful to disseminate, or cause to be disseminated, any false advertisement for the purpose of inducing the purchase of food, drugs, devices, or cosmetics (FTCA §52(a)). For the purposes of the FTCA, "false advertisement" means any advertisement, other than labeling, that is misleading in any material respect. When determining whether an advertisement is misleading, the government takes into account not only the representations made in the advertisement, but also the extent to which the advertisement fails to reveal material facts about the consequences that may result from the use of the device under the conditions prescribed in the advertisement, or under the conditions that are customary and usual for use of the device (FTCA §55(a)(1)).

ADULTERATION AND MISBRANDING

Section 301 of the FD&CA makes it a federal crime, punishable by fine or imprisonment or both, to "introduce or deliver for introduction into interstate commerce any food, drug, device, or cosmetic that is adulterated or misbranded."

Section 501 of the FD&CA deems a device adulterated for a variety of reasons related to both the contents and the conditions under which it has been prepared, packed, or held. With regard to labeling, a device—in particular, an IVD product—is considered adulterated if its strength differs from, or its purity or quality falls below, that which it purports or is represented to possess. A device is also considered adulterated if the manufacturer fails to comply with the Good Manufacturing Practice (GMP) regulations that deal with the quality assurance (QA) program requirements for labeling. These requirements are discussed in detail in Chapter 26.

Section 502 of the FD&CA deems a device misbranded:

- if its labeling is false or misleading in any particular;
- if, in a packaged form, its label fails to contain the name and place of business of the manufacturer, packer, or distributor;
- if its label fails to contain an accurate statement of the quantity of the contents in terms of weight, measure, or numerical count within reasonable variations permitted by the regulations;
- if any word, statement, or other information required by the law is not prominently displayed in comparison to other information on the label;
- if the required information is not stated in a way that makes it likely to be read and understood by the ordinary individual under the normal and customary conditions of purchase and use;
- if its labeling does not contain adequate directions for use including adequate warnings against use in certain pathological conditions or against unsafe dosage, methods, or duration of administration or application;
- if its labeling fails to warn against use by children where its use may be dangerous to their health; or
- if it is dangerous to health when used in the dosage, manner, or with the frequency or duration prescribed or suggested in the labeling.

These provisions of Section 502 apply equally to food, drugs, cosmetics, veterinary drugs, biologics, and medical devices. The Medical Device Amendments of 1976 expanded the provisions of the law with regard to medical devices. Section 520(e)(1)(A) established that a medical device may "be restricted to sale, distribution, or use only upon the written or oral authorization of a practitioner licensed by law to administer or use such a device." Section 502(q) extended the criminal provisions of the FD&CA to include distribution or offering for sale in any state a restricted device if its advertising is false or misleading, or if it is sold, distributed, or used in violation of regulations prescribed under Section 520(e).

The Medical Device Amendments of 1976 expanded the circumstances under which a medical device is considered misbranded. Under the amendments, a device is misbranded:

- if it is subject to a performance standard and does not bear all the labeling prescribed in the performance standard;
- if it is a restricted device and fails to bear the appropriate statement of restriction required by regulation;
- if it is restricted, and fails to bear "a brief statement of the intended uses of the device, and relevant warnings, precautions, side effects, and contraindications.";
- if there is a failure or refusal to comply with the requirements prescribed in Section 518 on notification and other remedies, or a failure to furnish any materials or information requested by or under Section 519 on reports and records;
- if the labeling or advertising bears any representation or suggestion that the device has the official approval of the FDA or creates the impression of FDA approval because the manufacturer possesses an FDA registration or premarket notification number; or
- if the labeling fails to bear the device's established name prominently and in type at least half as large as that used for any trade or brand name.

The Secretary of the DHHS may designate an official name for any drug or device if such action is necessary or desirable in the interest of usefulness and simplicity. The established name of a device is the official name designated by the Secretary of the DHHS. If an official name has not been designated, then the established name is the one that appears in the *United States Pharmacopeia*, the *Homeopathic Pharmacopeia of the United States*, the *National Formulary*, or any of their official supplements. A few medical devices are listed in these official compendiums. Two examples are poly (glycolide/l-lactide) surgical suture (PGL suture) and medical adhesive tape or adhesive bandage. Both are classified as devices under 21 CFR 878.4493 and 21 CFR 880.5240, respectively. Requirements, including labeling considerations, for both devices are described in official monographs in the *United States Pharmacopeia*. However, for most medical devices, neither condition is true. For these devices, the established name is the common or usual name of the device.

Section 721 of the FD&CA requires that a device (e.g., an IVD product) that employs color additives must comply with the provisions of the Color Additive Amendments of 1960* if the color additive comes into direct contact with the body of a human or other animal for a significant period of time. Failure to do so renders the device misbranded.

FALSE OR MISLEADING LABELING

The FD&CA declares that a device is misbranded if its labeling is false or misleading in any particular. The FDA takes a broad interpretation of the phrase "false and misleading." Labeling need

---

* *Color Additive Amendments of 1960*. US Public Law 86–618 US Coce vol 21 §376. July 12, 1960.

not be untrue, forged, fraudulent, or deceptive to be viewed as being in violation of the law. A word, statement, or illustration may be strictly true, but yet be misleading to the customer. Misleading means "the labeling is deceptive if it is such as to create or lead to a false impression in the mind of the reader" (Cardamone p. 4). A false impression may be created not only by false or deceptive statements, but also by ambiguity and misdirection. For example, a device labeled as a "Bust Developer" was determined to be misbranded because a booklet accompanying the device suggested that use of the device would increase breast size. The court upheld the FDA determination even though explanatory material in an accompanying booklet referred to bust size, defined as chest measurement, and not breast size (*U.S. v. Iso-Tensor*)*.

Failure to inform the customer of facts relevant to the statements made will render the device misbranded just as much as a blatantly false or exaggerated claim. A label that is silent on important considerations may be just as deceptive as one that makes exaggerated claims.

A manufacturer who knowingly makes false representations to the government to obtain FDA approval of the device can be held accountable not only for its statements, but also for medical costs incurred by the government as a result of complications associated with the falsely represented product (Justice pp. 1–2). Under the terms of the False Claims Act,** the government can seek compensatory and punitive damages from the manufacturer. Such actions clearly demonstrate the rapidly increasing interaction between food and drug law, healthcare law, government contract law, and product liability law. Manufacturers must be extremely cautious that the claims they make for their devices can be substantiated and do not overstate the advantages or minimize the risks associated with the device.

ADVERTISING AND PROMOTION

For many years, the world of device advertising and promotion has been relatively freewheeling when compared to drug advertising and promotion. However, the FDA signaled an increasing interest in this area when the Center for Devices and Radiological Health's (CDRH) Office of Compliance established a Device Labeling Compliance Branch (DLCB) in February 1992. After the formation of the DLCB, the number of warning letters and field corrective actions for misleading labeling and advertising of devices increased sharply. The Director of the CDRH at the time, Dr. Bruce Burlington, had said "that eliminating false or misleading marketing claims will be one of the center's top priorities" (Warning Letters p. 2). There are several things a manufacturer can do that are likely to draw a compliance action in this area.

First is describing the product or a particular use of the product as "FDA Approved," or making representations that create the impression of FDA approval because the manufacturer possesses an FDA registration or premarket notification number. Making such a claim is a clear violation of Section 301(l) of the FD&CA and renders the product misbranded. A person found guilty of such a violation is subject to the penalties outlined in the FD&CA.

Second is making unsupported claims about a product. The FDA applies the model that has been in use in the Center for Drug Evaluation and Research for years. The drug model for advertising regulation is based on a deceptively simple principle: Advertising and promotional claims must be based on the product's FDA-approved labeling, including indications and potential adverse effects, or on substantial evidence from adequate and well-controlled studies (GMP Letter p. 1). Manufacturers or distributors of medical devices may not promote their devices for indications that are not in the approved labeling. For example, if an orthopedic device is approved and labeled, "For

---

* *U.S. v. Iso-Tensor*, CA-9 (1977).
** *False Claims Act*. US Code vol 31 §§3729 *et seq.*

Cemented Use Only," the manufacturer or distributor must not represent that uncemented use is safe or effective (Orthopaedic Device p. 2).

Since 1993, CDRH's Office of Compliance has issued to manufacturers numerous warning letters alleging misleading labeling and advertising. The following are examples of the actions taken by the FDA against manufacturers for making misleading and unsupported claims in advertising and product promotion.

In September 1999, the FDA issued a warning letter to the manufacturer of a device approved for recording intracranial neural activity and for stimulation of subsurface levels of the brain during surgery. In the warning letter, the FDA alleged that promotional materials made unapproved claims for the device including a claim that this system "was developed for use in sterotactic functional neurosurgery/intraoperative microelectrode guidance during pallidotomy, thalamotomy, and the implantation of deep brain stimulators for the treatment of Parkinson's Disease and essential tremor." According to the warning letter, the manufacturer had included certain claims in the 510(k) application related to use of the device during pallidotomies, thalamotomies, epilepsy surgery and other procedures where appropriate. The FDA stated that unless the firm provided valid scientific evidence to demonstrate the effectiveness of each claim, these claims were to be removed from the labeling. The manufacturer subsequently responded that these specific claims had been removed from labeling, manuals, the "Indications for Use" page, and promotional material (Warning, September 1999 p. 1).

In the same letter, the FDA advised the manufacturer that their promotional material made improper claims about FDA approval. Statements such as "the first instrument of its kind to gain both US FDA approval," and "cleared by the FDA for distribution in the USA" are a violation of 21 CFR 807.97. Such statements are misleading and constitute misbranding (Warning, September 1999 p. 2).

In April 1993, the FDA issued warning letters to some of the largest hearing aid manufacturers and distributors in the United States (Hearing Aid p. 1). In the letters, the FDA warned the firms that their advertising, promotion, and labeling were misleading because they created unrealistic expectations for the performance of their products. The FDA told the companies to "eliminate all misleading promotional literature and advertising immediately," and warned that "continued distribution of the hearing aids with misleading claims could result in enforcement actions such as seizure, injunction, and civil penalties." Further, the FDA advised the firms to take steps to "correct the misconceptions they have created by their misleading promotion and advertising." According to the FDA:

- The promotional materials failed to disclose important information about the limits of the devices' effectiveness and implied that most or all people who use the hearing aids would benefit equally—a generalization that is not supported by scientific data. For example, although the hearing aids may increase the volume of sound for the user, they may not improve speech recognition. The FDA said these firms failed to disclose that some background noise would be amplified and that some speech sounds would not.
- The promotional materials overstated the quality and value of the hearing aids with such claims as, "If you have nerve deafness, hearing again is no big thing!"
- The promotional materials misled the public by including testimonials that were not supported by documentation or scientific evidence.

In June 1993, a daily cleaner for contact lenses was recalled because promotional material made a claim for use as a disinfectant. That claim exceeded the product's PMA-approved indication as a daily cleaner (FDA Enforcement Report p. 7). In this case, the promotional material in question included two article reprints from the *Journal of the American Optometric Association*.

In August 1993, the FDA accused two firms of "promoting X-ray bone densitometers for unapproved uses via undocumented claims that the devices can be used to predict fracture risk, diagnose osteoporosis, and determine total body bone mineral content" (Densitometer Makers p. 3). The FDA warned the manufacturers to cease distributing their promotional material, advertising, and labeling carrying these undocumented claims. In the same warning letter, the FDA asked the companies involved to propose (a) ways to make the device incapable of performing unapproved functions and (b) ways of correcting the "misleading statements and claims" that X-ray bone densitometers are safe and effective for predicting fracture risk. In addition, the FDA indicated in the warning letters that the intended, unapproved uses promoted by these manufacturers raised the classification of the devices from Class II to Class III, requiring a PMA.

In March 2000, the FDA issued a warning letter to a manufacturer alleging a number of significant inappropriate representations with respect to a particular product on the company's Internet Web site. One of the objectionable practices cited in the warning letter was the failure to include a "Use in Specific Populations" section in the warnings page even though such restrictions were part of the conditions of approval for the device (Warning, March 2000 p. 2).

In the same warning letter, FDA objected to the posting of a press release on the Web site. The FDA alleged that the press release selectively quoted parts of a study discussed in the *New England Journal of Medicine* and presented one study's results as being conclusive and established. According to the FDA, "Quoting selected portions of a study is not included within the scope of the recent litigation involving FDA's regulation of the distribution of peer-reviewed journal reprints" (Warning, March 2000 p.3).

Promoting a device for a seemingly nonmedical use can invite regulatory action by the FDA. For example, an FDA district office issued a warning letter to a Florida company for promoting the use of diagnostic ultrasound equipment for recording an image of the fetus for keepsake and entertainment purposes (Warning, May 2000 p. 1).

Other actions that may lead to problems with the FDA are providing misleading information on prescription statements, making misleading comparisons to competitors' products, failing to achieve fair balance, and using testimonials (Warning Letters, p. 2). FDA has taken the position that testimonial statements made by other individuals but appearing in promotional literature on the companies Internet website "essentially becomes claims of the company" (Warning, March 2001 p. 1).

Overreaching claims can backfire on a manufacturer even if they do not provoke action from the FDA. False or misleading statements in product advertising can invite attack through private litigation in two principal ways (Gibbs, Medical Device Promotion p. 296). The first way is through product litigation. False or misleading statements may establish a basis for negligence, strict liability, or breech of warranty. These may increase the likelihood of punitive damages if the manufacturer is found guilty. A second risk is that a competitor may seek relief under the Lanham Act.* The number of Lanham Act suits involving healthcare companies, including medical device manufacturers, has increased in recent years.

The Lanham Act was originally enacted to provide relief to companies against narrow forms of unfair competition, such as trademark infringement. However, many observers believed that the United States should fall in line with common practice in many foreign countries. These countries allowed an injured competitor to seek damages due to false claims about a product, even in the absence of the violation of a specific property right, such as a trademark (Gibbs, Medical Device Promotion p. 297). In recent years, the courts have held that a "false description or representation" of a competitor's goods or services is a violation. This standard has been interpreted broadly to

---

* *Lanham Act.* US Code, vol 15, §§1051–72, 1091–96, 1111–21, 1123-27. 1988.

encompass both "blatant falsehood" and deceptive "innuendo, indirect imitation and ambiguous suggestions," whether or not the implied claims were actually intended (Gibbs, Medical Device Promotion p. 298).

PROMOTION OF OFF-LABEL USES

The FDA approves a drug or device for use in the United States based on scientific data provided by the sponsor (usually the manufacturer). The FDA approves only what the sponsor may recommend about uses in its official labeling. The FDA, however, does not regulate the practice of medicine; the FDA does not approve or disapprove specific uses of legally marketed drugs and devices by the physician in his or her own practice (Archer, p. 1054). Former FDA Commissioner David A. Kessler, in testimony before the House Subcommittee on Human Resources and Intergovernmental Relations, reiterated the FDA position: "While long-standing FDA policy permits an individual physician taking care of his or her patient to make a decision whether to use an approved drug [or device] for uses beyond the labeling, the law is clear that the promotion of such uses is illegal." However, Dr. Kessler went on to state, "Physicians may not, under the guise of scientific exchange, involve themselves in the manufacturer's promotion of unapproved uses" (House, Government Operations pp. 156–157).

Such promotional activities can take several forms. In the case of the daily cleaner for contact lenses, for example, the FDA held that distribution by the manufacturer of reprints from a respected journal, which discussed an unapproved use of the product, constituted a violation of the FD&CA. Another example is promotional activities that are disguised as education or public relations. The FDA intends to hold these activities to the same standards as other more traditional promotional material. Citing a recent FDA case in his June 1991 testimony before the House Subcommittee on Human Resources and Intergovernmental Relations, Dr. Kessler said the "FDA recently took action against the publication of an apparently independent scientific journal which reported on the unapproved uses of various oncology products. The journal in fact was sponsored by the manufacturer of the products that were discussed favorably in the journal, and FDA informed the manufacturer that the journal would be considered promotional. FDA has reached an agreement with the manufacturer of those products under which the manufacturer will cease publication of the journal, mail a letter to all physicians who received the journal admitting the firm's sponsorship, and submit to FDA for pre-clearance all of its oncology promotional materials for two years" (House, Government Operations pp. 157–158).

The application of the FDA policy against dissemination of information on off-label uses has been contested in the courts. In a suit filed in the US District Court for the District of Columbia, the Washington Legal Foundation (WLF) contended that the FDA policy exceeds the agency's legal authority to regulate labeling and violates First Amendment speech rights. In *Washington Legal Foundation v. Friedman*,* the US District Court for the District of Columbia concluded that the restrictions imposed by the FDA failed to meet the four-part test governing the constitutionality of limits on commercial speech established in *Central Hudson Gas and Electric Corp. v. Public Service Commission of New York*.** *Central Hudson* held that restrictions on commercial speech are constitutionally permissible if (1) the speech at issue does not concern unlawful activity and is not inherently misleading, (2) the government has a substantial interest in controlling the speech, (3) the restrictions imposed on the speech advance the government's interest in a direct and material way, and (4) the government's chosen means are a reasonable match for the ends it seeks. In *Friedman*, the court found that the FDA had identified a legitimate objective but its policies were

---

* *Washington Legal Foundation v. Friedman*, 13 F. Supp 2d 51 (DDC 1998): 71.
** *Central Hudson Gas and Electric Corp. v. Public Service Commission of New York*, 447 US 557 (1980).

"considerably more extensive than necessary to further the substantial government interest in encouraging manufacturers to get new uses on-label"*.

The FDA appealed the decision to the United States Court of Appeals for the District of Columbia. The appellate court in *Washington Legal Foundation v. Henney*** vacated part of the district court decision and injunction insofar as it "declared unconstitutional (1) statutory provisions concerning the dissemination by manufacturers of certain written materials concerning new uses of approved products (21 U.S.C. 360aaa *et seq.*), and (2) an FDA guidance document concerning certain industry-supported scientific and educational activities known generally as industry-supported continuing medical education or 'CME'" (Decision p. 14286). The FDA contends that the appellate court decision establishes these statutory provisions as "a 'safe harbor' for manufacturers that comply with them; the CME guidance document details how the agency intends to exercise its enforcement discretion" (Decision p. 14286). The FDA views the appellate court decision as allowing it to proceed, in the context of case-by-case enforcement, to determine from a manufacturer's written materials and activities how it intends that its products be used. The FDA further grants that the Court of Appeals also recognized that if the agency brings an enforcement action, a manufacturer may raise a First Amendment defense (Decision p. 14286).

The WLF takes a different view of the outcome. In May 2001, the WLF filed a citizens' petition requesting that the FDA withdraw the *Federal Register* notice it published on March 16, 2000, entitled "Decision in *Washington Legal Foundation v. Henney*" (WLF p. 1). In its place, WLF proposed a statement indicating "manufacturers will not be subject to enforcement action for disseminating enduring materials that contain truthful information about off-label uses of FDA-approved products" (WLF p. 3). On January 28, 2002, the FDA denied the request stating, "The Federal Register notice is consistent with both parties' positions in the Court of Appeals and this does not present a facial violation of the First amendment" (FDA Reply, p. 2). At the time this chapter was prepared, it remains to be seen if the WLF will take any further legal action.

At the moment, the two "safe harbors" identified by the FDA are contained in 21 CFR 99 (regulations promulgated under §360aaa of the FD&CA) and in the "Guidance for Industry—Industry-Supported Scientific and Educational Activities," published in the *Federal Register* on December 3, 1997.

REPRINTS OF SCIENTIFIC ARTICLES

With respect to reprints of scientific articles, a manufacturer may only disseminate information on a new use of a product to a healthcare practitioner, a pharmacy benefit manager, a health insurance issuer, a group health plan, or a federal or state government agency (21 CFR 99.105). A manufacturer may disseminate written information concerning the safety, effectiveness, or benefit of a use of a device not described in the approved labeling or include such information in the statement of intended use for a cleared device, provided that the manufacturer complies with the relevant requirements of 21 CFR 99 – *Dissemination of information on unapproved/new uses or marketed drugs, biologics, and devices*. This information must (21 CFR 99.101(a)):

- Be about a device that has been approved, licensed, or cleared for marketing by the FDA.
- Be in the form of:
  - An unabridged reprint or copy of an article (1) peer-reviewed by experts qualified by scientific training or experience to evaluate the safety or effectiveness of the device

---

* *Washington Legal Foundation v, Friedman*, p. 71.
** *Washington Legal Foundation v. Henney*, 202 F. 3d 331 (DC Cir 2000).

involved and (2) was published in a scientific or medical journal. In addition, the article must be about a clinical investigation with respect to the device and must be considered to be scientifically sound by the experts described in this paragraph; or

- An unabridged reference publication (1) that includes information about a clinical investigation with respect to the device and (2) that experts qualified by scientific training or experience to evaluate the safety or effectiveness of the device that is the subject of the clinical investigation would consider to be scientifically sound.

- Not pose a significant risk to the public health.
- Not be false or misleading. The FDA may consider information disseminated under this part to be false or misleading if, among other things, the information includes only favorable publications when unfavorable publications exist, or if it excludes articles, reference publications, or other information required under 21 CFR 99.103(a)(4), or if the information presents conclusions that clearly cannot be supported by the results of the study.
- Not be derived from clinical research conducted by another manufacturer unless the manufacturer disseminating the information has the permission of such other manufacturer to make the dissemination.

The FDA will find that all journal articles and reference publications are scientifically sound except (21 CFR §99.101(b)):

- letters to the editor;
- abstracts of a publication;
- those regarding trials in healthy people;
- flagged reference publications that contain little or no substantive discussion of the relevant clinical investigation; and
- those regarding observations in four or fewer people that do not reflect any systematic attempt to collect data, unless the manufacturer demonstrates to the FDA that such reports could help guide a physician in his or her medical practice.

For purposes of the regulation, a scientific or medical journal is a scientific or medical publication that meets the following requirements (21 CFR 99.3(j)):

- The journal must be published by an organization that has an editorial board that uses experts who have demonstrated expertise in the subject of an article under review by the organization and who are independent of the organization. These experts must be free to review and objectively select, reject, or provide comments about proposed articles. The organization must have a publicly stated policy, to which it adheres, of full disclosure of any conflict of interest or biases for all authors or contributors involved with the journal or organization.
- The journal's articles must be peer-reviewed and published in accordance with the regular peer-review procedures of the organization.
- The journal must be generally recognized to be of national scope and reputation.
- The journal must be indexed in the *Index Medicus* of the National Library of Medicine of the National Institutes of Health.
- The article must not be published in the form of a special supplement that has been funded in whole or in part by one or more manufacturers.

A reference publication is a publication that (21 CFR 99.3(i)):

- was not written, edited, excerpted, or published specifically for, or at the request of, a device manufacturer;
- was not edited or significantly influenced by such a manufacturer;
- is not solely distributed through such a manufacturer, but is generally available in bookstores or other distribution channels where medical textbooks are sold;
- does not focus on any particular device of a manufacturer that disseminates information under this part and does not have a primary focus on new uses of devices that are marketed or are under investigation by a manufacturer supporting the dissemination of information; and
- does not present materials that are false or misleading.

A reprint or copy of an article or reference publication is "unabridged" only if it retains the same appearance, form, format, content, or configuration as the original article or publication. Such reprint or copy of an article or reference publication may not be disseminated with any information that is promotional in nature. A manufacturer may cite a particular discussion about a new use in a reference publication in the explanatory or other information attached to or otherwise accompanying the reference publication.

The FDA requires that any information disseminated under the provisions of this regulation include (21 CFR 99.103(a)):

1. A prominently displayed statement disclosing that:

THIS INFORMATION CONCERNS A USE THAT HAS NOT BEEN APPROVED OR CLEARED BY THE FOOD AND DRUG ADMINISTRATION.

If the information includes both an approved and unapproved use (or uses) or a cleared and uncleared use (or uses), the manufacturer must modify this statement to identify the unapproved or uncleared new use (or uses). The statement must be permanently affixed to the front of each reprint or copy of an article from a scientific or medical journal and to the front of each reference publication disseminated under this regulation.

If applicable, the statement must disclose that the information is being disseminated at the expense of the manufacturer. It must include a list of the names of any authors of the information who were employees of, or consultants to, or received compensation from the manufacturer, or who had a significant financial interest in the manufacturer during the time that the study comprising the subject of the article or reference publication began up through one year after the time the article/reference publication was written and published. It must also identify any person that has provided funding to conduct the study relating to the new use of a device for which such information is being disseminated. Finally, the statement must include a notice that there are products or treatments that have been approved or cleared for the use that is the subject of the information being disseminated, if applicable.

2. The official labeling for the device.

3. A bibliography of other articles (that concern reports of clinical investigations both supporting and not supporting the new use) from a scientific reference publication or scientific or medical journal that have been previously published about the new use of the device covered by the information that is being disseminated, unless the disseminated information already includes such a bibliography.

4. Any additional information required by the FDA under the provisions of 21 CFR 99.301(a)(2). This information must be attached to the front of the disseminated material or, if attached to the back of the disseminated material, a sticker or notation must appear on the front of the disseminated material informing the reader that FDA-required information appears on the back. This information may consist of:

- Objective and scientifically sound information that the FDA has determined is necessary to provide objectivity and balance pertaining to the safety or effectiveness of the new use of the device. This may include information that the manufacturer has submitted to the FDA or, where appropriate, a summary of such information and any other information that can be made publicly available; and
- An objective statement prepared by the FDA, based on data or other scientifically sound information, bearing on the safety or effectiveness of the new use of the drug or device.

Except as provided above, the statements, bibliography, and other information required by this regulation must be attached to such disseminated material (21 CFR 99.103(b)).

When determining whether a statement is "prominently displayed," the FDA may consider type size, font, layout, contrast, graphic design, headlines, spacing, and any other technique to achieve emphasis or notice. The required statements must be highlighted, outlined, boxed, or otherwise graphically designed and presented in such a way to achieve emphasis or notice and must be distinct from the other information being disseminated (21 CFR 99.103(c)).

At least 60 days before disseminating any written information concerning the safety, effectiveness, or benefit of a new use for a device, a manufacturer must submit to the FDA an identical copy of the information to be disseminated; including any information (e.g., the bibliography) and statements required under 21 CFR 99.103 (21 CFR § 99.210).

CONTINUING MEDICAL EDUCATION

In its "Guidance to industry—Industry-Supported Scientific and Educational Activities," the FDA recognizes the importance of two manufacturer-supported sources of information for healthcare professionals: (1) information resulting from the activities performed by, or on behalf of, manufacturers of products, and (2) information resulting from activities supported by manufacturers that are otherwise independent from the promotional influence of the supporting company. In the FDA's view, the former is subject to the labeling and advertising provisions of the FD&CA, while the latter is not (CME p. 64094).

When determining when an activity is independent of substantial influence of a manufacturer, FDA looks to determine the extent the company is in a position to influence the presentation of information related to its products. Influence can be both direct and indirect. Direct influence includes the selection of presenters or determining the treatment that topics will receive. Indirect influence would come through the relationship between the manufacturer and the provider. For example, the provider may have reason to believe that future support from the company depends on the treatment the provider gives to the companies products (CME p. 64095).

When evaluating an industry-supported program for "independence," the FDA will consider the following factors (CME pp. 64097–64099):

- The provider's control of presenter's and moderator's content and selections—Does the provider maintain full control over the content of the program and the selection of speakers and moderators? Has the manufacturer engaged in scripting, targeting points for emphasis, or other actions to influence program content, or has the manufacturer suggested speakers who are involved in promoting the company's products?

- Disclosure—Are all material facts concerning the extent of funding and significant relationships between the provider, presenters, or moderators and the supporting company disclosed when any unapproved uses of products would be discussed?
- Focus of the program—Is the intent of the program to produce an independent and non-promotional activity that focuses on education and is free from commercial influence and bias? Are reasonable and relevant alternative treatment options, if they exist, discussed in a fair and balanced way?
- Relationships between the provider and the supporting company—Are there legal, business, or other relationships between the company and the provider that could place the company in a position to exert influence over the content of the program?
- Provider involvement in sales and marketing—Is the provider or any of its employees who are involved in scientific or educational activities also involved in activities that assist the manufacturer in advertising or selling its products?
- Demonstrated failure of provider to meet standards—Has the provider a history of failing to meet standards of independence, balance, objectivity, or scientific rigor when putting on ostensibly independent programs?
- Multiple presentations—While recognizing that some repeat programs can serve public health interests, the FDA will consider whether multiple presentations on the same topic are held.
- Audience selection—Are invitations or mailings being sent to an audience targeted by the sponsoring company or is the audience being selected to support sales or marketing goals (e.g., as rewards or to influence opinion leaders)?
- Opportunities for discussion—During live programs, are there opportunities for questions and meaningful discussion during the program?
- Dissemination—Has information about the company's products presented in the scientific or educational activity been disseminated after the program, by or at the behest of the company, other than in response to an unsolicited request?
- Ancillary promotional activities—Have promotional activities such as presentation by sales representatives taken place in the meeting room?
- Complaints—Have the provider, presenters, or attendees raised any complaints regarding attempts by the supporting company to influence content?

The FDA does not consider this list to be exhaustive. Other factors may be appropriate for consideration in particular cases.

The FDA considers a written agreement between the provider and the supporting manufacturer to be one way to ensure the independence of an activity. This document should reflect that the provider will be solely responsible for designing and conducting the activity, and that the activity will be educational, nonpromotional, and free from commercial bias.

ADVERTISING ON THE INTERNET

The Internet offers a fast and cost-efficient way for medical device manufacturers to provide current and potential customers with timely and accurate information on their products and services. When properly done, advertising on the Internet can enable the customer to locate and filter the specific information they are seeking at the time most convenient for them.

For its many benefits, the Internet can be fraught with perils for the unwary manufacturer. The FDA considers product information placed on the Internet to be no different that any other labeling or advertising material. A review of warning letters issued by the FDA's Office of Compliance in the past few years indicates that many of the incidents cited by the FDA either had to do with significant inappropriate representations on Internet Web sites, or came to light because of information obtained by the FDA from Web sites. The Internet has enabled the Office of Compliance to be proactive in reviewing a manufacturer's promotional literature rather than waiting for complaints from competitors or the public. Therefore, the manufacturer must be as careful and exercise as much internal control over the information that appears on its Web site as it does with other labeling and advertising material.

Material must be controlled, reviewed, and approved before it is placed on the Web site. Obsolete material must be promptly removed. "Hyperlinking" Web pages allows the reader (and FDA) to easily move from one part of the site to another or even to another Web site. Great care must be taken to ensure that these links are correct and appropriate. In at least one of the warning letter cases cited earlier in this chapter, the manufacturer had a "spurious" link that took the reader (in this case, an FDA reviewer) to a page describing a new use for the device; that use had not received FDA approval.

Manufacturers must take special care in linking to other Web sites, particularly if those sites contain discussions of unapproved uses of the manufacturer's products. The manufacturer should monitor linked Web sites to ensure that content changes do not create a problem. Linking to the Web sites of well-respected, independent organizations or publications, such as the American Medical Association, American Heart Association, or the *New England Journal of Medicine*, should not create a problem even though those sites may contain information about unapproved or "off-label" uses of the manufacturer's products. However, linking directly to pages within those sites that contain off-label use information could be problematic. Certainly, linking to sites that are not independent of the manufacturer or that are dedicated to discussing off-label use of the manufacturer's products is likely to be considered by the FDA as promoting off-label use.

Because of the highly dynamic nature of the media, the manufacturer needs also to consider how to maintain a history of changes to the information content on its Web site so it can reconstruct the particulars in the event a question arises in the future. By the time a problem comes to light, for example, through an FDA warning letter, the information on the site may have changed several times. It may be necessary to go back and reconstruct the information content at the time of the alleged problem in order to develop corrective action or mount a defense if the manufacturer believes the allegations are unfounded.

When building Web sites, manufacturers are frequently faced with the dilemma that their products have different approved uses in different markets. The FDA recognizes that the Internet is a global resource and they do not have authority to regulate the information that is available to customers located outside the United States. To solve this problem, the FDA has taken the position that product information should be segregated into separate partitions that are only accessed through icons that indicate the intended audience. For example, a manufacturer could have a US icon and a European or International icon that would separate the FDA-cleared information from that intended for customers outside the United States. A clear and prominent statement of the intended audience at the entry to the partition helps to reinforce the separation. Translation of information into a language appropriate to the intended audience also helps reinforce the manufacturer's intent not to promote unapproved uses in any market.

Precisely because the Internet is a global resource, the FDA does not consider the location of the server where the information is stored to have any significance on its ability to regulate the site's information content for products approved in the United States. In addition, manufacturers head-quartered outside the United States have to meet the same requirements as any other company that promotes and advertises devices that have received marketing clearance in the United States.

THINGS TO REMEMBER

The FD&CA, as amended by Congress in 1976 and 1990, is the primary law under which the FDA derives authority to take action against regulated medical devices. The FDA also exercises authority to regulate labeling of medical devices under the provisions of the FPLA and the RCHSA.

The FDA draws a distinction between label and labeling. The term "label" covers all written, printed, or graphic information applied to a medical device, or any of its containers or wrappers. "Labeling," on the other hand, includes all labels as well as other written, printed, or graphic matter that accompanies the medical device. The FDA interprets "accompanying" liberally to include more than physical association with the medical device. The FDA includes in its definition of labeling posters, tags, pamphlets, circulars, booklets, brochures, instruction books, direction sheets, fillers, and so on, regardless of whether they are shipped with the device or brought together with the device after shipment.

The FD&CA makes it a federal crime to introduce or deliver for introduction into interstate commerce a medical device that is adulterated or misbranded. With regard to labeling, a device—in particular, an IVD product—is considered adulterated if its strength differs from, or its purity or quality falls below, that which it purports or is represented to possess. A device is misbranded if its labeling is false or misleading in any particular. Labeling need not be untrue, forged, fraudulent, or deceptive to be viewed as being in violation of the law. A false impression may be created not only by false or deceptive statements, but also by ambiguity and misdirection. Failure to inform the customer of facts relevant to the statements made will render the device misbranded just as much as a blatantly false or exaggerated claim. A label that is silent on important considerations may be just as deceptive as one that makes exaggerated claims.

The FDA considers advertising and promotion of medical devices—particularly the promotion of unapproved or "off-label" uses—to be regulated under the FD&CA. The FDA signaled an increasing interest in this area when the CDRH Office of Compliance established the DLCB in February 1992. Following the formation of the DLCB, the number of warning letters and field corrective actions for misleading labeling and advertising of devices has increased sharply.

However, the application of the FDA policy against dissemination of information on off-label uses has been contested in the courts. The appellate court, in *Washington Legal Foundation v. Henney*, vacated part of an earlier district court decision and injunction insofar as it declared unconstitutional (1) statutory provisions concerning the dissemination by manufacturers of certain written materials concerning new uses of approved products, and (2) an FDA guidance document concerning certain industry-supported scientific and educational activities known generally as industry-supported CME. The FDA contends that the appellate court decision establishes these statutory provisions as "a 'safe harbor' for manufacturers that comply with them; the CME guidance document details how the agency intends to exercise its enforcement discretion." The WLF sees the result somewhat differently and more legal challenges may be forthcoming.

The Internet offers a fast and cost-efficient way for medical device manufacturers to provide current and potential customers with timely and accurate information on their products and services. For its many benefits, the Internet can be fraught with perils for the unwary manufacturer. The manufacturer must be as careful and exercise as much internal control over the information that appears on its Web site as it does with other labeling and advertising material.

21 General Device Labeling in the United States

From the medical device manufacturer's point of view, one of the most significant changes to the Federal Food, Drug, and Cosmetic Act (FD&CA) contained in the Medical Device Amendments of 1976 was the introduction of a three-tiered classification system for medical devices. All devices intended for human use are placed in one of three classes based on the need for regulatory control to ensure safety and effectiveness. The Commissioner of Food and Drugs is responsible for classifying or reclassifying devices and is aided in this process by classification panels of experts.

Class I medical devices are those that are subject only to the general controls authorized by Sections 501 (adulteration), 502 (misbranding), 510 (registration), 516 (banned devices), 518 (notification and other remedies), 519 (records and reports), and 520 (general provisions) of the FD&CA.

In addition to the requirements of Class I, Class II devices are subject to performance standards promulgated under Section 514 (performance standards) of the FD&CA.

Class III devices require premarket approval to provide reasonable assurance of safety and effectiveness under Section 515 (premarket approval) of the FD&CA.

Devices that have already been classified by the Commissioner are listed in the Code of Federal Regulations (CFR) under Title 21—*Food and Drugs*, Parts 862 through 892. Appendix F provides a listing of the categories under which devices are grouped in Title 21 of the CFR.

In order to make a recommendation to the Commissioner on the classification of regulatory control appropriate to the device, the classification panel reviews the device for safety and effectiveness. In doing so, the panel considers the following, among other relevant factors:

- The persons for whom the use of the device is represented or intended
- The conditions of use for the device, including conditions of use prescribed, recommended, or suggested in the labeling or advertising of the device, and other intended conditions of use
- The probable benefit to health from the use of the device, weighed against any probable injury or illness from the use of the device
- The reliability of the device

These factors must be considered and addressed by the manufacturer in every aspect of a device including its labeling. The device's labeling must include the appropriate information to facilitate its use. The FDA often finds a relationship between the labeling and mishandling of the device. The FDA has reported that approximately 40 percent of the Medical Device Reports (MDRs) filed involve user errors (Cardamone p. 37). In addition to the cases where labeling is a primary factor leading to an MDR, labeling may be an "underlying" or secondary cause of misuse.

The FDA has encountered three general problem areas with labeling. They are (Cardamone, p 37):

1. misbranding—the failure of the labeling to meet the requirements of the labeling regulations;
2. poor labeling control—improper labeling of a device during manufacturing because of a failure in the quality control or Good Manufacturing Practice (GMP) program; and
3. inadequate labeling—labeling, while not misbranding, that could be improved to prevent misuse or mishandling of the device

Misbranding and poor labeling controls are violations of the letter of the law. The third point, inadequate labeling, or "labeling that is less than it can or should be," may not violate the law, but it can lead to the very problems that the Medical Device Amendments of 1976 was intended to prevent. This chapter, along with Chapter 22 and Chapter 23 describes the regulations that deal with misbranding of devices.

MISBRANDING

The FDA has developed a minimum set of labeling requirements that are applicable to all medical devices. These requirements are contained in 21 CFR 801, and include:

- general labeling provisions (Subpart A),
- labeling requirements for over-the-counter (OTC or nonprescription) devices (Subpart C),
- exemptions from adequate directions for use (Subpart D),
- other exemptions (Subpart E), and
- special requirements for specific devices (Subpart H).

Failure to follow or satisfy these requirements may lead to the device being deemed misbranded. The following sections discuss each of these requirements in detail.

Normally, substances subject to the FD&CA are exempted by Section 2(f)(2) of the Federal Hazardous Substances Act (FHSA) from regulation by the Consumer Product Safety Commission (CPSC). However, IVD products are specifically required by FDA regulations to include in their labeling any warnings or precautions for users required by the CPSC. The hazardous substance labeling requirements are discussed in Chapter 22. It must be noted that when a product regulated under the FD&CA contains a substance that presents a substantial risk of injury or illness from any handling or use that is customary or usual, it may be regarded as misbranded under the FD&CA for failing to reveal material facts if that label does not bear information to alert the consumer to this hazard (21 CFR 1500.81).

GENERAL LABELING PROVISIONS

Subpart A of 21 CFR 801 details a set of general labeling requirements that apply to all prescription, OCT, and IVD products. The section number (e.g., §801.1) of the corresponding section in 21 CFR is listed with each topic for reference.

NAME AND PLACE OF BUSINESS (§801.1)

The device label must contain the name and address of the manufacturer, packer, or distributor. In the case of a corporation, the name must be the actual corporate name, which may be preceded or followed by the name of a particular division of the corporation. In the case of an individual, partnership, or association, the name under which business is conducted must be used.

The address must include the street address, city, state, and zip code. The street address may be omitted if the firm's street address is listed in the current city directory or telephone directory. The requirement to include the zip code on the label applies only to consumer products. In the case of nonconsumer packaging, the zip code must appear on either the label or the labeling (including the invoice).

If the firm listed on the label is not the manufacturer of the device, the name must be qualified with a phrase (such as "Manufactured for..." or "Distributed by....") that indicates the connection between the firm and the device.

A firm who remanufactures, rebuilds, reconditions, or updates a previously owned medical device for commercial distribution is considered a manufacturer. However, since this firm is not the original manufacturer of the device, the name must be qualified by an appropriate phrase that reveals the connection of the firm with the device. For example, the name of a rebuilder or reconditioner could be qualified by "Rebuilt by..." or "Reconditioned by..." (Lowery, Puleo p. 1–6).

INTENDED USES (§801.4)

The term "intended uses" refers to the objective intent of the persons responsible for the labeling of a device. The intent is established by expressions of the responsible person or his/her representatives. For example, intent is established by labeling claims, advertising, or by other oral or written statements.

The term "intended uses" is often confused with "indications" or "indications for use." While these are complementary terms, they have distinctly separate meanings. "Intended uses" refers to the general functional use of a device. For example, the intended use of a medical laser could be described as (Medical Laser p. 21):

THE [PRODUCT TRADE NAME] LASER IS A GENERAL SURGICAL INSTRUMENT USED IN THE FREE BEAM MODE TO PHOTOCOAGULATE, VAPORIZE/ABLATE SOFT TISSUE (MUSCLE, CONNECTIVE TISSUE, ORGANS), OR FOR CUTTING, EXCISION, INCISION, AND COAGULATION OF SOFT TISSUE IN THE CONTACT MODE IN OPEN/CLOSED SURGICAL PROCEDURES IN GENERAL SURGERY.

The tissue effects listed in the previous example constitute the intended uses for surgical lasers.

The term "indications for use," on the other hand, refers to the specific therapeutic or diagnostic use, or group of similar uses, of the device, that is the disease, condition, or pathology for which the principal effect of the device is used to prevent, treat, cure, mitigate, or diagnose. For example, the indications for use of a medical laser could be (Medical Laser p. 21):

THE [PRODUCT TRADE NAME] IS INTENDED FOR PERCUTANEOUS LASER DISC DECOMPRESSION (PLDD).

The intended use of a device may be implied in the indications for use. In order to treat the indicated condition, the laser must have one of the tissue effects listed above. For more information on intended uses, see the "Indications for Use" section later in this chapter.

The intended uses statement usually appears near the front of the package insert or in a prominent place on the label, and is introduced by a phrase such as: "The [product trade name] is intended for ..." or "The [product trade name] is indicated for use ..." (Final Draft p. 11). For example, a statement incorporating both the intended use and indications for use for a cardiac pacemaker intended by its manufacturer to be used to pace the atria might read:

THE [PRODUCT TRADE NAME] PULSE GENERATOR IS INTENDED FOR PERMANENT ATRIAL PACING APPLICATIONS. ATRIAL INDICATIONS MAY INCLUDE SINUS NODE DYSFUNCTION OR SICK SINUS SYNDROME (E.G., SINUS BRADYCARDIA, SINUS ARREST AND/OR EXIT BLOCK, BRADYCARDIA-TACHYCARDIA SYNDROME, ETC.).

The intent may also be demonstrated by the circumstances surrounding the distribution of the device. For example, intent may be shown by offering or using a device, with the knowledge of the persons responsible for labeling the device or their representative, for a purpose for which it was neither labeled nor advertised.

If a manufacturer knows, or has information indicating, that a medical device is being used for purposes other than those for which it is offered, the manufacturer is required to provide adequate labeling for the uses to which it is being put. An example might be a manufacturer of dental X-ray equipment who routinely sells equipment to podiatrists (Cardamone p. 5).

The intended uses of a medical device may change after it has been introduced into interstate commerce by its manufacturer. In this case, it is the responsibility of the packer, distributor, or seller to supply adequate labeling in accordance with the new use intended by the packer, distributor, or seller.

ADEQUATE DIRECTIONS FOR USE (§801.5)

The term "directions for use" refers to the directions under which a medical device can be used safely and for the purposes for which it is intended. Adequate directions for use means directions sufficient for a layperson (e.g., patient or unlicensed healthcare provider) to use the device safely and for the intended purpose. Adequate directions for use includes:

- statements of all conditions, purposes, or uses for which the device is prescribed, recommended, or suggested in its oral, written, printed, or graphic advertising;
- the dosage for each use and the usual dosage for persons of differing ages and physical conditions;
- the frequency of administration or application;
- the duration of administration or application;
- the time of administration in relation to other time factors such as meals or the onset of symptoms;
- the routes or methods of administration or application; and
- any necessary preparations or adjustments prior to use.

Directions for use requirements for prescription, OTC and IVD products are discussed later in this chapter.

MISLEADING STATEMENTS (§801.6)

One representation that would render a device misbranded is a false or misleading statement in the device's labeling with respect to another device, drug, food, or cosmetic. Some examples of other false or misleading statements that would render a device misbranded are (Lowery §11):

- incorrect, inadequate, or incomplete identification;
- unsubstantiated claims of therapeutic value;
- inaccuracies concerning condition, state, treatment, size, shape, or style;
- substitution of parts or materials;
- subjective or unsubstantiated quality or performance claims; or
- use of the prefix US (United States) or other similar indication suggesting government or agency approval or endorsement of the product.

PROMINENCE OF REQUIRED STATEMENTS (§801.15)

Any word, statement, or other information required by the FD&CA must be prominently placed on the label or labeling. Prominently is defined by the act to mean conspicuous "(as compared to other words, statements, designs, or devices, in the labeling)." A required word, statement, or other information may lack prominence for any of the following reasons:

- If the required information does not appear on the part or panel of the label that is presented or displayed under customary conditions of purchase
- If there is sufficient space and the required information fails to appear on two or more panels, each of which is designed to render it to be displayed under customary conditions of purchase
- If the label fails to extend over the package space provided
- If insufficient space is provided for the required information because of the placement or conspicuousness of information that is not required
- Smallness or style of type, insufficient background contrast, designs that obscure labeling, or overcrowding of the labeling that renders it unreadable

Exemptions may be granted in those cases where the labeling lacks sufficient space for the required information (FD&CA §352(b)). To qualify for an exemption, the manufacturer must demonstrate that:

- Existing label space is not taken up by inclusion of any word, statement, design, or device not required in the labeling.
- Label space is not used to give prominence to information other than that required by the FD&CA.
- Existing label space is not used for any representation in a foreign language.

All required labeling must be in English. However, if a product is distributed solely within Puerto Rico or in a US territory where the predominant language is other than English, the predominant language may be substituted for English.

If a label contains any representation in a foreign language, then all of the information required by the FD&CA must also appear in that foreign language.

SPANISH LANGUAGE (§801.16)

A device distributed solely within Puerto Rico may be labeled in Spanish.

LABELING REQUIREMENTS FOR OVER-THE-COUNTER (NONPRESCRIPTION) DEVICES

Subpart C of 21 CFR 801 describes the particular requirements for devices that are sold without prescription through retail outlets. This subpart of the regulations implements the requirements of the Fair Packaging and Labeling Act (FPLA).

PRINCIPAL DISPLAY PANEL (§801.60)

The principal display panel means the part of the label that is most likely to be seen by the customer under customary conditions of display for retail sale. The principal display panel must be large enough to accommodate all of the information required by the FD&CA with "clarity and conspicuousness." Vignettes, crowding, and obscuring designs are not allowed.

If the package has alternative parts, or faces, that are likely to be seen under the customary conditions of display, then each of the alternative principal panels must contain all of the information required by the regulations.

For purposes of this part of the regulations, the term "area of the principal panel" means the side or surface that bears the principal display panel. This is considered to be:

- for a rectangular package, the height multiplied by the width of the side customarily displayed;
- for a cylindrical or nearly cylindrical package, 40 percent of the height multiplied by the circumference; or
- for other shapes, 40 percent of the total surface of the container, unless there is an obvious face (e.g., the top of a triangular package) that is customarily displayed.

STATEMENT OF IDENTITY (§801.61)

The statement of identity must be present on the principal display panel of an over-the-counter device. The statement of identity must contain the following:

- The common name of the device must be shown, followed by an accurate statement of the principal intended actions of the device.
- The indications for use must be included in the directions for use.
- The statement of identity must be presented in bold-faced type, must be reasonably related in size to the most prominent printed matter on the principal display panel, and must be in lines generally parallel to the base on which the package rests as it is designed to be displayed.

DECLARATION OF NET QUANTITY OF CONTENTS (§801.62)

The label of an over-the-counter device in packaged form must contain a statement of the net quantity of contents in terms of weight, measure, numerical count, or a combination of numerical count and weight, measure, or size. Section 801.62 requires that weights and measures be expressed in the customary inch-pound system. However, Congress, in the 1992 amendments to the FPLA, required, among other things, that the most appropriate units of both the customary inch-pound and the International System of Units (SI) be used for the declaration of net quantity of contents (Metric Labeling p. 67445). The remainder of this section describes the requirements using the customary inch-pound system. The reader should be aware that a separate statement of net quantity of contents in terms of the metric system of weight or measure is not regarded as a supplemental statement. Metric units are described in the "SI Declarations" Section later in this chapter.

The declaration of the net quantity shall express an accurate statement of the quantity of the contents of the package. Reasonable variations caused by loss or gain of moisture during distribution, or unavoidable deviations in GMP, will be recognized as long as the variation from the stated quantity is not unreasonably large. The net quantity is expressed as:

- Count—If the numerical count does not give accurate information as to the quantity of devices in the package, it must be supplemented by statements of weight, measure, or size.
- Weight—To be expressed in avoirdupois pounds and ounces.
- Measure—In the case of established customer usage, the quantity may be expressed in terms of linear measure or measure of area. If necessary for accuracy, these measures are augmented by a statement of the weight, measure, or size of the individual unit or the entire device. Liquid measure of the contents is expressed in terms of the US gallon (231 cubic inches) and quart, pint, and fluid-ounce subdivisions of the gallon. These measurements are made with the liquid contents at a temperature of 68 °F (20 °C).

The declaration may contain common or decimal fractions. Common fractions must be reduced to their lowest terms, and a decimal fraction is not to be carried out to more than two places. A statement that includes small fractions of an ounce shall be deemed to permit smaller variations than one that does not include such fractions.

The declaration must be located on the principal display panel. It must be visually separated from other information appearing on the label. The minimum separation is defined in the regulations to be:

- from the information above and/or below, by a space equal to the heights of the lettering used in the declaration and
- from the information left and right, by a space equal to twice the width of the letter "N" of the type font used in the declaration.

The declaration shall not use any terms qualifying a unit of weight, measure, or count (such as "giant pint" or "full quart") that tend to exaggerate. Unless the package has a principal display panel of 5 square inches or less, the declaration is to be placed within the bottom 30 percent of the area, in lines generally parallel to the base on which the package rests in its normal mode of display.

The declaration must appear in conspicuous, and easily legible bold-faced print or type in distinct contrast (by typography, layout, color, embossing, or molding) to other matter on the package. If all the label information is blown, embossed, or molded on the surface of glass or plastic, then the declaration of net quantity may also be blown, embossed, or molded on the surface. The requirements of conspicuousness and legibility include the following specifications:

- A letter shall not be more than three times as high as it is wide.
- The minimum letter size as a function of the area of the principal display panel is listed in Table 21.1. The minimum letter size requirement applies to upper-case or capital letters. When upper- and lower-case letters, or all lower-case letters are used, then it is the lower-case "o," or its equivalent, that must meet the minimum standard. When fractions are used in the labeling, each numeral must meet one-half of the minimum height standard.

The declaration may appear in more than one line on the principal display panel. The term "net weight" must be used when stating the net quantity of contents in terms of weight, e.g., "Net wt 6 oz." The term "net" or "net contents" is optional when stating the quantity in terms of fluid measure or numerical counts, e.g., "6 fl oz" or "net contents 6 fl oz." Avoirdupois ounces are distinguished from fluid ounces through association of terms. For quantities, only the abbreviations given in Table 21.2 may be employed. Periods and plural forms are optional.

TABLE 21.1
Type Size as a Function of Label Area for OTC Devices

Label Size (sq in.)	Printed Type Size	Blown, Embossed, or Molded in Glass or Plastic (minimum)
≤5	1/16 in.	1/8 in.
>5 but ≤25	1/8 in.	3/16 in.
>25 but ≤100	3/16 in.	1/4 in.
>100 but ≤400	1/4 in.	5/16 in.
>400	1/2 in.	9/16 in.

Source: 21 CFR 801.62(h)

TABLE 21.2
U.S. Abbreviations for Weights and Measures[a]

gallon	gal	milliliter	ml
quart	qt	cubic centimeter	cc
pint	pt	yard	yd
ounce	oz	feet or foot	ft
pound	lb	inch	in
grain	gr	meter	m
kilogram	kg	centimeter	cm
gram	g	millimeter	mm
milligram	mg	fluid	fl
microgram	mcg	square	sq
liter	l	weight	wt

[a] Periods and plural forms are optional.

Source: 21 CFR 801.62(l)

TABLE 21.3
OTC Device Labeling in Terms of Weight or Liquid Measure

Package Contains	Declaration	Examples
<1 pound	ounces	Net wt 12 oz
>1 but <4 pounds	ounces (pounds with the remainder in ounces) or ounces (pounds with the remainder in common or decimal fraction of a pound)	Net wt 24 oz (1 lb 8 oz) Net wt 24 oz (1.5 lb) or Net wt 24 oz (1.5 lb)
>4 pounds	pounds and ounces or pounds with the remainder in common or decimal fraction of a pound	Net wt 10 lb 8 oz Net wt 10.5 lb or Net wt 10.5 lb
<1 pint	fluid ounces	Net contents 8 fl oz
>1 pint but <1 gallon	fluid ounces (quarts; or quarts and pints; or pints, as appropriate, with the remainder in fluid ounces) or fluid ounces (quarts; or pints, as appropriate, with the remainder in common or decimal fraction of a pint or quart)	Net contents 56 fl oz (1 qt 1 pt 8 oz) Net contents 56 fl oz (1 qt 1.5 pt)
>1 gallon	gallons and common or decimal fraction of gallons or gallons and quarts; or quarts and pints, as appropriate, with the remainder in fluid ounces or gallons and quarts; or quarts and pints, as appropriate, with the remainder in common or decimal fraction of a pint or quart	Net contents 2.5 gal Net contents 2 gal 1 qt 1 pt 8 oz 2 gal 1 qt 1.5 pt

Source: 21 CFR 801.62(i), 801.62(j), and 801.62(k)

For packages where the net quantity of contents is labeled in terms of weight or fluid measure, the contents are expressed in the units shown in Table 21.3.

For packages where the net quantity of contents is labeled in terms of linear measure or area measure, the contents are expressed as shown in Table 21.4.

A separate statement of the net quantity in metric measurements is not regarded as a supplemental statement. An accurate statement of net quantity in the metric system of weight or measure may also appear on the principal display panel or on any other panel.

CUSTOMARY INCH-POUND DECLARATIONS

Declarations in the customary inch-pound system must be in the largest whole inch-pound unit, with the remainder expressed as a common or decimal fraction of the largest whole unit (e.g., "1 1/2 lb"

TABLE 21.4
OTC Device Labeling in Terms of Linear or Area Measure

Package Contains	Declaration	Examples
<1 foot	inches	6 in.
>1 foot	inches (yards; or yards and feet; or feet, as appropriate, with any remainder in inches) or inches (yards; or yards and feet; or feet, as appropriate, with any remainder in common or decimal fraction of a foot or yard)	86 in. (2 yd 1 ft 2 in.) 30 in. (2 1/2 ft) 30 in. (2.5 ft)
<1 square foot	square inches	100 sq in.
>1 square foot	square inches (square yards; or foot square yards and square feet; or square feet, as appropriate, with any remainder in square inches) or square inches (square yards; or square yards and square feet; or square feet, as appropriate, with any remainder in common or decimal fraction of a square foot or square yard)	158 sq in. (1 sq ft 14 sq in.) 1,944 sq in. (1 1/2 sq yd) or 1,944 sq in. (1.5 sq yd)

Source: 21 CFR 801.62(m) and 801.62(n)

or "1.5 lb"). The remainder may be expressed in terms of the next smaller whole unit and any common or decimal fraction of that unit (e.g., "1 qt 1 1/2 fl oz" or "1 qt 1.5 fl oz").

Packages that contain more than 1 pound but less than 4 pounds may bear a dual customary inch-pound declaration. The dual declaration, when provided, must give the total ounces followed by a parenthetical declaration giving the largest whole unit with any remainder expressed as described in the previous paragraph (e.g., "24 oz (1 1/2 lb)," "24 oz (1.5 lb)," or "24 oz (1 lb 8 oz)"). Packaging containing more than a pint but less than a gallon may also bear a dual declaration. In this case, the declaration must give the total fluid ounces followed by a parenthetical declaration of the largest whole unit with any remainder expressed as described above. The dual declaration must appear on one line and may precede or follow the SI declaration.

Common fractions are to be expressed in terms of halves, quarters, eighths, sixteenths, or thirty-seconds. When there is a firmly established general commercial usage and trade custom of using a different common fraction to express the net quantity of contents for a particular commodity, such common fractions may be used. Common fractions are always reduced to their lowest terms.

The symbols and abbreviations for the customary inch-pound units in Table 21.5 may be used when stating the net quantity of contents. No other symbols or abbreviations may be use.

SI DECLARATIONS

The SI declaration of net quantity of contents shall be expressed in terms of the appropriate multiple or submultiple that it will result in a number between 1 and 1000 (e.g., 500 g, not 0.5 kg; 1.96 kg, not 1,960 g). The SI declaration is never expressed in mixed units (e.g., 1.5 kg, not 1 kg 500 g). The SI units may precede or follow the customary inch-pound declaration (e.g., "Net wt 15 oz (425 g)" or "Net mass 425 g (15 oz)"; "Net 1 qt 1.8 fl oz (1 L)" or "Net 1 L (1 qt 1.8 fl oz)").

The symbols and abbreviations for the SI units in Table 21.5 may be used when stating the net quantity of contents. No other symbols or abbreviations may be used.

CONVERSION BETWEEN SYSTEMS

When calculating the conversion of customary inch-pound quantities to SI quantities and SI quantities to customary inch-pound quantities, the conversion factors in Table 21.6 are used. The conversion

TABLE 21.5
U.S. Symbols And Abbreviations for Weights and Measures

S.I. Unit	Symbol[a]	Customary Inch-Pound Unit	Symbol[b]
Microgram	μg	Ounce	Oz
Milligram	mg	Pound	Lb
Gram	g	Pint	Pt
Kilogram	kg	Quart	Qt
Centimeter	cm	Gallon	Gal
Square centimeter	cm²	Inch	In.
Cubic centimeter	cm³	Foot	Ft
Millimeter	mm	Yard	Yd
Meter	m	Fluid	Fl
Square meter	M²	Each	Ea
Cubic meter	M³	Piece	Pc
Milliliter	mL or ml	Square	Sq
Liter	L or l	Cubic	Cu

[a] Lower-case letters are to be used for the SI symbols, except for liter and milliliter where the capital "L" is preferred. Periods are not to be used after the SI symbols. It is unacceptable to add "s" to an SI symbol to express the plural of the symbol.

[b] Both upper- and lower-case letters may be used with the customary inch-pound abbreviations. An "s" should not be added to express the plural when the abbreviation is used.

Source: Metric Labeling p. 67462

factor or the quantity to be converted may not be rounded prior to calculating the converted quantity. Quantities calculated with the conversion factors in Table 21.6 may be rounded down to avoid overstating the net contents.

The manufacturer establishes the number of significant digits in the declaration. However, the accuracy of the contents declaration may never be sacrificed or exaggerated. A decimal fraction may never be carried out to more than three decimal places. When rounding results in a difference between the two systems, the FDA can be expected to use the largest of the two numbers for enforcement purposes to determine whether a package contains at least the declared amount of product.

USE OF THE TERM "NET"

The use of the terms "net," "net weight," or "net mass" when stating the net quantity of contents in terms of weight or mass is optional. When used, the terms may either precede or follow the declaration (e.g., "Net wt 1 lb (453 g)" or "1 lb (453 g) Net"). The use of the terms "net" or "net contents" with fluid measures or numerical counts is also optional. When used, the terms may either precede or follow the declaration of quantity. It is sufficient to distinguish between weight or mass and fluid contents by the appropriate units (e.g., "Net wt 6 oz (170 g)" or "6 oz (170 g)" versus "Net contents 6 fl oz (177 ml)" or "6 fl oz (177 ml)").

However, it is important to remember that the contents declaration must always state accurately the new quantity of contents even when the terms discussed in the previous paragraph are not used. Reasonable variation caused by loss or gain of moisture during the course of good distribution practice or by unavoidable deviations in current good manufacturing practice (cGMP) will be recognized. Variations from stated quantities must not be unreasonably large.

TABLE 21.6
SI and Customary Inch-Pound Conversion Factors[a]

Inch-pound			SI	
		Mass or Weight		
1 ounce	= 28.349 5 g	1 milligram	= 0.000 035 274 0 oz	
1 pound	= 453.592 g		= 0.015 432 4 grain	
	= 0.453 592 kg	1 gram	= 0.035 274 0 oz	
1 grain	= 64.798 9 mg	1 kilogram	= 2.204 62 lb	
		Volume or Capacity		
1 cubic inch	= 16.387 1 cm³	1 cubic centimeter	= 0.061 023 7 in³	
1 cubic foot	= 0.028 316 8 m³	1 cubic decimeter	= 0.035 314 7 ft³	
1 cubic yard	= 0.764 555 m³	1 cubic meter	= 35.314 7 ft³	
1 fluid ounce	= 29.573 5 mL		= 1.307 95 yd³	
1 liquid pint	= 473.177 mL	1 milliliter	= 0.033 814 0 fl oz	
	= 0.473 177 L	1 liter	= 1.056 69 liq qt	
1 liquid quart	= 946.353 mL	1 liter	= 0.264 172 gal	
	= 0.946 353 L			
1 dry pint	= 550.61 mL			
1 dry quart	= 1.101 22 L			
1 dry peck	= 8.809 68 L			
1 gallon[b]	= 3.785 41 L			
1 bushel[c]	= 35.239 L			
		Length		
1 inch	= 2.54 cm*	1 millimeter	= 0.039 370 1 in	
1 foot	= 30.48 cm*	1 centimeter	= 0.393 701 in	
	= 0.304 8 m*			
1 yard	= 0.914 4 m*			
		Area		
1 square inch	= 6.451 6 cm²*	1 square centimeter	= 0.155 000 in²	
1 square foot	= 929.030 cm²	1 square decimeter	= 0.107 639 ft²	
1 square yard	= 0.836 127 m²	1 square meter	= 10.763 9 ft²	

* Denotes an exact number.

[a] These conversion factors are given to six significant digits to provide such accuracy when necessary.

[b] Gallon means the US gallon of 231 cubic inches.

[c] Bushel means the US bushel of 2,150.42 cubic inches.

Source: Metric Labeling pp. 67461–67462.

USE OF THE TERM "WEIGHT"

The term "weight" may be abbreviated as "wt" in the net quantity of contents declaration. Both upper- and lower-case letters may be used for this abbreviation.

EXEMPTIONS FROM ADEQUATE DIRECTIONS FOR USE

Section 502(f)(1) of the FD&CA deems a device misbranded if its labeling fails to provide adequate directions for use. This is further defined in 21 CFR 801.5 to mean that the labeling must provide directions sufficient for a layperson to use the device safely and for the purpose intended. Subpart D of 21 CFR 801 describes the conditions where a device is exempted from this requirement.

PRESCRIPTION DEVICES (§801.109)

Some devices, because of (a) potential harmful effects, (b) the method of use, or (c) the collateral measures necessary to their use, are not safe except under the supervision of a licensed practitioner. For these devices, "adequate directions for use" cannot be prepared because the safe use of the device depends on the professional training and experience of the practitioner. These devices are exempt from Section 502(f)(1) of the FD&CA if the following conditions are met:

- A prescription device is exempt if it is in the possession of either a licensed practitioner or persons lawfully engaged in the manufacture or distribution of the device.
- The label of a prescription device, other than a surgical instrument, must bear the prescription statement:

  CAUTION: FEDERAL LAW RESTRICTS THIS DEVICE TO SALE BY OR ON THE ORDER OF A _____.

  The blank is filled with the word "physician," "dentist," or "veterinarian," or the descriptive designation of any practitioner licensed by the law of the state in which he or she practices to use or order the use of the device.
- The device label must bear a statement of the method of its application or use.
- The device labeling, on or within the package, must bear appropriate information for use, including indications, effects, routes, methods, and frequency and duration of administration, and any relevant hazards, contraindications, side effects, and precautions under which the device can be safely used.
- All labeling, other than labels and cartons, that bears any information for use of the device must also bear the date the labeling was issued or the date of the latest revision.

The information for use may be omitted from the dispensing package if, but only if, the directions, hazards, warnings, and other information are commonly known to the licensed practitioner. The FDA will, upon receipt of a written request, offer an opinion on a proposal to omit required labeling under this proviso.

Any labeling that contains or purports to contain information for use, whether or not it is on or within the package from which the device is dispensed, must contain adequate information for such use, including indications, effects, routes, methods, and frequency and duration of administration, plus any relevant hazards, contraindications, side effects, and precautions under which the device can be safely used. This information is not required on "reminder-piece" labeling, which calls attention to the name of the device but does not contain any instructions for its use.

Advertising, for example, falls under this requirement. A US appellate court has held that: "Most, if not all, advertising is labeling. The term labeling is defined in the FD&CA as including all printed matter accompanying any article. Congress did not, and we cannot, exclude from the definition printed matter which constitutes advertising" (Cardamone p. 3).

INDICATIONS FOR USE

Indications, or "Indications for Use," identify the target population in which there is sufficient valid scientific evidence to demonstrate that the device as labeled will provide clinically significant results without presenting an unreasonable risk of illness or injury. As appropriate, the labeling should state that the device (trade name) is "indicated" or "intended for use" (Device Labeling p. 4):

- in the treatment, mitigation, prevention, or diagnosis of a recognized disease or condition, or an important manifestation of a disease or condition;
- in the relief or mitigation of symptoms associated with a disease or condition, or an important manifestation of a disease or condition; and/or
- as an aid or adjunct to a mode of therapy or diagnosis.

If the device is indicated, or intended for use, in a subgroup of a population with a disease, symptom, or syndrome, the labeling should:

- describe the available evidence that supports the indication;
- state the limitations of the usefulness of the device;
- describe the specific tests needed for proper selection and monitoring of the patients;
- provide information on the appropriate kind, degree, and duration of the anticipated improvement, if available; and
- include relevant information on the recommended intervals between use, the usual duration of treatment, or any appropriate modifications of the recommended interval or duration.

The labeling should state when safety considerations indicate that the device should be reserved or restricted for use in certain situations—such as cases not responsive to other devices, drugs, or surgical procedures. The labeling should also state whether specific conditions should be met before the device is used on a long-term basis (e.g., demonstration of the responsiveness of a device in a short-term trial). If the indications for long-term use are different than for short-term use, the labeling must identify the specific indications for both.

In cases where there is a common belief that a device may be effective for a certain use, or there is a common use of the device for a condition, but the preponderance of evidence demonstrates that the device is ineffective when applied to that use or condition, the FDA may require that the labeling state that there is a lack of evidence that the device is effective for that use or condition (Device Labeling p. 5). During the classification procedure, the FDA may determine that special labeling about the indicated use of a device is required. The FDA is particularly concerned about the need to ensure that patients are informed about the uses, benefits, and risks of medical devices. Some examples of special indications-for-use labeling are discussed in Chapter 26.

Contraindications

Contraindications describe the situations in which the device should not be used because the risk of use clearly outweighs the benefit. For example, a device may, but not always, be contraindicated because of (a) hypersensitivity to an ingredient of a permanently implanted device; (b) substantial risk of a patient being harmed because of age, sex, concomitant therapy, disease state, or other condition; or (c) continued use in the face of an unacceptably hazardous adverse reaction.

The contraindications should immediately follow the indications for use in the labeling and should state only known hazards. Theoretical possibilities are not to be listed. If no contraindications are known, then the labeling should state in the contraindications section that there are "none known."

Warnings

The "warnings" section of the labeling follows the contraindications. It describes serious adverse reactions and potential safety hazards, any limitation in use imposed on the device by them, and steps to take should they occur. Appropriate warnings are included if there is reasonable evidence of an association of a serious hazard with the use of the device. A causal relationship need not have been proved (Device Labeling p. 6).

A warning is appropriate when a device is commonly used for a disease or condition for which there is insufficient valid scientific evidence to demonstrate effectiveness, and such usage is associated with a serious risk or hazard.

PRECAUTIONS

The "precautions" section of the labeling contains information about any special care to be exercised by the practitioner and/or patient in order for the device to be used safely for its intended purpose. Examples can include: (1) the need for protective clothing to be worn during use, (2) laboratory tests or other evaluation procedures that may be helpful in identifying adverse reactions, and (3) testing the patient's response, including the frequency of such tests or procedures before, during, and following use of the device. This section includes precautionary statements not appropriate for inclusion in other sections of the labeling.

It may be necessary to limit the use of the device because of:

- lack of long-term safety and effectiveness data;
- lack of safety and effectiveness data for specific patient populations such as pregnant women, infants, and so on;
- growth processes still occurring in the body; or
- anatomical or physiological limitations on the effectiveness of the device.

If the safety and effectiveness of the device for use in specific patient populations has not been established on the basis of valid scientific evidence, the Indications for Use section shall specifically identify the persons for whom use of the device is indicated, and the Precautions section shall include the following statement (Device Labeling pp. 6–7):

SAFETY AND EFFECTIVENESS IN [E.G., PREGNANT WOMEN, CHILDREN UNDER THE AGE OF ..., ETC.] HAVE NOT BEEN ESTABLISHED.

If a specific hazard is associated with the use of the device in certain patient populations, the hazard is to be described in the precautions section of the labeling. It may be appropriate, because of the nature of the hazard, to describe it in the warnings or contraindications sections of the labeling. In this case, the precautions section must refer to the hazard with a statement such as: "See warnings section for information on...."

ADVERSE REACTIONS

An "adverse reaction" is an undesirable effect reasonably associated with the use of the device. It may occur as part of the effect of the device or may be unpredictable in its occurrence.

All adverse reactions reasonably associated with the use of the device must be included in this section, even if these reactions have already been mentioned under contraindications, warnings, and/or precautions. A statement directing the reader to other sections of the labeling for more information regarding the adverse reaction and any steps to be taken, if appropriate, should follow the listing of the adverse reaction.

Adverse reactions should be listed in descending order based on their clinical significance as determined by their severity and frequency. The labeling should provide frequency data from adequately reported clinical studies when the data are not well known to the device user (practitioner and/or patient) and/or when needed in deciding between the use of the device and an alternative procedure or approach (Device Labeling p. 7).

RETAIL EXEMPTION FOR PRESCRIPTION DEVICES (§801.110)

A prescription device delivered to the ultimate user by a licensed practitioner in the course of his or her practice, or upon a prescription, or other order lawfully issued in the course of his or her professional practice, need only bear the name and address of the practitioner, the directions for use, and any required cautionary statements.

MEDICAL DEVICES HAVING COMMONLY KNOWN DIRECTIONS (§801.116)

A device is exempt from the requirement that the labeling bear adequate directions for use if adequate directions for common uses are known to the ordinary individual. A good example of this type of device is the adhesive bandage.* Such a device needs little in the way of instructions for use. Adhesive bandages are properly labeled if the outer protective container bears a statement that the contents may not be sterile if the individual package has been opened or damaged. The outer container must bear the name of any added antimicrobial agents. The individual package should indicate the dimensions of the compress. Both the individual package and the outer protective container must indicate the name of the manufacturer, packer, or distributor, and the outer protective package must include the address of the manufacturer, packer, or distributor (USP p. 198).

IN VITRO DIAGNOSTIC PRODUCTS (§801.119)

IVD products are intended for use in the diagnosis of disease or other conditions. IVD products will be in compliance with the general labeling provisions in 21 CFR 801 if they meet the specific requirements stated in 21 CFR 809.10. These requirements are described in Chapter 22.

MEDICAL DEVICES FOR PROCESSING, REPACKING, OR MANUFACTURING (§801.122)

A device intended for processing, repacking, or use in the manufacturing of another drug or device is exempt from Section 502(f)(1) (adequate directions for use) of the FD&CA if its label bears the statement:

CAUTION: FOR MANUFACTURING, PROCESSING, OR REPACKING.

These devices are subject to the other provisions in Section 502 of the FD&CA. Devices moving between manufacturing facilities or between a manufacturer and a contract sterilizer are exempted from Section 502(f)(1) and other provisions of Section 502 under another Subpart E of the regulations. This exemption is discussed later in this chapter.

MEDICAL DEVICES FOR USE IN TEACHING, LAW ENFORCEMENT, RESEARCH, AND ANALYSIS (§ 801.125)

A prescription device is exempt from the adequate directions for use if it is shipped or sold to, or in the possession of persons regularly and lawfully engaged in:

- instruction in pharmacy, chemistry, or medicine not involving clinical use;
- law enforcement;

---

* See 21 CFR 880.5240.

- research not involving clinical use; or
- chemical analysis or physical testing.

A device is exempt from adequate directions for use if it is to be used only for such instruction, law enforcement, research, analysis, or testing.

MEDICAL DEVICES: EXPIRATION OF EXEMPTIONS (§801.127)

The exemption from the adequate directions for use requirement is terminated if:

- the device is shipped or delivered to an individual in whose possession the device is not exempt, or
- the device is used for other than exempt purposes.

OTHER EXEMPTIONS

Section 21 CFR 801.122 describes the conditions under which a device intended for processing, repacking, or manufacturing is exempt from the adequate directions for use requirement. Subpart E of 21 CFR 801 deals with other exemptions to the labeling and packaging requirements of Sections 502(b) and (f) of the FD&CA for devices that will be processed, labeled, or repacked in substantial quantity at a site other than where they were processed or packed.

MEDICAL DEVICES: PROCESSING, LABELING, OR REPACKING (§801.150)

In-process devices that are being transported from one manufacturing site to another are exempt from the packaging and labeling requirements of Sections 502(b) (Name/address of manufacturer and accurate statement of quantity of contents) and 502(f) (adequate directions for use) of the FD&CA. This exemption is in force if:

- The person introducing the device into commerce is the operator of the firm where further processing takes place.
- The person introducing the device into commerce is not the operator, but delivery is made under a written agreement that includes the names and postal addresses of both parties, and the specifications for processing, labeling, or repacking are sufficient to ensure that, if followed, the device will not be adulterated or misbranded when the processing is completed.
- Unsterilized devices that are labeled as sterile are not misbranded while in transit to a contract sterilizer if both the following conditions are met:
  - There is a written agreement between the two parties that contains:
    - the names and postal addresses of the firms involved, and is signed by the person authorizing the shipment and the person in charge of the sterilization facility;
    - instructions for maintaining proper records to ensure that the number of units shipped is the same as the number sterilized;
    - acknowledgment that the devices are nonsterile and are being shipped for further processing; and
    - the specification of the sterilization process, sterilant media, equipment, and the testing method or quality controls to be used by the contract sterilizer.
  - Each pallet, carton, or other designated unit is conspicuously marked to show its nonsterile condition. Following sterilization, and until it is established that the device is sterile, each pallet, carton, or other designated unit must be conspicuously marked to show that it has not been released from quarantine.

While in-process devices are exempt from the requirements of Sections 502(b) and 502(f) of the FD&CA, they are subject to GMP regulations. These requirements are described in Chapter 26.

STERILE DEVICES

The labeling of sterile devices should receive special attention. Some sterile devices may only be for single use; they may not be designed or manufactured to withstand the rigors of the necessary recleaning and resterilization process. For these devices, the labeling should advise against resterilization and reuse.

Some medical devices are designed to have parts of the devices sterile. For these devices, the labeling must advise the user of the extent to which the device is "sterile." A possible limiting statement might be (Lowery, Puleo p. 11-5):

CAUTION: ONLY THE FLUID PATH OF THE SET IS STERILE AND NONPYROGENIC. DO NOT USE IN A STERILE OR ASEPTIC AREA WITHOUT PROPER PRECAUTIONS.

Some devices are designed so that they can be cleaned and resterilized. Other devices are intended to be sterilized by the user before the initial use. In both cases, the device labeling must provide adequate instructions to allow the user to carry out the necessary cleaning and sterilization process. Such instructions are also be appropriate for certain sterile, single-use only implants that may become contaminated before implantation (e.g., orthopedic implants). For any device intended to come in contact with the patient's skin, unless that device is provided sterile, the labeling should included initial processing instructions, even if the device is not reusable.

At a minimum, the reprocessing instructions should include a statement that the device should be thoroughly cleaned before reuse. Cleaning may be the only step required for safe and effective reprocessing, or it may only be the first step. In either case, adequate cleaning instructions are critical.

The instructions must indicate the appropriate level of microbicidal processing for the device. The FDA has identified four levels of processing. The appropriate level of processing depends on the intended use, the type of contact with the patient, and the risk of infection if the device is contaminated with any microorganisms. The four levels are: sterilization, or high-, intermediate-, or low-level disinfection (Reusable p. 5).

The instructions must be clear, legible, and in logical order (e.g., preprocessing, cleaning, rinsing, disinfection or sterilization, final rinsing, and postprocessing handling). The process outlined in the instructions must be written for the intended audience and be feasible in the intended location of reprocessing (e.g., healthcare facility or home). The user must have the knowledge and means of carry out the reprocessing steps.

The instructions must be comprehensive. Comprehensive instructions enable the user to understand precisely what is required to carry out the reprocessing of the device. The elements of comprehensive reprocessing instructions are listed below (Reusable pp. 6–9). Not all of the elements may apply to a specific device.

1. Any special accessories (e.g., tools, trays, test kits, and so on) needed to carry out the process should be described.
2. Any special preprocessing handling procedures that are needed should be described.
3. If the device consists of more than one removable part, the instructions for disassembly and reassembly should be included.
4. The labeling should recommend a method of cleaning the parts of the device.

5. Compatible cleaning and lubricating agents or a class of agents should be listed. The labeling of a reusable device should refer to the labeling of the cleaning and lubricating agent for preparation of and use instructions for those agents. If a specific agent or class of agent is not specified, then validation testing should qualify commonly used products.

6. Specific instructions for rinsing after cleaning and any liquid chemical disinfection or sterilization should be provided. The recommendations should include the quality of rinse water, volume, and duration of the rinse. If the instructions in the labeling of the cleaning and disinfection/sterilization agent are sufficient, the device labeling may refer to those instruction.

7. The labeling should specify at least one validated method for disinfection or sterilization. These instructions should include specific process parameters used to control the process (e.g., cycle parameters, aeration, loading of sterilizer, and so on). Any limits on the sterilization process that must be observed by the user should also be included (e.g., limits on maximum temperature or pressure during processing).

8. Special postprocessing procedures should be recommended where needed to eliminate or minimize recontamination of the device before use. The FDA requires a recommended postprocess aeration time be provided if labeling recommends ethylene oxide (EtO) sterilization.

9. The labeling should state a limit, when appropriate, on the number of times that a device can be resterilized and reused without affecting the safety and effectiveness of the device. For some devices (i.e., reusable surgical gowns), the labeling includes the number of allowed reuses and provides a mechanism so the user can keep track of the number of reuses. The labeling of other devices may describe a performance test that should be executed before each reuse, or identify an observable characteristic (e.g., corrosion) that enables the user to identify the point when the device is no longer fit for use.

10. When warranted, the labeling should include special warnings and precautions regarding the reprocessing process. These warnings may relate to user safety or to conditions that may significantly affect the effectiveness of the reprocessing or the performance of the device. For example, sterilization or disinfection processes that are inappropriate for the device should be identified (e.g., "Warning: Do not autoclave").

11. Devices intended to be maintained by lay users are of particular concern. Even certain tasks associated with reuse of prescription devices may be performed by the patient or lay healthcare providers.

12. When appropriate, the labeling may refer to recognized professional practices/guidance documents for procedures or other information rather than reproducing the details in the device labeling. In a similar manner, references may be made to the labeling of accessory devices used in processing when the labeling statements are consistent and complementary.

13. The labeling should contain a telephone number that the user can call for additional information on infection-control procedures.

14. The labeling may advise the user of his or her responsibility to validate any deviations from the recommend reprocessing procedure, and may state appropriate disclaimers if there are deviations.

Many of the products (i.e., devices such as sterilizers, liquid chemical sterilants, and disinfectants intended for use on medical devices) used to process medical devices for use/reuse are themselves currently subject to FDA clearance. The instructions for processing/reprocessing of a medical devices should not include reference to devices or agents that are not approved for commercial distribution.

If reuse of a product is a common practice, the manufacturer is encouraged by the FDA to include instructions for cleaning and sterilization. In some cases, the FDA may require the manufacturer to include such instructions. A case in point is reuse of kidney dialyzers. Even though dialyzers were often labeled for single use only, reuse had become a common practice. As a result, the FDA is requiring manufacturers to include instructions in their labeling for proper reprocessing (Dialysis p. 1).

REPROCESSED SINGLE-USE DEVICES

In general, Section 502 of the FD&CA declares a medical device is misbranded if it does not prominently and conspicuously display the name of the device's manufacturer, a generally recognized abbreviation of the manufacturer's name, or a unique and generally recognized symbol identifying the manufacturer. For reprocessed single-use devices, this requirement has created confusion and debate. The FDA recognizes the reprocessor as the manufacturer of the reprocessed device and may require the reprocessor to submit a 510(k) to obtain market clearance for the reprocessed device. The MDUFMA attempts to resolve this problem by introducing a new provision into the FD&CA. The MDUFMA modifies Section 502 of the FD&CA by adding Section 502(v), which declares that a reprocessed single-use device is misbranded unless all labeling of the device prominently and conspicuously bears the statement:

REPROCESSED DEVICE FOR SINGLE USE. REPROCESSED BY [NAME OF REPROCESSOR]

Section 502(v) goes into effect on January 27, 2004, which is 15 months after enactment of legislation. The requirement only applies to devices "introduced . . . into interstate commerce after such effective date" (MDUFMA §302(a)(2)).

ELECTROMAGNETIC COMPATIBILITY LABELING

Technological advances have improved the performance of many electronic devices by adding speed, increasing accuracy, permitting the inclusion of new, advanced functions, reducing size, and increasing portability. Nowhere has this trend been more evident than in the wireless communication industry. A rarity only a few years ago, wireless devices—cell phones—now are common in both public and private spaces. The proliferation and portability of these devices have greatly increased the potential for wireless communications and medical devices to come into proximity. With increased proximity, there is an increased possibility of the radiated electromagnetic fields produced by wireless communication devices interfering with medical devices. The increased potential for electromagnetic interference (EMI) has raised serious concerns regarding the mutual compatibility of medical devices and wireless communications devices.

There are a number of sources of relatively intense electromagnetic fields in the everyday environment. These sources include commercial radio and television stations, and amateur radio (HAM) transmitters. These sources are obvious and relatively easy to avoid. Others, such as hand-held portable transceivers (e.g., citizens band [CB] radios; "walkie-talkie;" security, fire, and police transceivers; cellular telephones; and other personal communication devices) and medium-range mobile transceivers (such as those used in police cars, fire trucks, ambulances, and taxis) can be difficult to identify and avoid.

At times, medical devices can even interfere with each other. Certain devices are designed to emit electromagnetic energy as part of their functional modes of operation, e.g., electrosurgery units, devices that use radio frequency energy to simultaneously cut and cauterize. When such devices are activated in proximity to other medical devices, the other devices may be effected. However, the effects are known, understood, and, most important, expected.

From a labeling standpoint, the directions for use should address the following points:

- The labeling should describe the effects of and list possible sources of both radiated and conducted EMI.
- A discussion of the effects and causes of electrostatic discharge should be included.
- The labeling should contain a list of other devices that pose potential electrical problems.
- A description of conditions (e.g., loosened electrodes) that can cause environmental EMI effects to be more pronounced should be included.
- The labeling should advise the operator on steps to identify and resolve environmental EMI with the safe and effective use of the device.
- When appropriate, the labeling should advise the user on procedures for determining whether the device is susceptible to levels of EMI occurring at the intended-use site, recommend periodic testing to check on susceptibility, and recommend a course of action if the device fails the test.

When EMI poses a potential hazard to users of a device, then some or all of the information listed above should be included in the warnings and precautions sections of the labeling as well. The FDA believes that a failure to provide adequate warnings constitutes a failure of the labeling to disclose material facts and, thus, renders the device misbranded (Alpert, Wheelchair p. 3). The FDA is requiring labeling that informs powered wheelchair users of the risks posed by EMI. The cardiac pacemaker industry, working with the FDA, has developed a generic warning for physician and patient manuals for implanted cardiac pacemaker and implanted defibrillators. Both applications are discussed in Chapter 26.

ELECTRONIC LABELING

Manufacturers have long sought approval from the FDA to provide medical device labeling in electronic form using media such as CD-ROM or over the Internet. The MDUFMA amends Section 502(f) of the FD&CA to allow required labeling for prescription devices intended for use in healthcare facilities to be made available solely by electronic means. The manufacturer may provide electronic labeling if (MDUFMA §206):

- the device is intended for use in healthcare facilities;
- the labeling complies with all applicable requirements of the FD&CA;
- the manufacturer provides healthcare facilities the opportunity to request the labeling in paper form; and
- the manufacturer can promptly provide the healthcare facility with the requested information at not additional cost.

SPECIAL REQUIREMENTS FOR SPECIFIC DEVICES

The topics discussed in the previous parts of this chapter are generally applicable to a wide range of medical devices. However, during the classification process, the FDA continues to establish very specific labeling requirements for specific devices. These requirements are contained in Subpart B of 21 CFR 800 and Subpart H of 21 CFR 801, and include:

- Requirements for labeling for contact-lens solutions and tablets (§800.12(c))
- Use-related statements (§§801.405—801.437). These parts describe specific labeling, which may include not only package labeling but informational literature, patient release forms, performance testing, and/or specific tolerances or prohibitions on certain ingredients. The following list includes devices or device components that have additional labeling requirements:
  - Denture repair or refitting kits (§801.405)
  - Use of impact-resistant lenses in sunglasses and eyeglasses (§801.410)
  - Maximum acceptable levels of ozone (§801.415)
  - Chlorofluorocarbon propellants (§801.417)
  - Professional and patient labeling for hearing aids (§801.420)
  - Conditions for sale of hearing aids (§801.421)
  - User labeling for menstrual tampons (§801.430)
  - Products containing or manufactured with chlorofluorocarbons or other ozone-depleting substances (§801.433)
  - User labeling for latex condoms (§801.435)
  - User labeling for devices that contain natural rubber (§801.437)

Chapter 26 provides a summary of the special labeling requirements for the specific devices listed in Subpart H of 21 CFR 801.

EXPORT OF MEDICAL DEVICES

The FDA derives its authority to regulate the export of medical devices from Section 801(e) of the FD&CA.

EXPORT OF APPROVED DEVICES

Under Section 801(e), FDA approval is not required for the export of any device that is approved for commercial distribution within the United States under either a premarket notification (510(k)) or premarket approval application (PMA). The device must be produced in a manufacturing site registered with the FDA in accordance with the GMPs contained in the Quality System Regulation (QSR) of 21 CFR 820. The device must meet the labeling requirements of 21 CFR 801 and any other applicable labeling requirements discussed in subsequent chapters. In addition to those devices with an approved 510(k) or PMA, preamendment and 510(k)-exempt devices may be exported without prior FDA approval. Upon request, the FDA will issue a Certificate for Foreign Government (CFG) for these devices in order to facilitate international marketing of the product, but the law does not require the certificate (Exporting Medical Devices p. 4).

EXPORT OF UNAPPROVED DEVICES

A device that has not been cleared for marketing in the United States through the 510(k), PMA, or investigational device exemption (IDE) process may be exported without FDA permission if the following condition are met (Exporting Medical Devices p 4):

- It is manufactured according to the specifications of the foreign purchaser.
- It is not in conflict with the laws of the country to which it is intended for export. A letter from the destination country approving importation is evidence that the device is not in conflict with the laws of that country. If the destination is a country in the European

Economic Area (EEA) and the device has been awarded the CE Conformity Marking, the FDA will accept the CE Conformity Marking certificate in lieu of a letter from the foreign government.

- It is not sold or offered for sale in domestic commerce.
- It is clearly labeled (on the outside of the shipping package) as intended for export only. Assuring compliance with these conditions is the responsibility of the exporting party, and no approval is required from the FDA. During routine inspections, the FDA will evaluate whether a firm has complied with Section 801(e)(1) of the FD&CA (Exports and Imports §5).

Using its enforcement discretion, the FDA has allowed firms to export unapproved devices that would be considered adulterated (because they lack 510[k] marketing clearance) without prior approval if they meet the following conditions (Exporting Medical Devices p. 5):

- It can be reasonable assumed that the device would be found to be substantially equivalent to a type of device classified by the FDA in Class I or II (i.e., there is an existing classification regulation applicable to the device in question) and it is not subject to a mandatory performance standard under Section 514 of the FD&CA.
- The device meets the requirements of Section 801(e)(1) described above.

This provision includes only devices that are identical in design, construction, and intended use to Class I or Class II devices, or to devices that reasonably would be substantially equivalent to a Class I or Class II device. Devices not included would be preamendment Class III devices where the FDA has called for submission of a PMA, postamendment Class III devices, and devices found not to be substantially equivalent to a 510(k) device. Prior to the FDA Export Reform and Enhancement Act of 1996 (ER&EA), firms were sometimes faced with a dilemma when they wished to export, under Section 801(e)(1), an unapproved device for which they reasonably could expect to obtain a 510(k) approval. While the FDA did not require prior approval to export, neither could it issue a CFG because these devices had not undergone marketing clearance. The ER&EA added a new provision that authorizes the FDA to issue a CFG if the exporter submits a signed statement that the product meets the requirements of Section 801(e)(1). A false statement is violation of the US law, subjecting the perpetrator to a substantial penalty (Exporting Medical Devices p 5).

Section 801(e)(2) of the FD&CA requires that FDA permission must be obtained to export an unapproved device that:

- does not comply with a performance standard promulgated under Section 514 of the FD&CA;
- requires, but does not have, an approved PMA under Section 515 of the FD&CA;
- is undergoing clinical investigation under the provisions of Section 520(g) of the FD&CA; or
- is a banned device under Section 516 of the FD&CA.

The FDA can grant an export license for such a product upon written application from the manufacturer provided that the FDA determines that the export of the device would not be contrary to public health and safety, and provided that the destination country has approved the device or has no objection to its import. A device approved under this procedure must not be sold or offered for sale in the domestic market, and the labeling on the outside of the shipping package must indicate that the device is intended for export only.

The ER&EA provides a convenient alternative to exporting medical devices under Section 801(e)(2). Under Section 802 of the FD&CA, Class III devices and devices required to meet a Section 514 performance standard may be exported without prior FDA approval if both the device and the manufacturer meet certain criteria. A manufacturer may export a device under Section 802 by submitting a "simple notification" per Section 802(g) prior to beginning export of the device.

Section 802(b)(1) allows qualified devices to be exported to certain listed ("Tier 1") countries if the device complies with the laws of that country and has valid marketing authorization by the appropriate authority. The listed (or Tier 1) countries are: Australia, Canada, Israel, Japan, New Zealand, Switzerland, South Africa, members of the European Union (the EU includes United Kingdom, Spain, Ireland, Denmark, Greece, Belgium, Portugal, Germany, France, Italy, Luxembourg, the Netherlands, Sweden, Finland, and Austria) or the EEA (which includes the EU countries and Norway, Iceland, and Liechtenstein). For Tier 1 countries, the manufacturer must provide the FDA with the name of the device and the date it begins to export to a listed country.

Section 802(b)(1) permits firms to export unapproved medical devices to countries other than Tier 1 countries if the device is authorized for sale in a Tier 1 country and if the importing country accepts that authorization. Some South American countries, for example, now permit marketing of any medical device with a CE mark (Exporting Medical Devices p. 7).

In order to qualify for export under Section 802, devices must (FD&CA, § 802(f)):

- meet the requirements of Section 801(e)(1) described above;
- substantially meet the GMP requirements of the QSR or an international standard recognized by the FDA (At the time this chapter was written, none were recognized by the FDA.);
- not be adulterated other than by the lack of marketing approval;
- not be the subject of a notice by the Department of Health and Human Services (DHHS) that reimportation would pose an imminent hazard, nor pose an imminent hazard to the receiving country; and
- not be mislabeled other than by possessing the language, units of measure, or any other labeling authorized by the recipient country. In addition, the labeling must comply with the requirements and conditions of use in the listed country that gave marketing authorization, and must be promoted in accordance with its labeling.

Although the FDA does not require a manufacturer to obtain written permission prior to export, a foreign purchaser requests proof of compliance with US law prior to export. The FDA can provide a Certificate of Exportability (COE) under Section 802 to facilitate export of a medical device under Section 802.

Once an adulterated or misbranded device has been sold or offered for sale in the domestic market, it may not be exported as an alternative for bringing the device into compliance with the requirements of the FD&CA (Exporting Medical Devices pp. 4–5).

Other Provisions of the ER&EA

The ER&EA provided three additional situations where export permission is automatically granted under Section 802 of the FD&CA.

Exporting for Investigational Use

A manufacturer who wishes to export an unapproved device for investigational use may export the device under an IDE, under Section 801(e)(2), or under 802(c) of the act depending on the country to which the device is being exported. For instance, Section 801(e)(2) of the FD&CA allows an unapproved device intended for investigational use to be exported to any country, if the device meets the requirements of 801(e)(1) and the exporter submits information to the FDA that would enable the agency to determine that exportation is not contrary to the public health or safety and that the foreign country approves of the exportation.

Alternatively, Section 802(c) allows an unapproved device intended for investigational use to be exported to a Tier 1 country (Australia, Canada, Israel, Japan, New Zealand, Switzerland, South Africa or member countries of the EEA) without FDA authorization if the unapproved device is exported in accordance with the laws of that country. Devices being exported under Section 802(c) are not required to meet the requirements of the IDE regulation. However, compliance with the basic export requirements of 802(f) described above and the record-keeping requirements in 802(g) is required. Exportation of an unapproved device for investigational use to any country other than a Tier 1 country requires authorization by the FDA under Section 801(e)(2) of the FD&CA (Exporting Medical Devices p 9).

Exporting for Marketing or in Anticipation of Foreign Marketing Approval

Section 802(d) of the FD&CA allows the exportation of an unapproved device "intended for formulation, filling, packaging, labeling, or further processing in anticipation of market authorization" in any Tier 1 country. The only expressed requirements for such exports are that the product complies with the laws of the foreign country and the requirements in Section 802(f) described above. Records for such exports must be kept in accordance with Section 802(g) of the act.

Devices Intended for Treatment of Non-US Diseases

Section 802(e)(1) of the FD&CA allows the export of a medical device intended for the diagnosis, treatment, or prevention of a tropical disease or another disease not prevalent in the US that does not otherwise meet the criteria of Section 802. FDA approval is required. Before approving the application, the FDA must be satisfied that the (a) device does not present unreasonable risk, (b) the benefits of using the device outweigh the risks, and (c) the risks of using available alternatives was considered.

IMPORT OF MEDICAL DEVICES

Devices imported into the United States must meet all of the applicable requirements of the FD&CA, including compliance with all of the labeling regulations. In addition, all wording required by the FD&CA on the label or labeling must be translated into English (Parr, Barcome, Import of Medical Devices p. 1-40).

To ensure that the FDA is notified of all regulated products imported into the United States, the importer, or the importer's representative, must file an entry notice and an entry bond with the US Customs Service. The FDA, working with the US Customs Service, is responsible for reviewing all entry papers before the device is allowed into the United States. This review may include a wharf examination, sample analysis, and/or label review, or the product may be allowed to proceed into commerce without sampling or examination.

If the sampling or examination indicates that the device may be in violation, the FDA, depending on the nature of the apparent violation, may release the product with comment, or may decide to issue a detention notice for it. If the device is detained, the owner may apply for permission to relabel or otherwise recondition the product to bring it into compliance. If permission is granted and the relabeling or reconditioning is successful, the FDA will release the device. If the FDA believes that the owner cannot or will not recondition the product properly, it may refuse to permit reconditioning. If permission is refused or the reconditioning is unsuccessful, the FDA will refuse entry.

The provisions of Tariff Act of 1930,* which is enforced by the US Customs Service, requires that all imported articles must be marked in English with the name of the country of origin. The intent is to inform the ultimate purchaser in the United States of the country where the imported article was made. Certain articles are exempted from this rule. They include (a) articles that are in transit through the United States, (b) articles that are under bond to insure export, and (c) articles that are specifically exempted by the law. Certain articles may also require special marking (Parr, Barcome, Import of Medical Devices p. 1-42).

The name of the country of origin must be permanently, conspicuously, and legibly displayed on the device or its container. The article should be marked as indelibly as the nature of the product or its container will allow. Containers for devices excepted from marking must be marked with the name of the country of origin of the device unless the container is also exempted from marking.

Any reasonable system of marking that fulfills the purpose of the law is generally acceptable. However, any marking that does not remain on the product during handling, or for any other reason except deliberate removal, is not acceptable. The best form of marking is one that becomes a part of the article itself, such as branding, stenciling, stamping, printing, molding, and similar methods. In the case of on earthenware or chinaware, glazing on the name of the country of origin in the process of firing is the preferred method of marking.

Adhesive labels may be used in some instances. If paper sticker or pressure-sensitive labels are used, they must be affixed so securely in a conspicuous place that unless deliberately removed they will remain on the article while it is in storage or on display and until it is delivered to the ultimate purchaser. Tags may also be used for this purpose. When tags are used, they must be attached in a conspicuous place and in a manner that assures that unless deliberately removed they will remain on the article until it reaches the ultimate purchaser (19 CFR 134.44).

The marking must be placed where it can be seen with casual handling of the article. The marking must not be in a position where it would be covered or concealed by subsequent attachments or additions. It must be visible without disassembling the article or removing or changing the position of any of the article's parts, and it must be of an adequate size and clear enough to be read easily by a person with normal vision (Parr, Barcome, Import of Medical Devices p. 1-43)

If the name of the country, standing alone, does not clearly indicate the country of origin, then the name of the country of origin must be preceded by a phrase such as "Made in…," "Product of…," or other words of similar meaning (19 CFR 134.46). This phrase must appear in close proximity to the country of origin and be in a comparable type size. Such a phrase would be required, for example, when additional words, marks, or other information appear with the name of the country in a way that might mislead the ultimate purchaser of the article as to the country of origin.

Sometimes an imported article is combined with another article after being imported but before being delivered to the ultimate purchaser. In this case, the imported article must be marked in such a was as to clearly indicate that the country of origin applies only to the imported article. For example, literature can be marked as "Printed in …," or words of a similar nature. Another example would be packaging material that is imported and marked "Container made in…."

Certain articles and their component parts are required to be marked legibly and conspicuously with their country of origin by die stamping, cast-in-the-mold lettering, etching (acid or electrolyte),

---

* *Tariff Act of 1930*, as amended. US Code vol. 19, §§1202–1677 (1930).

engraving, or by means of metal plates that bear the prescribed marking and that are securely attached to the article in a conspicuous place by welding, screws, or rivets. These articles include:

- clippers, shears and scissors, and blades for the same;
- knives having a folding or movable blade;
- knives with fixed blades, including forks , steels, and cleavers;
- safety razors and safety-razor blades;
- surgical needles including hypodermic needles, other surgical and dental instruments, scientific and laboratory instruments and apparatus, surveying and mathematical instruments, and drawing instruments made wholly or partially of metal; pliers, pincers, nippers, and hinged tools for handling wire, finished or unfinished; and
- thermostatic containers constructed with a vacuum or partial vacuum to maintain the temperature of the contents, or blanks and pistons of such articles.

Under the following circumstances, an article is exempted from marking by Section 1304 the Tariff Act if it is (19 CFR 134.32):

- an article that is incapable of being marked;
- an article that cannot be marked prior to shipment to the United States without injury;
- an article that cannot be marked prior to shipment to the United States except at an expense economically prohibitive to its importation;
- an article for which the marking of the containers will reasonably indicate the origin of the articles;
- an article that is a crude substance;
- an article imported for use by the importer and not intended for sale in its imported or any other form;
- an article to be processed in the United States by the importer or on the importer's behalf other than for the purpose of concealing the origin of such article and in such manner that any mark contemplated by this regulation would necessarily be obliterated, destroyed, or permanently concealed
- an article for which the ultimate purchaser must necessarily know, or in the case of an article from a North American Free Trade Agreement (NAFTA)* signatory country, must reasonably know, the country of origin by reason of the circumstances of its importation or by reason of the character of the article even though it is not marked to indicate its origin;
- an article that was produced more than 20 years prior to its importation into the United States;
- an article entered into or withdrawn from a warehouse for immediate exportation or for transportation and exportation;
- a product of American fisheries that is free of duty;
- a product of a possession of the United States;
- a product of the United States that was exported and returned;
- an article exempt from duty under 19 CFR 10.151 through 10.153, 145.31, or 145.32;

---

* North American Free Trade Agreement (NAFTA) entered into by the United States, Canada, and Mexico on August 13, 1992.

- an article that cannot be marked after importation except at an expense that would be economically prohibitive unless the importer, producer, seller, or shipper failed to mark the articles before importation to avoid meeting the requirements of the law;
- original works of art from a NAFTA country; and
- goods of a NAFTA country that are provided for in subheading 6904.10 or heading 8541 or 8542 of the Harmonized Tariff Schedule of the United States (HTSUS) (19 USC 1202).

BANNED DEVICES

A banned device is a device intended for human use that the FDA Commissioner has determined "presents substantial deception or an unreasonable and substantial risk of illness or injury." A device that is banned under the provisions of 21 CFR 895 is adulterated under section 501(g) of the FD&CA. A restricted device that is banned may also be misbranded under section 502(q) of the FD&CA.

LABELING (§895.25)

In determining whether a device presents substantial deception or an unreasonable risk of illness or injury, the FDA will consider all available data and information. During this process, the FDA may determine that labeling, a change in labeling, or, for a restricted device, a change in advertising, can correct or eliminate the substantial deception or unreasonable and substantial risk of illness or injury presented by a device. In this case, the FDA will provide the person(s) responsible for labeling or advertising the device with a written notice specifying:

- the nature of the deception, risk of illness or injury, or danger to the health of individuals presented by the device;
- the labeling, change in labeling, or, if the device is a restricted device, change in advertising, necessary to correct the deception or eliminate or reduce the risk or danger; and
- the period of time within which the labeling, change in labeling, or change in advertising must be accomplished.

The FDA-specified labeling or change in the labeling or advertising may include a specific statement, notice, or warning. This statement, notice, or warning must be in the manner and form prescribed by the FDA and will identify the deception, the risk of illness or injury, or the unreasonable, direct, and substantial danger to the health of individuals associated with the device as previously labeled. Such a statement, notice, or warning must be included for a time period specified by the FDA. The time period will be determined based on a number of factors including the degree of deception, risk, or danger associated with the use of the device, and any other factors that the FDA Commissioner considers pertinent.

The FDA will allow the person(s) responsible for labeling or advertising the device reasonable time, considering the degree of deception, risk, or danger associated with the use of the device, to accomplish the labeling or change in labeling or advertising. However, the FDA may request that the person(s) responsible for labeling the device stop distribution of the device until the changes are made.

If the voluntary action to correct or eliminate the substantial deception or unreasonable and substantial risk of illness or injury presented by a device is not taken, the FDA may move to prevent the introduction of the device into commerce. The FDA Commissioner may consider the failure of the person(s) responsible for the labeling of the device to accomplish the required labeling or changes in labeling or advertising as a basis for initiating proceedings to declare a device a banned device.

Veterinary Use (§895.1(d))

Although 21 CFR 895 does not cover devices intended for animal use, the person responsible for labeling of a device that is banned cannot avoid the ban by simply relabeling the device for veterinary use. A device that has been banned for human use, however, may have a valid veterinary use. Such a device may be marketed as a veterinary device provided that it complies with all requirements applicable to veterinary devices under the FD&CA and the applicable regulations in 21 CFR. The label for the device must bear the statement:

For Veterinary Use Only.
Caution: Federal law prohibits the distribution of this device for human use.

Even a device that bears the preceding label will be considered a banned device if the FDA determines that it is intended for human use. In making this determination, the FDA will consider, among other things, the ultimate destination of the device.

THINGS TO REMEMBER

The Medical Device Amendments of 1976 introduced a three-tiered classification system for medical devices based on the required level of regulatory controls necessary to protect public health and safety. Part of that classification system was the establishment of a set of factors to be considered when evaluating the safety and effectiveness of a device. These factors must be considered and addressed by the manufacturer when developing labeling for the device.

The basic requirements for device labeling are found in 21 CFR 801. This part of the CFR deals with general labeling provisions applicable to all devices, labeling for OTC (nonprescription) devices, exemptions from the adequate directions for use requirement (e.g., prescription devices), other exemptions for special purposes (e.g., devices in transit between manufacturing facilities), and special labeling requirements for special devices (e.g., hearing aids).

The FDA regulates the export of medical devices under Section 801(e) of the FD&CA. Under this provision, FDA approval is not required for the export of any device that is approved for commercial distribution within the United States under either a premarket notification (510(k)) or PMA. A device that would otherwise be considered to be adulterated or misbranded within the meaning of Sections 501 or 502 of the FD&CA if it were sold or offered for sale in commercial distribution in the United States, may be exported under Section 801(e)(1) of the FD&CA, provided the device is intended solely for export. Devices imported into the United States must meet all of the applicable requirements of the FD&CA, including compliance with all of the labeling regulations.

A device intended for human use that the FDA Commissioner has determined presents substantial deception or an unreasonable and substantial risk of illness or injury, may be banned under the provisions of 21 CFR 895. Such a device is adulterated. A restricted device that is banned may also be misbranded.

22 *In Vitro* Diagnostic Product Labeling

In vitro diagnostics (IVD) products include reagents, instruments, and systems that are intended for use in the diagnosis of disease or other conditions. The term diagnosis including a determination of the state of health in order to cure, mitigate, or prevent disease or its sequelae. These products are used for the collection, preparation, and examination of specimens taken from the human body. Included are products that are used in hospitals, clinical laboratories, and physician offices, as well as home-use testing kits and home-use mail-in specimen collections kits.

The FDA classifies IVD products as devices under Section 201(h) of the Federal Food, Drug, and Cosmetic Act (FD&CA). They may also be biological products subject to Section 351 of the Public Health Service Act (PHSA). The PHSA defines a biological product as any virus, therapeutic serum, toxin, antitoxin, vaccine, blood, blood compound or derivative, allergenic product or analogous product, or arsphenamine or its derivatives (or any other trivalent organic arsenic compound), applicable to the prevention, treatment, or cure of diseases or injuries of humans (Public Law 78-410 §262).

As devices intended for human use, IVD products are subject to the FDA's three-tier classification system based on the need for regulatory controls to ensure safety and effectiveness. The classifications for IVD products appear in Title 21 of the Code of Federal Regulation (CFR) in the classification regulations for (a) Clinical Chemistry and Clinical Toxicology Devices (21 CFR 862), (b) Hematology and Pathology Devices (21 CFR 864), and (c) Microbiology and Immunology Devices (21 CFR 866). (The FDA classification procedure is summarized in Appendix F.) More than 97 percent of IVD products are Class I or Class II. A few products, less than 3 percent have been classified as Class III. The Class III products are listed in Table F-1 in Appendix F, under parts 862, 864, and 866.

As devices, IVD products are subject to the general labeling requirements described in 21 CFR 801. However, because of the unique properties of IVD products, the FDA has established particular labeling requirements for them. These special requirements are contained in 21 CFR 809.10. An IVD product that meets these requirements is deemed to meet the requirements of 21 CFR 801 and section 501(f)(1) of the FD&CA.

The general labeling requirements for all IVD products are discussed in the following section. Because of the wide variation in design, methodology, and type of IVD products, and the range of scientific disciplines involved (e.g., chemistry, microbiology, immunology, hematology, pathology), the FDA has formulated three "product class models" that provide guidance for specific categories of IVD products. The product categories are clinical chemistry/toxicology, clinical microbiology/immunology, and clinical hematology/pathology (Bracey §1-1). Recommendations for organizing the labeling for these classes of devices are presented following the section on general labeling provisions.

In recent years, there has been a growing interest in IVD products intended for home use. The FDA has taken the view that the regulations provided in 21 CFR 809.10 are generally adequate for home-use IVD products even though the regulations were targeted at professional-use IVD products. However, some qualification of the requirements may be useful in preparing labeling for home-use IVD products. These qualifications are discussed later in this chapter.

Normally, substances subject to the FD&CA are exempt from regulation as hazardous substances. However, IVD products are specifically required to include in their labeling the warnings or precautions for users established by the Consumer Product Safety Commission (CPSC) in 16 CFR 1500. These requirements are discussed later in this chapter.

GENERAL IVD LABELING PROVISIONS

Subpart B of 21 CFR 809 details a set of general labeling requirements that apply to all IVD products. The section number (e.g., §809.10(a)) of the corresponding section in 21 CFR is listed with each topic for reference.

LABELING ON THE IMMEDIATE CONTAINER (§809.10(A))

The label on IVD products must contain the following information on the immediate container, unless the specific information is not applicable for a given device. All information must also appear on the outside container or wrapper, or be easily legible through the outside container or wrapper.

- The label must contain the name and place of business of the manufacturer, packer, or distributor. The address must include the street address, city, state, and zip code. The street address may be omitted if the firm's street address is listed in the current city directory or telephone directory. The requirement for name and place of business is discussed in detail in Chapter 21.
- The label must contain the proprietary name of the product (e.g., A^2 Medical hCG Standards).
- The label must contain the established name (common or usual name) of the product (e.g., human chorionic gonadotropin [hCG] standards kit).
- For reagents, the label must include an expiration date beyond which the product is not to be used. The expiration date is based on the storage conditions stated in the labeling.
- A lot or control number, identified as such, that is traceable to the complete manufacturing history of the product is required. Instrument lot numbers must permit tracing of all functional subassemblies of the instrument. For multiple unit products:
  - the lot or control number must permit the tracing of the identity of the individual units, and
  - if the product requires the units to be used together as a system, all units should bear the same lot or control number, if appropriate, or some other suitable uniform identification should be used.

If the presence of any of the label information listed above can be shown to interfere with the test, then the information can appear on the outside container or wrapper instead of the immediate container.

If possible, the information listed below should also appear on the label of the immediate container. However, if the immediate container is too small or is otherwise unable to accommodate a label with adequate space to contain the information, the following may appear on the outer container or wrapper only:

- The intended use or uses of the product (e.g., pregnancy detection).
- On the outer container or wrapper, any warnings or precautions for the user as established in the regulations contained in 16 CFR 1500 (hazardous substances). The requirements for labeling of hazardous substances are discussed later in this chapter.
- Any other warnings that are appropriate to any hazard presented by the product (e.g., "Do not pipette by mouth").
- The statement "For *In Vitro* Diagnostic Use" and any other limiting statement appropriate for the intended use of the product (e.g., "For use with A^2 Medical RIA Test Kit").

- For reagents, a list of the reactive ingredients, including:
  - the established name (common or usual name) of the ingredient, if any;
  - the quantity, proportion, or concentration of the ingredient; and
  - if the ingredient is derived from biological material, the source and measure of its activity (e.g., bovine serum albumin).
- The quantity, proportion, concentration, or activity is stated in the system generally used and recognized by the intended user (e.g., mg, mg/dl).
- For reagents, adequate instructions for proper storage to protect the stability of the product. Storage conditions must include, as appropriate, temperature, humidity, and light levels, as well as other pertinent factors.
- For reagents that must be manipulated by the user, such as reconstitution and/or mixing before use, the labeling must also provide storage instructions for the reconstituted or mixed product when stored in the original container.
- The basis for the storage instructions must be determined by reliable, meaningful, and specific test methods such as those described in 21 CFR 211.166. This section describes good practice for conducting stability testing for drug products as part of the current good manufacturing practices (GMPs) for finished pharmaceuticals. These practices are discussed briefly later in this chapter.
- For a reagent, a means by which the user may be assured that the product meets the appropriate standards of identity, strength, quality, and purity at the time of use must be presented. This information must be provided for the product as supplied and any resultant reconstituted or mixed product. In addition to the expiration date listed above, the label should include one or more of the following:
  - A statement of an observable indication of the adulteration of the product (e.g., turbidity, color change, precipitation)
  - Instructions for a simple method by which the user can reasonably determine that the product is useful for its intended purpose
- For a reagent, a declaration of the net quantity of contents, expressed in terms of weight or volume, numerical count, or any combination of these or other terms that accurately reflect the contents of the package must be provided. The use of metric designations is encouraged, wherever appropriate. If more than a single test can be performed using the product, any statement about the number of tests must be consistent with the instructions for use and the amount of material provided.

LABELING REQUIREMENTS FOR PACKAGE INSERTS (§809.10(B))

The labeling accompanying each product must contain the information listed below except where the information is not applicable, or as specified in a performance standard for a particular product class established under the authority of 21 CFR 861. The information must appear in one place in the labeling and be in the format and order listed below.

1. Begin with the proprietary and established name (common or usual name), if any, of the product.
2. Provide a concise statement of the intended use or uses of the product and whether it is a qualitative or quantitative type of procedure. The intended use should contain a statement indicating whether the product is for use in a clinical laboratory, doctor's office, or user's home, and whether it is for screening, monitoring, confirmating, or excluding, and/or aiding

in the diagnosis as an adjunct to other procedures. This section should include the clinical significance, if it can be stated in a few words. If the clinical significance statement is lengthy or complicated, it should appear under a separate heading entitled "Clinical Significance" (Glycohemoglobin §4). Examples of typical intended-use statements taken from FDA guidance documents include:

"THE A² MEDICAL CRP TEST SYSTEM IS AN *IN VITRO* DIAGNOSTIC DEVICE FOR THE QUANTITATION OF C-REACTIVE PROTEIN IN HUMAN SERUM BY RATE NEPHELOMETRY. THE MEASUREMENT OF CRP AIDS IN EVALUATION OF THE AMOUNT OF INJURY TO BODY TISSUES." (C-REACTIVE §4(A))

"A² MEDICAL'S hCG TEST IS A LABORATORY TEST INTENDED FOR THE QUANTITATIVE DETERMINATION OF PERCENT GLYCATED HEMOGLOBIN IN WHOLE BLOOD BY [METHODOLOGY] USING THE A² MEDICAL AUTOMATED SYSTEM TO MONITOR LONG TERM BLOOD GLUCOSE CONTROL IN INDIVIDUALS WITH DIABETES MELLITUS." (GLYCOHEMOGLOBIN § 4)

"THE A² MEDICAL CT TEST IS FOR THE DETECTION OF [C. TRACHOMATIS/CHLAMYDIAE] USING THE A² MEDICAL AUTOMATED SYSTEM FOR MALE URETHRAL AND URINE, FEMALE ENDOCERVICAL, AND INFANT CONJUNCTIVAL SWAB SPECIMENS." (CHLAMYDIAE P. 13)

Under conditions for use, describe any special applications of the device or specific contraindications or indications not addressed in the intended use statement. For example, such a statement for the A² Medical CT test might read: "for nasopharyngeal swabs only from infants as an aid in the diagnosis of chlamydial infant pneumonia" (Chlamydiae p. 13). These conditions for use may be addressed further in the sections on summary and explanation of the test, limitations of the procedure, or performance characteristics.

The intended use statement in the package insert must match exactly the "Indications for Use" statement in the summary of safety and effectiveness of any required premarket approval application (PMA).

3. Provide a summary and explanation of the test, including a short history of the methodology. The summary must include pertinent references and a balanced statement of the special merits and limitations of the test. If the product labeling refers to any other procedure, the appropriate literature citations must be included. The labeling must explain the nature of any differences from the original and their effect on the results.

4. The chemical, physical, physiological, or biological principles of the test procedure are to be explained. The explanation should be concise, stating the chemical reactions and techniques involved, if applicable.

5. For reagents, the labeling accompanying the product must include:
   • The established name (common or usual name) of each reactive ingredient, if any; the quantity, proportion, or concentration of the ingredient; and, if the ingredient derived from biological material, the source and measure of its activity.
   • A statement indicating the presence and character of any catalytic or nonreactive ingredients (e.g., buffers, preservatives, or stabilizers). The quantity, proportion, concentration, or activity is to be stated in units generally used and recognized by the intended user (e.g., mg, mg/dl).
   • Any warnings or precautions for the user as established in the regulations contained in 16 CFR 1500 (hazardous substances), and any other warnings appropriate to any hazard presented by the product.
   • The statement "For *In Vitro* Diagnostic Use" and any other limiting statement appropriate for the intended use of the product.

- Adequate instructions for reconstitution, mixing, dilution, and so on.
- Adequate instructions for proper storage to protect the stability of the product, including, as appropriate, temperature, humidity, and light conditions, as well as other pertinent factors. For reagents that must be manipulated by the users, such as reconstitution and/or mixing before use, the labeling must also provide storage instruction for the reconstituted or mixed product when stored in the original container. The basis for the storage instructions must be determined by reliable, meaningful, and specific test methods such as those described in 21 CFR 211.166.
- A statement of any purification or treatment required for use of the reagent.

6. For instruments, the labeling accompanying the product must include:
   - an explanation of the use or function of the instrument;
   - installation procedures and any special requirements;
   - a description of the principles of operation;
   - the performance characteristics and specifications of the instrument;
   - the operating instructions;
   - procedures for calibrating the instrument, including any material and/or equipment to be used;
   - any operational precautions and limitations;
   - any hazards as a result of the use of the instrument; and
   - service and maintenance information.

7. With regard to collection and preparation of specimens for analysis, the labeling should describe the type of specimen to be collected (e.g., whole blood, plasma, tumor tissue) and the amount of specimen required, both optimum and minimum. The labeling accompanying each product must include:
   - special precautions regarding collection of specimens, including any special preparation of the patient that might influence the validity of the tests (e.g., "The puncture site should be cleaned and thoroughly dried before obtaining the blood sample.");
   - any additives, preservatives, and so on that are necessary to maintain the integrity of the specimen;
   - any known substances that will interfere with or diminish the accuracy of the test; and/or
   - recommended storage, handling, and/or shipping instructions necessary for the protection and maintenance of the stability of the specimen.

8. The package insert should contain a step-by-step outline of the recommended procedures from receipt of the specimen to obtaining results. This outline should include;
   - a list of all materials provided (e.g., reagents, instruments, and equipment) with instructions for their use;
   - a list of all necessary materials that are not provided, including sufficient details (e.g., sizes, numbers, types, and quality) to enable the user to identify the needed materials;
   - a description of the amounts of reagents necessary, the times required for specific steps, proper temperature, and so on;
   - a statement describing the stability of the final reaction and the time limit for measuring the result to assure accuracy of the measurement;
   - a description of the required calibration, identifying and listing any necessary preparation of reference materials, samples, and blanks;
   - a description of the calibration range, including the highest and lowest values measured; and
   - the details of the necessary quality control materials and procedures to assure satisfactory performance. If both positive and negative controls are needed, this should be stated along with what are considered to be the satisfactory limits of performance.

- In addition to the previous information, the recommended procedures should include any points that may be useful in improving the precision and accuracy of the test.

9. The labeling must include an explanation of the procedure for calculating the value of the unknown. Each component of the formula used to calculate the unknown must be explained. A sample calculation with step-by-step instructions should be included. The results should be presented with the appropriate number of significant figures. If the result is qualitative rather than quantitative, a description adequate to enable the user to recognize the expected result must be provided (e.g., by color change, from blue to red).

10. State any limitations of the procedure, such as known extrinsic factors or interfering substances that might affect the results. If certain results indicate that further testing, either more specific or more sensitive, is needed, this need must be stated in the labeling. The accompanying labeling must state the range(s) of expected values as obtained with the product from studies of various population(s). It must indicate how the range was established and identify the population(s) on which it was established. Any variance in test results as a function of geographical location, age or sex of population studied, season of the year, type of test employed, specimen collection and handling procedures, and so on should be described.

    The labeling should instruct the user on interpreting positive, negative, and equivocal, indeterminate, or borderline results, and describe the clinical significance of each result.

11. Specific performance characteristics must be described, including, as appropriate, information on such things as accuracy, precision, specificity, and sensitivity. These performance characteristics must be based on generally accepted methods using biological specimens from normal and abnormal populations. A summary of the data upon which the specific performance characteristics (e.g., accuracy, precision, specificity, and sensitivity) are based is to be included in the labeling. If positive and negative predictive values are furnished, this calculation of predictive values requires knowledge about the prevalence of the disease in the population sampled (Estrogen §5.8 (a)).

12. A bibliography listing pertinent references keyed to the text in the labeling

13. The name and place of business of the manufacturer, packer, or distributor must be included. The address must include the street address, city, state, and zip code. The street address may be omitted if the firm's street address is listed in the current city directory or telephone directory. If the "person" listed is not the manufacturer of the IVD product, the name must be qualified with a phrase that indicates the connection, such as "Manufacturer for ..." or "Distributed by ...."

14. Finally, the labeling must bear the date of last revision.

SPECIAL CASES

The regulations governing the information required on the labeling accompanying each product recognize two special cases. The first is the multipurpose instrument. The second is the reagent supplied as a replacement in a diagnostic system. The labeling accompanying a multipurpose instrument used for diagnostic purposes, but not committed to specific diagnostic procedure or system, is required to contain only the following:

- The proprietary and established name of the instrument
- The intended use or uses
- The specific instrument requirements listed in the previous section

- The name and address of the manufacturer, packer, or distributor
- The date of the last revision of the labeling

The labeling for a reagent intended for use as a replacement in a diagnostic system may include only those items necessary to adequately identify the reagent and to describe its proper use in the system.

EXEMPTIONS FROM IVD PRODUCT LABELING REQUIREMENTS (§809.10(C))

The shipment or other delivery of an IVD product is exempt from the requirements in 21 CFR 809.10(a) and 809.10(b) and from any performance standards promulgated under 21 CFR 861 if one of the following conditions is met:

- If the shipment or delivery of an IVD product is for an investigation being conducted under an approved investigational device exemption (IDE), but only if all of the requirements of the IDE have been met.
- If the shipment or delivery of an IVD product is for an investigation that is not subject to an IDE, and if the following conditions are met:
  - For a product in the laboratory research phase of development that is not represented to be an effective IVD product and all labeling prominently bears the statement:

    FOR RESEARCH USE ONLY. NOT FOR USE IN DIAGNOSTIC PROCEDURES.

  - For a product being shipped or delivered for product testing prior to full commercial market release and all labeling prominently bears the statement:

    FOR INVESTIGATIONAL USE ONLY. THE PERFORMANCE CHARACTERISTICS OF THIS PRODUCT HAVE NOT BEEN ESTABLISHED.

  - An example of this type of testing would be use of the product on specimens acquired from humans in order to compare the usefulness of the product with other products or procedures that are currently in use or are already recognized as useful.
- A description of the conditions under which a diagnostic device is exempt from the requirements for an IDE is located in Chapter 24.

LABELING FOR GENERAL-PURPOSE LABORATORY REAGENTS AND EQUIPMENT (§809.10(D))

General-purpose laboratory reagents and equipment include routine laboratory reagents and equipment whose uses are generally known to persons trained in their use. These items need not include the directions for use required by 21 CFR 809.10(a) and 21 CFR 809.10(b) (labeling on the immediate container and labeling requirements for package insert, respectively) if their labeling meets the following requirements:

- For reagents, the label on the immediate container and the outer container, if one exists, must include the following information:
  - The proprietary name of the reagent
  - The established name (common or usual name), if any, of the reagent (e.g., hydrochloric acid)
  - The name and place of business of the manufacturer, packer, or distributor

- A lot or control number, identified as such, that is traceable to the manufacturing history of the product
- The following information should also be included on the immediate container label. However, if the immediate container is too small or otherwise unable to accommodate a label with sufficient space to contain the required information, the information may appear only on the outer container label. This exemption is allowed only if the reagent is packaged within an outer container from which it is removed for use.
    - The established name (common or usual name), if any, of the reagent ingredients, including the quantity, proportion, or concentration of each ingredient (e.g., hydrochloric acid: formula weight 36.46, assay 37.9 percent, specific gravity 1.192 at 60 °F).
    - If the reagent is derived from biological material, the source and, if applicable, a measure of the activity.
    - The quantity, proportion, concentration, or activity must be stated in units that are recognized by the intended user.
    - A statement of the purity and quality of the reagent, including a quantitative declaration of any impurities present. This requirement can be satisfied by reference to an accepted and generally available standard that contains the same information. Standards established by the American Chemical Society, or the National Research Council, or set forth in the U.S. Pharmacopeia or National Formulary.
    - A statement of warnings or precautions for users as established in 16 CFR 1500 (hazardous substance) and any other warnings as appropriate to the hazards presented by the product.
    - The statement "For Laboratory Use."
    - Appropriate storage instructions that are adequate to protect the stability of the product. When applicable, the storage instructions should specify temperature, humidity, and light conditions, as well as other pertinent factors. This information should be based on meaningful, reliable test methods such as those described in 21 CFR 211.166.
    - A declaration of the net quantity of contents in terms of weight or volume, numerical count, or any combination of these or other terms that accurately reflect the contents of the package. Use of metric designations is encouraged wherever appropriate.
- The label for general-purpose laboratory equipment (e.g., a beaker or pipette) need only include a description of the product, its composition, and physical characteristics, if these are necessary for its proper use.

PRODUCT CLASS MODELS FOR IVD PRODUCTS

The FDA has developed, as guidance for preparing product submissions, product-class models that apply to specific categories of IVD products. As previously stated, the three product categories are clinical chemistry/toxicology, clinical microbiology/immunology, and clinical hematology/pathology (Bracey §1-1). Each of these models lists the information that is appropriate to include in the labeling for the product class. The product labeling must satisfy the general requirements in 21 CFR 809.10.

CLINICAL CHEMISTRY/TOXICOLOGY

These are systems intended to detect or measure the level or concentration of various chemicals or toxic substances in specimens of human origin. Normally, the labeling for these products includes the package insert, vial label, and kit-box label. The labeling should include the following particular information (Bracey §13):

- Name and intended use of test
- Summary and explanation of test
- Chemical and biological principles of test procedure
- Reagents
- Specimen collection, preparation, analysis, and storage
- Assay procedure
- Equipment and materials required
- Test results interpretation
- Limitations of test procedure
- Specific performance characteristics
- Selected bibliography

CLINICAL MICROBIOLOGY/IMMUNOLOGY

These are systems intended to identify and deal with microorganisms or to measure specific substances in serum or other body fluids by immunochemical techniques. The labeling should include the following particular information (Bracey §13):

- Intended use (include clinical application or purpose of test)
- Principle of test
- Material required but not supplied
- Specimen collection and preparation
- Sample record sheet
- Data record sheet
- Precautions
- Interpretation of results
- Limitations of procedure
- Quality control
- Performance characteristics, e.g., precision, sensitivity (detection level)
- Expected results
- References

CLINICAL HEMATOLOGY/PATHOLOGY

These are systems intended to process blood or collect, process, and store human cells and tissue. Normally, the labeling for these products includes the outside container label, vial label, and package insert. The labeling should include the following particular information (Bracey §13)

- Name and intended use of test
- Summary and explanation of test
- Hematological or pathological principles of test procedure (e.g., morphology, dye adsorption)
- Reagent characterization, including the source (e.g., rabbit brain)
- Limitations of procedure
- Specimen collection, preparation, analysis, and storage
- Performance specifications
- Selected bibliography

HOME-USE IVD PRODUCTS

Traditionally, healthcare professionals in hospitals, clinical laboratories, and physician offices have primarily used IVD products. In recent years, there has been a growing interest in IVD products intended for home use. Home-use IVD products must meet the criteria for "adequate directions for use" as provided in Section 502(f)(1) of the FD&CA. Like other IVD products, home-use IVD products are deemed to be in compliance if they meet the requirements of 21 CFR 809.10. Labeling for home-use IVD products that are sold directly to the consumer must also comply with the requirements of 21 CFR 801, Subpart C—Labeling requirements for over-the-counter devices. These requirements are discussed in Chapter 21.

To satisfy the "adequate directions for use" requirement, the labeling for a home-use IVD product must be simple, concise, and easy to understand. The labeling should make liberal use of illustrations and drawings, use bold print or other methods to highlight warnings and precautions, and provide color coding of reagent containers where practicable (Home-Use IVDs §3(C)).

The FDA has taken the view that the regulations provided in 21 CFR 809.10 are generally adequate for home-use IVD products even though the regulations were targeted at professional-use IVD products. However, some qualification of the requirements may be useful in preparing labeling for home-use IVD products (Home-Use IVDs §3(C)).

The label on the IVD product should address the following:

- The intended use statement (§809.10(a)(2)) that is visible on the outer package should clearly indicate the type of diagnostic procedure offered by the IVD product. Procedures include screening, monitoring, or diagnostic tests. The label should also clearly describe the specific conditions, disorder, or risk factor for which the test is to be used. The intended use description should be accompanied by a clear indication of who should use the test and the conditions for its use. Any necessary contraindication statements should follow the intended use description.
- A listing of the quantity, proportion, concentration, or activity of each reactive ingredient (§809.10(a)(3)) may be eliminated unless the information is necessary for proper use of the test by a layperson.

The labeling accompanying the product should take account of the following:

- The intended use statement (§809.10(b)(2)) as described above should be reiterated in the package insert.
- The summary and explanation of the test (§809.10(b)(3)) can be combined with the description of the principle of the procedure (§809.10(b)(4)) to provide a simple, understandable explanation of how the test works. The combined section should discuss both the medical benefits and the limitations of the test with regard to its intended use. Presentation of the chemical reactions or formulas is not necessary but may be referenced in the bibliography.
- A separate section in the labeling should describe how to interpret the test results and include any limitations of the procedure. This section should clearly indicate the significance of the test result in light of the test's intended use. Specific information should be provided to let the user know what follow-up action is appropriate based on the results. The meaning of false-positive and false-negative results should be explained, including the possible sources of false results and their implications.

- The labeling should provide a section that describes any foods, medications, or other substances that would interfere with the test. This should include a list of what substances should be avoided and for how long before the test is performed.
- The information required to describe the reagents (§809.10(b)(5)), instruments (§809.10(b)(6)), specimen collection procedures (§809.10(b)(7)), and the step-by-step outline of recommended procedures (§809.10(b)(8)) may be combined into a single test-procedures section. While combined into a single section, this part of the labeling should include all of the information required in these sections of the CFR except for the quantity, proportion, concentration, or activity of each reactive ingredient included in the test kit. This information need not be provided unless it is necessary for the proper use of the test by a layperson. The description of the specimen collection, preparation, and analysis steps should be bolstered by the use of pictures and illustrations, preferably in color.
- The section on interpreting the test results (§809.10(b)(9)) should include information on troubleshooting the test.
- The information required in the section on expected values (§809.10(b)(11)) and specific performance characteristics (§809.10(b)(12)) may be combined into a single information section (i.e., test performance characteristics). This section should present a summary of data from both laboratory and consumer field evaluation of the test. The data from the two evaluations should be presented separately and should include a concise discussion of the study protocols used to evaluate the test's performance. Also, this section should include a discussion and/or presentation of data relating to the test's accuracy. This discussion should include information about the level of false-positive and false-negative results, and should be presented for both the laboratory and consumer field evaluations.
- The remaining requirements for a bibliography (§809.10(b)(13)), name and place of business of the manufacturer, packer, or distributor (§809.10(b)(14)), and label revision date (§809.10(b)(15)) are addressed in the same way as for professional-use IVDs.

The FDA recommends that the manufacturer provide a toll-free telephone number and/or an address so that the user can obtain answers to any questions concerning use of the device.

The FDA is also concerned that consumers be aware of significant changes in the labeling for home-use IVDs, particularly labeling that accompanies frequently used devices (e.g., home-use glucose test devices). Manufacturers are advised to alert consumers by including special inserts highlighting significant labeling changes.

As part of a premarket evaluation of home-use IVD products, FDA recommends that a manufacturer conduct a consumer field evaluation to determine the device's performance when used by lay users, unassisted, following instructions provided in the labeling (Home-Use IVDs §3(D)). To assess the ability of the lay user to read and understand the labeling, the FDA recommends that the manufacturer provide a simple questionnaire as part of the test. The questionnaire could be used to determine whether the user understands the purpose of the test, the conditions of use, the limitations of the test, the meaning of the results, and the appropriate follow-up. The FDA would like to see the questionnaire and a tabulation of responses included in the premarket submission.

HAZARDOUS SUBSTANCES

A hazardous substance is defined by the Federal Hazardous Substances Act (FHSA) as any substance or mixture of substances that is toxic, corrosive, an irritant, a strong sensitizer, flammable or combustible, or that generates pressure through decomposition, heat, or other means (FHSA

§2(f)(1)(A)). The use of such substances is controlled under regulations promulgated by the CPSC if such substance or mixture of substances may cause substantial personal injury or illness during or as a proximate result of any customary or reasonably foreseeable handling or use, including reasonably foreseeable ingestion by children. CPSC regulations governing the use of hazardous substances are found in 16 CFR 1500.

Normally, substances subject to the FD&CA are exempted by Section 2(f)(2) of the FHSA from regulation by the CPSC. However, where a food, drug, or cosmetic offers a substantial risk of injury or illness from any handling or use that is customary or usual, it may be regarded as misbranded under the FD&CA if its label fails to bear information to alert the householder to the consequences that may result from use of the article (16 CFR, §1500.81).

However, IVD products are specifically required by FDA regulations to include in their labeling any warnings or precautions established in 16 CFR 1500 for users. This includes labeling on the immediate container (§809.10(a)(4)), in the labeling accompanying each product (§809.10(b)(5)(ii)), and on the labeling for general-purpose laboratory reagents (§809.10(d)(1)(iv)).

LABELING REQUIREMENTS

In 16 CFR 1500.3(14)(i), the regulations state that a product containing a hazardous substance is misbranded if its labeling does not include the following:

- The name and address of the manufacturer, packer, or distributor.
- The common or usual name, if one exists, or the chemical name of the hazardous substance or of each component that contributes substantially to its hazard.
- The word "DANGER" if the substance is extremely flammable, corrosive, or highly toxic. These terms are all defined in the CPSC regulations.
- The word "WARNING" or "CAUTION" for all other hazardous substances.
- An affirmative statement of the principal hazard (e.g., "Flammable," "Combustible," "Vapor harmful," "Causes burns," "Absorbed through skin").
- Precautionary measures describing the actions to be followed or avoided (e.g., "Do not get in eyes, on skin, or on clothing," "Use waterproof gloves and face shield or goggles when handling concentrate").
- Instructions, when necessary or appropriate, for first-aid treatment (e.g., "In case of eye contact, flush immediately with plenty of fresh water and get medical attention").
- The word "POISON" and the skull-and-crossbones symbol (Table 17.3, Symbol 4) for any hazardous substances that are defined as highly toxic. (The procedure for determining if a substance is highly toxic is given in 16 CFR § 1500.3(b)(6).)
- Instructions for handling and storage of packages requiring special care.
- The statement "Keep out of the reach of children" or its practical equivalent unless the product is intended for use by children. If the product is intended for use by children, then the label must include adequate directions for the protection of children from the hazards posed by the product.

PROMINENCE, PLACEMENT, AND CONSPICUOUSNESS OF LABELING

The labeling of products containing hazardous substances must bear certain cautionary statements on their labels. These statements are described in the previous section. Section 2(p)(2) of the FHSA specifies that all such statements must appear prominently on the product label in a conspicuous and legible type that contrasts by typography, layout, or color with other printed matter on the label.

In 16 CFR §1500.121, the Commissioner of the CPSC has established regulatory requirements to satisfy section 2(p)(2) of the FHSA. Labels that do not comply with this regulation may be considered misbranded under the FHSA.

Prominent Label Placement

For purposes of this regulation, a display panel is any surface of the immediate container, and of any outer container or wrapping that bears labeling. The principal display panel is "the portion(s) of the surface of the immediate container, and of any outer container or wrapping, which bear(s) the labeling designed to be most prominently displayed, shown, presented, or examined under conditions of retail sale" (16 CFR 1500.121(a)(2)(iv)). Depending on its design, a package may have more than one principal display panel. If so, each principal display panel must contain all the information the regulation requires to appear on the principal display panel. If the principal display panel is on a lid, cap, or other item that may be separated from the immediate container and discarded, the container is deemed to have a second principal display panel elsewhere on the immediate container.

All items of cautionary labeling required by the FHSA may appear on the principal display panel. However, the signal word, the statement of principal hazard(s), and, when appropriate, instructions to read carefully any cautionary material that is placed elsewhere on the label must appear on the principal display panel. This information is to be blocked together within a square or rectangular area, with or without a border, on the principal display panel of the immediate container. All cautionary statements placed on the principal display panel must be separated on all sides from all other printed or graphic matter by a borderline or by a space no smaller than the minimum allowable height of the type size for the cautionary material required by the FHSA (exclusive of signal words and statements of hazard) to appear on the principal display panel. The only exception to this requirement is the declaration of net contents required under the Fair Packaging and Labeling Act.

All cautionary material required by the FHSA to appear on the principal display panel must appear in lines that are generally parallel to any base on which the package rests as it is designed to be displayed for sale. This requirement does not apply to labeling on collapsible tubes, cylindrical containers with a narrow diameter, or containers where both the "front'" and "back" of the container are principal display panels (16 CFR 1500.121(b)(1)).

Any cautionary labeling required by the FHSA that does not appear on the principal display panel must be placed together on a display panel elsewhere on the container. The name and place of business of the manufacturer, packer, distributor, or seller may appear on any display panel. When required cautionary material appears on a display panel other than the principal display panel, the principal display panel must bear the statement, "Read carefully other cautions on the _____ panel," or its practical equivalent. The blank space contains a description of the location of the other panel (16 CFR 1500.121(b)(3)).

If a hazardous substance is "highly toxic," the label must bear the word "POISON" in addition to the signal word "DANGER," and must also bear the skull-and-crossbones symbol. Under the hazardous substance regulation (16 CFR 1500.14(b)), the label of some products may, in addition to any required signal word, be required to bear the word "POISON" and the skull-and-crossbones symbol because of the special hazard associated with their ingredients. In both instances, the word "POISON" and the skull-and-crossbones symbol need not appear on the principal display panel on the container unless all other cautionary labeling required by the FHSA appears on the principal display panel. The word "POISON" and the skull-and-crossbones symbol, when required, must appear either together with other cautionary labeling on a display panel other than the principal display panel or together with the signal word and statement(s) of principal hazard on the principal display panel (16 CFR 1500.121(b)(5)(i)).

Under certain conditions set forth in the FHSA, the label of a hazardous substance will require the word "POISON" instead of a signal word. The word "POISON" must appear in capital letters on the principal display panel, together with the statement(s) of the principal hazard. Certain substances for which the word "POISON" is required instead of any signal word are listed in 16 CFR 1500.129. This list includes the following in the concentrations listed (16 CFR 1500.129):

- Hydrochloric acid (HCl) and any preparation containing free or chemically unneutralized HCl in a concentration of 10 percent or more
- Sulfuric acid (H_2SO_4) and any preparation containing free or chemically unneutralized H_2SO_4 in a concentration of 10 percent or more
- Nitric acid (HNO_3) or any preparation containing free or chemically unneutralized HNO_3 in a concentration of 5 percent or more
- Carbolic acid (C_6H_5OH), also known as phenol, and any preparation containing C_6H_5OH in a concentration of 5 percent or more
- Oxalic acid ($H_2C_2O_4$) and any preparation containing free or chemically unneutralized $H_2C_2O_4$ in a concentration of 10 percent or more
- Any salt of oxalic acid and any preparation containing any such salt in a concentration of 10 percent or more
- Acetic acid ($HC_2H_2O_2$) or any preparation containing free or chemically unneutralized $HC_2H_2O_2$ in a concentration of 20 percent or more
- Hypochlorous acid, either free or combined, and any preparation containing the same in a concentration that will yield 10 percent or more by weight of available chlorine
- Potassium hydroxide (KOH) and any preparation containing free or chemically unneutralized KOH, including caustic potash and vienna paste (vienna caustic), in a concentration of 10 percent or more
- Sodium hydroxide (NaOH) and any preparation containing free or chemically unneutralized NaOH, including caustic soda and lye in a concentration of 10 percent or more
- Silver nitrate ($AgNO_3$), sometimes known as lunar caustic, and any preparation containing $AgNO_3$ in a concentration of 5 percent or more
- Ammonia water (NH_3) and any preparation containing free or chemically uncombined NH_3, including ammonium hydroxide and "hartshorn," in a concentration of 5 percent or more

AREA OF PRINCIPAL DISPLAY PANEL

The area of the principal display panel is the area of the side or surface of the immediate container, or of the side or surface of any outer container or wrapping, that bears the labeling designed to be most prominently displayed when the package is presented for retail sale. This area is not limited to the portion of the surface covered with labeling. It includes the entire surface excluding flanges at the tops and bottoms of cans, conical shoulders of cans, handles, and shoulders and necks of bottles and jars. Collapsible metal tubes containing hazardous substances must be labeled so that all cautionary labeling appears as close to the dispensing end of the container as possible.

To determine the proper type size for cautionary labeling, the area of the principal display panel (or other panel bearing cautionary labeling) is computed as follows (16 CFR 1500.121(c)(1)):

- For a rectangular package where one entire side is the principal display panel, the area of the principal display panel is the product of the height times the width of that side.
- For a cylindrical or nearly cylindrical container or tube on which the principal display panel appears on the side, the area of the principal display panel is 40 percent of the product of the height of the container times its circumference.
- For any other shape of container, the area of the principal display panel is 40 percent of the total surface of the container, excluding flanges at the tops and bottoms of cans, conical shoulders of cans, handles, and so on. However, if such a container has an obvious principal display panel (such as an oval or hourglass-shaped area on the side of a container), the area to be used to determine the proper type size is the entire area of the obvious principal display panel.

TYPE-SIZE REQUIREMENTS

The cautionary labeling statements must appear in a conspicuous and legible type that will contrast by typography, layout, or color with the other printed matter on the label. The size of cautionary labeling should be reasonably related to the type size of any other printing appearing on the same panel. However, the type size of cautionary labeling must meet the minimum size requirements in Table 22.1 unless there are specific requirements or exemptions in the regulations. All upper-case, or capital, letters must be at least equal in height to the required type size, and all other letters must be the same style as the upper–case, or capital, letters. The height of a capital or upper case letter should be no more that three times its width.

The signal word, or the word "POISON" when required in place of a signal word, and the statement of the principal hazard must be in all capital letters.

Where color is the primary method used to achieve the required contrast, the color of any cautionary labeling statement must sharply contrast with the background color. Combinations that may not give the appropriate contrast are black letters on a dark blue or dark green background, dark red letters on a light red background, light red letters on a reflective silver background, and white letters on a light gray or tan background (16 CFR 1500.121(d)(1)).

The presence of vignettes or the proximity of other labeling or lettering that renders any cautionary labeling statement obscured or inconspicuous is prohibited.

If all of the required cautionary labeling does not appear on the principal display panel, the statement to "Read carefully other cautions on the _____ panel," or its practical equivalent, must appear in at least the same type size as that required for the other cautionary material that appears

TABLE 22.1
Minimum Type Size Requirements

Area of principal display panel (a) in square inches	Minimum height of printed image of capital or upper case letters (in inches).					
	$0 < a \leq 2$	$2 < a \leq 5$	$5 < a \leq 10$	$10 < a \leq 15$	$15 < a \leq 30$	$a > 30$
Signal word[a]	3/64	1/16	3/32	7/64	1/8	5/32
Statement of hazard	3/64	3/64	1/16	3/32	3/32	7/64
Other cautionary material[b]	1/32	3/64	1/16	1/16	5/64	3/32

[a] Including the word "POISON" when required instead of a signal word by Section 3(b) of the FHSA (16 CFR 1500.129).

[b] Size of lettering for other cautionary material is based on the area of the display panel on which such cautionary material appears.

Source: 16 CFR 1500.121(v)(2)(ii), Table 1.

elsewhere on the label of a hazardous substance. The size of the cautionary labeling that appears on other display panels is determined by the area of the panel on which it appears.

If there is more than one principal hazard associated with the product, then each hazard must appear in the same size and style of type. Each must appear with the same degree of boldness or appear in the same color.

ACCOMPANYING DOCUMENTS

When the package of a hazardous substance is accompanied by literature containing the directions for use, the FHSA requires the literature to bear cautionary labeling. The cautionary labeling must appear in reasonable proximity to any direction for use. The type size used for the cautionary labeling must be reasonably related to the type size of any other printed matter in the accompanying documents. As with the labels, the cautionary labeling must be conspicuous and legible. The signal word and statement of principal hazard or hazards must appear in capital letters.

OUTER CONTAINER OR WRAPPINGS

All cautionary labeling that appears on the immediate container of a hazardous substance must also appear on any outer container or wrapping used in the retail display of the substance. The information must appear in the same manner as it does on the immediate container. If the cautionary labeling statements on the immediate container are clearly legible through any outer container or wrapper used in retail display, then they need not be repeated on the outer container or wrapping itself (16 CFR 1500.121(b)(4)).

CAUSTIC POISONS

IVD products may also fall under the jurisdiction of the Federal Caustic Poison Act (FCPA).* The provisions of the FCPA apply to any container suitable for household use and employed exclusively to hold any dangerous caustic or corrosive substance defined in Section 2(c) of the FCPA. For example, a product containing ammonia (NH_3) in sufficient concentration would be subject to the FCPA.** FDA regulations governing labeling of products covered by the FCPA are found in 21 CFR 1230.

Such a container must have a firmly attached label or sticker that will remain attached while the container is being used. The label or sticker must be placed on the container so that it is readily visible to the user. The label or sticker is required to bear information as follows:

- The common name of the dangerous, caustic, or corrosive substances as defined in Section 2(a) of the FCPA or any other name commonly employed to designate and identify such a substance must appear on the label. The common name must appear even if a trade name appears on the label.
- If the name on the label is other than that of the manufacturer, it must be qualified by such words as "distributed by," "packed for," "packed by," or "sold by," or similar wording that reveals the relationship between the firm named on the label and the manufacturer.
- The signal word "POISON" must appear in uncondensed sans serif (Gothic) capital letters. The proportions of the lettering of the word "poison" based on 24-point (typeface) size are shown in 21 CFR 1230.13.

---

* *Federal Caustic Poison Act.* US Code, vol. 15, §§402 *et seq.* (1927)
** See 21 CFR 2.110 – Definition of ammonia under *Federal Caustic Poison Act.*

- The word "POISON" must be followed by directions for treatment in case of ingestion of the substance. In addition, if the substance can cause external injury such as a chemical burn, the label or sticker must give directions for the appropriate treatment. The treatment(s) must be sanctioned by a competent medical authority, and the required treatment materials should, whenever practical, be ones that are usually available in the household.
- The manufacturer or wholesaler is exempted from the requirement to label the container with the directions for treatment, as long as the other information required by the FCPA is provided, the regulations in 21 CFR 1230 are met, and the container is for other than household use. A person who receives such a container from a manufacturer or wholesaler must add the directions for treatment of injury before he or she offers the container for retail sale.

All required labeling must be conspicuous in contrast to other printed material. The labeling must be in English unless the product is distributed solely in Puerto Rico, or a US territory where English is not the predominant language. In the case of products distributed solely in Puerto Rico, 21 CFR 801.15 permits the required labeling to be in the predominant language, Spanish.

STORAGE INSTRUCTIONS AND EXPIRATION DATES

The labeling for IVD reagents must include adequate instructions for proper storage to protect the stability of the product, and an expiration date beyond which the product is not to be used. The basis for the storage instructions and expiration dating must be determined by reliable, meaningful, and specific test methods. In the regulations, the FDA recommends that the IVD manufacturer use the stability-testing program described in the cGMPs for finished pharmaceuticals (21 CFR 211.166).

This section of the regulation specifies that there be a written testing program designed to assess the stability characteristics of drug products and that the program be followed. The program includes:

- a statistically sound sample size and test interval for each attribute examined to assure valid estimates of stability,
- specification for storage conditions for samples retained for testing,
- reliable, meaningful, and specific test methods for each attribute to be examined,
- testing conducted in the same container-closure system as that in which the product is marketed, and
- testing of reagents before and after reconstitution and/or mixing as directed in the labeling.

An adequate number of batches of each reagent must be tested to determine an appropriate expiration date, and the manufacturer must maintain a record of the data. Accelerated studies, combined with basic stability information on the components, reagents, and container-closure system, may be used to support tentative expiration dates provided full shelf-life studies are not available but are being conducted. Where data from accelerated studies are being used to project a tentative expiration date that is beyond a date supported by actual shelf-life studies, there must be stability studies conducted, including reagent testing at appropriate intervals, until the tentative expiration date is verified or the appropriate expiration date is determined.

THINGS TO REMEMBER

IVD products are those reagents, instruments, and systems intended for use in the diagnosis of disease or other conditions in humans. The FDA classifies these products as devices under 201(h) of the FD&CA. As devices, IVD products are subject to the general labeling requirements in 21 CFR 801. However, because of the unique properties of IVD products, the FDA has established particular labeling requirements for them. These special requirements are contained in 21 CFR 809.10. An IVD product that meets these requirements is deemed to meet the labeling requirements of 21 CFR 801.

Most products covered by the FD&CA are exempt from regulation by the CPSC. However, IVD products are specifically required by FDA regulations to include in their labeling any warnings or precautions for users required by the regulations promulgated under the FHSA.

The labeling for IVD reagents must include adequate instructions for proper storage to protect the stability of the product as well as an expiration date beyond which the product is not to be used. The FDA recommends that the IVD manufacturer use the stability-testing program described in the cGMPs for finished pharmaceuticals (21 CFR 211.166).

23 Radiation-Emitting Device Labeling

Congress declared in the preamble to the Radiation Control for Health and Safety Act of 1968 (RCHSA)* that the public health and safety must be protected from the dangers of electronic product radiation (RCHSA §263b). To accomplish this objective, the Secretary of the Department of Health and Human Services (DHHS) is authorized to establish an electronic-radiation control program. This program includes the authority to develop and administer performance standards to control the emission of electronic product radiation.

The electronic-radiation control program, administered by the Food and Drug Administration (FDA), covers any product that emits ionizing or nonionizing electromagnetic or particulate radiation, and products that emit sonic, infrasonic, or ultrasonic radiation as the result of operation of an electronic circuit. Radiation-emitting devices include those products that emit radiation either by design (e.g., X-ray equipment) or as a consequence of operation (e.g., television sets). Section 358 of the RCHSA authorizes the development of federal standards for these types of products. These standards are contained in the Code of Federal Regulations, 21 CFR Subchapter J—Radiological Health.

A product that emits radiation as a result of the decay of a radioactive element or isotope (e.g., an ionization-type smoke detector) is exempted from regulation under the RCHSA (Cardamone p. 25).

For the most part, the FDA does not choose to actively regulate common consumer products that emit either electromagnetic or sonic radiation. Examples include such common household products as radios and incandescent light bulbs. However, if the FDA concludes that such products compromise public health and safety it can move to regulate these products. An example cited by the FDA to illustrate this point is cordless telephones. Complaints from consumers led the FDA to confirm that injuries had occurred due to the placement, volume, and frequencies of ringers in certain brands of cordless telephones. As a result, the FDA set and enforced guidelines to protect the hearing of the users (Cardamone p. 25).

Labeling of radiation-emitting devices intended for medical application must also meet the general medical device labeling requirements in Chapter 21. The nature and placement of labeling for radiation-emitting devices may vary from the general labeling requirements discussed in Chapter 21. The concepts of "label" and "labeling," however, remain the same. The remainder of this chapter is devoted to those portions of the performance standards in 21 CFR that deal with product labeling. The section number (e.g., §1010.3) of the corresponding section in 21 CFR is listed with each topic for reference.

GENERAL LABELING REQUIREMENTS FOR ELECTRONIC PRODUCTS

Part 21 CFR 1010 specifies general requirements that are applicable to all electronic products covered by the performance standards in 21 CFR Subchapter J.

---

* This statute, which was originally part of the *Public Health Service Act* (PHSA) at US Code, vol. 42, §263b *et seq.*, was recodified by Section 19(3) of Public Law 101–629 – which repealed Section 354 and redesignated Sections 355 through 360F of the PHSA as Sections 531 through 542 of the *Federal Food, Drug, and Cosmetic Act* (FD&CA), US Code, vol. 21, §360hh *et seq., supra.*

Product Certification (§1010.2)

The manufacturer of a product covered by a performance standard must provide the dealer or distributor, at the time of delivery, with a certification, based on an approved testing program, that the product conforms to the requirements of the performance standard. The certification must be in the English language and must be on a tag or label permanently affixed to or inscribed on the product. The information must be readily accessible to view when the product is fully assembled for use, unless the particular performance standard specifies some other manner of certification.

In those cases where it is not feasible to affix the certification labeling to the device, as specified in the previous paragraph, the manufacturer may apply to the Director of the Center for Devices and Radiological Health (CDRH) for approval of an alternate means of providing the certification.

Product Identification (§1010.3)

Every manufacturer of an electronic product covered by a performance standard must provide the information listed below. This information must be in the English language and must be on a tag or label permanently affixed or inscribed on the product.

- The full name and address of the manufacturer of the product
  - If the product is sold under a name other than that of the manufacturer, the full name and address of the individual or firm under whose name the product is sold may appear provided the full name and address of the manufacturer of the device has been provided to the Director of the CDRH.
  - Abbreviations such as "Co.," "Inc.," or their equivalent in the country where the device is manufactured may be used. Also, the first and middle initials of individuals may be used.
- The place, month, and year of manufacture.
  - The place of manufacture may be expressed in a code provided the code and its key have been previously supplied to the Director of the CDRH.
  - The month and year of manufacture cannot be coded or abbreviated. The month and four-digit year must appear as follows:

MANUFACTURED: (Insert Month and Year of Manufacture)

Further, the information must be readily accessible to view when the product is fully assembled for use. In addition, a particular performance standard may specify additional requirements for display of this identification information.

In those cases where it is not feasible to affix identification labeling to the device, as specified above, the manufacturer may apply to the Director of the CDRH for approval of an alternate means of identification.

Every manufacturer of an electronic product covered by a performance standard must provide the Director of the CDRH with a complete listing of the brand names applied to the product, and the names and addresses of the individuals or companies for whom the devices are manufactured.

Variances (§1010.4)

Under 21 CFR 1010.4, a manufacturer may apply for a variance from one or more provisions of any performance standard. The conditions under which an application for variance will be considered are listed in Section 1010.4. If the FDA grants a variance, the manufacturer must modify the product certification to state:

- that the product conforms to the applicable standard except for the characteristics covered by the variance,
- that the product conforms to the provision of the variance, and
- the assigned number and effective date of the variance.

EXEMPTIONS FOR PRODUCTS INTENDED FOR US GOVERNMENT USE (§1010.5)

A department or agency of the United States (US) government may procure products that do not meet the provisions of a performance standard for the purposes of research, investigations, studies, demonstrations, or training, or for reasons of national security. A manufacturer may apply for an exemption from the performance standard in order to meet the procurement specification following the procedure described in 21 CFR 1010.5.

The manufacturer of a product for which an exemption is granted must provide the following statement on a tag or label permanently affixed to or inscribed on the product.

CAUTION

THIS ELECTRONIC PRODUCT HAS BEEN EXEMPTED FROM FOOD AND DRUG ADMINISTRATION RADIATION SAFETY PERFORMANCE STANDARDS PRESCRIBED IN THE CODE OF FEDERAL REGULATIONS, TITLE 21, CHAPTER I, SUBCHAPTER J, PURSUANT TO EXEMPTION NO._____, GRANTED ON _____.

The information must be readily accessible to view when the product is fully assembled for use, unless another manner is prescribed in the exemption.

EXPORT OF ELECTRONIC PRODUCTS (§1010.20)

The performance standards specified in 21 CFR Subchapter J do not apply to products intended solely for export, provided:

- the product and the outside of the shipping container are labeled or tagged to indicate that the product is intended for export; and
- the product meets all the applicable requirements, including labeling requirements, of the country to which the product is intended for export.

IONIZING RADIATION-EMITTING PRODUCTS

Products that emit ionizing radiation are subject to the performance standard in 21 CFR 1020. Products covered by this standard are:

- television receivers (§1020.10),
- cold-cathode gas discharge tubes (§1020.20),
- diagnostic X-ray systems and their major components (§1020.30), and
- cabinet X-ray systems (§1020.40).

The labeling requirements for each type of product are described in the following sections.

TELEVISION RECEIVERS (§1020.10)

A television receiver is an electronic product designed to receive and display a television picture through broadcast, cable, or closed-circuit television. Digital monitors such as computer screens are excluded from the standard (Cardamone p. 27).

The manufacturer of a product that could exceed the allowable radiation exposure rates specified in the standard because of the failure, improper adjustment, or improper replacement of a circuit or shield component shall include a warning label. The label must be permanently affixed or inscribed and must be clearly legible under conditions of service. The warning label must include the high-voltage specification and instructions for adjusting the high voltage to the specified value.

COLD-CATHODE GAS DISCHARGE TUBES (§1020.20)

A cold-cathode gas discharge tube is an electronic device in which an electron flow is produced and sustained by the ionization of gas atoms and ion bombardment of the cathode. The manufacturer must:

- provide appropriate safety instructions together with instructions for use, including the specifications for the proper power source with each tube;
- permanently affix tags or labels identifying the intended polarity of the terminals to each tube or enclosure;
- provide a warning that application of power in excess of the specifications to tubes designed to demonstrate heat, florescence, or magnetic effects may result in X-rays in excess of allowable limits;
- provide a warning that tubes designed to demonstrate the production of X-rays will produce X-rays when energized; and
- locate the required tag or label on the tube or enclosure so that it is readily visible and legible when the product is fully assembled for use.

DIAGNOSTIC X-RAY SYSTEMS AND THEIR MAJOR COMPONENTS (§1020.30)

A diagnostic X-ray system is one designed for irradiation of any part of the human body for purposes of diagnosis or visualizing. It incorporates one or more of the following components certified according the provisions of this section of the performance standard. These components include:

- tube housing assemblies, X-ray controls, X-ray high-voltage generators, tables, cradles, film changers, vertical cassette holders mounted in a fixed location, cassette holders with front panels, and beam-limiting devices manufactured after August 1, 1974;
- fluoroscopic imaging assemblies manufactured after August 1, 1974, and before April 26, 1977
- spot-film devices and image intensifiers manufactured after April 26, 1977;
- cephalometric devices manufactured after February 25, 1978;
- image-receptor support devices for mammographic X-ray systems manufactured after September 5, 1978; and
- computed Tomography (CT) gantries manufactured after September 3, 1985.

The general provisions of this performance standard apply to CT X-ray systems manufactured before November 29, 1984. Those manufactured after this date are also subject to the requirements in 21 CFR 1020.33, which was added to the performance standard in 1984.

Identification of Components (§1020.30(e))

In addition to the identification requirements in 21 CFR 1010.3, manufacturers of components subject to this performance standard must permanently inscribe or affix the model number and serial number to the component. The model and serial number must be legible and accessible to view. The word "model" or "type" must appear as part of the manufacturer's identification of certified X-ray components. High-voltage generators contained in X-ray tube housings and beam-limiting devices that are integral parts of the tube housing are exempt from this requirement.

When the certification of a system or subsystem consisting of two or more components has been authorized in writing by the Director of the CDRH, a single inscription, tag, or label bearing the model number and serial number may be used to identify the product.

A manufacturer of a tube housing must inscribe or affix to the housing the name of the manufacturer, model number, and serial number of the X-ray tube incorporated into the housing.

Replacement of an X-ray tube in a previously manufactured tube housing assembly certified under this performance standard constitutes manufacture of a new tube housing assembly. The person replacing the X-ray tube is considered a manufacturer under the provision of the regulation. This person must inscribe or affix to the housing the name of the manufacturer, model number, and serial number of the replacement X-ray tube. The individual must also remove, cover, or deface any previously affixed inscriptions, tags, or labels that are no longer applicable.

Replacement of the X-ray tube in tube housing assemblies that are designed and designated by their original manufacturer as containing quick-change X-ray tubes is not considered a manufacturing operation. The manufacturer of quick-change X-ray tubes must provide a label with the tube manufacturer's name, the model, and serial number of the X-ray tube. The manufacturer must instruct the assembler to attach the label to the tube housing assembly and remove, cover, or deface any labels that the tube manufacturer describes as no longer applicable.

Information Provided to the Assembler (§1020.30(g))

The manufacturers of X-ray systems covered by this performance standard must provide assemblers of those systems with the information necessary so that they can assemble, install, adjust, and test the components adequately to assure that the product will comply with the provisions of the standard.

Such instructions must include specifications of other components compatible with those to be installed when compliance of the system or subsystem depends on their compatibility. The specifications may describe pertinent physical characteristics of the components and/or may list by manufacturer model number the components that are compatible.

For X-ray controls and generators manufactured after May 3, 1994, the manufacturer must:

- Provide a statement of the rated line voltage and the range of line-voltage regulation for operation at maximum line current.
- Provide a statement of the maximum line current of the X-ray system based on the maximum input voltage and current characteristics of the tube housing assembly compatible with the rated output voltage and the rated output-current characteristics of the X-ray control and associated high-voltage generator. If the manufacturer of the X-ray control and associated high-voltage generator is unaware of the rated input voltage and current characteristics of the tube housing assembly, then the manufacturer must provide the assembler with the necessary information to determine the maximum line current for the particular tube housing assembly(ies).
- Provide a statement of the technique factors that constitute the maximum line-current conditions described above.

Information Provided to the User (§1020.30(h))

The manufacturers of X-ray systems covered by this performance standard must provide the purchaser with manuals (or instruction sheets) that include the information required in 21 CFR

1020.30(h). These manuals must include the technical and safety information required for safe operation of the equipment.

1. For all X-ray equipment, including that covered in Sections 1020.31, 1020.32, and 1020.33, adequate instructions concerning any radiological safety procedures and precautions that may be necessary because of unique features of the equipment must be included in the manual. A schedule of maintenance necessary to keep the equipment in compliance with this performance standard must also be included.

2. For each tube housing assembly:
   - Include statements about the leakage technique factors for all combinations of tube housing assemblies and beam-limiting devices for which the tube housing manufacturer states compatibility.
   - State the minimum filtration permanently in the useful beam expressed as millimeters of aluminum equivalent. In this performance standard, an aluminum equivalent is the thickness of aluminum (type 1100 alloy)* that affords the same attenuation, under specified conditions, as the material in question.
   - Declare the peak tube potential at which the aluminum equivalent was obtained.
   - Provide cooling curves for the anode and tube housing.
   - Include tube-rating charts.
   - If the tube is designed to operate from different types of X-ray high-voltage generators, or under modes of operation such as alternate focal-spot size or speed of anode rotation that affects its rating, specific identification of the differences in ratings must be noted.

3. For the X-ray control device that controls input power to the X-ray high-voltage generator and/or the X-ray tube, and for the associated X-ray high-voltage generator, the manufacturer must:
   - Provide a statement of the rated line voltage and the range of line-voltage regulation for operation at maximum line current.
   - State the maximum line current of the X-ray system based on the maximum input voltage and current characteristics of the tube housing assembly compatible with the rated output voltage and rated output-current characteristics of the X-ray control and associated high-voltage generator. If the manufacturer of the X-ray control and associated high-voltage generator does not know the rated input voltage and current characteristics of the tube housing assembly, sufficient information must be provided to allow the purchaser to determine the maximum line current for his or her particular tube housing assembly(ies).
   - Describe the technique factors** that constitute the maximum line-current conditions described above.
   - In the case of a battery-powered generator, include a specification of the minimum charge necessary for proper operation
   - State the generator rating and duty cycle.
   - Describe the maximum deviation from the preindication given by labeled technique-factor control settings or indications during any radiographic or CT exposure where the equipment is connected to a power supply described in the instructions for use. In the case of fixed technique factors, the maximum deviation from the nominal fixed value of each factor must be stated.

---

* The nominal chemical composition of type 1100 aluminum alloy is a minimum of 99.00 percent aluminum with 0.12 percent copper.
** Technique factors mean any of a variety of conditions of operation defined in 21 CFR 1020.30(b)(36) combining peak tube potential in kV and some other parameters such as tube current in mA.

- State the maximum deviation from the continuous indication of X-ray tube potential and current during any fluoroscopic exposure when the equipment is connected to a power supply described in the instructions for use.
- Explain the measurement criteria for all technique factors used above—for example, the beginning and end points of the exposure time measured with respect to a certain percentage of the voltage waveform.

4. For each variable-aperture beam-limiting device, the manufacturer must:
   - Provide leakage technique factors for all combinations of tube housing assemblies and beam-limiting devices for which the beam-limiting device manufacturer states compatibility.
   - State the minimum aluminum equivalent of that part of the device through which the useful beam passes and include the X-ray tube potential at which the aluminum equivalent was obtained. When two or more filters are provided as part of the device, the statement must include the aluminum equivalent of each filter.

Warning Label (§1020.30(j))

The control panel containing the main power switch must have the following warning statement clearly visible:

> **WARNING: THIS X-RAY UNIT MAY BE DANGEROUS TO PATIENT AND OPERATOR UNLESS SAFE EXPOSURE FACTORS AND OPERATING INSTRUCTIONS ARE OBSERVED.**

Repair of Components (§1020.30(d)(2)(iv))

Components installed temporarily in an X-ray system in place of certified components removed temporarily for repair must be certified. The temporarily installed component must bear the following labeling:

> TEMPORARILY INSTALLED COMPONENT
>
> THIS CERTIFIED COMPONENT HAS BEEN ASSEMBLED, INSTALLED, ADJUSTED, AND TESTED BY ME ACCORDING TO THE INSTRUCTIONS PROVIDED BY THE MANUFACTURER.
>
>   (SIGNATURE)
>   COMPANY NAME
>   STREET ADDRESS, P.O. BOX
>   CITY, STATE, ZIP CODE
>   DATE OF INSTALLATION

The person who installs the temporary component is not required to file a report of assembly with the CDRH if the temporary component is identified by a tag or label giving the information listed above. However, the replacement of the temporarily installed component by any component other than the one originally removed for repair must be reported to the CDRH.

Radiographic Equipment (§1020.31)

This part of the performance standard applies to equipment for the recording of images, except equipment involving the use of an image intensifier, or CT X-ray systems manufactured on or after November 29, 1984.

The system may provide a capability to override the positive beam-limitation feature in the case of system failure, and for servicing the equipment. A key is required for any override capability that is accessible to the operator. Each key or key switch shall be clearly and durably labeled:

FOR X-RAY FIELD LIMITATIONS SYSTEM FAILURE

THE OVERRIDE CAPABILITY IS CONSIDERED ACCESSIBLE TO THE OPERATOR IF IT IS REFERENCED IN THE OPERATOR'S MANUAL OR IN OTHER MATERIAL INTENDED FOR THE OPERATOR OR IF ITS LOCATION IS SUCH THAT THE OPERATOR WOULD CONSIDER IT PART OF THE OPERATIONAL CONTROLS.

A capability may be provided to override the automatic X-ray-field size adjustment in case of system failure. If provided, a signal visible from the operator's position must indicate when the automatic X-ray-field size adjustment override is engaged. Each system failure override switch must be clearly labeled:

FOR X-RAY FIELD LIMITATIONS SYSTEM FAILURE

Fluoroscopic Equipment (§1020.32)

This part of the performance standard covers equipment used for observing, by means of some fluorescent substance, the shadows of objects enclosed in media opaque to ordinary light, but transparent to X-rays. It is applicable to a system with or without an image intensifier.

The system may provide the capability to override the automatic adjustment of the X-ray-field size in case of a system failure. If provided, a signal visible at the fluoroscopist's position must indicate whenever the automatic field adjustment is overridden. Each system-failure override switch must be clearly labeled:

FOR X-RAY FIELD LIMITATIONS SYSTEM FAILURE

Computed Tomography (CT) Equipment (§1020.33)

The performance standard for CT X-ray systems became effective for systems manufactured or remanufactured on or after November 29, 1984. Beginning on this date, manufacturers were required to provide the user with the technical and safety information specified in 21 CFR 1020.33(c)(1) (conditions for operation) and 21 CFR 1020.33(c)(2) (dose information). The performance standard was later revised to add additional requirements for CT X-ray systems manufactured or remanufactured on or after September 3, 1985. For the purposes of this section, remanufactured means modifying a CT system in such a way that the resulting dose and imaging performance is substantially equivalent to any CT system originally manufactured on or after November 29, 1984.

The manufacturer of a CT X-ray system manufactured or remanufactured on or after September 3, 1985, must provide to the purchaser of the system the technical and safety information described in 21 CFR 1020.33(c) in addition to the general information required for all diagnostic X-ray systems by 21 CFR 1020.30(h). This information must be provided in a separate section of the user's instruction manual or in a separate manual devoted only to this information.

1. The manual must contain a statement of the CT conditions of operation* used to provide the information in this section of the user's instruction manual.
2. The following dose information obtained using the CT dosimetry phantom** is to be included. If the CT system is designed to image both the head and body, separate dosage information is to be provided for each application. All dose measurements must be performed with the CT dosimetry phantom placed on the patient couch or support without additional attenuating material present.
   - List the Computed Tomography Dose Index (CTDI) at the following locations in the dosimetry phantom:
     a. Along the axis of rotation of the phantom.
     b. Along a line parallel to the axis of rotation and 1.0 centimeter interior to the surface of the phantom with the phantom positioned so that the CTDI is the maximum obtainable at this depth.
     c. Along lines parallel to the axis of rotation and 1.0 centimeter interior to the surface of the phantom at positions 90, 180, and 270 degrees from the position in paragraph (b). The CT conditions of operation shall be the typical values suggested by the manufacturer for CT of the head or body. The location of the position where the CTDI is the maximum, as specified in paragraph (b), must be given by the manufacturer with respect to the housing of the scanning mechanism or other readily identifiable feature of the CT X-ray system so as to permit placement of the dosimetry phantom in this orientation.
   - List the CTDI in the center location of the dosimetry phantom for each selectable CT condition of operation that varies either the rate or the duration of X-ray exposure. This CTDI must be presented as a value that is normalized to the CTDI in the center location of the dosimetry phantom described above, with the CTDI described above having a value of one. As each individual CT condition of operation is changed, all other independent CT conditions of operation must be maintained at the typical values described above. These data must encompass the range of each CT condition of operation stated by the manufacturer as appropriate for CT of the head or body. When more than three selections of a CT condition of operation are available, the normalized CTDI must be provided for the minimum, maximum, and midrange value of the CT condition of operation. The normalized CTDI may be provided for additional selectable values of the CT condition selections.
   - List the CTDI at the location coincident with the maximum CTDI at 1.0 centimeter interior to the surface of the dosimetry phantom for each selectable peak tube potential. When more than three selections of peak tube potential are available, the CTDI shall be provided for the minimum, maximum, and a typical value of peak tube potential. The CTDI may be provided for additional selectable values of the peak tube potential. The CTDI must be presented as a value that is normalized to the maximum CTDI located at 1.0 centimeter interior to the surface of the dosimetry phantom, with the maximum CTDI having a value of one.
   - Provide the dose profile in the center location of the dosimetry phantom for each selectable nominal tomographic section thickness. When more that three selections or nominal

---

* CT conditions of operation means all selectable parameters governing the operation of a CT X-ray system including normal tomographic section thickness, filtration, and the technique factors as defined in 21 CFR 1020.30(b)(36).

** A CT dosimetry phantom is a cylinder of polymethl-methacrylate of specified dimensions and density used to determine the dose delivered by a CT X-ray system.

tomographic section thicknesses are available, the information must be provided for the minimum, maximum, and mid-range value of nominal tomographic section thickness. The dose profile may be provided for additional selectable values of the nominal tomographic section thickness. The dose profile must be presented on the same graph and to the same scale as the corresponding sensitivity profile required below.

- Provide a statement of the maximum deviation from the CTDI and dose profile values obtained using the CT dosimetry phantom. Deviation of actual values may not exceed these limits.

3. The following imaging performance data must be provided for the CT conditions of operation used to provide the information required by item 2 in this list. All other aspects of data collection, including the X-ray-attenuation properties of the material in the tomographic section, must be similar to those used to provide the dose information required by item 2. If the CT system is designed to image both the head and body, separate imaging performance information is to be provided for each application. Imaging performance data must:

- include a statement of the noise;
- provide a graphical presentation of the modulation transfer function for the same image processing and display mode as that used in the statement of the noise;
- describe the nominal tomographic section thickness(es);
- provide a graphical presentation of the sensitivity profile, at the location corresponding to the center location of the dosimetry phantom, for each selectable nominal tomographic section thickness for which the dose profile is given in item 2; and
- describe the dosimetry phantom or device and the test protocol or procedure used to determine the specifications and include a statement of the maximum deviation from the imaging performance specifications provided in the manual. Deviation of actual values may not exceed these limits.

4. The manufacturer of any CT X-ray system must provide the following quality assurance (QA) information with each system. All information must be provided in a separate section of the user's instructional manual.

- The manufacturer must provide a phantom or phantoms capable of providing an indication of contrast scale, noise, nominal tomographic section thickness, the spatial resolution capability of the system for low- and high-contrast objects, and measuring the mean CT number* of water or a reference material.
- The user's instruction manual must contain instructions on the use of the phantom(s) including a schedule of testing appropriate for the system, allowable variations for the indicated parameters, and a method to store, as records, (QA) data.
- The manual must provide representative images obtained with the phantom(s) using the same processing mode and CT conditions of operations as in item 3 for a properly functioning system of the same model. The representative images must be provided in two forms.
  - The first required form is photographic copies of the images obtained from the image display device.
  - The second required form is images stored as digital data on a storage medium compatible with the CT X-ray system. The CT X-ray system must be provided with the means to display these images on the image display device.

---

* CT number is the number used to represent the X-ray attenuation associated with each elemental area of the CT image.

The special requirements for controls and indicators for CT X-ray systems are described in 21 CFR 1010(f), (g), and (h). All of the parameters that control the scan or scan sequences must be indicated prior to the initiation of the scan. If some of the conditions of operation are at fixed values, they may be indicated by permanent marking on the equipment. All of these indicators must be visible from any point where the scan can be initiated.

For a system that produces a single tomogram during a scan, a way must be provided to visually determine the tomographic plane or a reference plane offset from the tomographic plane. For systems capable of producing multiple tomograms during a scan, a way must be provided to permit visual determination of the location of the reference plane. The manufacturer must provide information in the literature on the relationship of the reference plane to the planes of the tomogram. For any offset alignment system, the manufacturer must provide specific instructions for patient positioning.

The manufacturer must provide a method to calculate the mean and standard deviation of CT numbers for an array of picture elements about any location in the image. The user must be able to control the number of elements in this array. The manufacturer must include in the manual specific instructions on how to use the method provided for calculation of the mean and standard deviation of the CT number.

Cabinet X-Ray Systems (§1020.40)

A cabinet X-ray system is a device with the X-ray tube installed in an enclosure that is independent of the surrounding structure. The cabinet is designed to contain the material being X-rayed and to exclude personnel from its interior during generation of X-rays. Common examples include the X-ray systems designed to inspect carry-on baggage at airline, railroad, and bus terminals.

Every cabinet X-ray system must be equipped with two independent means of indicating when X-rays are being generated. One of the indicators may be a milliammeter labeled to indicate X-ray tube current. All other indicators must be legibly labeled, "X-RAY ON."

The device must provide an indicator that is clearly visible from each and every door, access panel, and port where it is possible to gain access to the interior of the cabinet. The indicator must be clearly labeled, "X-RAY ON."

A clearly visible and legible warning label must be permanently affixed or inscribed at the location of the controls that initiate X-ray generation. The label must bear the statement:

CAUTION: X-Rays Produced When Energized

Adjacent to every port on the device, there must be a clearly visible and legible label bearing the statement:

CAUTION: Do Not Insert Any Part of the Body When System is Energized—X-Ray Hazard

If the cabinet is designed to admit humans, it must have audible and visible warning signals within the cabinet that are activated for at least 10 seconds immediately prior to the first initiation of X-ray generation after closing any door or port. The system must also be equipped with a visible warning system inside the cabinet that is active when X-rays are being generated. A control must be accessible within the cabinet that will terminate or prevent generation of X-rays.

Signs indicating the meaning of the warning signals and containing instructions for use of the emergency control must be provided inside the cabinet. These signs must be legible, accessible to view, and illuminated when the main power control is in the "on" position.

The manufacturer of a cabinet X-ray system must provide for the purchaser, and others upon request at a cost not to exceed the cost of preparation and distribution, manuals and instructions that include the following technical and safety information:

- Potential, current, and duty cycle ratings of the X-ray generation equipment
- Adequate instructions concerning any radiological safety procedures and precautions that may be necessary because of unique features of the system
- A schedule of maintenance necessary to keep the system in compliance with the performance standard in 21 CFR 1020.40

If the cabinet X-ray system is intended for assembly or installation by the purchaser, the manufacturer must provide instructions for assembly, installation, adjustment, and testing of the cabinet x-ray system adequate to assure that the system is in compliance with applicable provisions of 21 CFR 1020.40 when assembled, installed, adjusted, and tested as directed.

MICROWAVE AND RADIO-FREQUENCY-EMITTING PRODUCTS

Products that emit microwave and radio frequency (RF) radiation are subject to the performance standard in 21 CFR 1030. At present, only microwave ovens are covered by this standard. A microwave oven, as defined in 21 CFR 1030.10(b), is a device manufactured for use in homes, in restaurants, in food vending or service establishments, on interstate carriers, and in similar facilities that is designed to heat, cook, or dry food through the application of electromagnetic energy at frequencies assigned by the Federal Communications Commission (FCC) in the normal industrial, scientific, and medical (ISM) heating bands ranging from 890 megahertz to 6,000 megahertz. Because microwave ovens are not classified as medical devices, their labeling requirements are not included in this chapter.

LIGHT-EMITTING PRODUCTS

Products that emit light energy are subject to the performance standard in 21 CFR 1040. Products covered by this standard are:

- laser products (§1040.10),
- specific-purpose laser products (§1040.11),
- sunlamp products and ultraviolet lamps intended for use in sunlamp products (§1040.20), and
- high-intensity mercury vapor discharge lamps (§1040.30).

The labeling requirements for each type of product are described in the following sections.

LASER PRODUCTS (§1040.10)

Lasers are devices capable of producing intense radiation at a specific wavelength for medical, scientific, and industrial purposes. The performance standard in 21 CFR 1040.10 is applicable to all laser products manufactured or assembled after August 1, 1976, except for lasers sold or used as components in another electronic product. Lasers used as components are exempt from the requirements of the performance standard if the laser is accompanied by a general warning notice that adequate instructions for safe installation of the laser product are provided in the servicing information available from the complete laser product manufacturer. The laser must be labeled with a statement that it is designed for use solely as a component and, therefore, does not comply with the appropriate requirements of the performance standard.

Lasers are classified based on the degree of hazard to skin and eyes from exposure to radiation, combined with the possibility of exposure of any part of the human body to laser or collateral radiation.* There are a total of six classifications (I, IIa, II, IIIa, IIIb, and IV). Class I levels of laser radiation are not considered hazardous. Class IV levels of laser radiation are considered an acute hazard to the skin and eyes from direct and scattered radiation. The accessible emission limits for laser radiation in each class are specified in Tables I, II-A, II, III-A, and III-B in 21 CFR 1040.10(d). The accessible emission limits for collateral radiation from laser products are specified in Table VI in 21 CFR 1040.10(d).

Labeling Requirements (§1040.10(g))

In addition to the certification and identification requirements in 21 CFR 1010.2 and 1010.3, each laser product is subject to the labeling requirements described in this section.

Each Class IIa laser product must have a label affixed bearing the following wording:

CLASS IIA LASER PRODUCT—AVOID LONG-TERM VIEWING OF DIRECT LASER RADIATION.

Each Class II laser product shall have a label affixed bearing the Warning Logotype A (Figure 23.1) and including the following wording:

[IN POSITION 1 ON THE LOGOTYPE]

LASER RADIATION—DO NOT STARE INTO BEAM

[IN POSITION 3 ON THE LOGOTYPE]

CLASS II LASER PRODUCT

Each Class IIIa laser product with an irradiation less than or equal to 2.5×10^{-3} W cm^{-2} must have a label affixed bearing the Warning Logotype A (Figure 23.1) and including the following wording:

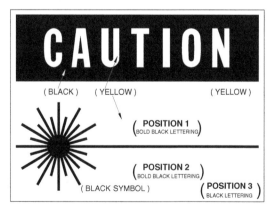

Source: 21 CFR 1040.10, Figure 1

FIGURE 23.1 Warning logotype A.

---

* Collateral radiation is defined in 21 CFR 1040.10(b)(12) as any electronic product radiation, except laser radiation, produced as a result of the operation of the laser.

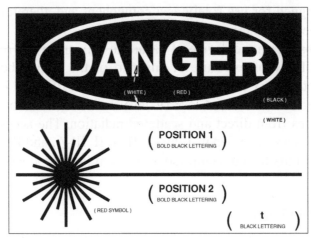

Source: 21 CFR 1040.10, Figure 2

FIGURE 23.2 Warning logotype B.

[IN POSITION 1 ON THE LOGOTYPE]

LASER RADIATION—DO NOT STARE INTO BEAM OR VIEW DIRECTLY WITH OPTICAL INSTRUMENTS

[IN POSITION 3 ON THE LOGOTYPE]

CLASS IIIa LASER PRODUCT

Each Class IIIa laser product with an irradiation greater than $2.5 \times 10^{-3}\,\text{W cm}^{-2}$ must have a label affixed bearing the Warning Logotype B (Figure 23.2) and including the following wording:

[IN POSITION 1 ON THE LOGOTYPE]

LASER RADIATION—AVOID DIRECT EYE EXPOSURE

[IN POSITION 3 ON THE LOGOTYPE]

CLASS IIIa LASER PRODUCT

Each Class IIIb laser product must have a label affixed bearing the Warning Logotype B (Figure 23.2) and including the following wording:

[IN POSITION 1 ON THE LOGOTYPE]

LASER RADIATION—AVOID DIRECT EXPOSURE TO BEAM

[IN POSITION 3 ON THE LOGOTYPE]

CLASS IIIb LASER PRODUCT

Each Class IV laser product must have a label affixed bearing the Warning Logotype B (Figure 23.2) and including the following wording:

[IN POSITION 1 ON THE LOGOTYPE]

LASER RADIATION—AVOID EYE OR SKIN EXPOSURE TO DIRECT OR SCATTERED RADIATION

[IN POSITION 3 ON THE LOGOTYPE]

CLASS IV LASER PRODUCT

Each Class II, IIIa, IIIb, and IV laser product must state in appropriate units, at position 2 of the required warning logotype, the maximum output of laser radiation, the pulse duration when appropriate, and the laser medium (e.g., CO_2) or the emitted wavelength(s).

Each laser product, except medical lasers and Class IIa lasers, must have a label affixed near each aperture though which accessible laser or collateral radiation in excess of the limits specified in 21 CFR 1040.10 is emitted. The label must bear one of the following warnings:

- If the radiation emitted through the aperture is laser radiation:

AVOID EXPOSURE—LASER RADIATION IS EMITTED FROM THIS APERTURE.

- If the radiation emitted through the aperture is collateral radiation described in Table VI, item 1, in 21 CFR 1040.10(d):

AVOID EXPOSURE—HAZARDOUS ELECTROMAGNETIC RADIATION IS EMITTED FROM THIS APERTURE.

- If the radiation emitted through the aperture is collateral radiation described in Table VI, item 2, in 21 CFR 1040.10(d):

AVOID EXPOSURE—HAZARDOUS X-RAYS ARE EMITTED FROM THIS APERTURE.

Special labeling is required if no safety interlocks are present that would prevent operation of the laser when some portion of the protective housing is displaced or removed, thereby allowing human access to laser or collateral radiation in excess of the limits of Class I and Table VI in 21 CFR 1040(d). Labels must be placed on any part of the housing that can be displaced or removed during operation, maintenance, or service. The labels must be visible on the protective housing prior to displacement or removal. In addition, the labels must be near the opening created by displacement or removal of the protective housing. The labels must bear one of the following warnings:

- If the accessible laser radiation is Class II:

CAUTION—LASER RADIATION WHEN OPEN. DO NOT STARE INTO BEAM.

- If the accessible laser radiation is Class IIIa with an irradiance less than or equal to 2.5 x 10^{-3} W cm^{-2}:

CAUTION—LASER RADIATION WHEN OPEN. DO NOT STARE INTO BEAM OR VIEW DIRECTLY WITH OPTICAL INSTRUMENTS.

- If the accessible laser radiation is Class IIIa with an irradiance greater than 2.5 x 10^{-3} W cm^{-2}:

DANGER—LASER RADIATION WHEN OPEN. AVOID DIRECT EYE EXPOSURE.

- If the accessible laser radiation is Class IIIb:

    DANGER—Laser radiation when open. AVOID DIRECT EXPOSURE TO BEAM.

- If the accessible laser radiation is Class IV:

    DANGER—Laser radiation when open. AVOID EYE OR SKIN EXPOSURE TO DIRECT OR SCATTERED RADIATION.

- If the collateral radiation is in excess of the accessible emission limits described in Table VI, item 1, in 21 CFR 1040.10(d):

    CAUTION—Hazardous electromagnetic radiation when open.

- If collateral radiation is in excess of the accessible emission limits described in Table VI, item 2, in 21 CFR 1040.10(d):

    CAUTION—Hazardous X-rays when open.

Special labeling is also required if the laser product is equipped with safety interlocks that can be defeated during operation of the laser, thereby allowing human access to laser or collateral radiation in excess of the limits of Class I and Table VI in 21 CFR 1040(d) when some portion of the protective housing is displaced or removed. Labels must be placed on any part of the housing that can be displaced or removed during operation, maintenance, or service. The labels must be visible on the product prior to and during the time the interlock is defeated. In addition, the labels must be near the opening created by displacement or removal of the protective housing. The labels must bear one of the following warnings:

- If the accessible laser radiation is Class II:

    CAUTION—Laser radiation when open and interlock defeated. DO NOT STARE INTO BEAM.

- If the accessible laser radiation is Class IIIa with an irradiance less than or equal to 2.5×10^{-3} W cm^{-2}:

    CAUTION—Laser radiation when open and interlock defeated. DO NOT STARE INTO BEAM OR VIEW DIRECTLY WITH OPTICAL INSTRUMENTS.

- If the accessible laser radiation is Class IIIa with an irradiance greater than 2.5×10^{-3} W cm^{-2}:

    DANGER—Laser radiation when open and interlock defeated. AVOID DIRECT EYE EXPOSURE.

- If the accessible laser radiation is Class IIIb:

    DANGER—Laser radiation when open and interlock defeated. AVOID DIRECT EXPOSURE TO BEAM.

- If the accessible laser radiation is Class IV:

    DANGER—Laser radiation when open and interlock defeated. AVOID EYE OR SKIN EXPOSURE TO DIRECT OR SCATTERED RADIATION.

- If the collateral radiation is in excess of the accessible emission limits described in Table VI, item 1, in 21 CFR 1040.10(d):

CAUTION—Hazardous electromagnetic radiation when open and interlock defeated.

- If collateral radiation is in excess of the accessible emission limits described in Table VI, item 2, in 21 CFR 1040.10(d):

CAUTION–Hazardous X-rays when open and interlock defeated.

If the laser product produces only visible laser radiation, then the phrase "laser light" may replace the phrase "laser radiation" on the labels listed above. If the product produces only invisible radiation, the word "invisible" must appropriately precede the word "radiation." If the product produces both visible and invisible radiation, the phrase "visible and invisible" or "visible and/or invisible" must precede the word "radiation."

All labels affixed to laser products must be positioned so that they can be read without exposing the reader to laser radiation in excess of the accessible limits of Class I or the limits of collateral radiation established in Table VI in 21 CFR 1040.10(d).

The labels required by 21 CFR 1040.10 and 1040.11 must be permanently affixed to or inscribed on the laser product. These labels must be legible and visible during operation, maintenance, or service, as appropriate. A manufacturer that discovers that the size, configuration, design, or function of a laser product would preclude compliance with the requirements for any required label, or would render the required wording of such label inappropriate or ineffective, can apply to the Director of the CDRH for approval of an alternative. Upon written application, the CDRH Director can approve alternate means of providing such label(s) or alternate wording for such labels, as applicable.

User Information (§1040.10(h)(1))

The manufacturer of a laser product must provide the following information with each laser product. This information may be contained as an integral part of any user instruction or operation manual that is regularly supplied with the product.

- The user instructions must contain adequate directions for assembly, operation, and maintenance, including clear warnings concerning precautions to be taken to avoid exposure to laser or collateral radiation in excess of the accessible emission limits in Tables I, II-A, II, III-A, III-B, and VI in 21 CFR 1040.10(d). The instructions must include a schedule of the maintenance necessary to keep the product in compliance with 21 CFR 1040.10 and 1040.11.
- The manufacturer must state the magnitude, in appropriate units, of the pulse duration(s), maximum radiant power, and, where applicable, the maximum radiant energy per pulse of the accessible laser radiation detectable in each direction that is in excess of the accessible emission limits in Table I in 21 CFR 1040.10(d) as determined according to the test methods in 21 CFR 1040.10(e).
- The user instructions should contain legible reproductions of all labels and hazard warnings that 21 CFR 1040.10(g) and 1040.11 require to be affixed or provided with the laser product. Color reproductions are optional. However, the reproductions must include the information required for positions 1, 2, and 3 of the applicable warning logotypes. The corresponding position of each label affixed to the product must be indicated or, if provided with the product, there must be a statement that such labels could not be affixed to the product but

were provided with the product. If the labels are provided with the product, a statement must be included giving the form and manner in which the labels were supplied.

- A listing must be provided of all controls, adjustments, and procedures for operation and maintenance, including the warning:

> CAUTION—USE OF CONTROLS OR ADJUSTMENTS OR PERFORMANCE OF PROCEDURES OTHER THAN THOSE SPECIFIED HEREIN MAY RESULT IN HAZARDOUS RADIATION EXPOSURE.

- In the case of laser products other than laser systems, a statement of the compatibility requirements for a laser energy source that will assure compliance with the requirements of 21 CFR 1040.10 and 1040.11 must be provided.
- In the case of a laser product classified with a 7-millimeter-diameter aperture stop as provided in 21 CFR 1040.10(e)(3)(i), if the use of a 50-millimeter aperture stop would result in a higher classification, the following warning must be included in the user information:

> CAUTION—THE USE OF OPTICAL INSTRUMENTS WITH THIS PRODUCT WILL INCREASE EYE HAZARD.

Purchasing and Servicing Information (§1040.10(h)(2))

Manufacturers of laser products must comply as follows:

- All catalogs, specification sheets, and descriptive brochures pertaining to each laser product must contain a legible reproduction of the class designation and warning that are required to be affixed to the product by 21 CFR 1040.10(g). Color reproductions are optional. The information provided must include the information required for positions 1, 2, and 3 of the applicable warning logotype.
- The manufacturer must provide servicing dealers and distributors with adequate instructions for service adjustments and service procedures for each laser product model. These instructions must include clear warnings and precautions to be taken to avoid possible exposure to laser and collateral radiation in excess of the accessible emissions limits in Tables I, II-A, II, III-A, III-B, and VI in 21 CFR 1040.10(d).
- The instructions to servicing dealers and distributors must include a schedule of the maintenance necessary to keep the product in compliance with 21 CFR 1040.10 and 1040.11.
- The instructions to servicing dealers and distributors must include a listing of those controls and procedures that could be utilized by persons other than the manufacturer or the manufacturer's agents to increase accessible emission levels of radiation. The instructions must also include a clear description of the location of displaceable portions of the protective housing that could allow human access to laser or collateral radiation in excess of the accessible emission limits in Tables I, II-A, II, III-A, III-B, and VI in 21 CFR 1040.10.
- The instructions must include, for each procedure or sequence of procedures to be accomplished, protective procedures for service personnel to avoid exposure to levels of laser and collateral radiation known to be hazardous. The servicing instructions must include legible reproductions of required labels and hazard warnings. Color reproductions are optional.

SPECIFIC-PURPOSE LASER PRODUCTS (§1040.11)

Specific-purpose laser products are defined in the performance standard as those products intended for: (a) medical uses, (b) surveying and similar activities, and (c) demonstrations.

Medical Laser Products (§1040.11)

A medical laser product is a medical device as defined in 21 USC 321(h) that is manufactured, designed, intended, or promoted for *in vivo* laser irradiation of any part of the human body for the purpose of: (a) diagnosis, surgery, or therapy, or (b) relative positioning of the human body.

Each medical laser product must comply with all the applicable requirements of 21 CFR 1040.10 for laser products of its class. In addition, the manufacturer must affix, near each aperture through which accessible laser radiation in excess of the limits for Class I is emitted, a label bearing the wording:

LASER APERTURE

The manufacturer must provide instructions with each Class III or IV medical laser product specifying a procedure and schedule for calibration of the measurement system as required by 21 CFR 1040.11(a)(1).

As a medical device, a medical laser product must meet the general medical device labeling requirements in Chapter 21. This includes providing the information necessary for the safe use of the laser by a trained practitioner. Manufacturers are encouraged to provide the information described below in a special part of the user manual provided with each medical laser. This part of the user manual should include the following sections (Medical Laser pp. 17–19):

1. Device description—This section should include a brief description of the laser, including the model number or other designation, principle of operation, design features, performance specifications, and other similar information.
2. General biological and physical characteristics—A section on general biological and physical characteristics should follow the description of the laser. This section should include unique characteristics of the laser in terms of its physical and biological effects on tissue. This information needs to be sufficient to enable the practitioner to anticipate the clinical effects of the laser, to understand aberrations, and to compare one laser to another.
3. Indications for use—A clear statement of the indications for use should follow the section on biological and physical characteristics, and must contain the following information:
   - The indications for use should address whether the laser has a broad range of surgical utility in some medical specialty area or is dedicated to specific uses (e.g., is used for a specific disease).
   - The surgical or medical specialty area(s) for the laser should be listed for both general- and specific-use lasers. Depending on the extent of the indications, the indications for use may be listed by the surgical or medical specialty area.
   - The indications-for-use section should describe, in aggregate terms, the type of target tissues.
   - The procedural parameters should be described in sufficient detail to characterize the conditions for safe and effective use. These terms should include those related to mode (contact or free beam), surgical approach (open or closed), associated instrumentation, and operational conditions.
   - Other factors important to understanding the indications for use should be described. For example, a laser may have both a general use in one specialty area and a specific use in another specialty area.
4. Contraindications—A description of the contraindications for use of the laser should follow the indications for use. Contraindications describe those situations where the laser should

not be used because the risks clearly outweigh the possible benefits. Contraindications could include unacceptable risks to tissues adjacent to the target such as critical nerves or blood vessels. The contraindications could include medical conditions, such as invasive cancer, that would preclude the use of the laser in selected situations. Contraindications may be general or they may be limited to specific surgical situations. Known hazards, not theoretical possibilities, are to be listed. If there are no known contraindications to the use of the laser, then this section should state, "None known."

5. Warnings—The contraindications should be followed by any warnings about serious adverse reactions and potential safety hazards, any limitations in use imposed by them, and the steps to be taken if they occur. A warning should be included if there is reasonable evidence that a serious hazard is associated with the use of the laser. A causal relationship between the use of the laser and the hazard need not be proven. Examples of serious hazards could include risks of igniting methane gas during rectal procedures, and the risks of air (gas) emboli if coaxial gas is used during certain procedures.

6. Precautions—The warnings should be followed by a list of precautions to be taken before, during, and after use of the laser. The precautions should include information to protect the physician and operating-room personnel as well as the patient. This information may include the need for protective clothing or glasses, for smoke-aerosol evacuation, or for wet packing to reduce the risk of fire and tissue damage. It may be appropriate in this section to caution the physician about the maximum power to be used with certain tips and accessories, and about the depth of penetration of the laser in comparison to other lasers. It should include a statement about the need for training in the use of the laser. Unlike the contraindications and warnings sections, the precautions section must include theoretical concerns.

7. Adverse reactions—The section following the precautions should identify all undesirable effects reasonably associated with the use of the laser. An adverse reaction may occur as part of the effect of the laser, or it may be unpredictable in its occurrence. This section must include all adverse reactions mentioned in the contraindications, warnings, and precautions sections. A statement directing the reader to another part of the labeling for more information regarding the effect of the adverse reaction and any steps to be taken if it occurs may follow the listing of an adverse reaction. The adverse reactions should be listed in descending order according to their clinical significance as determined by their severity and frequency.

8. Directions for use—The next section should provide key or important use information. This information should augment the professional skills of the practitioner and serve as a ready reference source. For example, the directions for use may include advice to the practitioner to start with lower energies and with shorter bursts, and to gradually increase these as needed. For specific intended uses, this section should provide precise use instructions.

9. Bibliography—This section is intended to support the labeling by providing the practitioner with a ready reference to the literature supporting the labeling statements.

The information in the labeling of components and accessories of the laser is derived from data on the laser system as a whole and should provide professional use instructions on the system. Accessories or components may be offered as replacements or as separate devices. They should be labeled to advise the practitioner to refer to the user manual that is supplied with the laser for full clinical-use information.

The labeling for accessories or components must not add to or modify the labeling of the laser system that is provided in the user manual. An exception may be a situation in which an accessory affects the nominal power required for use, or the accessory or component limits the amount of power that may be used without risk to the equipment or patient (Medical Laser p. 19).

Surveying, Leveling, or Alignment Laser Products

A surveying, leveling, or alignment laser product is defined in 21 CFR 1040.10(b)(39) as any laser product manufactured, designed, intended, or promoted for one or more of the following uses:

- Determining and delineating the form, extent, or position of a point, body, or area by taking angular measurements
- Positioning or adjusting parts in proper relation to one another
- Defining a plane, level, elevation, or straight line

There are no special labeling requirements for surveying, leveling, or alignment laser products beyond those specified in 21 CFR 1040.20.

Demonstration Laser Products

A demonstration laser product is defined in 21 CFR 1040.10(b)(13) as any laser product manufactured, designed, intended, or promoted for purposes of demonstration, entertainment, advertising display, or artistic composition. A laser that may be used for one of these purposes is not considered a demonstration laser unless it is specifically designed, manufactured, intended, or promoted for one or more of these uses. There are no special labeling requirements for demonstration laser products beyond those specified in 21 CFR 1040.20.

SUNLAMP PRODUCTS AND ULTRAVIOLET LAMPS INTENDED FOR USE IN SUNLAMP PRODUCTS (§1040.20)

A sunlamp* is a device designed to induce skin tanning and that incorporates one or more ultraviolet lamps intended to irradiate any part of the living human body by ultraviolet radiation with wavelength in air between 200 and 400 nanometers. In addition to the labeling requirements in 21 CFR 801 and the certification and identification requirements in 21 CFR 1010.2 and 1010.3, sunlamps and ultraviolet lamps intended for use in sunlamps are subject to the labeling requirements described in this section.

Sunlamp and Ultraviolet Lamp Labels (§1040.20(d))

Each sunlamp must have a label or labels that contain the following:

- A warning statement with the following words:

DANGER—ULTRAVIOLET RADIATION. FOLLOW INSTRUCTIONS. AVOID OVEREXPOSURE. AS WITH NATURAL SUNLIGHT, OVEREXPOSURE CAN CAUSE EYE AND SKIN INJURY AND ALLERGIC REACTIONS. REPEATED EXPOSURE MAY CAUSE PREMATURE AGING OF THE SKIN AND SKIN CANCER. WEAR PROTECTIVE EYEWEAR; FAILURE TO MAY RESULT IN SEVERE BURNS OR LONG-TERM INJURY TO THE EYES. MEDICATIONS OR COSMETICS MAY INCREASE YOUR SENSITIVITY TO THE ULTRAVIOLET RADIATION. CONSULT PHYSICIAN BEFORE USING SUNLAMP IF YOU ARE USING MEDICATIONS OR HAVE A HISTORY OF SKIN PROBLEMS OR BELIEVE YOURSELF ESPECIALLY SENSITIVE TO SUNLIGHT. IF YOU DO NOT TAN IN THE SUN, YOU ARE UNLIKELY TO TAN FROM THE USE OF THIS PRODUCT.

---

* A sunlamp is a Class I medical device under 21 CFR 878.4635.

- The recommended exposure position(s) expressed in terms of a distance specified both in meters and in feet (or in inches) or through the use of other means to indicate clearly the recommended exposure position(s)
- Directions for achieving the recommended exposure position(s) and a warning that the use of other positions may result in overexposure
- A recommended exposure schedule including duration and spacing of sequential exposures, and maximum exposure time(s) in minutes
- A statement of the time it may take before the expected results appear
- A designation of the type of ultraviolet lamp to be used in the product

Each ultraviolet lamp must have a label that contains the following:

- The words:

SUNLAMP—DANGER—ULTRAVIOLET RADIATION. FOLLOW INSTRUCTIONS.

- The words:

USE ONLY IN A FIXTURE EQUIPPED WITH A TIMER.

- The model identification

Any label prescribed in this performance standard for sunlamp products must be permanently affixed or inscribed on an exterior surface of the product when the product is fully assembled for use. The label must be readily accessible to view immediately before use of the product by the person being exposed. Any label prescribed in this performance standard for ultraviolet lamps must be permanently affixed or inscribed so as to be legible and readily accessible to view.

If the size, configuration, design, or function of the sunlamp or ultraviolet lamp would preclude compliance with the required labeling or would render the required wording inappropriate, ineffective, or unnecessary, the manufacturer may apply to the Director of the CDRH for a variance. Upon receipt of a written application from the manufacturer, the CDRH Director may approve alternate wording for the label(s), or deletion, as applicable.

In lieu of permanently affixing or inscribing on the ultraviolet lamp tags or labels containing the certification required by 21 CFR 1010.2(b) and the product identification required by 21 CFR 1010.3(a), the manufacturer may affix or inscribe the required tags or labels on the lamp package uniquely associated with the lamp. Such labeling is possible if the name of the manufacturer and the month and year of manufacture are permanently affixed or inscribed on the exterior surface of the ultraviolet lamp so as to be legible and accessible to view. The name of the manufacturer and the month and year of manufacture on the exterior surface of the lamp may be expressed in code or symbols, if the manufacturer has previously provided the Director of the CDRH with the key to the code or symbols and the location of the information on the ultraviolet lamp. The label or tag on the lamp package may provide either the month and year of manufacture without abbreviation, or information to allow the date to be readily decoded.

Information Provided to the User (§ 1040.20(e))

The manufacturers of sunlamp products or ultraviolet lamps covered by this performance standard must provide the purchaser with adequate instructions for use to avoid or to minimize potential injury to the user as specified in 21 CFR 1040.20(e). The user's instructions for sunlamp products must include the following technical and safety information, as applicable:

- Include a reproduction of the label(s) for sunlamp products required above. These label reproductions must be prominently displayed at the beginning of the instructions.
- Provide a statement of the maximum number of people who may be exposed to the product at the same time and a warning that only that number of protective eyewear has been provided.
- Clearly state the instructions for the proper operation of the product including the function, use, and setting of the timer and other controls, and the use of protective eyewear.
- Provide easy-to-understand instructions for determining the correct exposure time and schedule for persons according to skin type.
- Inform the user of where to obtain repairs and list recommended replacement components and accessories that are compatible with the product. This includes compatible protective eyewear, ultraviolet lamps, timers, reflectors, and filters that will, if installed or used as instructed, result in continued compliance with the requirements of this performance standard.

When an ultraviolet lamp is sold separately from a sunlamp product, the user's instructions with the ultraviolet lamp must:

- contain a reproduction of the labels required on ultraviolet lamps as specified above, prominently displayed at the beginning of the instructions;
- display a warning that the instructions accompanying the sunlamp product should always be followed to avoid or minimize potential injury; and
- provide a clear identification by brand and model designation of all lamp models for which replacement lamps are promoted, if applicable

HIGH-INTENSITY MERCURY VAPOR DISCHARGE LAMPS (§1040.30)

The provision of this section applies to any high-intensity mercury vapor discharge lamp that is designed, intended, or promoted for illumination purposes. As high-intensity mercury vapor discharge lamps are not classified as medical devices, their labeling requirements are not included in this chapter.

SONIC, INFRASONIC, AND ULTRASONIC RADIATION-EMITTING PRODUCTS

This performance standard covers products that emit sound energy. At present, only ultrasound therapy products are covered.

ULTRASOUND THERAPY PRODUCTS (§1050.10)

An ultrasonic therapy product is any device intended to generate and emit ultrasonic radiation for therapeutic purposes at ultrasonic frequencies above 16 kilohertz (kHz), or a generator or applicator designed or specifically designated for use in such a device. In addition to the labeling requirements in 21 CFR 801 and the certification and identification requirements in 21 CFR 1010.2 and 1010.3, ultrasonic therapy products are subject to the labeling requirements described in this section.

- Each operating control* must be clearly labeled, identifying the function controlled, and, where appropriate, the units of measure of the function. If a separate control and indicator

---

* An operating control is defined in 21 CFR 1050.10(b)(15) as any control used during operation that affects the ultrasonic radiation emitted by the applicator.

are associated with the same function, then labeling the appropriate units of measure of that function is required for the indicator but not for the control.

- Each service control* that is accessible without displacing or removing any part of the ultrasonic therapy product must be clearly labeled, identifying the function controlled. The label must include the phrase:

FOR SERVICE ADJUSTMENT ONLY

- Ultrasonic generators must bear a label that provides the following particulars:
  - The brand name, model designation, and unique serial number or other unique identification so that each is individually identifiable
  - Ultrasonic frequency (unless there is an operation control for varying this quantity)
  - Type of waveform (continuous wave or amplitude modulated)
- Ultrasonic generators employing amplitude-modulated waveforms must also bear a label that provides the following information:
  - Pulse duration and pulse repetition rate (unless there are operation controls for varying these quantities)
  - An illustration of the amplitude-modulated waveform
  - The ratio of the temporal-maximum effective intensity to the temporal-average effective intensity (If this ratio is a function of any operation-control setting, then the range of the ratio must be specified, and the waveform illustration must be provided for the maximum value of this ratio.)

Each ultrasonic applicator** must bear a label indicating the following particulars:

- The brand name, model designation, and unique serial number or other unique identification so that the applicator is individually identifiable
- A designation of the generator(s) for which the applicator is intended
- The ultrasonic frequency, effective radiating area, maximum beam nonuniformity ratio, type of applicator (focusing, collimating, diverging), and, for a focusing applicator, the focal length and focal area

The labels required by 21 CFR 1050.10 must be permanently affixed to or inscribed on the ultrasonic therapy product. These labels must be legible and clearly visible. A manufacturer discovering that the size, configuration, or design of an ultrasonic therapy product would preclude compliance with the requirements for any required label can apply to the Director of the CDRH for approval of an alternative. Upon written application, the CDRH Director can approve alternate means of providing such label(s).

DISCOVERY OF A PRODUCT DEFECT OR FAILURE TO COMPLY

A manufacturer who discovers that any electronic product produced, assembled, or imported by the firm has a defect or fails to comply with an applicable federal standard must notify the FDA. The manufacturer must also notify dealers or distributors and the purchasers of the equipment as specified

---

* A service control is defined in 21 CFR 1050.10(b)(22) as any control provided for the purpose of adjustment that is not used during operation and can affect the ultrasonic radiation emitted by the applicator, or can alter the calibration or accuracy of an indicator or control.

** An applicator is defined in 21 CFR 1050.10(b)(2) as that portion of a fully assembled ultrasonic therapy product that is designed to emit ultrasonic radiation, and that includes one or more ultrasonic transducers and any associated housing.

in 21 CFR 1003.10. The notification to the affected persons must include, among other information identified in the regulations, the following:

- Instructions for the use of the product pending the correction of the defect
- A clear evaluation in nontechnical terms of the hazards reasonably related to any defect or failure to comply with the applicable Federal standards
- The following statement:

THE MANUFACTURER WILL, WITHOUT CHARGE, REMEDY THE DEFECT OR BRING THE PRODUCT INTO COMPLIANCE WITH EACH APPLICABLE FEDERAL STANDARD IN ACCORDANCE WITH A PLAN TO BE APPROVED BY THE SECRETARY OF HEALTH AND HUMAN SERVICES, THE DETAILS OF WHICH WILL BE INCLUDED IN A SUBSEQUENT COMMUNICATION TO YOU.

- If, at the time that the notification is sent, the FDA has approved a plan for the repair, replacement, or refund of the product, the notification may include the details of the approved plan in lieu of this notice.

A number 10 white envelope with the name and address of the manufacturer in the upper left corner must be used to contain the notice from the manufacturer to affected persons. The following statement is to appear in the far left-hand third of the envelope centered in a red rectangle 3 3/4 inches wide and 2 _ inches high.

**IMPORTANT—
ELECTRONIC PRODUCT
RADIATION WARNING**

The statement shall be in reverse printing in three lines, all capitals, and centered in the rectangle. "Important" shall be in 36-point Gothic Bold type. "Electronic Product" and "Radiation Warning" shall be in 36-point Gothic Condensed type. The envelope containing the notice must not contain advertising or other extraneous material. Envelopes with markings similar to those described above may not be used by the manufacturer for any other mailings except those required in this part of the regulations.

RADIO FREQUENCY EMITTING DEVICES

Any device that generates RF energy is subject to regulation by the FCC under the authority of the Federal Communications Act of 1934.* The FCC is concerned with preventing harmful interference to authorized radio communication services. To carry out its mandate, the FCC may establish performance standards, test and certify devices, inspect installations, and license individual operators.

RADIO-FREQUENCY RADIATORS

Most medical devices that generate RF energy are classified as devices that may be operated without an individual operators license. Within this classification, the devices are grouped into three categories:

---

* *Communications Act of 1934*, as amended. US Code, vol. 47, §§151 *et seq.* (1934).

- Intentional radiator—a device that intentionally generates and emits RF energy by radiation or induction.
- Unintentional radiator—a device that intentionally generates RF energy for use within the device, or that sends RF signals to associated equipment via connecting wiring, but is not intended to emit RF energy by radiation or induction.
- Incidental radiator—a device that generates RF energy during the course of its operation, although the device is not intentionally designed to generate or emit RF energy. An example of an incidental radiator is a dc motor.

The labeling requirements for these devices are listed in 47 CFR Part 15—*Radio Frequency Devices*.

Intentional Radiators (§15.201)

Before being marketed, most intentional radiators that are operated under the provisions of 47 CFR 15 must be certified by the FCC following the procedures in 47 CFR 2, Subpart J. An FCC-certified device must be labeled with an FCC identifier consisting of two elements in the exact order indicated below, and the identifier must be preceded by the term "FCC ID:" in capital letters on a single line (47 CFR 2.926).

FCC ID: XXXYYYYYYYYYYYYY

The first element, XXX, is the three character grantee code assigned permanently by the FCC to a manufacturer. It consists of Arabic numerals, capital letters, or a combination of Arabic numerals and capital letters.

The second element, YYYYYYYYYYYYYY, is an equipment product code consisting of a maximum of 14 Arabic numerals, capital letters, or a combination of Arabic numerals and capital letters, and may include the dash or hyphen (-). The equipment product code must be one that has not been previously used in conjunction with the same grantee code. The combination produces a unique identifier for the device.

Separate FCC identifiers may be assigned to a device consisting of two or more sections assembled in a common enclosure, if the sections are constructed on separate subunits or circuit boards with independent frequency-controlling circuits. The FCC identifier assigned to the transmitter section is designated by the term "TX FCC ID," the FCC identifier assigned to the receiver section is designated by the term "RX FCC ID," and the FCC identifier assigned to any remaining sections is designated by the term "FCC ID."

The label bearing the FCC identifier must be etched, engraved, stamped, indelibly printed, or otherwise permanently marked on a permanently attached part of the equipment enclosure. The FCC identifier must be readily visible from the outside of the equipment enclosure. It is recommended, but not required, that the label be visible at all times during normal use of the equipment. The term "FCC ID:" and the code identification assigned by the FCC must be in a size of type large enough to be readily legible, consistent with the dimensions of the equipment and its nameplate. However, the type size of the FCC identification is not required to be larger than eight-point (47 CFR 2.925).

In addition to the requirements listed above, an intentional radiator, except receivers associated with the operation of a licensed radio service or a stand-alone cable input selector switch, must have the following statement in a conspicuous location on the device (47 CFR 15.19(a)(3)):

THIS DEVICE COMPLIES WITH PART 15 OF THE FCC RULES. OPERATION IS SUBJECT TO THE FOLLOWING TWO CONDITIONS: (1) THIS DEVICE MAY NOT CAUSE HARMFUL INTERFERENCE, AND (2) THIS DEVICE MUST ACCEPT ANY INTERFERENCE RECEIVED, INCLUDING INTERFERENCE THAT MAY CAUSE UNDESIRED OPERATION.

When a device is so small or for such use that it is not practical to place this statement on it, the information shall be placed in a prominent location in the instruction manual or pamphlet supplied to the user. An alternative, if no instruction manual exists, would be to place the required wording on the container in which the device is marketed.

The user's manual or instruction manual must caution the user that changes or modifications not expressly approved by the party responsible for compliance with FCC regulations could void the user's authority to operate the equipment.

Unintentional Radiators (§15.101)

Unless they are exempted in 47 CFR 15.23, 15.103, or 15.113, unintentional radiators must be authorized by the FCC or verified by the manufacturer prior to marketing. A table listing the equipment authorization required by type of device can be found in 47 CFR 15.101. The devices that are authorized by the FCC must be labeled as specified in the previous section, "Intentional Radiators."

Unintentional radiators may be subject to verification. Verification is a process defined in 47 CFR 2.902 that allows the manufacturer to make the measurements or take the necessary steps to insure that the equipment complies with the appropriate technical standards. Unless the FCC makes a specific request, the manufacturer is not required to submit a sample unit or representative data. Since no application is made to the FCC, there is no FCC identifier associated with the device. The labeling of such equipment may include model or type numbers, but the information may not be presented in a way that purports to be an FCC identifier (47 CFR 2.926).

Specialized medical digital devices that are classified as unintentional radiators are exempted from the labeling requirements of 47 CFR 15. Specialized medical digital devices are generally used at the direction of, or under the supervision of, a licensed healthcare practitioner. They may be used either in a healthcare facility or in the patient's home. However, devices marketed through retail channels for use by the general public are not exempt. The exemption does not apply to digital devices used for record keeping or for any purpose not directly related to medical treatment (47 CFR 15.103(e)).

Incidental Radiators (§15.13)

Manufacturers of incidental radiators must employ good engineering practices to minimize the risk of harmful interference.

INDUSTRIAL, SCIENTIFIC, AND MEDICAL EQUIPMENT

The FCC has the authority, under the Federal Communications Act of 1934 to prevent harmful interference to authorized radio communication services by regulating the operation of ISM equipment that emits electromagnetic energy within the RF spectrum. ISM equipment is designed to generate and use localized RF energy for ISM, domestic, or similar purposes, excluding applications in the field of telecommunications. Typical ISM applications include production of physical, chemical, or biological effects such as heating, ionization of gasses, mechanical vibration, hair removal, and acceleration of charged particles. Examples of medical devices covered by this regulation are medical diathermy equipment and ultrasonic equipment.

Source: 47 CFR 18.209(b)

FIGURE 23.3 Declaration of conformity logo.

ISM equipment may be operated on any frequency above 9 kHz except on certain frequency bands reserved for safety, search, and rescue equipment. The prohibited frequency bands are listed in 47 CFR 18.303. ISM equipment may radiate unlimited energy within the frequency bands reserved for ISM applications (47 CFR 18.301). The field strength levels of emissions that lie outside the ISM bands may not exceed the limits specified in 47 CFR 18.305.

The manufacturer or user of ISM equipment is granted an authorization to market or operate the equipment by following the procedures in 47 CFR, Chapter 1, Part 2, Subpart J. Each device must be labeled as specified in the section, "Intentional Radiators," unless the equipment is subject to the verification procedure described in 47 CFR 2.951 through 2.957 (47 CFR 18.209). Since the verification process does not involve an application to the FCC, there is no FCC identifier associated with the device. The labeling of such equipment may include model or type numbers, but the information may not be presented in a way that purports to be an FCC identifier (47 CFR 2.926).

Certain ISM equipment may be authorized for use under a Declaration of Conformity procedure. Devices authorized under the Declaration of Conformity procedure are labeled with the logo shown in Figure 23.3. The label may not be a stick-on paper label. It must be permanently affixed to the product in a location that is readily visible to the purchaser at the time of purchase. In this context, permanently affixed means that the label is etched, engraved, stamped, silk-screened, indelibly printed, or otherwise permanently marked on a permanently attached part of the equipment or on a nameplate of metal, plastic, or other material fastened to the equipment by welding, riveting, or a permanent adhesive. The label must be designed to last the expected lifetime of the equipment in the environment in which the equipment may be operated and must not be readily detachable (47 CFR 18.209(b)).

THINGS TO REMEMBER

Congress declared in the preamble to the RCHSA that the public health and safety must be protected from the dangers of electronic product radiation. To accomplish this objective, the Secretary of the DHHS is authorized to establish an electronic radiation-control program. The electronic radiation control program administered by the FDA covers any product that emits ionizing or nonionizing electromagnetic or particulate radiation and products that emit sonic, infrasonic, or ultrasonic radiation as the result of operation of an electronic circuit. Radiation-emitting devices include those products that emit radiation either by design (e.g., X-ray equipment) or as a consequence of operation (e.g., television sets).

For the most part, the FDA does not choose to actively regulate common consumer products that emit either electromagnetic or sonic radiation. Examples include such common household products as radios and incandescent light bulbs. However, if the FDA concludes that such products compromise public health and safety, they can move to regulate these products.

The electronic-radiation control program is established through a series of product standards contained in the Code of Federal Regulations, 21 CFR Subchapter J—*Radiological Health*.

The nature and placement of labeling for radiation-emitting devices may vary from the general labeling requirements discussed in Chapter 21. The concepts of "label" and "labeling," however, remain the same. This chapter is devoted to those portions of the performance standards in 21 CFR that deal with product labeling.

Any device that generates RF energy is subject to regulation by the FCC under the authority of the Federal Communications Act of 1934. The FCC is concerned with preventing harmful interference to authorized radio-communication services. Most medical devices that generate RF energy are classified as devices that may be operated without an individual operator's license. Within this classification, the devices are grouped into three categories: (a) intentional radiators, (b) unintentional radiators, and (c) incidental radiators. The labeling requirements for these devices are listed in 47 CFR Part 15—*Radio Frequency Devices*.

24 Bringing Devices to Market in the United States

Any person who proposes to bring to market a device intended for human use is required by Section 510 of the Federal Food, Drug, and Cosmetic Act (FD&CA) to report his or her intention to the Food and Drug Administration (FDA) at least 90 days before introducing the device into interstate commerce. Failure to comply is a violation of Section 301(p) of the FD&CA and exposes the manufacturer to both civil and criminal penalties under Section 302 of the FD&CA.

The Medical Device Amendments of 1976 directed the FDA to classify all medical devices that were in commercial distribution on May 28, 1976, into one of three regulatory categories depending on the degree of regulation necessary to provide reasonable assurance of their safety and effectiveness. Those devices already on the market in the United States prior to May 28, 1976, are often referred to as preamendment devices. The class into which the device is placed determines the requirements that must be met before a device can be introduced into interstate commerce. The classification also determines, in large measure, the form that the notification of intent to market a device must take.

Notification is required for all devices unless the device is specifically exempted by the regulations. Prior to the adoption of the Food and Drug Administration Modernization Act of 1997 (FDAMA), this exemption was generally applied to Class I devices. For example, the tongue depressor* is a Class I device that is exempt from premarket notification procedures as long as the device is not labeled or otherwise represented as sterile. The FDA had exempted almost all Class I devices (with the exception of "Reserved Devices") from the premarket notification requirement. A "Reserved Device" is a "Class I device that is intended for a use that is of substantial importance in preventing impairment of human health, or to any Class I device that presents a potential unreasonable risk of illness or injury" (FD&CA §510(l)).

FDAMA required the FDA to publish in the *Federal Register* a list of Class II devices that the agency believes do not require a premarket notification in order to provide reasonable assurance of safety and effectiveness. Devices on this list are exempt from the premarket notification requirement under Section 510(k) of the FD&CA. The remaining devices in Class I and Class II still require premarket notifications.

Class III devices are generally subject to the more extensive provisions of Section 515 of the FD&CA, which requires FDA approval of the safety and effectiveness of the device. Devices that are substantially equivalent to preamendment Class III devices can still be introduced to the market using the premarket notification process unless the FDA has published a notice calling for premarket approval application (PMA) under Section 515 of the FD&CA.

There is a special exemption to these provisions of the FD&CA for investigational devices. This exemption allows a device to be used in human subjects under controlled conditions in order to develop the data necessary to support the manufacturer's contention that the device is safe and effective for its intended use. Custom devices and humanitarian use devices, which are discussed later in this chapter, are also exempt from premarket notification.

---

* A tongue depressor is a device intended to displace the tongue to facilitate examination of the surrounding organs and tissues. See 21 CFR 880.6230.

PREMARKET NOTIFICATION (510(K))

Section 510(k) of the FD&CA requires that any person who proposes to introduce into interstate commerce a device intended for human use must notify the FDA at least 90 days in advance of placing the device on the market.

During the 90-day period, the FDA must determine whether the new or modified device is substantially equivalent in terms of safety and effectiveness to a predicate device, which can be either (a) a preamendment device or (b) a postamendment device that has been reclassified from Class III into Class I or II. The premarket notification program is meant to (Parr, Barcome, Regulatory Requirements, p. 5-1):

- Identify new devices. Devices that are not substantially equivalent to preamendment devices are automatically classified as Class III. These devices undergo either premarket approval (the approval of a PMA) or reclassification before they are marketed.
- Classify new devices. A new device that is substantially equivalent to a preamendment device is in the same regulatory class as the preamendment device.
- Provide equity by allowing manufacturers of new devices that are substantially equivalent to preamendment devices to market their devices without facing greater regulatory burdens than were faced by the manufacturer of the preamendment devices.

Premarket notification is required under regulations in Title 21 of the Code of Federal Regulations (CFR) Section 807.81 when:

- A device is first introduced into the market.
- A device is introduced into the market that may already be marketed by another manufacturer.
- The manufacturer is about to make a significant change or modification to a device currently in commercial distribution, or one that is being reintroduced into the market. In 21 CFR 807.81(3), the FDA defines a significant change as a "change or modification in design, material, chemical composition, energy source, or manufacturing process" that could significantly affect the safety and effectiveness of the device. The manufacturer is responsible for determining whether a proposed change to a device warrants submission of a premarket notification. The FDA has long taken the position that a 510(k) application is not required for every change in design, material, chemical composition, energy source, or manufacturing process, but only where such changes could significantly affect the safety or effectiveness of the device. The key issue is identifying what is a significant effect. The manufacturer is in the best position to make this determination based on good judgment, adequate supporting data, and sufficient documentation. To assist the manufacturer and help minimize disagreements between manufacturers and the agency, the FDA has developed guidance for when a change, including a change to labeling, exceeds the regulatory threshold for submission of a new 510(k).
- The manufacturer proposes a change in the intended use of the device.

Premarket Notification Review Program

A 510(k) notification must contain sufficient information, including all proposed labeling, to enable the FDA to evaluate the manufacturer's claim of substantial equivalence. This includes the labels on the device, on the packaging, the instruction manual, and advertising that describes the intended use. It should also include the service manual if the device is to be serviced by someone other than the manufacturer. The information content and format requirements for the 510(k) notification are set out in 21 CFR 807.87 through 807.94. In addition to the required information, the FDA advises including the labeling for the device to which substantial equivalence is claimed (Rice §3-2).

In verifying the claim, the FDA typically determines whether the device performs the same function and falls within the same established type as the predicate device. This process begins with a review of the intended use of the "new" device. If the device has a new intended use (what constitutes a new intended use is discussed in the following section), it will be considered as not substantially equivalent. If the new device is judged to have the same intended use as the predicate device, then the technology characteristics will be evaluated. The review of the premarket notification compares the characteristics of the new device (design, material, energy source, and so on) to the characteristics of the predicate device. If the differences do not raise new types of safety and effectiveness questions for the same intended use, the new device can be found substantially equivalent (Rice §2-1).

The final determination of substantial equivalence rests with the FDA. If the FDA finds that a device is not substantially equivalent to a predicate device, the new device is automatically placed into Class III. Such a device, unless reclassified by the FDA into a lower classification, requires the submission of a PMA. A manufacturer may petition the FDA under Section 513(f) of the FD&CA to have a device reclassified into Class I or Class II.

Once a device has been found to be substantially equivalent to its predicate device, the FDA will notify the manufacturer of its determination in writing. In the process of reviewing the premarket notification, the FDA may require warnings, qualifications, or other labeling restrictions. These will be set out in the response to the premarket notification (510(k) §3). For example, the FDA has determined that the labeling of orthopedic bone screws for certain uses must bear the warning: "WARNING: This device is not approved for screw attachment or fixations to the posterior elements (pedicles) of the cervical, thoracic, or lumbar spine" (Spinal System p. 10). The manufacturer must ensure that the labeling includes the intended use(s) identified in the letter of substantial equivalency. If there are no intended uses listed in the substantial equivalency letter, the manufacturer should ensure that the labeling contains those intended uses identified in the device classification of the predicate device.

SUBMITTING A PREMARKET NOTIFICATION (510(K)) FOR A CHANGE TO AN EXISTING DEVICE

Once a product has entered interstate commerce under the premarket notification procedure in Section 510(k) of the FD&CA, the manufacturer must notify the FDA of any significant change or modification in design, components, methods of manufacture, or intended use by submitting a new premarket notification. A significant change is defined in 21 CFR 807.81(a)(3) as one that (a) could significantly affect the safety or effectiveness of the device, or (b) is a major change or modification in the intended use of the device.

The use of the adjectives "significant" and "major" in the regulations create the "gray areas" that require interpretation. The manufacturer is responsible for determining whether a proposed change to a device warrants submission of a premarket notification.

For some specific devices, the FDA and industry have worked to establish definitive guidance for modifications (e.g., *Premarket Notification (510(k)) Guidance Document for Daily Wear Contact Lenses.** To aid manufacturers of devices where specific guidance documents do not exist, the FDA has developed a model to assist in deciding when a new premarket notification must be submitted for a change (Deciding When to Submit p. 2).

---

* This guidance document is available on the FDA Web site at http://www.fda.gov/cdrh/ode/conta.html.

The FDA model suggests that there are certain changes that will almost always lead to a decision to submit a new premarket notification. A modification in the intended use of the device, a change in warnings, precautions, or contraindications, or additional claims for enhanced safety or effectiveness are labeling changes that fall into this category (Deciding When to Submit pp. 9–12). Labeling changes that involve reformatting or other changes, such as correcting typographical errors, that do not result in a change in the directions for use will almost always lead to a decision not to submit a new premarket notification.

THE MEANING OF INTENDED USE

For a device to be judged as substantially equivalent to a preamendment device (or a postamendment device that has been reclassified from Class III to Class I or II), it must have essentially the same intended use as the predicate device.

The FDA does not require that the labeling of a new device contain the same identical therapeutic or diagnostic statements as those that appear on the predicate device labeling in order for the new device to have the same intended use (Parr, Barcome, Regulatory Requirements p. 5-23). Some elements of the labeling may not be critical to establishing the intended use of the device. An example cited in the FDA's Regulatory Requirements for Medical Devices workshop manual illustrates this point:

> *Non-implanted blood access device:* This is a preamendment Class II device intended to provide access to the patient's blood for such purposes as hemodialysis. Devices available before 1976 are labeled for insertion into the femoral vein. New devices that are labeled for insertion into the subclavian vein may have the same intended use if the labeling of both devices indicates that the aim of both devices is to provide access to the patient's blood for similar purposes. (Parr, Barcome, Regulatory Requirements, p. 5-23)*

The differences in the labeling, in this case, relate to the method of use, not the intended use of the device. The FDA has determined that the method of insertion did not raise any new safety or effectiveness questions. If there are no other significant changes in the device (i.e., technology, design, an so on), the new device would be substantially equivalent.

Another example from the same FDA manual illustrates the latitude that the FDA has in making a determination of substantial equivalence:

> *Conventional dialyzer:** This is also a preamendment Class II device. The preamendment devices are labeled for use as part of an artificial kidney system for patients with renal failure. The principal purpose of the device is to remove excess water from the blood system. A new device was labeled for use as part of a heart-lung machine for removing excess water from the vascular system at the end of surgery. The FDA concluded that these devices are substantially equivalent. (Parr, Barcome, Regulatory Requirements, p. 5–23)*

The FDA concluded that the differences in labeling related to nonessential conditions and did not substantially affect the safety or effectiveness of the devices. Barring significant changes in the design or technology used in the device, these devices are substantially equivalent. The key question in both these examples is whether or not the label indications of the new device introduces questions about safety and effectiveness that are different from those of the predicate device.

---

* See 21 CFR 876.5540(a)(2).
** See 21 CFR 876.5820(a)(2).

In the previous examples, the FDA found that the differences in label indications were not significant enough to warrant a finding that the devices had different intended uses. For some new devices, however, modifications in label indications may result in a finding that the devices are not substantially equivalent, even though the intended effect of the new device is very similar to the predicate device. This is illustrated in the following example:

Long-term percutaneous intravascular catheter:* This preamendment Class II device consists of a slender tube labeled for insertion into the vascular system for extended periods in order to sample blood, monitor blood pressure, or administer drugs. The new device, labeled for use as a spinal access catheter, was found by the FDA to not be substantially equivalent. (Parr, Barcome, Regulatory Requirements, p. 5-24)

The FDA concluded that even though both devices were intended to deliver drugs into the body, they related to different body systems. The introduction of the device into the spinal canal raised significantly different safety concerns (i.e., risk to the spinal cord) that were not present in the predicate device.

For a device to be found substantially equivalent, the 510(k) application must demonstrate that its intended use does not differ from the intended use of the predicated device. It must also demonstrate that any differences in the labeling are irrelevant in terms of safety and effectiveness.

PMA APPROVAL

Class III devices are those for which general controls and performance standards are deemed to be insufficient to provide a reasonable assurance of safety and effectiveness. Generally speaking, Class III devices (a) are represented in their labeling to be life sustaining or life supporting, (b) are implanted into the body, or (c) present potential unreasonable risk of injury or illness. Under Section 515 of the FD&CA, all devices placed in Class III are subject to PMA requirements.

Manufacturers of preamendment Class III devices (i.e., those introduced before May 28, 1976) are not required to submit a PMA. However, the FD&CA provides that the FDA may call for a PMA on any preamendment Class III device by publishing an appropriate classification regulation—an option the FDA exercised for replacement heart valves,** as an example. Replacement heart valves are Class III devices for which PMAs were required to be filed with the FDA on or before December 9, 1987.

Manufacturers of postamendment Class III devices (i.e., those introduced after May 28, 1976) that are substantially equivalent to a preamendment device are subject to the same regulatory requirements as the preamendment device. The FDA determines substantial equivalence after reviewing the manufacturer's PMA, which is discussed earlier in this chapter. Any devices that have been proposed for marketing after May 28, 1976, that are judged not to be substantially equivalent to a preamendment device are automatically placed in Class III until they are reclassified by the FDA.

There are no specific regulatory requirements for labeling that are peculiar to devices that require PMA approval. However, copies of all proposed labeling for the device must be submitted as part of the PMA. This includes such things as installation instructions and any literature or advertising that is attached to the device, its packaging or wrappers, or that constitutes labeling under Section 201(m) of the FD&CA (Park, Poneleit p. 2-10).

---

* See 21 CFR 870.1250.

** A replacement heart valve is a device intended to perform the function of any of the heart's natural valves. See 21 CFR 870.3925.

As a condition of approval, however, the FDA may impose restrictions on the device, including its labeling. These restrictions may come in the PMA approval order, in regulations issued at the time of approval, or in regulations issued subsequent to approval. Such restrictions or conditions could include prominent display, in the labeling of the device and in its advertising, of warnings, hazards, or precautions important for the safe and effective use of the device. The restrictions could also include patient information on alternative modes of therapy and on the risks and benefits of using the device. The FDA can require the inclusion of identification codes on the device or its labeling. In the case of implantable devices, this could include an identification card given to the patient. If the device is sold to a medical practitioner, it is sufficient that the manufacturer supplies the cards directly to the practitioner for distribution to the patients and that the manufacturer makes a reasonable effort to obtain patient identity information from the practitioner (Park, Poneleit p. 2-22).

PMA Supplements (§814.39)

After a PMA is approved, a manufacturer must submit a PMA supplement for review and approval by the FDA before making a change affecting the safety or effectiveness of the device. The PMA holder bears primary responsibility for determining if a proposed change to a device warrants submission of a PMA supplement. The manufacturer is in the best position to make this determination, based on good judgment, adequate supporting data, and sufficient documentation. In 21 CFR 814.39, the FDA identifies some of the changes that require a PMA supplement if the change affects safety or effectiveness. This list includes changes to device labeling.

A manufacturer may make certain changes in a device after FDA approval of the PMA without submitting a PMA supplement. Changes that do not affect the safety or effectiveness of the device can be made without prior approval of the FDA. However, these changes must be reported to the FDA in the periodic (usually annual) postapproval report required as a condition of approval. An example of such a change would be an editorial change in labeling that does not affect the safety or effectiveness of the device. Trivial changes, such as changing the color of label stock, do not have to be reported in the periodic postapproval report (Park, Poneleit p. 3-13).

Special PMA Supplement (§814.39(d))

Normally, approval of a PMA supplement is required before a change affecting the safety or effectiveness of a device can be implemented. These include changes to the indications for use of the device and labeling changes. However, in certain circumstances, changes that enhance the safety of the device can be placed into effect prior to receipt of a written FDA order approving the PMA supplement. The "Special PMA Supplement" allows the manufacturer to quickly implement labeling changes, as well as other changes listed in the regulations, provided the labeling changes:

- add to or strengthen a contraindication, warning, precaution, or information about an adverse reaction;
- add to or strengthen an instruction that is intended to enhance the safe use of the device; or
- delete misleading, false, or unsupported indications.

Licensing of a PMA Approval

The owner of a PMA approval may enter into an agreement to permit another firm to manufacture and distribute a device under a private label. Before such a license is legal, the FDA must approve either a PMA supplement submitted by the original PMA holder or an original PMA submitted by the licensee that includes, or includes by authorized reference, the original approved PMA. Either

PMA submission must include copies of all required labeling (draft or final) and a description of all differences between the original approved PMA holder's and the licensee's labeling. This can be in the form of a markup of the original labeling, identifying the differences incorporated into the licensee's labeling.

Final PMA Labeling Review

A manufacturer's original PMA or PMA supplement will often contain draft labeling. The FDA will review the draft labeling to determine if it satisfies the regulatory requirements. As a condition of approval, the manufacturer is required to submit a PMA amendment containing the final printed labeling prior to commercial distribution of the device. At this stage, the submission of the printed labeling is a formality to assure that FDA has a copy of the final labeling in the PMA file. The FDA expects that the final printed labeling will be virtually identical to the final draft labeling reviewed prior to approval. Any substantive changes must be submitted for approval through a PMA supplement.

Some manufacturers have attempted to submit final printed labeling that is inconsistent with the draft labeling approved by the FDA. As a result, the FDA's Office of Device Evaluation (ODE) has established a policy to deal with manufacturers who attempt to "slip" substantive changes in device labeling through in hopes that the FDA will not notice the difference between the draft and final printed labeling. If such an inconsistency is found, the FDA, in addition to the application of any other remedies that may be appropriate, will require the PMA applicant to (Final Draft p. 2):

- Submit final printed labeling for all future PMAs or PMA Supplements, whether for the same or different devices, before the FDA will issue a letter of approval. Draft labeling will not be accepted.
- Submit, as a condition of approval of a PMA or PMA supplement, a report of all labeling changes that do not require a PMA supplement within five days from the date the changes are made.

INVESTIGATIONAL DEVICE EXEMPTIONS (IDE)

Section 520(g) of the FD&CA encourages, to the extent consistent with the protection of the public health and safety and with ethical standards, the discovery and development of useful devices for human use. This section of the FD&CA authorizes the FDA to grant an exemption to the requirements of Sections 502 (Misbranded drugs and devices), 510 (Registration of producers of drugs and devices), 514 (Performance standards), 515 (Premarket approval), 516 (Banned devices), 519 (Records and reports on devices), 520(e) (Restricted devices), 520(f) (Good manufacturing practice requirements), or 721 (Listing and certification of color additives) to permit investigational use of a device by experts qualified by scientific training and experience to investigate the safety and effectiveness of such devices.

All clinical investigations of devices to determine safety and effectiveness are covered by the regulations in 21 CFR 812. The regulations, however, do exempt certain investigations from the requirements in 21 CFR 812. These exemptions are briefly discussed later in this chapter.

Those studies that are not exempt from the IDE regulation by 21 CFR 812.2(c) are subject to different levels of regulatory control. The regulations differentiate between significant-risk device studies and nonsignificant-risk device studies. The procedures for approval of an IDE differ accordingly. A study involving a significant-risk device* requires advanced approval by the FDA

---

* A significant-risk device is defined in 21 CFR 812.3(m) to mean an investigational device that for various reasons may present a potential for serious risk to the health, safety, or welfare of a subject.

of an IDE application. Approval of the Institutional Review Board (IRB) is required before the study can begin at that institution. The investigational use of a banned device is always considered to pose a significant risk.

Nonsignificant-risk devices are considered to have an approved IDE application, unless the FDA has notified the sponsor that approval of an application is required. The responsibility for making the initial determination of whether the device presents a significant risk belongs to the device's sponsor, who submits a proposal for the study to the IRB for review. If the IRB agrees with the sponsor that the study does not pose a significant risk, no IDE application to the FDA is necessary.

Regardless of the type of study, the device must be labeled in accordance with the requirements in 21 CFR 812.5. Further, the sponsor of the study may not engage in any of the prohibited promotional or other practices listed in 21 CFR 812.7.

LABELING OF INVESTIGATIONAL DEVICES (§812.5)

The label on the investigational device or its immediate container must contain the following information:

- The name and place of business of the manufacturer, packer, or distributor. The address must include the street address, city, state, and zip code. The street address may be omitted if the firm's street address is listed in the current city directory or telephone directory. The requirement for name and place of business is discussed in detail in Chapter 21.
- If appropriate, the quantity of contents.
- The following statement:

> CAUTION—INVESTIGATIONAL DEVICE. LIMITED BY FEDERAL (OR UNITED STATES) LAW TO INVESTIGATIONAL USE.

- A description of all relevant contraindications, hazards, adverse effects, interfering substances or devices, warnings, and precautions.

An investigational device shipped solely for research on or with laboratory animals must bear the following statement:

> CAUTION—DEVICE FOR INVESTIGATIONAL USE IN LABORATORY ANIMALS OR OTHER TESTS THAT DO NOT INVOLVE HUMAN SUBJECTS.

PROHIBITED PRACTICES (§§812.5(B) AND 812.7)

As with any other regulated device, the labeling of an investigational device may not make any statement that is false or misleading in any particular. Further, the labeling of an investigational device cannot represent the device as safe or effective for the purpose for which it is being investigated.

A study sponsor, an investigator, or any person acting on their behalf may not:

- Promote or test market an investigational device until after the device has been approved by the FDA for commercial distribution. The FDA recognizes that a sponsor may need to recruit investigators and has established guidelines for advertising to obtain investigators. These guidelines are described in the following section.
- Commercialize an investigational device by charging the subjects or investigators a price higher than necessary to recover the costs of manufacture, research, development, and handling of the device.

- Unduly prolong an investigation.
- Make any representation that the investigational device is safe or effective for the purpose for which it is being investigated.

ADVERTISING OF AN INVESTIGATIONAL DEVICE (§812.7)

Although the sponsor of an investigational device is prohibited by 21 CFR 812.7 from "promoting" or "test marketing" an investigational device, the FDA recognizes that most sponsors need to obtain investigators under whose immediate direction the device can be administered, dispensed, or used involving a human subject.

As a guideline, the procedures listed below are not legal requirements, but they are acceptable to the FDA. Sponsors who follow the guidelines may be reasonably assured that their notices will not violate the requirements in 21 CFR 812.7. However, this guidance does not bind the FDA or the public. An alternative approach may be used if such an approach satisfies the requirements of the FD&CA and the regulations (Availability p. 7).

Advertising for Investigators

To obtain investigators, sponsors of investigational devices may place notices in health professional publications, send letters to potential investigators, or, through exhibits or oral presentations at professional meetings, announce that they are seeking investigators (Availability p. 1). To help sponsors of clinical investigations obtain investigators, the FDA has produced a guidance document entitled "Preparing Notices of Availability of Investigational Medical Devices and for Recruiting Study Subjects." Prepared by the FDA's Center for Devices and Radiological Health (CDRH), the guidance document contains the FDA's current thinking on preparing notices of the availability of investigational medical devices and for recruiting study subjects. It describes ways acceptable to the FDA for sponsors of clinical investigations to prepare and disseminate notices, mailings, exhibits, and oral presentations to inform potential investigators of the availability of an investigational device. The sponsor of a clinical study involving human subjects must (Availability pp. 1–4):

- Announce the availability of an investigational device only in medical or scientific publications or at medical or scientific conferences whose readership or audience is composed primarily of experts who are qualified by training and experience to investigate the safety and effectiveness of such a device.
- State in clear terms that the purpose of the notification is only to obtain investigators and not to make the device generally available. Enrolling more investigators or subjects than is necessary to establish the safety and effectiveness of the device will be considered promotion or commercialization of the investigational device. The appropriate number of investigators and subjects necessary to demonstrate the safety and effectiveness of the device is established in the investigational plan approved as part of the IDE application.
- Limit the information presented in any notice of availability to the following:
  - The proposed use of the device
  - The name and address of the sponsor
  - How to apply to be an investigator
  - How to obtain the device for investigational use
- The notice should further list the investigator's responsibilities during the course of the investigation, namely the need to:
  - Await IRB and FDA approval before allowing any subject to participate

- Obtain informed consent from the subject
- Permit the use of the device only on subjects under the investigator's supervision
- Report adverse reactions
- Keep accurate records
- Conduct the investigation in accordance with the signed agreement with the study sponsor, the investigational plan, FDA regulations, and whatever conditions of approval are imposed by the reviewing IRB or the FDA
- Use direct mailing for the sole purpose of soliciting qualified experts to conduct the investigation. The FDA considers an undirected mass mailing an inappropriate vehicle for soliciting clinical investigators. Such a mailing will be considered promotional.
- Print in the notice, in type at least as large as the largest printing in the notice, the investigational device caution statement ("CAUTION—Investigational device. Limited by Federal (or United States) law to investigational use.") An oral presentation must include a clear, unequivocal statement that the device is available only for investigational use.
- Make only objective statements concerning the physical nature of the investigational device.
- Ensure that no claims are made that state or imply, directly or indirectly, that the device is reliable, durable, dependable, safe, or effective for the purpose under investigation. The sponsor may not claim that the investigational device is in any way superior to any other device.
- Not present comparative descriptions of the investigational device with any other device, although the announcement may include reasonably sized drawings or photographs of the device.
- Not include information regarding pricing data, but information may be included stating where such data may be obtained.
- A sponsor or investigator should not offer volume discounts for an investigational device. The FDA would regard such discounts as promotional.

Recruiting Study Subjects

Direct recruiting of subjects to participate in a clinical study is acceptable under FDA guidelines. However, the sponsor and investigators should take the following into consideration when preparing direct recruiting advertisements (Availability pp. 4–6).

- The FDA sees direct recruiting advertisement as a part of the informed-consent and subject-selection process. The IRB should review the advertising to ensure that the information is not misleading. Such a review is especially critical when the study involves subjects that are likely to be vulnerable to undue influence.
- When direct advertising is used, the IRB should review both the information contained in the advertisement and the mode of its communication to determine that the procedure for recruiting subjects is not coercive and does not state or imply a certainty of favorable outcome or other benefits beyond what is outlined in the informed-consent document and the study protocol.
- The advertising must not make any claims, explicitly or implicitly, that the device is safe and effective for the purpose under investigation or that it is equivalent or superior to any other device. The phrase "new treatment" should not be used without explaining that the device is under investigation. A phrase like "receive new treatment" implies that all study subjects will receive a product of proven worth.
- The advertisement should not promise "free medical treatment" when the intent is only to say that the subject will not be charged for taking part in the study. Adverting may state that subjects will be paid, but should not emphasize the payment or state the amount to be paid.

The FDA takes the position that advertising to recruit subjects should be limited to the information the prospective subject needs to determine their eligibility and interest. The FDA recommends, but does not require, that the direct advertising include the following information (Availability pp. 6–7):

- The name and address of the clinical investigator and/or research facility
- The condition under study and/or the purpose of the research
- A summary of the criteria that will be used to determine eligibility for the study
- A brief list of participation benefits, if any (e.g., a no-cost health examination)
- The time or other commitments required of the subjects
- Who to contact for further information

MANUFACTURING PRACTICES FOR INVESTIGATIONAL DEVICES (§812.20(B)(3))

Investigational devices are exempt from the Good Manufacturing Practice (GMP) regulations contained in 21 CFR 820 unless the sponsor states an intention to comply with these requirements. However, investigational devices must be manufactured, processed, packaged, labeled, stored, and, where appropriate, installed under an effective quality assurance (QA) system. A description of the methods, facilities, and controls that make up the quality system must be described in the IDE application. The description must be sufficiently detailed so that a reviewer generally familiar with GMP can make a knowledgeable judgment about the quality control used in the manufacture of the investigational device.

While investigational devices are exempt from the GMP regulations, they are not exempt from the requirements to meet the standards of identity, strength, quality, and purity specified in section 501(c) of the FD&CA. Neither are they exempt from product liability actions (Park, Melvin p. 4-74).

Once an investigational device is approved for commercial release, it will be subject to the Quality System Regulation (QSR). Therefore, the FDA recommends that the original QA system be based on or exceed the device GMP requirements. The device GMP regulations, as they relate to labeling, are discussed in Chapter 26.

EXEMPTED IDE INVESTIGATIONS (§812.2(C))

This section lists categories of devices that are exempt from the IDE requirements in 21 CFR 812. Included are:

1. A device in commercial distribution before May 28, 1976, when used or investigated in accordance with the indications for use in the labeling that was in effect at that time.*
2. A device that has been determined by the FDA to be substantially equivalent to a device in commercial distribution before May 28, 1976, and that is used or investigated for the indications in the labeling reviewed by the FDA when making the substantial equivalence determination
3. A diagnostic device, if the sponsor complies with the labeling requirements in 21 CFR 809.10(c) and if the testing:
   - is noninvasive;
   - does not require an invasive sampling procedure that presents significant risk;
   - does not, by design or intention, introduce energy into the subject; and
   - is not used as a diagnostic procedure without confirmation of the diagnosis by another medically established diagnostic product or procedure.

---

* For Class II and Class III devices, this and the exemption in Item 2 expire on the date published in an FDA regulation or order that calls for submission of a premarket approval application (PMA) or establishes a performance standard for a Class II device.

This exemption has been used to allow the investigation of *in vitro* diagnostic (IVD) products without an IDE. The labeling requirements to qualify for this exemption are discussed in Chapter 22.

The FDA has been concerned for several years that some manufacturers have used this exemption to "commercialize" IVDs without appropriate approvals (Commercialization of IVDs, p. 3). The FDA has put manufacturers on notice that it intends to use its enforcement discretion to fix a time for manufacturers to either obtain appropriate approval or withdraw from the market IVDs that are labeled for investigational or research use.

4. A device undergoing consumer-preference testing, testing of a modification, or testing of the combination of two or more commercially distributed devices, if the purpose of the testing is not to determine safety or effectiveness and does not put the subject at risk.

5. A device intended solely for veterinary use.

6. A device shipped solely for research on or with laboratory animals and labeled in accordance with 21 CFR 812.5(c).

7. A custom device, unless that device is being used to determine safety or effectiveness. Custom devices are discussed later in this chapter.

While these devices are exempt from the requirements of 21 CFR 812, they are not exempt from the requirement for IRB review and approval under 21 CFR 55, and are subject to informed consent under 21 CFR 50 if they involve human subjects.

INVESTIGATIONAL USE OF MARKETED PRODUCTS

Good medical practice requires physicians to use commercially available medical devices according to their best knowledge and judgment. If a physician uses a marketed device for an indication not in the approved labeling, he or she has a responsibility to the patient to (a) be well-informed about the device, (b) base his or her decision to use the device on sound scientific rationale and medical evidence, and, (c) maintain appropriate records of the device's use and effects. Such a use of the device is a recognized part of the "practice of medicine" and does not require the submission of an IDE or the review of an IRB, unless the institution requires an IRB review where the device is to be used.

The investigational use of a marketed product, however, suggests the use of the device in the context of a study protocol. When the principal intent of the investigational use of a marketed device is to develop information about the safety or efficacy of uses other than those in the approved labeling, the submission of an IDE is generally required (Park, Melvin p. 4-70).

INVESTIGATIONAL EXEMPTION FOR INTRAOCULAR LENSES (IOLS)

On March 31, 1997, Section 21 CFR 813 was removed from the CFR. Henceforth, investigation of intraocular lenses (IOLs) will be governed by the IDE regulation in 21 CFR 812. There are several differences between Parts 812 and 813. To assist sponsors of IOL studies, FDA prepared a letter to IOL sponsors detailing the changes and offering its suggestions on how to address these differences.

There are two requirements that were unique to Part 813. The following will continue to be required under Sections 812.20(b)(11) and 812.20(b)(1) respectively (Revocation p. 2):

- Section 813.155(b)(2) required the sponsor to provide an identification card directly or indirectly to the subject of a clinical study. The identification card will continue to be required.

- Section 813.5(a) stated that the label of an IOL should bear the sterility shelf life of the lens. This labeling requirement will continue to be required.

Custom Devices (§812.3(b))

Section 520(b) of the FD&CA exempts devices that, in order to comply with an order of an individual physician or dentist, deviate from the otherwise-applicable performance standards promulgated under section 515, or the premarket approval requirements in section 515.

A custom device is defined as a medical device that:

- necessarily deviates from devices that are generally available or from applicable performance standards or premarket approval requirements in order to comply with an order of an individual physician or dentist,
- is not generally available or generally used by other physicians,
- is not generally available in finished form for purchase or dispensing by prescription,
- is not offered for commercial distribution through labeling or advertising, and
- is intended (a) for use by the individual patient named in the order and is made in a specific form for that patient or (b) to meet the special needs of the physician or dentist in the course of professional practice.

A custom device is also granted special status under 21 CFR 812.2(c)(7) in that its use is exempted from the requirements for an investigational device.

HUMANITARIAN DEVICE EXEMPTION (HDE)

In 1996, the FDA promulgated regulations to implement Section 530(m) of the FD&CA. This section of the FD&CA is intended to encourage the discovery and use of devices that would benefit patients in the treatment and diagnosis of diseases or conditions that affect fewer than 4,000 individuals in the United States. A humanitarian device exemption (HDE) is an application that is similar to a PMA, but exempt from the effectiveness requirements of a PMA (HDE §1).

The FDA may grant a request for an exemption from the effectiveness requirements of Sections 514 and 515 of the FD&CA for a device that (21 CFR 814.100):

- is designed to treat or diagnose a disease or condition that affects fewer than 4,000 individuals in the United States (the FDA makes this determination at the time of the request for humanitarian use device [HUD] designation.);
- would not be available to a person with a disease or condition referred to above unless the FDA grants such an exemption;
- will not expose patients to an unreasonable or significant risk of illness or injury, and the probable benefit to health from the use of the device outweighs the risk of injury or illness from its use, taking into account the probable risks and benefits of currently available devices or alternative forms of treatment, and
- there is no comparable device, other than under this exemption, available to treat or diagnose such disease or condition.

A device that meets these criteria can be designated a humanitarian use device (HUD).

Obtaining marketing approval for an HDE involves two steps:

1. obtaining designation of the device as an HUD from the FDA's Office of Orphan Products Development, and
2. submitting an HDE application to the FDA's ODE.

A person granted an exemption under section 520(m) of the FD&CA must submit periodic reports as described in 21 CFR 814.126(b). The FDA may suspend or withdraw approval of an HDE after providing notice and an opportunity for an informal hearing.

A manufacturer submitting an HDE application must provide the same labeling as required for a PMA as described in 21 CFR 814.20(b)(10). In addition, the labeling must bear the statement (21 CFR 814.104(b)(4)(ii)):

> HUMANITARIAN DEVICE. AUTHORIZED BY FEDERAL LAW FOR USE IN THE [TREATMENT OR DIAGNOSIS] OF [SPECIFY DISEASE OR CONDITION]. THE EFFECTIVENESS OF THIS DEVICE FOR THIS USE HAS NOT BEEN DEMONSTRATED.

The FDA will approve an HDE if none of the reasons in 21 CFR 814.118 for denying approval of the application applies. The FDA can approve an application using draft final labeling, but only if all the deficiencies in the application are editorial or similar minor deficiencies in the draft final labeling. Such an approval is conditional and depends on the applicant incorporating the specified labeling changes exactly as directed. The applicant is required to submit a copy of the final printed labeling before marketing the device.

THINGS TO REMEMBER

Any person who proposes to bring a device intended for human use to market is required by Section 510 of the FD&CA to report to the FDA his or her intention at least 90 days before introducing the device into interstate commerce.

The Medical Device Amendments of 1976 directed the FDA to classify all medical devices into one of three regulatory categories depending on the degree of regulation necessary to provide reasonable assurance of their safety and effectiveness. Most devices are in Class I and Class II, and are subject to the general regulatory controls requiring premarket notification under Section 510(k) of the FD&CA. Class III devices are subject to the more extensive provisions of Section 515 of the FD&CA, which require FDA approval of the safety and effectiveness of the device. There is a special exemption for investigational devices. This exemption allows a device that is properly labeled to be used in a human subject under controlled conditions in order to develop the data necessary to support the manufacturer's contention that the device is safe and effective for its intended use.

Section 510(k) of the FD&CA contains the premarket notification requirements. A manufacturer of a new or significantly modified device must notify the FDA in order to establish substantial equivalence in terms of safety and effectiveness to a preamendment device or a postamendment device that has been reclassified from Class III to Class I or II. In verifying the manufacturer's claim, the FDA typically determines, by reviewing the device's labeling, if the device performs the same function and falls within an established type as a preamendment device. If the differences do not raise new safety and effectiveness questions for the same intended use, the new device is found substantially equivalent.

Class III devices are those for which general controls and performance standards are deemed to be insufficient to provide reasonable assurance of safety and effectiveness. Under Section 515 of the FD&CA, all devices placed in Class III are subject to PMA requirements.

Manufacturers of preamendment Class III devices (i.e., those introduced before May 28, 1976) are generally not required to submit a PMA. Manufacturers of postamendment Class III devices (i.e., those introduced after May 28, 1976) that are substantially equivalent to a preamendment device are subject to the same regulatory requirements as the preamendment device.

Section 520(g) of the FD&CA encourages, to the extent consistent with the protection of the public health and safety and with ethical standards, the discovery and development of useful devices for human use. This section of the FD&CA authorizes the FDA to grant an exemption to permit the investigational use of devices by experts qualified by scientific training and experience to investigate the safety and effectiveness of such devices. All clinical investigations of devices to determine safety and effectiveness are covered by the regulations in 21 CFR 812.

Those studies that are not exempt from the IDE regulation by 21 CFR 812.2(c) are subject to different levels of regulatory control based on the degree of risk to the patient. A study involving a significant-risk device requires the advanced approval by the FDA of an IDE application. Nonsignificant-risk devices are considered to have an approved IDE application, unless the FDA has notified the sponsor that approval of an application is required.

Regardless of the type of study, the device must be labeled in accordance with the requirements in 21 CFR 812.5. Further, the sponsor of the study may not engage in any of the prohibited promotional or other practices listed in 21 CFR 812.7.

Two other kinds of exempt devcies are custom devices and HUDs. A custom device is intended for use on a single patient or to meet the special needs of the physician or dentist in the course of professional practice. A HDE is similar to a PMA, but is exempt from the effectiveness requirements of a PMA. An approved HDE authorizes the marketing of a HUD.

25 Good Manufacturing Practice in Labeling

Section 520(f) of the Federal Food, Drug, and Cosmetic Act (FD&CA) requires that domestic or foreign manufacturers of medical devices intended for commercial distribution in the United States have a quality assurance (QA) program. The purposes of the QA program are to ensure that devices conform to their specifications and to help prevent the production of defective products that can endanger consumers. Section 520(f) of the FD&CA further states that the Secretary of the Department of Health and Human Services (DHHS) may prescribe regulations requiring that the methods used in, and the facilities and controls used for, the manufacture, pre-production design validation (including a process to assess the performance of a device but not including an evaluation of the safety or effectiveness of a device), packing, storage, and installation of a device conform to current good manufacturing practice (FD&CA, § 520(f)(1)(A)). The Good Manufacturing Practice (GMP) regulations, which were promulgated under the authority of this section of the FD&CA, require that various specifications and controls are established for medical devices and that finished devices meet these specifications.

The authority for applying GMP to medical devices was added to the FD&CA as a part of the Medical Device Amendments of 1976. These regulations, which are located in 21 CFR 820, specify the methods used in, and the facilities and controls used for, the manufacture, packaging, storage, and installation of a device. Except for editorial changes to update organizational references in the regulations and revisions to the list of critical devices included in the preamble of the regulation, the device GMP requirements were not revised for more than 15 years.

The Safe Medical Devices Act of 1990 (SMDA) amended section 520(f)(1) of the FD&CA to provide clear authority to add preproduction-design-validation controls to the scope of the GMP regulations. Under this expanded authority, the Food and Drug Administration (FDA) undertook a major overhaul of the GMP regulations to (1) add design, purchasing, and servicing control; (2) modify the critical device requirements; (3) revise certain existing requirements; and (4) harmonize the GMP regulations for medical devices with the quality system specifications in the 1994 edition of ISO 9001, *Quality management systems – Requirements*.

The new regulation was published as a final rule in October 1996 under the title "Quality System Regulation" (QSR). This regulation sets out the current good manufacturing practice (cGMP) requirements. These requirements establish "the methods used in, and the facilities and controls used for, the design, manufacture, packaging, labeling, storage, installation, and servicing of all finished devices intended for human use" (21 CFR, § 820.1(a)(1)).

The QSR covers QA programs and organizations, buildings, equipment, components, production and process controls, packaging and labeling controls, distribution and installation, device evaluation, and records. The 1976 regulation established a few additional controls, including controls on labeling for "critical devices." A critical device was defined as "a device that is intended for surgical implant into the body or to support or sustain life and whose failure to perform when properly used in accordance with instructions for use provided in the labeling can be reasonably expected to result in a significant injury to the user." The QSR does not use the term "critical devices." However, 21 CFR 820.65 requires that a manufacturer of a device that meets this definition establish and maintain a traceability system for the device. For evaluating when this is needed, the FDA uses the 1988 publication, *Advisory List of Critical Devices – 1988*. This list was published in the *Federal Register* on March 17, 1988.

Section 21 CFR 820.5 requires that "each manufacturer shall establish and maintain a quality system that is appropriate for the specific medical device(s) designed or manufactured, and that meets the requirements of this part." Within the framework of the regulations, the manufacturer has some flexibility when establishing "appropriate" requirements for each type or family of device. It is the manufacturer's responsibility to determine the specific quality-system elements necessary and to develop and implement procedures tailored to the particular device and its manufacturing process. When a manufacturer decides to not implement a provision of the QSR that is qualified by the term "where appropriate," it must document its justification for not implementing the requirement. The justification should show that not implementing a requirement is not reasonably expected to result in a product that does not meet specifications or a failure to carry out any necessary corrective action (21 CFR 820.1(a)(3)). An example cited in the FDA's GMP manual states, "For example, if it is impossible to mix up labels at a manufacturer because there is only one label or one product, then there is no necessity for the manufacturer to comply with all of the GMP requirements under device labeling" (Lowery §1).

GMP APPLICATIONS

Any finished device intended for human use that is commercially distributed in the United States is covered by the GMP regulations unless there is an approved exemption. However, the regulations apply equally to domestic and foreign manufacturers.

The GMP regulations also apply equally to all medical device manufacturers regardless of size. The FDA does not grant any exception to the GMP regulations to firms because they are small. The FDA has recognized that small firms may not need all of the same systems as large firms to achieve control over their manufacturing processes. For example, a small manufacturer may need less documentation than a large manufacturer in order to achieve a state-of-control (Lowery, §1). Manufacturers must use good judgment when applying the GMP regulations to their specific products and operations (QSR §1).

Some system components, such as blood tubing and major diagnostic X-ray components, may be considered finished devices because they are accessories to finished devices. The manufacturers of such components are subject to the GMP regulations when the accessories are labeled and sold separately from the primary device for health-related purposes to hospitals, physicians, or other users (Lowery §1).

Some firms are exempt from the GMP regulations or are not routinely inspected. These firms are, however, subject to the FD&CA. If they produce devices that are adulterated and/or misbranded they are subject to the penalties outlined in the FD&CA. The principal exemptions to the GMP regulations are briefly discussed in the following paragraphs. It should be noted that a device labeled as sterile is never exempted from the GMP regulations (Lowery §1).

GMP Exemptions

The FDA has determined that certain establishments are exempt from some or all of the QSR. An exemption or variance may be granted by the FDA in response to a citizen's petition, through an FDA policy statement, through an exemption published in the *Federal Register*, or by FDA classification regulations published in the *Federal Register*. The tongue depressor* is an example of a device that is exempted by its classification regulation from most aspects of the GMP regulations because minimal manufacturing controls are required in order to produce a device that meets its specifications. An investigational device that meets the requirements of the investigational device exemption (IDE) regulation (21 CFR 812) is exempt from the requirements in 21 CFR 820.

Custom-Device Manufacturers

A custom device is exempt from some of the statutory requirements of the FD&CA. The conditions that must be present for a device to be recognized by the FDA as a custom device are described in Chapter 24. While a custom device is not required to comply with some aspects of the FD&CA, a custom-device manufacturer is subject to the QSR including those aspects of the regulation dealing with control of labeling and labeling operations. The FDA has not made a practice of inspecting custom-device manufacturers (Lowery §1). However, a custom-device manufacturer is required to comply and could be subject to penalties if found to be out of compliance.

Contract Manufacturers

A firm that manufactures, packages, labels, and/or sterilizes a finished device under the terms of a contract with another firm is a contract manufacturer. The arrangement between the firms must be documented in a written contract. The contract manufacturer must comply with the applicable requirements of the QSR and must register its facilities with the FDA.

The firm whose name appears on the product is legally liable for all the shortcomings of its subcontractors. However, both firms may be subject to regulatory action if the subcontractor is found to be out of compliance with the applicable QSR requirements. The FDA has used its enforcement authority to increase scrutiny of 510(k) notices and premarket approval applications (PMAs) for firms whose contractors are having QSR problems (Contract Manufacturers p. 6).

Contract Testing Laboratories

Contract laboratories that design or test components or finished devices for a manufacturer are considered an extension of the manufacturer's quality system (Lowery §1). These laboratories may provide services to a number of customers, many of which are not medical device manufacturers. Contract laboratories are not subject to routine GMP inspections. However, the finished-device manufacturer is responsible for assuring that equipment and procedures used by a lab are adequate and appropriate (21 CFR 820.50). A manufacturer's internal test laboratory, even if it does testing for various facilities within the firm, is subject to FDA inspection (Lowery §1).

Component Manufacturers

A component is defined in 21 CFR 820.3(c) as "any raw material, substance, piece, part, software, firmware, labeling, or assembly which is intended to be included as part of the finished, packaged, and labeled device." Manufacturers who supply components or parts of finished devices are not subject to the QSR. However, the FDA encourages these manufacturers to use appropriate provisions of the QSR as guidance. In the preamble to the original GMP regulations, the FDA amplified on this requirement. The FDA agreed with comments stating that the GMP regulations should apply only to finished-device manufacturers, because manufacturers of components may supply only a

---

* A tongue depressor is a device intended to displace the tongue to facilitate examination of the surrounding organs and tissues. The classification regulation in 21 CFR 880.6230 exempts this Class I device from all GMP requirements except general requirements for records (21 CFR 820.180) and complaint files (21 CFR 820.198).

fraction of their production to finished-device manufacturers. Further, it was suggested that components for many finished devices are readily available in the marketplace and are not manufactured exclusively for use in medical devices (GMP p. 31512). A printer, for example, who supplies labels for a medical device would not be subject to GMP regulations. The FDA generally holds the final-device manufacturer responsible for establishing assessment criteria for suppliers that specify the requirements, including quality requirements, that the suppliers must meet.

A manufacturer who produces components specifically for use in a finished device that the firm produces, whether in the same or a different location, is subject to the QSR for the component-manufacturing operation. For example, a firm that produces a printed circuit board in one location and ships it to another facility for assembly into a medical device would be subject to QSR requirements at the printed-circuit-board manufacturing operation. The FDA considers the component manufacturing to be part of the finished device manufacturing operation (Lowery §1).

REMANUFACTURERS

A remanufacturer is defined as "any person who processes, conditions, renovates, repackages, restores, or does any other act to a finished device that significantly changes the finished device's performance or safety specifications, or intended use" (21 CFR 820.3(w)). The FDA considers remanufacturers to be manufacturers subject to the QSR. They must establish and implement quality systems to assure the safety and effectiveness of the devices that are distributed. Remanufacturers are required to comply with the labeling requirements of 21 CFR 801.1(c). However, since the firm is not the original manufacturer of the device, the name must be qualified by an appropriate phrase that reveals the connection of that firm with the device. For example, the name of a remanufacturer could be qualified by the phrase "Remanufactured by …." (Lowery §1).

REPACKERS AND RELABELERS

A firm that repackages or otherwise changes the container, wrapper, or labeling of any device packaging in furtherance of the distribution of the device from the original place of manufacture to the person who makes the final delivery or sale to the ultimate consumer is considered a manufacturer and is subject to the applicable requirements of the QSR. A firm is a repacker or relabeler if it (Lowery §1):

- packages and/or labels previously manufactured finished devices or accessories;
- receives finished devices in bulk (e.g., surgical tubing, syringes), repacks them into individual packages, and labels them; or
- receives previously manufactured devices that have been packaged and labeled by another manufacturer, and combines them into a kit with other unpackaged devices that are received in bulk.

For applying GMP regulations, a firm is not considered a repacker or relabeler if it only places previously packaged and labeled devices into another package for convenience. From the FDA point of view, such activity "is essentially the same as a drug store employee placing packaged items into a bag labeled with the name of the drug store." (Lowery, Puleo §1).

A distributor that adds a label bearing its name and address is not considered a relabeler and would not be required to maintain records demonstrating compliance with the labeling control requirements of the QSR. A manufacturer who simply affixes a sticker label bearing a distributor's name and address would not require record keeping demonstrating compliance with labeling controls requirements for that operation (Lowery §1).

SPECIFICATION DEVELOPER

A specification developer is a firm that provides specifications to another manufacturer who, in turn, produces a finished device that meets the specifications. The firm that produces the device under terms of a contract with the specification developer is called a contract manufacturer. The contract manufacturer may package and label the device or it may be sent to the specification developer for packaging and labeling. Contract manufacturers of finished devices must comply with the applicable requirements of the GMP regulations. The specification developer and the contract manufacturer may be held jointly responsible for the manufacturing processes performed (Lowery §1).

If a product carries the specification developer's label, the specification developer is responsible for maintaining a complaint file and processing complaints, and for maintaining the specifications and other appropriate parts of the device master record.*

QSR LABELING CONTROLS

A medical device manufacturer's QA program must incorporate a number of elements that relate to labeling in order to comply with the QSR. Labeling includes equipment labels, control labels, package labels, directions for use, maintenance manuals, and so on. Displays on electronic-message panels and cathode ray tube (CRT) screens are considered labeling if instructions, prompts, cautions, or parameter-identification information are provided (Lowery §11).

The QA program must assure that the labeling meets the requirements in the device master record with respect to legibility, adhesion, and so on, and assure that labeling operations are controlled so that the correct labeling is always used.

The original GMP regulations dealt primarily with control of labeling within the manufacturing process. Included were inspection, handling, storage, and distribution requirements for device labeling. The regulations required that the manufacturer have physical facilities of suitable design that provided sufficient space for packaging and labeling operations. Also covered was the design requirement that labels remain legible during the customary conditions of processing, storage, handling, distribution, and use. Failure to meet these requirements resulted in adulteration of the device.

The QSR expanded the scope of the GMPs to include preproduction-design processes, including the design of labeling. Failure to comply with the applicable requirements, such as proofreading and change control, could lead to errors in content. In such cases, the device would be misbranded and adulterated.

A number of the sections of the QSR directly affect labeling, including Sections 820.80 and 820.70(f). Section 820.120 deals with specific requirements for device labeling. These sections deal with physical design applications of labeling to assure legibility under normal conditions of use over the expected life of the device. They help assure the proper inspection, handling, storage, and distribution of labeling.

The requirements in 820.30(c) address the intended use of the device and the needs of the user and patient. Proper design and procurement of the labels and labeling is necessary to ensure "adequate

---

* A device master record is a term used in the QSR to describe all of the routine documentation required to manufacture a device that will consistently meet the firm's requirements. In 21 CFR 820.3(j), a device master record is described as "a compilation of records containing the procedures and specifications for a finished device." The contents of the device master record are described in 21 CFR 820.181.

labeling" for a device. Labeling design includes generating labeling content to make sure the content meets FDA requirements as well as the needs of the customer.

The rest of this section contains a discussion of some of the requirements of the QSR. The section numbers (e.g., §820.30) of the corresponding section of the QSR is listed with each topic for reference.

CONTENT DEVELOPMENT AND APPROVAL (§820.30)

The FDA notes that the establishment of labeling requirements is an important element of design input (cGMP p. 52618). A manufacturer must consider a variety of factors when establishing design and labeling requirements. These include the safety needs of the user, the environment in which the device will be used, reliability, safeguards against misuse, and, where applicable, maintainability and serviceability. Adequate maintenance instructions must be provided, so that the user can maintain the safety and effectiveness of the device. Points to consider in the design of good labeling are discussed in Chapter 27 and Chapter 28.

All labeling, except labels and cartons, that includes information for use of the device must include the date that the labeling was issued or last revised (21 CFR 801.109(e)).

The total finished design output consists of the device, its packaging and labeling, the associated specifications and drawings, and the production and quality-system specifications and procedures. All of these elements are included in the device master record. In addition, each manufacturer must establish and maintain a design history record* for each device.

DOCUMENT CONTROLS (§820.40)

To meet the QSR, a manufacturer must establish and maintain document-control procedures covering all the documents required to demonstrate compliance with this regulation. These procedures must provide for document approval and distribution and change control.

Document Approval and Distribution (§820.40(a))

The manufacturer must designate one or more individuals who review all documents required to meet the QSR for accuracy and approve them prior to issuance. This includes the specifications for labels and labeling. The approval, including the date and signature of the individual(s) approving the document, must itself be documented. The documents needed to meet the requirements of the QSR must be available at all locations for which they are designated, used, or otherwise necessary, and all obsolete documents must be promptly removed from all points of use or otherwise prevented from unintended use.

Document Changes (§820.40(b))

The QSR requires that changes to specifications, methods, or procedures for components, finished devices, manufacturing materials, production, installation, servicing, or quality systems be documented, reviewed, and approved by individuals in the same functions/organizations that performed the original review and approval, unless specifically designated otherwise. In addition, any change to a specification, method, or procedure that may affect the quality** must be validated as adequate

---

* A design history record is a term used in the QSR to describe all of the routine documentation required to demonstrate that the design was developed in accordance with the approved design plan. In 21 CFR 820.3(e), a design history record is described as "a compilation of records which describes the design history of a finished device."

** Quality is defined in 21 CFR 820.3(s) as "the totality of features and characteristics that bear on the ability of a device to satisfy fitness-for-use, including safety and performance."

for its intended use before it is approved and issued. The FDA has concluded from a review of recall data that many recalls are the result of a failure to validate changes to ensure that the changes are adequate for their intended purpose.

The manufacturer must also have a procedure in place to determine whether the change to the device or to the manufacturing-process specifications, including packaging and labeling specifications, is significant enough to require the submission of a premarket notification (510(k)) under 21 CFR 807.81(a)(3) or the submission of a supplement to a PMA under 21 CFR 814.39(a). A record of this evaluation and its results must be maintained.

A record of changes to documents must be maintained. The information required to be included in the documentation-change record is detailed in 21 CFR 820.40(b).

LABEL INTEGRITY (§820.120(A))

A manufacturer must ensure that the labels are designed, printed, and, where applicable, applied in such a way that they remain legible and affixed to the device during the customary conditions of distribution, storage, and use. Design, in this context, refers to the materials chosen for the labeling, the placement of the labels, and the mounting methods—and not the content of the labels. For example, labels printed onto plastic *in vitro* diagnostic (IVD) media plates would be considered unacceptable if they can smear. Another example would be "wet" inks, used in many magazines, which smear when touched by sweaty or oily fingers. This type of ink would be unacceptable for package inserts, instruction manuals, and similar types of device labeling. Temperature and humidity exposure during shipping and storage can cause some glues used to affix labels to packaging to loosen, resulting in labeling becoming detached from the device or its packaging. Finally, labels must be placed so they will be seen but not abraded during use. An example of improper placement would be locating a safety label on the footrest area of a ladder, where it would likely be scuffed off after a few uses.

PURCHASING (§820.50)

Each manufacturer must establish and maintain procedures to ensure that all components, including packaging and labeling material, conform to specifications in the device master record. These procedures include establishing and maintaining assessment criteria for suppliers. Potential suppliers must be able to meet requirements, including quality requirements, based on the documented assessment criteria.

The manufacturer must establish purchasing forms that clearly describe or reference the specifications for the components. The purchasing forms must include an agreement that the supplier will notify the manufacturer of any changes in the product so that the manufacturer may determine whether the change may affect the quality of the finished device. The approval, including the date and signature of the individual(s) approving the form, must be documented.

RECEIPT AND INSPECTION (§§820.80 AND 820.120(B))

The manufacturer must establish an appropriate mix of supplier and incoming controls to ensure that purchased components, including packaging and labeling materials, are acceptable for their intended use. The language in the regulations is purposely structured to provide the manufacturer with some latitude in deciding whether to carry out acceptance procedures at the location of the supplier under contract, in-house, or using some combination of these approaches. The manufacturer may rely on the component supplier to perform testing if the manufacturer specifies or is knowledgeable about the supplier's quality system, particularly the inspection and test programs, and if the supplier has specifications that properly define the manufacturer's acceptable limits for the component or material parameters (Lowery §6). However, incoming components, manufacturing materials, and finished devices may not be used or processed until they have been verified as conforming to specified requirements, and their acceptance or rejection by the designated individual(s) is documented.

Labeling must be examined for accuracy, including, where applicable, the correct expiration date, control number, storage instructions, handling instructions, and additional processing instructions. The labels may not be placed in storage or used until they have been examined by the designated individual(s) and their release has been approved. The release, including the date, name, and signature of the individual performing the examination, must be documented in the device history record.*

If a labeling mistake cannot occur because a firm makes only one device and there are no control numbers or expiration dates, then a receiving inspection is adequate to check compliance with the specifications in the device master record. If, however, there is any possibility that a mistake could occur, additional inspections are needed at the time the labeling is issued for use to verify that the necessary information is present and correct. An example would be inspecting to see that the correct expiration date is used.

If expiration dating is used, the expiration date must reflect the period after final packaging during which the device is fit for its intended use when stored and used according to the instructions included in its labeling. The manufacturer should have stability-test data that establish the period during which the device remains fit for use.

HANDLING (§§820.60 AND 820.140)

Each manufacturer must establish and maintain procedures to ensure that mix-ups, damage, deterioration, or other adverse effects to components, finished devices, and manufacturing materials do not occur during any stage of handling.

To ensure orderly handling, the manufacturer must establish and maintain procedures for identifying components, manufacturing materials, and finished devices until they leave the manufacturer's control.

STORAGE (§§820.120(c) AND 820.150)

Printed packaging and labeling material, including preprinted containers, must be stored in an area and in a manner that provides proper identification of the material and is designed to prevent mix-ups. Any control number or other identification used must be legible and clearly visible while the material is in storage.

The manufacturer must establish and maintain procedures for authorizing the movement of material in and out of designated storage areas. Access to these areas should be limited to authorized personnel.

If materials, including packaging and labeling, deteriorate over time, they must be stored in such a way as to facilitate proper stock rotation. Procedures must be in place to ensure that all obsolete, rejected, or deteriorated material in storage is not inadvertently used.

Storage controls should be appropriate for the numbers and kinds of devices manufactured in a facility. A firm that manufactures a single product would not need an extensive control system for labeling storage. Similarly, a firm that manufactures several devices having dissimilar labeling would not need as elaborate a system as one that manufactures products where the labeling could easily be mixed up.

---

* A device history record is a term used in the QSR to describe all of the documentation required to construct the history of a completed device. In 21 CFR 820.3(i), a device history record is described as "a compilation of records containing the production history of a finished device."

Labeling and Packaging Operations (§§820.70 and 820.120(d))

A manufacturer must provide an environment that provides the proper conditions for all operations, including packaging and labeling. Conditions to be considered include lighting, ventilation, space, temperature, humidity, air pressure, filtration, airborne contamination, static electricity, and other environmental considerations. The manufacturer must periodically inspect its facilities and review its control systems to verify that the systems are adequate and functioning properly. Records of such inspections must be made and reviewed.

Labeling and packaging operations must be controlled to prevent labeling mix-ups between similar products or labels. The extent of production-area controls will be determined by the likelihood that a labeling mix-up could occur. A firm that manufactures a single product would need to take far fewer precautions to avoid mix-ups than would a firm that manufactures similar products in the same facility. One accepted technique for preventing labeling mix-ups is to physically separate labeling and packaging operations, or to perform the operations at different times for different devices. If a mix-up could occur in the production area, it is important that the area be carefully examined to make sure that any devices and labeling left over from the previous operation are removed before beginning a different operation. The surrounding area, tables, packaging lines, printing machines, and other equipment must be cleared of labels and other material used in the previous operation.

The FDA recommends that labeling containing precoded serial numbers, manufacturing dates, expiration dates, or other control codes be destroyed and not returned to storage. The QSR does not require the reconciliation of the number of labels used with the number issued, although this control is recommended in some instances, such as when different sizes of the same device are being packaged or otherwise labeled (Lowery §11).

Nonconforming product (§820.90)

A manufacturer must establish and maintain procedures to control product that does not conform to specified requirements, including requirements for labels and labeling. These procedures must address the identification, documentation, evaluation, segregation, and disposition of nonconforming product. The manufacturer must establish and maintain procedures for rework of nonconforming products. In the case of labels and labeling, these procedures must include reevaluation of the nonconforming product after the rework is completed to ensure that the product meets its current approved specifications. The rework and reevaluation activities are documented in the device history record.

Distribution (§§820.80(d) and 820.160)

A manufacturer must establish and maintain procedures for accepting finished devices to ensure that each production run, lot, or batch of finished devices meets acceptance criteria. Finished devices are to be held in quarantine or otherwise adequately controlled until released. The conditions that must be met for a device to be released for distribution are listed in 21 CFR 820.80(d).

Just as with components, the manufacturer must provide storage facilities for finished devices that prevent mix-ups and ensure that only those devices approved for release are distributed. When a device's fitness-for-use or quality deteriorates over time, the manufacturer's procedures must ensure that the oldest approved devices are distributed first and that expired devices are not distributed.

Once again, the extent of controls will depend on the numbers and types of finished devices being distributed. One case that requires special attention is the control of devices prelabeled as "sterile" but not yet sterilized. The manufacturer must be absolutely certain that such a prelabeled but unsterilized device is not released from a controlled area. Shipping of devices labeled as sterile for sterilization is discussed later in this chapter.

Special caution must be used for such devices that are used as samples. These market promotion devices either must be sterile or must be conspicuously marked with a caution statement because otherwise the recipient might use a packaged and labeled market-promotion device. 21 CFR 820.25(b) requires quality-awareness training for all personnel. This includes sales and marketing personnel, who must be informed of labeling-control requirements and the consequences of a violation (Lowery §11).

Traceability (§§820.65 and 820.120(e))

The manufacturer of a device that is intended for surgical implant into the body or to support or sustain life and whose failure to perform when properly used in accordance with instructions for use provided in the labeling can be reasonably expected to result in a significant injury to the user must establish and maintain traceability procedures for the device. These procedures are intended to facilitate corrective action should a defect or other problem with the device, including problems with its labeling, be identified.

The labeling of such a device must contain a control number* for traceability. The control number identifies each unit, batch, or lot of finished devices, and, where appropriate, critical components of the finished device. The control number is recorded in the device history record. The control number must be on or accompany the device through distribution.

The control number for traceability is not required on every label on the device. It must, however, appear on the label that reaches the ultimate user (Lowery §11). The label on a shipping carton, for example, would not meet the traceability requirement because the carton is likely to be discarded before the device reaches the user.

The control number applies to the finished device and not to the labeling. However, most labels will contains a separate number, such as a drawing number, which is used to identify and control labeling materials in the manufacturing process.

Complaint Handling (§820.198)

Every manufacturer must maintain complaint files and have established procedures for receiving, reviewing, and evaluating complaints. The manufacturer must review, evaluate, and investigate all complaints involving possible failure of labeling or packaging to meet any of its specifications. Complaints pertaining to death, injury, or any hazard to safety shall be immediately reviewed, evaluated, and investigated by a designated individual(s). Events that result in a death or serious injury are reportable under the medical device reporting regulation in 21 CFR §803. Labeling can be a factor and has to be considered when evaluating the complaint (21 CFR 803(d)).

When an investigation is conducted, a written record of the investigation is made and retained at a location formally designated by the manufacturer. The information required to be included in the record of investigation is detailed in 21 CFR 820.198(e). Records of investigations of events that are determined to be reportable under the medical device reporting requirements of 21 CFR 803 must also include the information required by §803. When this information cannot be obtained, a record of the reason must be maintained in the record of investigation.

---

* A control number, as defined in 21 CFR 820.3(d), is "any distinctive symbols, such as a distinctive combination of letters or numbers, or both, from which the history of the manufacturing, packaging, labeling, and distribution of a unit, lot, or batch of finished devices can be determined."

OVERLABELING

Overlabeling is the placing of a new label over an old label. In general, the FDA discourages this practice (Lowery §11). However, it is an acceptable practice as long as the new label and the processes for using it meet the QSR requirements for attachment, legibility, reprocessing, and change control.

SHIPPING FOR PROCESSING

In-process devices that are being transported from one manufacturing site to another are exempt from the packaging and labeling requirements of Sections 502(b) (name/address of manufacturer and accurate statement of quantity of contents) and 502(f) (adequate directions for use) of the FD&CA. The conditions necessary for this exemption to be in force are described in Chapter 21. However, these devices are subject to QSR controls to the extent that the specifications for processing, labeling, or repacking are sufficient to ensure that, if followed, the device will not be adulterated or misbranded.

One type of in-process device that poses a unique labeling problem is the finished device that is being shipped from the manufacturer to a contract sterilizer. It is a common industry practice for manufacturers to ship finished devices labeled as sterile even though they are not sterile during shipment. Such a practice is permissible under 21 CFR 801.150 provided all of the conditions described in §801.150(e) have been satisfied. These are discussed in detail in Chapter 21.

Each pallet, carton, or other designated unit must be conspicuously marked to show its nonsterile nature while in transit between the manufacturer and the contract sterilizer and while being held prior to sterilization. Rigorous attention must be paid to ensuring that unsterilized product is not mistaken for sterilized product. The FDA recommends the use of visual indicators to distinguish between products before and after sterilization. Examples of such indicators are labels that change color when exposed to steam or ethylene oxide, or adhesive "dots" that change color when exposed to radiation (Cardamone p. 24). These labels, however, only indicate that the device has been exposed to a sterilant. They do not guarantee that the product is sterile. Additional controls are necessary to ensure that the sterilization process has achieved the required sterility assurance level.

Devices that have been sterilized may be held or may be shipped to the manufacturer's warehouse or other controlled distribution point prior to final release. These devices must be properly labeled to clearly indicate that they are in quarantine awaiting final approval. The pallet, carton, or other designated unit must be conspicuously marked to indicate the status of the device, such as "sterilized, awaiting test results," or an equivalent statement. The firm must be able to demonstrate that it is in control of the device until final release. If necessary, the device could be recovered for reprocessing, or destroyed. For this reason, the FDA does not consider a distributor's warehouse or facility to be a controlled distribution point (Cardamone p. 25).

THINGS TO REMEMBER

Section 520(f) of the FD&CA requires that domestic or foreign manufacturers of medical devices intended for commercial distribution in the United States have a QA program. The purpose of the QA program is to ensure that devices conform to their specifications, and to help prevent the production of defective products that can endanger consumers.

The GMP regulations established under the authority of the Medical Device Amendments of 1976 deal primarily with control of labeling within the manufacturing process. Included are inspection, handling, storage, and distribution requirements for device labeling. The regulations require that the manufacturer have physical facilities of suitable design that provide sufficient space for packaging and labeling operations. Also covered is the design requirement that labels remain legible during the customary conditions of processing, storage, handling, distribution, and use. Failure to meet these requirements results in adulteration of the device. The SMDA extended the GMP regulation to cover content development and change control for labeling.

The QSR covers any finished device intended for human use that is commercially distributed in the United States unless there is an approved exemption. The regulation applies equally to domestic and foreign manufacturers. The QSR applies equally to all medical device manufacturers regardless of size. However, the FDA does not grant any exception to the QSR to firms because they are small. The FDA has recognized that small firms may not need all of the same systems as larger firms to achieve control over their manufacturing processes. Manufacturers must use good judgment when applying the QSR to their specific products and operations.

Some firms are exempt from some or all of the QSR or are not routinely inspected. These firms are, however, subject to the FD&CA. If they produce devices that are adulterated and/or misbranded, they are subject to the penalties outlined in the FD&CA. The principal exemptions to the QSR include custom-device manufacturers, contract manufacturers, contract laboratories, component manufacturers, and other firms that have been granted an exemption or variance by the FDA.

A medical device manufacturer's QA program must incorporate a number of elements that relate to labeling in order to comply with the QSR. Labeling includes equipment labels, control labels, package labels, directions for use, maintenance manuals, and so on. Displays on electronic-message panels and CRT screens are considered labeling if instructions and parameter-identification information are given.

In-process devices that are being transported from one manufacturing site to another are exempt from the packaging and labeling requirements of Sections 502(b) and 502(f). One type of in-process device that poses a unique labeling problem is the finished device that is being shipped from the manufacturer to a contract sterilizer. It is a common industry practice for manufacturers to ship finished devices labeled as sterile even though they are unsterile during shipment. Special care must be taken to ensure that these devices are properly labeled to prevent a device from being released until the device has been sterilized, and the sterilization process has achieved the required sterility assurance level.

26 Special Labeling Requirements for Specific Devices in the United States

During the process of classifying devices (see Appendix F) the Food and Drug Administration (FDA) may establish labeling requirements for specific device categories. These requirements are usually broader in scope than the device-specific requirements established during the 510(k)/premarket approval application (PMA) review process described in Chapter 24. Occasionally, during the review of a product application, the FDA will identify a labeling issue of sufficient importance to make a general rule requiring specific wording in the labeling. For example, the FDA has determined that orthopedic bone-screw labeling must bear the warning: "WARNING: This device is not approved for screw attachment or fixations to the posterior elements (pedicles) of the cervical, thoracic, or lumbar spine" (Spinal System p. 10). These requirements appear in guidance documents or other communications targeted at the affected manufacturers. Eventually, some of these requirements are incorporated directly into the Code of Federal Regulations (CFR). This chapter summarizes these special labeling requirements.

LABELING FOR CONTACT LENS SOLUTIONS AND TABLETS (§800.12(C))

This part contains requirements for labeling for contact-lens solutions and tablets that are available for retail sale, which must be packaged in a tamper-resistant package. A tamper-resistant package is one that has an indicator or barrier to entry that, when breached or missing, can reasonably be expected to provide visual evidence to the consumer that tampering has occurred.

Each retail package is required to bear a prominently placed statement alerting the consumer to the tamper-resistant feature of the package. The labeling statement must be placed so that it would be unaffected if the tamper-resistant feature of the package is breached or missing. If the tamper-resistant feature of the package is one that has an identifiable characteristic, such as a shrink band, the characteristic must be referred to in the statement. For example, a package with a shrink band could say: "For your protection, this bottle has an imprinted seal around the neck."

USE-RELATED STATEMENTS

This part contains specific use-related labeling requirements for the devices listed in the following sections. The section number (e.g., §801.405) of the corresponding section in 21 CFR is listed with each topic for reference.

LABELING OF ARTICLES INTENDED FOR LAY USE IN REPAIRING AND/OR REFITTING OF DENTURES (§801.405)

The American Dental Association and leading dental authorities have advised the FDA of their concern regarding the safety of denture reliners, repair kits, pads, cushions, and other articles marketed and labeled for lay use in the repair, refitting, or cushioning of ill-fitting, broken, or irritating dentures. The FDA has concluded that laypersons cannot be expected to properly repair or refit dentures and that the long-term wearing of improperly repaired or refitted dentures may cause irreparable damage to the oral cavity. Such articles designed for lay use should be limited to emergency or temporary situations pending the services of a licensed dentist. Therefore, the FDA regards such devices as unsafe and misbranded under the Federal Food, Drug, and Cosmetic Act (FD&CA) unless the labeling:

- Limits the use of denture repair kits to emergency repairs pending unavoidable delay in obtaining professional reconstruction of the denture, with:
  - Each indications-for-use statement being preceded and modified by the word "emergency."
  - The inclusion of the conspicuous warning statements:

    "WARNING—FOR EMERGENCY REPAIRS ONLY. LONG TERM USE OF HOME-REPAIRED DENTURES MAY CAUSE FASTER BONE LOSS, CONTINUING IRRITATION, SORES, AND TUMORS. THIS KIT FOR EMERGENCY USE ONLY. SEE DENTIST WITHOUT DELAY."

  - The inclusion of the following information in the directions for use:

    SPECIAL TRAINING AND TOOLS ARE NEEDED TO REPAIR DENTURES TO FIT PROPERLY. HOME-REPAIRED DENTURES MAY CAUSE IRRITATION TO THE GUMS AND DISCOMFORT AND TIREDNESS WHILE EATING. LONG TERM USE MAY LEAD TO MORE TROUBLES, EVEN PERMANENT CHANGES IN BONES, TEETH, AND GUMS, WHICH MAY MAKE IT IMPOSSIBLE TO WEAR DENTURES IN THE FUTURE. FOR THESE REASONS, DENTURES REPAIRED WITH THIS KIT SHOULD BE USED ONLY IN AN EMERGENCY UNTIL A DENTIST CAN BE SEEN. DENTURES THAT DON'T FIT PROPERLY CAUSE IRRITATION AND INJURY TO THE GUMS AND FASTER BONE LOSS, WHICH IS PERMANENT. DENTURES THAT DON'T FIT PROPERLY CAUSE GUM CHANGES THAT MAY REQUIRE SURGERY FOR CORRECTION. CONTINUING IRRITATION AND INJURY MAY LEAD TO CANCER IN THE MOUTH. YOU MUST SEE YOUR DENTIST AS SOON AS POSSIBLE.

- Limits the use of denture reliners, pads, and cushions to temporary refitting pending unavoidable delay in obtaining professional reconstruction of the denture, with:
  - each indications-for-use statement being preceded and modified by the word "temporary"
  - the inclusion of the conspicuous warning statement":

    "WARNING—FOR TEMPORARY USE ONLY. LONG TERM USE OF THIS PRODUCT MAY LEAD TO FASTER BONE LOSS, CONTINUING IRRITATION, SORES, AND TUMORS. FOR USE ONLY UNTIL A DENTIST CAN BE SEEN."

  - the inclusion of the following information in the directions for use:

    USE OF THESE PREPARATIONS OR DEVICES MAY TEMPORARILY DECREASE THE DISCOMFORT; HOWEVER, THEIR USE WILL NOT MAKE THE DENTURE FIT PROPERLY. SPECIAL TRAINING AND TOOLS ARE NEEDED TO REPAIR A DENTURE TO FIT PROPERLY. DENTURES THAT DO NOT FIT PROPERLY CAUSE IRRITATION AND INJURY TO THE GUMS AND FASTER BONE LOSS, WHICH IS PERMANENT AND MAY REQUIRE A COMPLETELY NEW DENTURE. CHANGES IN THE GUMS CAUSED BY DENTURES THAT DO NOT FIT PROPERLY MAY REQUIRE SURGERY FOR CORRECTION. CONTINUING IRRITATION AND INJURY MAY LEAD TO CANCER IN THE MOUTH. YOU MUST SEE YOUR DENTIST AS SOON AS POSSIBLE.

- If the denture relining or repair material forms a permanent bond with the denture, inclusion of the warning:

  THIS RELINER BECOMES FIXED TO THE DENTURE AND A COMPLETELY NEW DENTURE MAY BE REQUIRED BECAUSE OF ITS USE.

Labeling claims that exaggerate the usefulness or the safety of these materials or fail to disclose all facts relating to the claims of usefulness will be regarded as false and misbranded under Sections 201(n) and 502(a) of the FD&CA.

Maximum Acceptable Level of Ozone (§801.415)

Ozone is a toxic gas with no known useful medical application in specific, adjunctive, or preventative therapy. For ozone to be effective as a germicide, it must be present in a concentration far greater than humans or animals can safely tolerate.

A device that generates ozone by design or as a by-product will be considered misbranded under Section 502 of the FD&CA if:

- the device generates ozone, releases it into the atmosphere, and does not indicate in its labeling the maximum acceptable concentration of ozone that may be generated (not to exceed 0.05 part per million by volume of air circulated through the device), and does not indicate the smallest area in which such a device can be used so as not to produce an ozone concentration in excess of 0.05 part per million;
- the device labeling claims any medical condition for which there is no proof of safety and effectiveness; or
- the device generates ozone at a level less than 0.05 part per million by volume of air circulating through the device and it is labeled for use as a germicide or deodorizer

Hearing-Aid Devices: Professional and Patient Labeling (§801.420)

Hearing aids shall be clearly and permanently marked with the following information:

- The name of the manufacturer or distributor, the model name or number, the serial number, and the year of manufacture
- A "+" symbol indicating the positive connection for the battery insertion, unless it is physically impossible to insert the battery in the reverse position

A "user instructional brochure" must accompany the hearing aid, and must be provided to the prospective user by the dispenser of the hearing aid. The following information, to the extent applicable to the particular requirements and characteristics of the hearing aid, must be included in a user instructional brochure:

1. An illustration(s) of the hearing aid, indicating operating controls, user adjustments, and the battery compartment
2. Information on the function of all controls intended for user adjustment
3. A description of any accessory that may accompany the hearing aid (e.g., accessories for use with a television or telephone)
4. Specific instructions for:
   - Use of the hearing aid
   - Maintenance and care of the hearing aid, including the procedure for washing the earmold, replacing the tubing on those hearing aids that use tubing, and for storing the hearing aid when it will not be used for an extended time
   - Replacing or recharging the battery, including a generic designation of replacement batteries
   - Information on how and when to obtain repair service, including at least one specific address where the user can go or a mailing address for repair service
5. A description of commonly occurring avoidable conditions that could adversely affect or damage the hearing aid, such as dropping it, immersing it in water, or exposing it to excessive heat

6. Identification of any known side effects associated with the use of a hearing aid that may warrant consultation with a physician (e.g., skin irritation or accelerated accumulation of earwax)

7. A statement that a hearing aid will not restore normal hearing and will not prevent or improve hearing impairment resulting from organic conditions

8. A statement that, in most cases, infrequent use of a hearing aid does not permit the user to obtain full benefit from it

9. A statement that the use of a hearing aid is only part of hearing habilitation and may need to be supplemented by auditory training and instruction in lipreading

10. The following warning:

WARNING TO HEARING AID DISPENSERS

A HEARING AID DISPENSER SHOULD ADVISE A PROSPECTIVE HEARING AID USER TO CONSULT PROMPTLY WITH A LICENSED PHYSICIAN (PREFERABLY AN EAR SPECIALIST) BEFORE DISPENSING A HEARING AID IF THE HEARING AID DISPENSER DETERMINES THROUGH INQUIRY, ACTUAL OBSERVATION, OR REVIEW OF ANY OTHER AVAILABLE INFORMATION CONCERNING THE PROSPECTIVE USER, THAT THE PROSPECTIVE USER HAS ANY OF THE FOLLOWING CONDITIONS:

(I) VISIBLE CONGENITAL OR TRAUMATIC DEFORMITY OF THE EAR.

(II) HISTORY OF ACTIVE DRAINAGE FROM THE EAR WITHIN THE PREVIOUS 90 DAYS.

(III) HISTORY OF SUDDEN OR RAPIDLY PROGRESSIVE HEARING LOSS WITHIN THE PREVIOUS 90 DAYS.

(IV) ACUTE OR CHRONIC DIZZINESS.

(V) UNILATERAL HEARING LOSS OF SUDDEN OR RECENT ONSET WITHIN THE PREVIOUS 90 DAYS.

(VI) AUDIOMETRIC AIR-BONE GAP EQUAL TO OR GREATER THAN 15 DECIBELS AT 500 HERTZ (Hz), 1,000 Hz, AND 2,000 Hz.

(VII) VISIBLE EVIDENCE OF SIGNIFICANT CERUMEN ACCUMULATION OR A FOREIGN BODY IN THE EAR CANAL.

(VIII) PAIN OR DISCOMFORT IN THE EAR.

SPECIAL CARE SHOULD BE EXERCISED IN SELECTING AND FITTING A HEARING AID WHOSE MAXIMUM SOUND PRESSURE LEVEL EXCEEDS 132 DECIBELS BECAUSE THERE MAY BE RISK OF IMPAIRING THE REMAINING HEARING OF THE HEARING AID USER. (This provision is required only for those hearing aids with a maximum sound pressure capability greater than 132 decibels (dB).)

11. The following notice for prospective hearing-aid users:

IMPORTANT NOTICE FOR PROSPECTIVE HEARING AID USERS

GOOD HEALTH PRACTICE REQUIRES THAT A PERSON WITH A HEARING LOSS HAVE A MEDICAL EVALUATION BY A LICENSED PHYSICIAN (PREFERABLY A PHYSICIAN WHO SPECIALIZES IN DISEASES OF THE EAR) BEFORE PURCHASING A HEARING AID. LICENSED PHYSICIANS WHO SPECIALIZE IN DISEASES OF

THE EAR ARE OFTEN REFERRED TO AS OTOLARYNGOLOGISTS, OTOLOGISTS OR OTORHINOLARYNGOL-OGISTS. THE PURPOSE OF MEDICAL EVALUATION IS TO ASSURE THAT ALL MEDICALLY TREATABLE CONDITIONS THAT MAY AFFECT HEARING ARE IDENTIFIED AND TREATED BEFORE THE HEARING AID IS PURCHASED.

FOLLOWING THE MEDICAL EVALUATION, THE PHYSICIAN WILL GIVE YOU A WRITTEN STATEMENT THAT STATES THAT YOUR HEARING LOSS HAS BEEN MEDICALLY EVALUATED AND THAT YOU MAY BE CON-SIDERED A CANDIDATE FOR A HEARING AID. THE PHYSICIAN WILL REFER YOU TO AN AUDIOLOGIST OR A HEARING AID DISPENSER, AS APPROPRIATE, FOR A HEARING AID EVALUATION.

THE AUDIOLOGIST OR HEARING AID DISPENSER WILL CONDUCT A HEARING AID EVALUATION TO ASSESS YOUR ABILITY TO HEAR WITH AND WITHOUT A HEARING AID. THE HEARING AID EVALUATION WILL ENABLE THE AUDIOLOGIST OR DISPENSER TO SELECT AND FIT A HEARING AID TO YOUR INDIVIDUAL NEEDS.

IF YOU HAVE RESERVATIONS ABOUT YOUR ABILITY TO ADAPT TO AMPLIFICATION, YOU SHOULD INQUIRE ABOUT THE AVAILABILITY OF A TRIAL-RENTAL OR PURCHASE-OPTION PROGRAM. MANY HEARING AID DISPENSERS NOW OFFER PROGRAMS THAT PERMIT YOU TO WEAR A HEARING AID FOR A PERIOD OF TIME FOR A NOMINAL FEE AFTER WHICH YOU MAY DECIDE IF YOU WANT TO PURCHASE THE HEARING AID.

FEDERAL LAW RESTRICTS THE SALE OF HEARING AIDS TO THOSE INDIVIDUALS WHO HAVE OBTAINED A MEDICAL EVALUATION FROM A LICENSED PHYSICIAN. FEDERAL LAW PERMITS A FULLY INFORMED ADULT TO SIGN A WAIVER STATEMENT DECLINING THE MEDICAL EVALUATION FOR RELIGIOUS OR PERSONAL BELIEFS THAT PRECLUDE CONSULTATION WITH A PHYSICIAN. THE EXERCISE OF SUCH A WAIVER IS NOT IN YOUR BEST HEALTH INTEREST AND ITS USE IS STRONGLY DISCOURAGED.

CHILDREN WITH HEARING LOSS

IN ADDITION TO SEEING A PHYSICIAN FOR A MEDICAL EVALUATION, A CHILD WITH A HEARING LOSS SHOULD BE DIRECTED TO AN AUDIOLOGIST FOR EVALUATION AND REHABILITATION SINCE HEARING LOSS MAY CAUSE PROBLEMS IN LANGUAGE DEVELOPMENT AND THE EDUCATIONAL AND SOCIAL GROWTH OF A CHILD. AN AUDIOLOGIST IS QUALIFIED BY TRAINING AND EXPERIENCE TO ASSIST IN THE EVALUATION AND REHABILITATION OF A CHILD WITH A HEARING LOSS.

12. The following technical data, unless the data are provided in separate labeling accompanying the device:*
    - Saturation-output curve (SSPL 90 curve)
    - Frequency-response curve
    - Average saturation output (HF-Average SSPL 90)
    - Average full-on gain (HF-Average full-on gain)
    - Reference test gain
    - Frequency range
    - Total harmonic distortion
    - Equivalent input noise
    - Battery-current drain
    - Induction-coil sensitivity (telephone coil aids only)
    - Input-output curve (ACG aids only)
    - Attack and release times (ACG aids only)

---

* The determination of technical data values for the hearing-aid labeling shall be conducted in accordance with the test procedures of the American National Standard "Specification of Hearing Aid Characteristics," ANSI S3.22-1996.

If the hearing aid is used or rebuilt, this fact must be clearly stated on the package and on a tag attached to the hearing aid. It may also be stated in the user instructional brochure.

The user instructional brochure may contain statements other than those required by the regulations, provided (a) the statements are not false or misleading and do not diminish the impact of the required statements, and (b) the statements are not prohibited by 21 CFR 801, or by the regulations of the Federal Trade Commission.

HEARING AIDS: CONDITIONS FOR SALE (§801.421)

A hearing-aid dispenser may not sell a hearing aid unless the prospective user has presented a written statement signed by a licensed physician stating that the patient's hearing loss has been medically evaluated and the patient may be considered a candidate for a hearing aid. The medical evaluation must have taken place within the preceding six months.

A prospective hearing-aid user who is 18 years of age or older may waive the medical evaluation requirement provided that the hearing-aid dispenser (21 CFR 801.421(a)):

- informs the prospective user that the exercise of the waiver is not in the user's best health interest;
- does not in any way actively encourage the prospective user to waive such a medical evaluation; and
- affords the prospective user the opportunity to sign the following statement:

I HAVE BEEN ADVISED BY _____ (HEARING AID DISPENSER'S NAME) THAT THE FOOD AND DRUG ADMINISTRATION HAS DETERMINED THAT MY BEST HEALTH INTEREST WOULD BE SERVED IF I HAD A MEDICAL EVALUATION BY A LICENSED PHYSICIAN (PREFERABLY A PHYSICIAN WHO SPECIALIZES IN DISEASES OF THE EAR) BEFORE PURCHASING A HEARING AID. I DO NOT WISH A MEDICAL EVALUATION BEFORE PURCHASING A HEARING AID.

Before a prospective hearing-aid user signs any statement and before the sale of a hearing aid, the hearing-aid dispenser must (21 CFR 801.421(b)):

- provide the prospective user a copy of the user instructional brochure for a hearing aid that has been or may be selected for the prospective user;
- review the content of the user instructional brochure with the prospective user orally or in the predominate method of communication used during the sale;
- afford the prospective user an opportunity to read the user instructional brochure.

Upon request, a dispenser must provide a copy of the user instructional brochure for the hearing aid being considered or the name and address of the manufacturer or distributor from whom a user instructional brochure for the hearing aid may be obtained.

In addition to assuring that a user instructional brochure accompanies each hearing aid, a manufacturer or distributor must (21 CFR 801.421(c)(2)):

- provide sufficient copies of the user instructional brochure to sellers for distribution to users and prospective users; and
- provide a copy of the user instructional brochure to any hearing-aid professional, user, or prospective user who requests a copy in writing.

Group auditory trainers, defined as a group amplification system purchased by a qualified school or institution for the purpose of communicating with and educating individuals with hearing impairments, are exempt from the requirements of this regulation (21 CFR 801.421(e)).

User Labeling for Menstrual Tampons (§801.430)

Data have shown that toxic shock syndrome (TSS), a rare but serious and sometimes fatal disease, is associated with the use of menstrual tampons. To protect the public and minimize the effects of TSS, the FDA requires that menstrual tampons be labeled as follows:

- The labeling must prominently and legibly display consumer information in a way to make it likely to be read and understood by the ordinary individual under normal conditions of purchase and use, including:
  - the warnings signs of TSS (e.g., sudden fever [usually 102° or more] and vomiting, diarrhea, fainting or near fainting when standing up, dizziness, or a rash that looks like sunburn);
  - what to do if these or other signs of TSS appear, including removal of the tampon at once and seeking medical attention immediately;
  - the risk of TSS to all women using tampons during their menstrual period, especially the reported higher risk to women under 30 years of age and teenage girls, the estimated incidence of TSS of 1 to 17 per 100,000 menstruating women and girls per year, and the risk of death from contracting TSS;
  - the advisability of using tampons with the minimum absorbency needed to control menstrual flow in order to reduce the risk of contracting TSS;
  - avoiding the risk of getting tampon-associated TSS by not using tampons, and reducing the risk of getting TSS by alternating tampon use with sanitary napkins use during menstrual periods; and
  - the need to seek medical attention before again using tampons if TSS warning signs have occurred in the past, or if women have any questions about TSS or tampon use.
- If the information required above is to be included as a package insert, the following alert statement must appear prominently and legibly on the package label:

Attention: Tampons are associated with Toxic Shock Syndrome (TSS). TSS is a rare but serious disease that may cause death. Read and save the enclosed information.

- Menstrual-tampon package labels shall bear one of the absorbency terms in Table 26.1. These terms rate the absorbency of the production run, lot, or batch as measured by the test in 21 CFR 801.430(f)(2).The package label must include an explanation of the ranges of absorbency and a description of how the consumer can use the range to select the tampon with the minimum absorbency to control menstrual flow in order to reduce the risk of contracting TSS.

Any menstrual tampon that is not labeled according the requirements listed above that was initially introduced or initially delivered into commerce after March 1, 1990, is misbranded under Sections 201(n), 502(a), and 502(f) of the FD&CA.

Section 21 CFR 810.430(g) exempts any menstrual tampon intended to be dispensed by a vending machine from the requirements of 21 CFR 801.430. Specifically, the statements on absorbency need not be provided.

TABLE 26.1
Menstrual Tampon Terms of Absorbency

Ranges of Absorbency in Grams [a]	Corresponding Terms of Absorbency
6 and under	Junior absorbency
6 to 9	Regular absorbency
9 to 12	Super absorbency
12 to 15	Super plus absorbency
15 to 18	Ultra absorbency
Above 18	None

[a] These ranges are defined, respectively, as follows: less than or equal to 6 grams (g); greater than 6 g up to and including 9 g; greater than 9 g up to and including 12 g; greater than 12 g up to and including 15 g; greater than 15 g up to and including 18 g; and greater than 18 g.

Source: 21 CFR 801.430(e)(1)

PRESCRIPTION AND RESTRICTED DEVICES CONTAINING OR MANUFACTURED WITH OZONE-DEPLETING SUBSTANCES (§§801.417 AND 801.433)

Under the provision of the Clean Air Act,* any product containing or manufactured with chlorofluorocarbons, halons, carbon tetrachloride, methyl chloride, or any other Class I substance designated by the Environmental Protection Agency (EPA) must bear the following warning unless it is exempted by the regulation (40 CFR. 82.106):

> **WARNING: CONTAINS [OR MANUFACTURED WITH, IF APPLICABLE] [INSERT NAME OF SUBSTANCE], A SUBSTANCE WHICH HARMS PUBLIC HEALTH AND ENVIRONMENT BY DESTROYING OZONE IN THE UPPER ATMOSPHERE.**

A product need not bear the warning if it contains trace quantities of a controlled substance remaining as a residue or impurity due to a chemical reaction, and where the controlled substance serves no useful purpose in or for the product itself. However, if such product was manufactured using the controlled substance, the product is required to be labeled as a "product manufactured with" the controlled substance, unless otherwise exempted (40 CFR, 82.106(b)).

The warning statement must be conspicuous and legible, and must appear in sharp contrast to the background. Examples of combinations of colors that may not satisfy the proposed requirement for sharp contrast are (a) black letters on a dark blue or dark green background, (b) dark red letters on a light red background, (c) light red letters on a reflective silver background, and (d) white letters on a light gray or tan background.

The name of the Class I substance that is inserted into the warning statement is the standard chemical name of the substance as listed in 40 CFR 82, Appendix A to Subpart A, except that:

- the acronym "CFC" may be substituted for "chlorofluorocarbon";
- the acronym "HCFC" may be substituted for "hydrochlorofluorocarbon"; and
- the term "1,1,1-trichloroethane" may be substituted for "methyl chloroform."

If a product contains or is manufactured with more than one substance controlled by the regulation, the warning statement may include the names of all of the substances in a single warning

---

* *Clean Air Act.* US Code, vol. 42, §§7401 *et seq.* (1963).

TABLE 26.2
Minimum Type Size Requirements

	Minimum Height of Printed Image of Capital or Upper-Case Letters.					
Area of principal display panel (a) in square inches	$0 < a \leq 2$	$2 < a \leq 5$	$5 < a \leq 10$	$10 < a \leq 15$	$15 < a \leq 30$	$a > 30$
Signal word [a]	3/64	1/16	3/32	7/64	1/8	5/32
Statement	3/64	3/64	1/16	3/32	3/32	7/64

Source: 40 CFR 82.110(f)(1), Table 1

statement, provided that the combined statement clearly distinguishes which substances the product contains and which were used in the manufacturing process.

The warning statement must be blocked within a square or rectangular area, with or without a border. The warning statement must appear in lines that are parallel to the surrounding text on the product's principle display panel, supplemental printed material, or promotional printed material.

The signal word "WARNING" must appear in all capital letters. The ratio of the height of a capital letter to its width shall be such that the height of the letter is no more than three times its width.

Table 26.2 specifies the minimum type-size requirements for the warning statement and is based upon the area of the display panel of the product. Where the statement is on outer packaging as well as the display panel area, the statement shall appear in the same minimum type size as on the display panel. The warning statement must appear at least as large as the type sizes specified in Table 26.2. The type size refers to the height of the capital letters. A larger type size materially enhances the legibility of the statement and is desirable.

The use of chlorofluorocarbon in devices as a propellant is generally prohibited except for the following essential uses:*

- Metered-dose corticosteroid human drugs for oral inhalation
- Metered-dose short-acting adrenergic bronchodilator human drugs for oral inhalation
- Metered-dose ergotamine tartrate drug products administered by oral inhalation for use in humans
- Anesthetic drugs for topical use on accessible mucous membranes of humans where a cannula is used for application
- Metered-dose cromolyn sodium human drugs administered by oral inhalation
- Metered-dose ipratropium bromide human drugs administered by oral inhalation
- Metered-dose atropine sulfate aerosol human drugs administered by oral inhalation
- Metered-dose nedocromil sodium human drugs administered by oral inhalation
- Metered-dosed ipratropium bromide and albuterol sulfate, in combination, administered by oral inhalation for human use
- Sterile aerosol talc administered intrapleurally by thoracoscopy for human use

Any prescription and restricted device containing or manufactured with chlorofluorocarbons, halons, carbon tetrachloride, methyl chloride, or any other substance designated as Class I by the EPA must bear a warning statement. The warning statement must be clearly legible and conspicuous on the product, its immediate container, its outer packaging, or its other labeling in accordance with the requirements of 40 CFR 82. The warning must appear with such prominence and conspicuousness as to render it likely to be read and understood by consumers under normal conditions of purchase.

---

* See 21 CFR Chapter 1, Subpart G – *Provision Applicable to Specific Products, Subject to the Federal Food, Drug, and Cosmetic Act*, Section 2.125 – *Use of chlorofluorocarbon in self-pressurized containers.*

However, the following alternative warning statement may be used for prescription and restricted device products (21 CFR 801.433(b)(1)):

THIS PRODUCT CONTAINS [OR IS MANUFACTURED WITH, IF APPLICABLE] [INSERT NAME OF SUBSTANCE], A SUBSTANCE WHICH HARMS THE ENVIRONMENT BY DESTROYING OZONE IN THE UPPER ATMOSPHERE.

YOUR PHYSICIAN HAS DETERMINED THAT THIS PRODUCT IS LIKELY TO HELP YOUR PERSONAL HEALTH. **USE THIS PRODUCT AS DIRECTED, UNLESS INSTRUCTED TO DO OTHERWISE BY YOUR PHYSICIAN.** IF YOU HAVE ANY QUESTIONS ABOUT ALTERNATIVES, CONSULT WITH YOUR PHYSICIAN.

The warning statement must be clearly legible and conspicuous on the product, its immediate container, its outer packaging, or its other labeling in accordance with the requirements of 40 CFR 82. It must appear with such prominence and conspicuousness as to render it likely to be read and understood by consumers under normal conditions of purchase.

If the alternative warning statement is used, the following warning statement must be placed on the package labeling intended to be read by the physician (physician package insert) after the "How supplied" section that describes special handling and storage conditions on the physician labeling:

WARNING: CONTAINS [OR MANUFACTURED WITH, IF APPLICABLE] [INSERT NAME OF SUBSTANCE], A SUBSTANCE WHICH HARMS PUBLIC HEALTH AND ENVIRONMENT BY DESTROYING OZONE IN THE UPPER ATMOSPHERE.

A NOTICE SIMILAR TO THE ABOVE **WARNING** HAS BEEN PLACED IN THE INFORMATION FOR THE PATIENT [OR PATIENT INFORMATION LEAFLET, IF APPLICABLE] OF THIS PRODUCT UNDER ENVIRON- MENTAL PROTECTION AGENCY (EPA) REGULATIONS. THE PATIENT'S WARNING STATES THAT THE PATIENT SHOULD CONSULT HIS OR HER PHYSICIAN IF THERE ARE QUESTIONS ABOUT ALTERNATIVES.

The requirements described in 21 CFR 801.433 do not replace or relieve the manufacturer from any requirements imposed under 40 CFR 82.

USER LABELING FOR LATEX CONDOMS (§801.435)

The requirements in this section apply to condoms* and condoms with spermicidal lubricant** that are formed from latex films. Performance data show that the material integrity of these types of condoms will degrade over time. To protect the public health and minimize the risk of device failure, latex condoms must bear an expiration date that is supported by the testing as described in 21 CFR 801.435(d) and (h).

The expiration date must be displayed prominently and legibly on the primary packaging (i.e., individual package) and higher levels of packaging (e.g., boxes of condoms) in order to ensure the visibility to consumers of the expiration date. The time period upon which the expiration date is based begins with the date of packaging of the condom.

USER LABELING FOR DEVICES THAT CONTAIN NATURAL RUBBER (§801.437)

Data in the Medical Device Reporting system and the scientific literature indicate that some individuals are at risk of severe anaphylactic reactions to natural latex proteins. To minimize the risk to individuals sensitive to natural latex proteins and to protect the public health, the FDA has

---

* See 21 CFR 884.5300.
** See 21 CFR 884.5310.

implemented minimum labeling requirements for devices composed of or containing, or having packaging or components that are composed of or contain, natural rubber that contacts humans. The term "natural rubber" includes natural-rubber latex, dry natural rubber, and synthetic latex or synthetic rubber that contains natural rubber in its formulation.

For the purposes of this regulation, "natural-rubber latex" means rubber that is produced by the natural-rubber latex process that involves the use of natural latex in a concentrated colloidal suspension. The regulations apply to products that are intended to contact or are likely to contact the user or patient. This includes contact when the device that contains natural rubber is connected to the patient by a liquid path or an enclosed gas path; or the device containing the natural rubber is fully or partially coated with a powder, and such powder may carry natural-rubber proteins that may contaminate the environment of the user or patient.

Devices containing natural-rubber latex that contacts humans must bear the following statement in bold print on the device labeling:

CAUTION: THIS PRODUCT CONTAINS NATURAL RUBBER LATEX WHICH MAY CAUSE ALLERGIC REACTIONS.

This statement must appear on all device labels and other labeling. It must appear on the principal display panel of the device packaging, the outside package, container, or wrapper, and the immediate device package, container, or wrapper.

If a device contains dry natural rubber that contacts humans and is not covered by the previous requirement, it must bear the following statement in bold print on the device labeling:

THIS PRODUCT CONTAINS DRY NATURAL RUBBER.

This statement must appear on all device labels and other labeling. It must appear on the principal display panel of the device packaging, the outside package, container or wrapper, and the immediate device package, container, or wrapper.

Devices that have packaging that contains natural-rubber latex that contacts humans must bear the following statement in bold print on the device labeling:

CAUTION: THE PACKAGING OF THIS PRODUCT CONTAINS NATURAL RUBBER LATEX WHICH MAY CAUSE ALLERGIC REACTIONS.

This statement shall appear on the packaging that contains the natural rubber, and the outside package, container, or wrapper.

Devices that have packaging containing dry natural rubber that contacts humans must bear the following statement in bold print on the device labeling:

THE PACKAGING OF THIS PRODUCT CONTAINS DRY NATURAL RUBBER.

This statement shall appear on the packaging that contains the natural rubber, and on the outside package, container, or wrapper.

Any device that contains natural rubber that contacts humans may not use the term "hypoallergenic" in its labeling.

CONTRACEPTIVES AND SEXUALLY TRANSMITTED DISEASES (STDS)

Over a number of years, the FDA has developed recommendations for the labeling of all contraceptive devices regarding protection from pregnancy and sexually transmitted diseases (STDs). It is the agency's intent to encourage a uniform approach to this type of labeling information for the consumer. Where appropriate, the FDA recommends that the same information be included in the professional labeling for these devices.

PROTECTION FROM PREGNANCY

Contraceptive devices include male and female condoms, diaphragms, cervical caps, intrauterine contraceptive devices (IUCDs), and tubal occlusion devices (TODs). The labeling for these devices should contain an easy-to-read table with the pregnancy rates associated with all methods. Such a table is intended to provide the consumer with the pregnancy rate on his or her selected contraceptive, as well as comparative information on the pregnancy rates of various other contraceptive products. The intent is to provide appropriate information in an understandable form to help people appreciate the effectiveness of their contraceptive of choice with respect to the many other products available on the market. Those products that require professional labeling should include the same information in an appropriate section of the labeling. The FDA considers the communication of information on pregnancy rates to contraceptive device users to be essential for the safe and effective use of those devices (Contraceptive Labeling p. 1).

PROTECTION FROM SEXUALLY TRANSMITTED DISEASES (STDs)

In addition to including contraceptive efficacy data, the labeling of contraceptive devices should include information on the ability of the contraceptive device to provide protection from STDs. The FDA has developed statements that manufacturers are strongly encouraged to include in their patient/consumer labeling regarding the transmission of STDs. The recommended statement depends on the particular device (Contraceptive Labeling pp. 2–4).

- For IUCDs, TODs,, and natural-skin condoms:

  THIS PRODUCT IS INTENDED TO PREVENT PREGNANCY. IT DOES NOT PROTECT AGAINST HIV INFECTION AND OTHER SEXUALLY TRANSMITTED DISEASES (STDs).

- For natural-skin condoms, the following additional statement should be included:

  IN ORDER TO HELP REDUCE THE RISK OF TRANSMISSION OF MANY SEXUALLY TRANSMITTED DISEASES (STDs), INCLUDING HIV INFECTIONS (AIDS), USE A LATEX CONDOM.

- On the condom wrapper/principal display panel for latex condoms for men:

  IF USED PROPERLY, LATEX CONDOMS WILL HELP TO REDUCE THE RISK OF TRANSMISSION OF HIV INFECTION (AIDS) AND MANY SEXUALLY TRANSMITTED DISEASES (STDs).

- In the directions for use for latex condoms for men:

  IF USED PROPERLY, LATEX CONDOMS WILL HELP TO REDUCE THE RISK OF TRANSMISSION OF HIV INFECTION (AIDS) AND MANY SEXUALLY TRANSMITTED DISEASES (STDs), INCLUDING CHLAMYDIA, GENITAL HERPES, GENITAL WARTS, GONORRHEA, HEPATITIS B, AND SYPHILIS.

- For male condoms made from new materials, the following interim labeling, based on slippage and breakage data, should be included while a contraceptive effectiveness study is underway):

  YOU MAY USE THIS [INSERT NAME] CONDOM IF YOU OR YOUR PARTNER ARE ALLERGIC TO LATEX.

YOU SHOULD KNOW:

THE RISKS OF PREGNANCY AND SEXUALLY TRANSMITTED DISEASES (STDs), INCLUDING AIDS
(HIV INFECTION), ARE NOT KNOWN FOR THIS CONDOM. A STUDY IS BEING DONE.

THERE ARE LABORATORY TESTS ON THIS [NAME OF MATERIAL] MATERIAL. THESE TESTS SHOW
THAT ORGANISMS EVEN AS SMALL AS SPERM AND VIRUSES LIKE HIV CANNOT PASS THROUGH IT.

*LATEX CONDOMS [IN BOLDFACE] FOR MEN, IF USED CORRECTLY WITH EVERY ACT OF VAGINAL
INTERCOURSE, ARE HIGHLY EFFECTIVE AT PREVENTING PREGNANCY, AS WELL AS STDs, INCLUDING
AIDS (HIV INFECTION).*

- For female condoms, the following key elements are placed on the primary package as well as the package insert:

LATEX CONDOMS FOR MEN ARE HIGHLY EFFECTIVE AT PREVENTING SEXUALLY TRANSMITTED DISEASES
(STDs), INCLUDING AIDS (HIV INFECTION), IF USED PROPERLY.

IF YOU ARE NOT GOING TO USE A MALE LATEX CONDOM, YOU CAN USE [INSERT NAME] TO HELP
PROTECT YOURSELF AND YOUR PARTNER.

[INSERT NAME] ONLY WORKS WHEN YOU USE IT. USE IT EVERY TIME YOU HAVE SEX.

BEFORE YOU TRY [INSERT NAME], BE SURE TO READ THE DIRECTIONS IN THE BOX AND LEARN HOW
TO USE IT PROPERLY.

For diaphragms and cervical caps, the FDA is aware of some clinical work looking at these devices used with vaginal spermicides as a means of protection from a few STDs. However, the FDA believes that these studies are not conclusive. Therefore, the FDA has not developed definitive statements that can be used for uniform labeling of these devices. The FDA holds that no claims for STD protection afforded by diaphragms or cervical caps may be made without submission of an appropriate 510(k) premarket notification or PMA supplement.

ELECTROMAGNETIC INTERFERENCE

In recent years, the FDA has become increasing concerned about the effects of electromagnetic interference (EMI) on active medical devices of all kinds. This concern has manifested itself in new or strengthened performance requirements and new labeling directed at both healthcare professionals and patients. The following are examples taken from FDA guidance documents that illustrate the kind of information FDA recommends.

ELECTRICALLY POWERED WHEELCHAIRS

The FDA has received numerous reports of erratic, unintentional movement of powered wheelchairs. Some of these instances have resulted in serious injuries. The number of incidents that can be linked to EMI is unclear. However, testing at the FDA's laboratory coupled with information from manufacturers and powered-wheelchair users has lead the FDA to conclude the EMI is one of the causes (Alpert, Wheelchair p. 1).

Because of the evidence that EMI can adversely effect the behavior of powered wheelchairs, the FDA has concluded that additional steps to inform the wheelchair user about the potential hazard are required.

Label on the Product

The following EMI information should be affixed to the powered wheelchair for reference by the user. It should be contained preferably on a durable label and secured to the powered wheelchair within clear view of the user. It should be included in the product literature as well. The suggested labeling is (Alpert, Wheelchair – Attachment A pp. 1–2):

<u>**WARNING:** RADIO WAVE SOURCES MAY AFFECT POWERED WHEELCHAIR CONTROL</u>

RADIO WAVE SOURCES, SUCH AS RADIO STATIONS, TV STATIONS, AMATEUR RADIO (HAM) TRANS-MITTERS, TWO-WAY RADIOS, AND CELLULAR PHONES, CAN AFFECT POWERED WHEELCHAIRS AND MOTORIZED SCOOTERS (IN THIS TEXT, BOTH WILL BE CALLED POWERED WHEELCHAIRS). FOLLOWING THE WARNINGS LISTED BELOW SHOULD REDUCE THE CHANCE OF UNINTENDED BRAKE RELEASE OR POWERED WHEELCHAIR MOVEMENT WHICH COULD RESULT IN SERIOUS INJURY.

I) DO NOT TURN ON HAND-HELD PERSONAL COMMUNICATION DEVICES, SUCH AS CITIZENS BAND (CB) RADIOS AND CELLULAR PHONES, WHILE THE POWERED WHEELCHAIR IS TURNED ON;

2) BE AWARE OF NEARBY TRANSMITTERS, SUCH AS RADIO OR TV STATIONS, AND TRY TO AVOID COMING CLOSE TO THEM;

3) IF UNINTENDED MOVEMENT OR BRAKE RELEASE OCCURS, TURN THE POWERED WHEELCHAIR OFF AS SOON AS IT IS SAFE;

4) BE AWARE THAT ADDING ACCESSORIES OR COMPONENTS, OR MODIFYING THE POWERED WHEELCHAIR, MAY MAKE IT MORE SUSCEPTIBLE TO INTERFERENCE FROM RADIO WAVE SOURCES (NOTE: THERE IS NO EASY WAY TO EVALUATE THEIR EFFECT ON THE OVERALL IMMUNITY OF THE POWERED WHEELCHAIR); AND

5) REPORT ALL INCIDENTS OF UNINTENDED MOVEMENT OR BRAKE RELEASE TO THE POWERED WHEELCHAIR MANUFACTURER, AND NOTE WHETHER THERE IS A RADIO WAVE SOURCE NEARBY.

<u>IMPORTANT INFORMATION</u>

1) **20** VOLTS PER METER (V/M) IS A GENERALLY ACHIEVABLE AND USEFUL IMMUNITY LEVEL AGAINST INTERFERENCE FROM RADIO WAVE SOURCES (THE HIGHER THE LEVEL, THE GREATER THE PROTECTION);

2) THIS PRODUCT HAS AN IMMUNITY LEVEL OF [STATE THE IMMUNITY LEVEL OF THIS POWERED WHEELCHAIR MODEL AND INCLUDE THE CONFIGURA-TION TESTED IN THE ACCOMPANYING LITERATURE; i.e. SPECIFY THE ACCESSORIES AND/OR COMPONENTS INCLUDED FOR TESTING, OR SPECIFY THAT THEY WERE NOT INCLUDED FOR TESTING IN THE OPER-ATOR'S MANUAL]; or alternatively state

2) THE IMMUNITY LEVEL OF THIS PRODUCT IS NOT KNOWN.

Accompanying Product Literature

This portion of the labeling will provide the user with basic information that describes the problems with EMI, the known sources of EMI, the protective measures either to lessen the possibility of exposure or to minimize the degree of exposure, and any suggested action should unexpected or erratic movement occur. The FDA suggested labeling is listed below (Alpert, Wheelchair – Attachment A pp. 2–5):

CAUTION: IT IS VERY IMPORTANT THAT YOU READ THIS INFORMATION REGARDING THE POSSIBLE EFFECTS OF ELECTROMAGNETIC INTERFERENCE ON YOUR POWERED WHEELCHAIR.

Electromagnetic Interference (EMI) From Radio Wave Sources

Powered wheelchairs and motorized scooters [in this text, both will be referred to as powered wheelchairs] may be susceptible to electromagnetic interference (EMI), which is interfering electromagnetic energy (EM) emitted from sources such as radio stations, TV stations, amateur radio (HAM) transmitters, two-way radios, and cellular phones. The interference (from radio wave sources) can cause the powered wheelchair to release its brakes, move by itself, or move in unintended directions. It can also permanently damage the powered wheelchair's control system. The intensity of the interfering EM energy can be measured in volts per meter (V/m). Each powered wheelchair can resist EMI up to a certain intensity. This is called its "immunity level." The higher the immunity level, the greater the protection. At this time, current technology is capable of achieving at least a 20 V/m immunity level, which would provide useful protection from the more common sources of radiated EMI. This powered wheelchair model as shipped, with no further modification, has an immunity level of **[STATE THE IMMUNITY LEVEL OF THIS POWERED WHEELCHAIR MODEL AND INCLUDE THE CONFIGURATION TESTED IN THE ACCOMPANYING LITERATURE; I.E. SPECIFY THE ACCESSORIES AND/OR COMPONENTS INCLUDED FOR TESTING, OR SPECIFY THAT THEY WERE NOT INCLUDED FOR TESTING IN THE OPERATOR'S MANUAL]**, or alternatively state: The immunity level of this powered wheelchair model is not known.

There are a number of sources of relatively intense electromagnetic fields in the everyday environment. Some of these sources are obvious and easy to avoid. Others are not apparent and exposure is unavoidable. However, we believe that by following the warnings listed below, your risk to EMI will be minimized.

The sources of radiated EMI can be broadly classified into three types:

1) Hand-held portable transceivers (transmitters-receivers) with the .antenna mounted directly on the transmitting unit. Examples include: citizens band (CB) radios, "walkie talkie," security, fire, and police transceivers, cellular telephones, and other personal communication devices. NOTE: Some cellular telephones and similar devices transmit signals while they are ON, even when not being used;

2) Medium-range mobile transceivers, such as those used in police cars, fire trucks, ambulances, and taxis. These usually have the antenna mounted on the outside of the vehicle; and

3) Long-range transmitters and transceivers, such as commercial broadcast transmitters (radio and TV broadcast antenna towers) and amateur (HAM) radios.

NOTE: OTHER TYPES OF HAND-HELD DEVICES, SUCH AS CORDLESS PHONES, LAPTOP COMPUTERS, AM/FM RADIOS, TV SETS, CD PLAYERS, AND CASSETTE PLAYERS, AND SMALL APPLIANCES, SUCH AS ELECTRIC SHAVERS AND HAIR DRYERS, SO FAR AS WE KNOW, ARE NOT LIKELY TO CAUSE EMI PROBLEMS TO YOUR POWERED WHEELCHAIR.

POWERED WHEELCHAIR ELECTROMAGNETIC INTERFERENCE (EMI)

BECAUSE EM ENERGY RAPIDLY BECOMES MORE INTENSE AS ONE MOVES CLOSER TO THE TRANSMITTING ANTENNA (SOURCE), THE EM FIELDS FROM HAND-HELD RADIO WAVE SOURCES (TRANSCEIVERS) ARE OF SPECIAL CONCERN. IT IS POSSIBLE TO UNINTENTIONALLY BRING HIGH LEVELS OF EM ENERGY VERY CLOSE TO THE POWERED WHEELCHAIR'S CONTROL SYSTEM WHILE USING THESE DEVICES. THIS CAN AFFECT POWERED WHEELCHAIR MOVEMENT AND BRAKING. THEREFORE, THE WARNINGS LISTED BELOW ARE RECOMMENDED TO PREVENT POSSIBLE INTERFERENCE WITH THE CONTROL SYSTEM OF THE POWERED WHEELCHAIR.

WARNINGS

ELECTROMAGNETIC INTERFERENCE (EMI) FROM SOURCES SUCH AS RADIO AND TV STATIONS, AMATEUR RADIO (HAM) TRANSMITTERS, TWO-WAY RADIOS, AND CELLULAR PHONES CAN AFFECT POWERED WHEELCHAIRS AND MOTORIZED SCOOTERS. FOLLOWING THE WARNINGS LISTED BELOW SHOULD REDUCE THE CHANCE OF UNINTENDED BRAKE RELEASE OR POWERED WHEELCHAIR MOVEMENT WHICH COULD RESULT IN SERIOUS INJURY.

1) DO NOT OPERATE HAND-HELD TRANSCEIVERS (TRANSMITTERS-RECEIVERS), SUCH AS CITIZENS BAND (CB) RADIOS, OR TURN ON PERSONAL COMMUNICATION DEVICES, SUCH AS CELLULAR PHONES, WHILE THE POWERED WHEELCHAIR IS TURNED ON;

2) BE AWARE OF NEARBY TRANSMITTERS, SUCH AS RADIO OR TV STATIONS, AND TRY TO AVOID COMING CLOSE TO THEM;

3) IF UNINTENDED MOVEMENT OR BRAKE RELEASE OCCURS, TURN THE POWERED WHEELCHAIR OFF AS SOON AS IT IS SAFE;

4) BE AWARE THAT ADDING ACCESSORIES OR COMPONENTS, OR MODIFYING THE POWERED WHEELCHAIR, MAY MAKE IT MORE SUSCEPTIBLE TO EMI (NOTE: THERE IS NO EASY WAY TO EVALUATE THEIR EFFECT ON THE OVERALL IMMUNITY OF THE POWERED WHEELCHAIR); AND

5) REPORT ALL INCIDENTS OF UNINTENDED MOVEMENT OR BRAKE RELEASE TO THE POWERED WHEELCHAIR MANUFACTURER, AND NOTE WHETHER THERE IS A SOURCE OF EMI NEARBY.

IMPORTANT INFORMATION

1) 20 VOLTS PER METER (V/M) IS A GENERALLY ACHIEVABLE AND USEFUL IMMUNITY LEVEL AGAINST EMI (THE HIGHER THE LEVEL, THE GREATER THE PROTECTION);

2) THIS PRODUCT HAS AN IMMUNITY LEVEL OF [STATE THE IMMUNITY LEVEL OF THIS POWERED WHEELCHAIR MODEL AND INCLUDE THE CONFIGURATION TESTED IN THE ACCOMPANYING LITERATURE; I.E. SPECIFY THE ACCESSORIES AND/OR COMPONENTS INCLUDED FOR TESTING,

OR SPECIFY THAT THEY WERE NOT INCLUDED FOR TESTING IN THE OPERATOR'S MANUAL]; OR ALTERNATIVELY STATE

2) THE IMMUNITY LEVEL OF THIS PRODUCT IS NOT KNOWN.

IMPLANTABLE PACEMAKERS/DEFIBRILLATORS

Implantable pacemakers or implantable defibrillators can be affected by the electromagnetic fields produced by cellular telephones and certain electronic theft-detection system. The FDA working with industry has developed guidance regarding warning or caution statements that should be included in labeling for these devices.

Cellular Telephones

Testing conducted by industry, the FDA, and the University of Oklahoma Center for the Study of Wireless Electromagnetic Compatibility indicates that some implantable cardiac pacemakers and implantable defibrillators can be influenced by cellular telephones in close proximity to the implant. Digital cellular telephones, because of their higher power level and signal-modulation characteristics, appear more likely to interfere with the operation of some implants than analog telephones. The following language is for physician and patient manuals to provide guidance for those implants where interference may be a problem.

CELLULAR PHONES

RECENT STUDIES HAVE INDICATED THERE MAY BE A POTENTIAL INTERACTION BETWEEN CELLULAR PHONES AND [PACEMAKER OR IMPLANTABLE DEFIBRILLATOR] OPERATION. POTENTIAL EFFECTS MAY BE DUE TO EITHER THE RADIO-FREQUENCY SIGNAL OR THE MAGNET WITHIN THE PHONE AND COULD INCLUDE INHIBITION OR [ASYNCHRONOUS PACING OR DELIVERY OF ADDITIONAL THERAPIES] WHEN THE PHONE IS IN CLOSE PROXIMITY (WITHIN 6 INCHES) TO THE PULSE GENERATOR. IT IS IMPORTANT TO NOTE, BASED ON TESTING TO DATE, THAT ANY EFFECT RESULTING FROM AN INTERACTION BETWEEN CELLULAR PHONES AND [PACEMAKERS OR IMPLANTED DEFIBRILLATORS] IS TEMPORARY. SIMPLY MOVING THE PHONE AWAY FROM THE IMPLANTED DEVICE WILL RETURN IT TO ITS PREVIOUS STATE OF OPERATION. BECAUSE OF THE GREAT VARIETY OF CELLULAR PHONES AND THE WIDE VARIANCE IN PATIENT PHYSIOLOGY, AN ABSOLUTE RECOMMENDATION TO COVER ALL PATIENTS CANNOT BE MADE. THE FOLLOWING INFORMATION PROVIDES A GENERAL GUIDELINE TO PATIENTS HAVING AN IMPLANTED [PACEMAKER OR DEFIBRILLATOR] WHO DESIRE TO OPERATE A CELLULAR PHONE.

- MAINTAIN A MINIMUM SEPARATION OF 6 INCHES (15 CENTIMETERS) BETWEEN A HAND-HELD PERSONAL CELLULAR PHONE AND THE IMPLANTED DEVICE. PORTABLE AND MOBILE CELLULAR PHONES GENERALLY TRANSMIT AT HIGHER POWER LEVELS COMPARED TO HAND-HELD MODELS. FOR PHONES TRANSMITTING ABOVE 3 WATTS, A MINIMUM SEPARATION OF 12 INCHES (30 CENTIMETERS) BETWEEN THE ANTENNA AND THE IMPLANTED DEVICE IS ADVISED.

- PATIENTS SHOULD HOLD THE PHONE TO THE EAR OPPOSITE THE SIDE OF THE IMPLANTED DEVICE. PATIENTS SHOULD NOT CARRY THE PHONE IN A BREAST POCKET OR ON A BELT OVER OR WITHIN 6 INCHES OF THE IMPLANTED DEVICE AS SOME PHONES EMIT SIGNALS WHEN THEY ARE TURNED ON BUT NOT IN USE (I.E., IN THE LISTEN OR STANDBY MODE). STORING THE PHONE IN A LOCATION OPPOSITE TO THE SIDE OF THE IMPLANT IS RECOMMENDED.

Electronic Article Surveillance (Theft-Prevention) Systems

Testing conducted by industry and FDA has demonstrated that some active implantable medical devices can be influence by the electromagnetic fields produced by electronic article surveillance (EAS) systems. These fields can be quite strong in close proximity to the EAS "gates." In those situations where the potential for interference exists, the industry and FDA have developed the "don't linger, don't lean" guidance:

> CAUTION: ELECTRONIC ARTICLE SURVEILLANCE (EAS) EQUIPMENT SUCH AS RETAIL THEFT PRE-VENTION SYSTEMS MAY INTERACT WITH DEVICES. PATIENTS SHOULD BE ADVISED TO WALK DIRECTLY THROUGH AND NOT REMAIN NEAR AN EAS SYSTEM LONGER THAN NECESSARY.

THINGS TO REMEMBER

During the process of classifying devices, the FDA may establish specific labeling requirements for specific device categories. These requirements are usually broader in scope than the device-specific requirements established during the 510(k)/PMA review process. Occasionally, during the review of a product application, the FDA will identify a labeling issue of sufficient importance to make a general rule requiring specific wording in the labeling. These requirements appear first in guidance documents or other communications targeted at the affected manufacturers. Eventually, many of these requirements are incorporated directly into the CFR. The manufacturer is responsible for complying with these requirements. The Division of Small Manufacturers Assistance within the Center for Devices and Radiological Health (CDRH) is available to help manufacturers identify appropriate guidance documents and other resources to help identify the published requirements for a specific device.

Part IX

Development of Device Labels

27 Reducing Labeling Problems

By common practice, all printed, written, or graphic materials on or accompanying a medical device are termed "labeling." In addition to identifying a device (i.e., name of the device, name and address of the manufacturer/distributor, model, serial/lot number, and so on), the labeling must convey the directions for how to use and care for the medical device. The labeling must impart the information that the intended user needs to understand in order to use the device safely for the purpose(s) intended by the manufacturer. This includes such important topics as risks, precautions, warnings, and potential adverse reactions.

High-quality labeling is a vital, but often overlooked, component of any successful medical device. In a study conducted by the United States (US) Food and Drug Administration (FDA), fully 40 percent of the Medical Device Reports (MDRs) filings were found to involve user error in applying the device. In many of these cases, the labeling, while meeting the strict requirements of the applicable regulations, failed to present in an appropriate way the information needed by the user. Accurate, well-written labeling is important to the safe, reliable operation of a medical device, whether the intended user is a professional or a layperson.

The previous chapters have focused on the regulatory requirements for labeling, which must be properly addressed before the product is allowed on the market. This chapter will present some concepts and practices that the reader may find useful in planning and writing labeling materials in order to reduce problems with use of the device. Chapter 28 will highlight some important considerations and give some helpful suggestions to those who must plan the design and typography of medical device labeling.

It is important to remember that there is no one right way to develop labeling. The specific content and format will depend on the device and the people who use it. There can be no guarantee that the user will not make errors. However, well-organized material that is easy to read and understand will promote the safe and effective use of the device. This is true whether the intended user is a highly trained healthcare professional or a layperson.

CONSIDER THE AUDIENCE

A good writer begins a new project by considering the audience. Developing labels and instruction manuals for medical devices should be no different. Before developing any labeling, particularly the instruction manual, the writer must determine who the user will be. Will the typical user be a physician, allied medical professional, or a layperson? Clearly the information required and the assumptions that the writer should make will be vastly different depending on the audience.

LAY USERS

Developing high-quality material for lay users that is easy to read, understand, and follow is a challenging task. Lay users differ widely in age, education, reading level, primary language, and life experience. In most cases, it will be impossible to develop a profile of the "typical" user. The best that the writer may be able to achieve is to pinpoint the general characteristics of the user population. Questions that should be asked include (Backinger p. 3):

- Are the users elderly? Are they children?
- Do the users have vision or hearing problems?
- Do the users have serious illnesses or disabilities that could interfere with their ability to manipulate the device?
- Are the users likely to be on medications that may interfere with memory, understanding, or their ability to carry out procedures?
- What is the user's likely level of reading comprehension? (For lay users, this is generally assumed to be a sixth-grade reading level.)
- What is the users' primary language?
- Where will users be using the device (e.g., in the hospital, at home)?

Thoroughly understanding the characteristics of the user population will enable the writer to provide the needed information in a way that is understandable to the people who must use the device. The objective should be to promote the safe and effective use of the device by all who require it.

Professional Users

Many medical devices are intended by their manufacturers to be used only by professionals who are qualified by their training and experience to use, or supervise the use of, the device. These devices are generally exempt from the requirement that the labeling contain "adequate directions for use" for the device, provided the labeling meets the minimum requirements in the regulations of the country where the devices are being sold. The labeling for a surgical suture does not, for example, have to explain the procedure that the surgeon uses in sewing up a wound.

However, when developing labeling for professionals, the writer must remember that professionals are not all the same. A device may be used by both specialists and general practitioners, who may have different levels of detailed knowledge about the appropriate uses of the device or the techniques involved in using it. When faced with this situation, the writer may find it necessary to provide a section containing summary information for quick reference by the specialist who needs to know about the novel features of the device, backed up by more detailed material for the professional who needs to be trained in the use of the device.

ORGANIZE THE INFORMATION

Organization is the placing of information into an orderly, functional, structured whole. Chaos is so frightening and repugnant to humans that they often associate it with their most terrible experiences. This may account for the extreme irritation and dislike that readers experience when writing lacks a clear plan. No other defect can arouse so much anger and aversion (Tichy p. 33). Writers generally want to avoid arousing the antipathy of the reader, but they can end up inviting it by neglecting to plan.

At times, the organization of the device labeling is specified in detail in the regulations. For example, the warning logotypes required in the United States for laser equipment (see Figure 23.1 and Figure 23.2) are precisely specified in 21 CFR 1040.10(g). In a similar fashion, the format and order of information appearing in the package insert of an *in vitro* diagnostic (IVD) product sold in the United States is specified in 21 CFR 809.10(b).

In most cases, however, the writer is free to choose the organization that best fits the needs of the intended user and the material that must be presented. The FDA booklet *Write It Write: Recommendations for Developing User Instruction Manuals for Medical Devices Used in Home Healthcare* suggests 18 elements that should be considered when developing a manual for medical devices used in the home. They are (Backinger pp. 5–15):

⚠ WARNING

Using this device can cause injury to your unborn baby. Do not use if you are pregnant.

FIGURE 27.1 Example of a warning statement.

1. User assistance—Provide the user with the company name, address, device name, model number, and the telephone number of the customer-assistance department. If applicable, provide the address of a Web site where the user can obtain additional information.
2. Table of contents—Use the same headings that are used in the text.
3. General warnings and precautions—Include contraindications, warnings, precautions, adverse reactions, and, when applicable, limitations on the population of patients for whom the device is indicated.
4. Purpose of the device (indications for use)—Identify the target population for which the device provides clinically significant benefit at a reasonable risk.
5. Description of the device—Briefly describe the device and its parts and accessories, and the function and purpose of all user-adjustable controls. Accessories may be covered in the main manual or may be described in a separate section.
6. Environmental conditions that affect use—Explain both the conditions necessary for the proper functioning of the device, and the conditions under which the device should not be used.
7. Setup instructions—Include in this section detailed instructions for any steps that the user must perform before the device can be used.
8. Device check-out—Describe procedures that the user should follow to verify that the device is operating correctly, including when to check it (e.g., weekly, monthly, before each use), the step-by-step procedures to be followed, and what to do if the device is not working correctly.
9. Operating instructions—Describe any necessary preparations, the step-by-step procedures that the user should follow. Include the following:
   - Special preparation required before operating the device (e.g., washing hands)
   - Warnings and safety instructions listed immediately prior to the related step or action
   - Step-by-step procedures, presented in a logical order, for operating the device
   - The results of incorrect operation and ways the user can recognize if the device is not behaving properly
   - A place to record user-specific instructions
   - If appropriate, how to read and interpret the results of the procedure
   - Special steps to be taken once the procedure is over, including proper disposal of any waste or residues created as a result of using the device
   - Whom to contact in the event of a problem (this may be by reference to the user-assistance section of the manual)
10. Cleaning—include instructions for when and how to clean the device, giving the method and materials (e.g., clean with soap and water), and inform the user of limitations in cleaning and/or sterilization methods (e.g., do not immerse in water, do not autoclave), and describe what is likely to happen if the device is not cleaned.

11. Maintenance—Clearly describe the maintenance that the user must perform. If maintenance is required by others, then describe the proper maintenance actions, who is responsible, and when the maintenance is needed. A chart can be a convenient way of providing this information. Provide the user with the address and telephone number of service centers where the required maintenance can be obtained, or provide a place where the user can write in this information.

12. Storage—Describe the proper preparation of the device for storage, the conditions for proper storage, the expected results of improper storage, and a notice if extended storage may affect the performance of the device.

13. Troubleshooting—Provide a section that helps the user identify the cause of problems listing the most serious problems first.

14. Accessories and supplies—In many cases, a separate section on accessories and supplies will not be required. The information can be included in the appropriate sections of the manual along with information on the main device. In those cases where there are a lot of accessories or supplies, it may enhance the usefulness of the manual to have a separate section devoted to these items. Regardless of where the information is placed, describe and illustrate all of the supplies, materials, and accessories needed to operate and maintain the device. Specify quantity, size, and type of all supplies and accessories needed in sufficient detail that the user can easily obtain them.

15. Technical information—For home-use devices this section should contain only the information that the user needs to safely use the device. More involved information can be presented (a) in a separate manual for the professional user, (b) in separate sections for the lay user and the professional who will dispense or service the device, or (c) in an appendix or supplement to the user instructions.

16. Summary—Provide a brief recap of the essential information in the manual. The summary should include only the information that the user really needs. The summary should be placed at the end to encourage the user to read the entire manual before reaching the summary.

17. Index—Provide an alphabetized listing of important subjects in the manual with page numbers so that the user can find needed information quickly and the manual can achieve its maximum usefulness. The preparation of an index requires careful study so that it is more than an expanded, alphabetical table of contents. For large and complicated manuals, the services of a professional indexer may be valuable.

18. Date—The date that the manual was issued or last revised should appear in a conspicuous place. The back cover of the manual is a popular location for this information. The date of last revision is required by various nations' regulations on some manuals. It is recommended that the date of the last revision appear on the manual for all devices.

This structure is recommended for manuals intended for the lay user. In preparing material for the professional, the writer should take account of the training and experience of the reader. This will generally mean more technical and scientific information, a greater reliance on technical and medical terms, and less emphasis on step-by-step operating procedures. However, the basic elements are still applicable and should be covered in the manual.

WRITE TO THE READER

One of the most serious mistakes that writers can make is writing to themselves. When material is clear to them, they mistakenly think it is clear to others. The writer must estimate closely the reader's interests, reading skill, and knowledge of the subject. The good writer has the ability to "forget himself or herself and become the reader" (Tichy p. 21). The ability to know and understand the reader is key to writing functional prose.

CHOOSE WORDS CAREFULLY

Effective language involves the correct, succinct, and clear use of words to communicate information between human beings. For medical device labeling, good language conveys instructions simply, directly, and unambiguously. The writer should strive to minimize the effort the reader will have to invest in understanding what is being communicated. One way is to minimize unnecessary technical terminology, polysyllabic words, and complicated expressions.

When writing for the lay reader, it is advisable to avoid using technical terms altogether. When technical terms must be employed, be sure to provide a nontechnical explanation or definition. Even those that are simple technical terms for the professional can be mysterious and confusing for the lay person. For example, the package insert accompanying one brand of test strips for a home blood-glucose monitor explained in detail how extremes in "hematocrit" can affect the accuracy of the instrument. The package insert informed the user that a hematocrit below 25 percent or above 60 percent would result in inaccurate readings. Unfortunately, the writer failed to explain what a hematocrit was, or how the user should find out if his or her hematocrit was high or low.

On the other hand, technical terminology is indispensable in conveying ideas to readers in the same specialization. It is the easiest, simplest, cleanest, and quickest road to precision. However, a trap that many writers fall into is using technical terminology where it is unnecessary and even confusing. A careful writer will replace technical terminology with nontechnical language when it is possible to do so without sacrificing brevity and clarity.

Another trap that writers may fall into is using terms from other specializations. When writing for the healthcare professional, the writer should avoid adding expressions from engineering to the technical language of medical expressions. The language from one specialty is enough.

Always introduce each subject, control, indicator, or device before it is used in the text. If abbreviations or uncommon terms must be used, define them before they are used so the reader doesn't lose concentration by having to wonder, "What is this?"

AVOID CLUTTER

Clutter is the result of using falsely elegant words and phrases when simpler ones would do. Readers are very busy trying to learn how to use a new device or carry out a procedure. Cluttered language just gets in the way—distracting the user and obscuring the important points. According to the noted author and editor, William Zinsser, "fighting clutter is like fighting weeds—the writer is always slightly behind" (Zinsser p. 14). Some examples of clutter and possible simplifications are shown in Table 27.1. There are a number of books available on technical writing that contain helpful techniques for weeding out clutter.

WRITING INSTRUCTIONS

In many cases, the instructional sections are the heart of the manual. They convey the information that the user must understand in order to use the device as intended. Instructions must be carefully planned and organized. The following three methods that are used frequently in documenting the instructions for use (IFU) of a medical device (Backinger pp. 18–30):

TABLE 27.1
Examples of Cluttered Language

Original	Possible Simplifications
abovementioned	the, this, that, those, these
accounted for by the fact that	due to, caused by
after this is accomplished	then
an example of this is the fact	for example, as an example, thus
another aspect of the situation to be considered	as for
approximately	about
as regards	about
assistance	help
at the time of	when
at this point in time	now
care should be used so as not to	be careful
carried out in	perform
causes power to be applied to	switches power to
comes up to	reaches
due to the fact that	because, due to
if at all possible	if possible
in conjunction with	and, together with
in light of the fact that	because
in view of the above, in view of the foregoing circumstances, in view of the fact that	therefore
subsequent	next
notwithstanding the fact that	although
of very minor importance	unimportant
take appropriate measure	act
taking this factor into consideration, it is apparent that	therefore, therefore it seems
transpire	happen, occur
will also serve as a chance to	may
with this in mind, it is clear that	therefore

Sources: Cardamone p. 41; Tichy pp. 396–400.

- Text—using words in complete sentences that form paragraphs
- Flowchart—a diagram using symbols or brief verbal phrases to represent the order of an operation
- List—an item-by-item series of words, phrases, or sentences

Each of these methods has advantages and disadvantages that may make one or another better suited for a particular device and user population. Some of the advantages and disadvantage are described in the following sections. Whatever system is chosen, avoid mixing formats in the same section of the manual. Mixing formats is likely to confuse the reader. However, when used carefully, different formats can serve well in different parts of the manual (e.g., a list of supplies needed preceding an operating procedure documented in a flowchart).

TEXT

Text is a technique that uses complete sentences to form paragraphs. Most people are familiar with this organizational structure, which increases the reader's comfort. Text is also the most adaptable technique. It is capable of conveying information ranging from very simple to the most complex and involved instructions. The advantages of text as a technique for organizing instructions include:

- Text is easy to develop, update, and handle.
- Particularly for lay users, text with accompanying graphics is well suited to simple instructions where few decisions are required.
- Text can be constructed to tell the user what actions must be taken when more than one procedure is required at a time.

The disadvantages of text as a format for instructions include:

- The writer may be tempted to include information that is not necessary to the proper execution of the procedure.
- Text usually requires more space to convey the information.
- With text, the reader will have more difficulty than with either the flowchart or list techniques in finding specific items within the procedure.
- With text, it may be more difficult for the user to picture the whole procedure quickly.

FLOWCHART

A flowchart is a diagram that uses symbols or brief verbal phrases to convey the order of steps in a procedure. Flowcharts are useful when the procedure requires the user to make frequent decisions between various alternative courses of action. The advantages of a flowchart are:

- By using symbols or verbal clues, the flowchart avoids containing excessive information.
- The flowchart visually illustrates the logical order of the procedure.
- The flowchart can provide an overview of the entire process.
- The flowchart leads the user through a series of sequential decisions.
- The flowchart simplifies conditional action, which would have to be described in text as "if then or else" statements.

The disadvantages of the flowchart as a means of conveying instructions include:

- The very compactness of the information in a flowchart may be difficult for some users to understand without training.
- Complex decision processes involving more than two alternatives or combinations of several input variables may be difficult to diagram simply and accurately.
- Flowcharts are most effective when they can be viewed all at once. Therefore, they should be contained entirely on one page, which may not be possible for complex tasks.

LIST

A list is an item-by-item series of words, phrases, or sentences. The items in a list should be equal or nearly equal divisions of a subject. Lists, for example, are good for informing the user of supplies needed for a procedure. The advantages of lists include:

- Lists can be a good compromise between text and flowcharts for lengthy tasks.
- Lists are well adapted to procedures that require checking off steps that are completed.
- Lists can be helpful when the user is learning a procedure.

The main disadvantage of lists is that they are not as well suited to tasks that contain conditional logic, which involves jumping around within the procedure.

WARNINGS AND CAUTIONS

Providing the user with the information needed to use a device in the safest manner possible is one of the fundamental purposes of labeling. A significant part of that responsibility is to instruct the user on any hazards that may be associated with use or misuse of the device. The signal words "warning," "caution," and "danger" are normally associated with such statements to alert the user to the significance of the information. A recognized symbol or icon, such as those shown in Table 17.3, may also be associated with the statement for further emphasis.

In general, a warning is a statement that alerts the user to the possibility of death, injury, or other serious adverse reactions associated with the use or misuse of the device.

A caution is a statement that alerts the user to the possibility of a problem associated with the use or misuse of the device. Such problems include device malfunction, device failure, damage to the device, or damage to other property. The caution statement should include the steps that should be taken to avoid an adverse outcome.

Warnings and cautions can be divided into two categories:

- General warnings and cautions provide essential information required before using the device. These warnings and cautions should be placed in a separate section at the beginning of the manual where the user will see them right away. Frequently, general warnings and cautions are listed in separate sections with the headings **WARNINGS** and **PRECAUTIONS**.
- Specific warnings and cautions apply to specific instructions or actions and should immediately precede the instructions to which they apply. Research indicates that combining all warnings into a section does not help the user avoid the hazard unless the warnings also appear with the steps in the instructions to which they apply (Backinger p. 32).

There is a special kind of caution statement associated with conditions under which the device should not be used. These "contraindications" describe situations where the risk of device use outweighs the benefits to the patient. Examples would include directions such as "Not to be used on pregnant women," or "Do not use with patients who suffer with chronic atrial fibrillation." Contraindications for device use should be listed along with the general warnings and cautions, frequently in a separate section headed **CONTRAINDICATIONS**.

It may be appropriate to provide the user with information on the probability that a given hazard will occur. However, the writer must be careful not to overestimate or underestimate the likelihood of a hazard occurring. The use of complex statistics to explain the likelihood of a hazard should be avoided when writing for the lay user.

Symbols or icons can be very useful in drawing the reader's attention to a given warning. Color can also be used effectively for this purpose. Use of color is discussed in Chapter 28.

The following elements should be present in a warning or caution (Backinger p. 33):

- A signal word (e.g., WARNING or CAUTION) should come first, usually above or to the left of the remaining text.
- An appropriate symbol or icon may be used for emphasis. In some cases, the symbol or icon may be required by a particular regulation, as in the case of warnings involving hazardous substances. The recommended location for the symbol or icon is above or to the left of the signal word (ISO 3864-1, p. 4).
- A primary hazard statement explaining the possible problem should follow the signal word.

- The consequences of not following the instructions should be clearly explained to the user.
- The final element is an explanation of the do's and don'ts necessary in order for the user to avoid the hazard.

An example of a warning statement combining all of these elements is illustrated in Figure 27.1.

All warning and caution statements should stand out from the other material in the labeling. Using the highlighting techniques described in Chapter 28, as well as an appropriate symbol or icon, can draw the user's attention to the information.

A warning or caution statement should never include procedures or other information not essential to the warning. Instructions to be followed if the hazard occurs should immediately following the warning statement in the body of the text. If the labeling contains a troubleshooting section, the information should be included there as well, because the user may turn to that section when a problem occurs.

LABELING EVALUATION

All labeling should undergo some premarket testing and evaluation. Pretesting can identify specific strengths and weaknesses of labeling that can be used to improve the product and reduce problems later on.

Pretesting should look at user comprehension, user performance, acceptability, and credibility. Focus on the characteristics of the intended user group to make the labeling most useful to them. Too often, labeling is written without the target user clearly in mind. The result is labeling that the user cannot understand or follow.

Several methods can be used to test medical device labeling, including focus-group interviews, in-depth individual interviews, questionnaires, readability testing, and operator-performance studies. Some techniques, such as the focus-group and individual interviews, are particularly useful during the planning and development stages of a project, when the writer is looking for the best ways to organize and present the instructions. Other techniques, such as questionnaires, readability testing, reviews by professionals familiar with the device, and operator-performance studies, where potential users are asked to operate the device, are well suited to testing once a draft of the manual is completed. Most often, a combination of methods must be used to develop the most effective labeling possible.

LIMITING LIABILITY

A set of instructions, no matter how carefully designed and well written, cannot guarantee that the user will never make a mistake. Instructions cannot prevent the intentional misuse of a device. However, device manufacturers can limit liability by properly using product warnings.

Effective warnings use combinations of words and pictures in labels, package inserts, and instruction/user guides. Warnings should pertain not only to the product, but also to its surroundings and users. For example, a warning should describe what effect a radiological or electronically controlled device might have on other equipment or people in the same room. Warnings should detail possible hazards with use, maintenance, and disposal of the device. They should include the words "caution," "warning," or "danger" in large print above a picture that shows the hazard and its effects. A message should describe how to avoid potential dangers. Advertising should be limited to approved uses and should avoid absolute words such as "eliminates," "never," "always," "safe," and "prevents." Instead, it should clearly and truthfully communicate the information consumers need to make decisions. Advertising should be unambiguous and make claims that are "performance based" (Labels pp. 3–4).

THINGS TO REMEMBER

High-quality labeling is a vital, but often overlooked, component of any successful medical device. Accurate, well-written labeling is important to the safe, reliable operation of a medical device regardless of whether the intended user is a professional or a layperson.

It is important to remember that there is no one right way to develop labeling. The specific content and format will depend on the device and the people who use it.

A good writer begins a new project by considering the audience. Thoroughly understanding the characteristics of the user population will enable the writer to provide the information needed. The objective should be to promote the safe and effective use of the device.

Develop a clear plan for the work so that the information is placed into an orderly, functional, and structured whole. At times, the organization of the device labeling is specified in detail in the regulations. In most cases, however, the writer is free to choose the organization that best fits the material that must be presented and the needs of the intended user.

Choose your words carefully so they convey instructions simply, directly, and unambiguously. The writer should strive to minimize the effort the reader will have to invest in understanding what is being communicated. One way is to minimize unnecessary technical terminology, polysyllabic words, and complicated expressions. Try to avoid using falsely elegant words and phrases when simpler ones would do.

Understand and write to the reading level of the expected audience. For lay users, this is a sixth-grade reading level. Even when writing for the trained professional, a good writer will take pains to come crisply to the point. Clear, clean sentences or lists are acceptable in instruction manuals and in other labeling. Bury an idea in a lot of words and the reader will probably miss the point.

Providing the user with the information needed to use a device in the safest manner possible is one of the fundamental purposes of labeling. A significant part of that responsibility is to instruct the user on any hazards that may be associated with use or misuse of the device. The signal words "warning," "caution," and "danger" are normally associated with such statements to alert the user to the significance of the information. A warning or caution statement should never include procedures or other information not essential to the warning. Instructions to be followed if the hazard occurs should immediately following the warning statement in the body of the text.

All labeling should undergo some premarket testing and evaluation looking at user comprehension, user performance, acceptability, and credibility. Focus on the characteristics of the intended user group to make the labeling most useful to them.

28 Designing Good Labeling

The design of the labeling should complement and enhance the subject of the labeling. It is universally true that good design cannot improve poorly written material, but a weak design can seriously compromise the best writing.

The material in this chapter is intended to highlight some important considerations and to give some helpful suggestions to those who must plan the design and typography of medical device labeling. The material highlights the bare essentials and, therefore, is not intended to serve all of the needs of professional designers, typographers, or production people.

CONDITIONS FOR USE

The designer of medical device labeling must consider the conditions under which the device will be stored and used. Labels applied to the device or its containers must remain in place and legible during the customary conditions of distribution, storage, and use. Similarly, the manuals or other documents accompanying the device must be durable enough to tolerate expected use for as long as necessary in order to achieve the device's intended purpose.

DEVICE LABELS

Labels applied to the device or its containers must remain in place and legible during the customary conditions of processing, distribution, storage, and use. This means that the designer must consider:

- the environmental conditions (e.g., temperature, relative humidity, and atmospheric pressure) in which the device and its packaging will be transported, stored, and used;
- the maintenance procedures (e.g., cleaning, disinfecting, and/or sterilization) that are likely to be used with the device;
- whether liquids, solvents, or grease will be used on or around the device;
- the placement of the labels to avoid damage to them during normal use of the device;
- the lighting conditions in which the device is likely to be used;
- whether the user is likely to have poor vision or other sensory problems; and
- whether the device is likely to be subjected to vibration (e.g., used in an ambulance).

Proper consideration of these factors should lead the designer to appropriate choices of label substrate, dimensions, print size, ink, finish, and mounting methods so that the labels remain attached and legible in the customary and expected conditions. In determining the customary and expected conditions, the designer should consider the mistakes that a user might reasonably be expected to make. For example, if the user is instructed to clean the device with a sponge or cloth moistened in water or 70 percent isopropyl alcohol, the designer should consider what would happen to the labels if the user used pure isopropyl or methyl alcohol.

ACCOMPANYING DOCUMENTATION

As with the labels, the designer should consider the environment where the documentation accompanying the device, such as the manual, will be used. Most of the conditions for use that apply to the labels also apply to the accompanying documentation. In addition to those issues listed in the previous section, the designer of the accompanying documentation should consider:

- whether the user will need to refer to the documents every time the device is used, or only while learning how to use the device, and for troubleshooting after that;
- whether the user will need to have both hands free while using the documentation;
- whether the documents will need to be sized to fit in the carrying case of a portable device; and
- whether the documents will need to be updated by the user of the device (i.e., inserting and removing pages).

Consideration of these factors, and those described in the previous section, will lead the designer to appropriate choices of paper size, paper weight (or alternative material), finish, and binding method so that the accompanying documentation will be durable enough to tolerate the expected usage and last as long as necessary to achieve the intended purpose of the device.

LAYOUT

A layout is the designer's blueprint for a document. The layout specifies the dimensions, materials, type fonts and sizes, colors, and so on, to be used in the label or document. For large documents, such as technical manuals, it will show how the document will be structured—by sections, chapters, and subheads. The layout specifies page size, how major divisions within the document will be identified, how pages will be numbered, and what type fonts and sizes will be used for various purposes. In short, it contains all of the information necessary to turn the writer's manuscript into a final printed document.

The usefulness of large documents can be enhanced by separating sections to make them easy to find. The use of colored paper, colored paper edges, and labeled tabs are common techniques.

All pages in a document should be numbered. There are common approaches to page numbering. Numbering the pages consecutively is the traditional and most widely accepted method used in publishing (Chicago p. 40). However, pages can be numbered by chapter and page (e.g., 4-20 refers to page 20 in Chapter 4). A variation on this method is to use the chapter heading, or a shortened version of the chapter heading, and the page number (e.g., Index-5 refers to page 5 of the Index). Use of either of the latter systems, rather than page numbers only, makes updating of the manual easier because pages can be inserted or removed (for example, from a loose-leaf binder) without having to renumber the entire volume.

PHYSICAL ATTRIBUTES

The physical attributes of labeling will have a marked influence on both the ease of use and subjective appeal. Labels should be carefully placed on the device and its packaging so they are clearly visible and can be easily read by the intended user. Warning statements should always be prominently located so they are visible to the operator in normal use. Documents should be compact, accessible, and easy for the user to manipulate in the environment where the device will be used. Failure to adequately address these physical issues will materially affect the degree to which the labeling is read, comprehended, followed, and retained (Callan p. 19).

Technical manuals and similar reference material are often large documents and should be sized accordingly. In metric countries, international paper sizes, also known as the International Standards Organization (ISO) sizes, are widely used. The three ISO paper series are designated A, B, and C. The A series is for general printed matter, including stationary and publications. The B series is

TABLE 28.1
International (ISO) Paper Sizes

Size	In Millimeters [a]	Approximate Inches
A Series		
4A	$1{,}682 \times 2{,}378$	$66\frac{1}{4} \times 93\frac{3}{8}$
2A0	$1{,}189 \times 1{,}682$	$46\frac{3}{4} \times 66\frac{1}{4}$
A0	841×1189	$33\frac{1}{8} \times 46\frac{3}{4}$
A1	594×841	$23\frac{3}{8} \times 33\frac{1}{8}$
A2	420×594	$16\frac{1}{2} \times 23\frac{3}{8}$
A3	297×420	$11\frac{3}{4} \times 16\frac{1}{2}$
A4	210×297	$8\frac{1}{4} \times 11\frac{3}{4}$
A5	148×210	$5\frac{7}{8} \times 8\frac{1}{4}$
A6	105×148	$4\frac{1}{8} \times 5\frac{7}{8}$
A7	74×105	$2\frac{7}{8} \times 4\frac{1}{8}$
A8	52×74	$2 \times 2\frac{7}{8}$
A9	37×52	$1\frac{1}{2} \times 2$
A10	26×37	$1 \times 1\frac{1}{2}$
B Series		
B0	$1{,}000 \times 1{,}414$	$39\frac{3}{8} \times 55\frac{5}{8}$
B1	707×1000	$27\frac{7}{8} \times 39\frac{3}{8}$
B2	500×707	$19\frac{5}{8} \times 27\frac{7}{8}$
B3	353×500	$12\frac{7}{8} \times 19\frac{5}{8}$
B4	250×353	$9\frac{7}{8} \times 12\frac{7}{8}$
B5	176×250	$7 \times 9\frac{7}{8}$
B6	125×176	5×7
B7	88×125	$3\frac{1}{2} \times 5$
B8	62×88	$2\frac{1}{2} \times 3\frac{1}{2}$
B9	44×62	$1\frac{3}{4} \times 2\frac{1}{2}$
B10	31×44	$1\frac{1}{4} \times 1\frac{3}{4}$

[a] ISO standards are based on a rectangle whose sides have a ratio of one to the square root of 2 (1:1.414). No matter how many times a sheet of these proportions is halved, each half retains the same constant proportions.

Source: Kuhn p. 2 and p. 11

primarily for posters, wall charts, and other items. The B series is about halfway between two A sizes. Although intended for posters, it is sometimes used as an alternative to the A size when the difference between sizes (e.g., A4 to A5) represents too large a jump. The C series is for envelopes. The dimensions for the A and B series are given in Table 28.1. Manuals are typically A4 (210 mm x 297 mm) or B5 (176 mm x 250 mm) sizes. Standard sizes for printed manuals in the United States are $8\frac{1}{2}$ inches by 11 inches and $5\frac{1}{2}$ inches by $8\frac{1}{2}$ inches.

Manuals should be designed so that they will lie flat open without assistance. This frees the reader to use both hands to perform tasks.

Manuals may need to be updated from time to time by adding or deleting pages. A ring binder is the preferred solution for meeting this requirement. However, the pages in ring binders have a tendency to pull out when used frequently. A heavy, good-quality paper should be used for this purpose. Spiral binding is preferred for documents that are not intended to be updated by the user (Callan p. 20).

Some reference materials (e.g., quick reference guides, methods sheets) are intended to be used while the described equipment is in operation. These documents should be designed for ease of access and use. Appropriate materials (e.g., card stock, plastic laminate) should be used to provide the necessary durability in normal use.

Glossy paper should be avoided to reduce glare. This may be a particular problem for the elderly. Paper should always be heavy enough to prevent text and graphics from bleeding or showing through. Bleed is distracting and reduces the clarity of printed words and illustrations.

TYPE FONTS AND SIZE

The most commonly available and widely used typefaces (i.e., Baskerville, Bembo, Garamond, Janson, Palatino, and Times Roman) are about equally legible, although Times Roman is the most widely available (Chicago pp. 767–769). This book is printed in Times.

Serif type is generally considered easier to read than sans serif type. A serif is the small horizontal stroke that finishes off the main stroke of the letter. The eye uses the serif as an anchor to stay on the correct line, which makes serif easier and less fatiguing to read than sans serif in text (Backinger p. 37). A heavy sans serif type (also called "Gothic" type) is frequently used for hazard-signal words (e.g., DANGER, WARNING, CAUTION, POISON), and other required wording in hazard warnings. The use of a different type font, along with other highlighting techniques, helps to draw the reader's attention.

The type size selected for medical device labeling must be large enough to be easily read by the intended audience. Twelve-point type is recommend for the general audience. It is a good compromise between legibility and the need to conserve space. Fourteen-point type is recommended for the visually impaired and the elderly reader (Backinger p. 37).

Other type sizes may have limited application in labeling. Ten-point type (in which many texts and general-interest books are set) is an acceptable minimum for the general audience, but not for the elderly. It may be useful for footnotes and other reference material that is not part of the body of the manual. Nine-point and smaller type is hard for most people to read and encourages the reader to skip the material or risk eyestrain (Callan p. 20). Large type (18-point and larger) should be used sparingly.

Frequent changes in font style or type size should be avoided. These can be a major distraction for the reader.

HIGHLIGHTING

Highlighting is a technique for emphasizing important words, thoughts, or phrases. Commonly used highlighting techniques include using boldfaced, underlined, or italicized type, and using capital letters. Highlighting also includes color, reverse printing, varied font styles, boxing-in text, offsetting border and background patterns, and white space. All of these techniques can be used to provide visual relief and to stress important points.

To be effective, highlighting must be used sparingly. Overuse creates visual clutter and lessens the impact of the message. Avoid particularly the overuse of capitals and italics. Text printed in all capitals reduces reading speed because the shapes of the letters do not vary greatly. Used appropriately, however, the use of all capitals can be a powerful technique for drawing the reader's attention to important information, particularly if the reader is scanning the text. Italics should be used sparingly because they also retard reading speed.

Use different types of highlighting for different purposes. For example, use bold warnings; box in important information; and use all capitals to emphasize important words (see Figure 27.1). Whatever methods are chosen, be consistent in their application.

WHITE SPACE

Judicious use of white space can be an important technique for improving the appearance of a document, making it easier to read and emphasizing major divisions. Of particular importance is the amount of white space between lines of text, referred to in the printing trade as the "leading" or "lead." Too much lead impairs reading speed, increases printing cost, and makes a set of instructions unnecessarily lengthy. Too little lead makes the lines blur together, especially for those with poor vision. In general, the larger the type, or the longer the lines of text, the more lead is needed.

Use of white space at the end of individual lines is a technique that can improve the readability of labeling. In printing, the traditional technique produces a column of type with its left and right margins neatly aligned. Although this produces a very clean look, there are problems associated with the variable letter and word spacing within a line and the use of hyphenation to "justify" the line that can affect reading speed and comprehension. An increasingly acceptable technique for dealing with this problem in books as in other kinds of printed material, is simply not to justify the lines of type. The right edge is allowed to "run ragged." Word spacing is the same in every line, and each line ends with the word falling nearest, but not beyond, the maximum length of the type line. The ragged right margin makes it easier for the eye to distinguish one line from another (Callan p. 20).

ILLUSTRATIONS AND GRAPHICS

Illustrations and graphics consist of photographs, drawings, cartoons, symbols, icons, tables, and graphs. The use of illustrations can greatly simplify the instructions for operating of the device by augmenting text descriptions. The old axiom, "one picture is worth a thousand words" is still true. Also, illustrations are usually remembered better than words, and they can greatly help the semiliterate reader understand the operation of the device (Callan p. 12).

Good illustrations should be clean, simple, uncluttered, and accurate. Each illustration should convey one idea, and it should not simply repeat the text. The virtue of graphics lies in the ability to show aspects of the device that are difficult to convey verbally. Photographs show the exact appearance of an object. Drawings can be used to emphasize specific details. Exploded views and cut-away views may be useful in explaining complex details. Exploded views, however, should be used only if the user is required to assemble or take apart the device (Backinger p. 42).

Illustrations and graphics should be located near the relevant text. This keeps the reader's attention focused and makes the flow of the instructions easier to follow. Side-by-side columns of text and graphics make both easy to follow (Backinger p. 41).

A symbol is a sign or picture that has been developed to represent an idea, as opposed to an icon, which is a drawing that looks like the idea it is meant to represent (Backinger p. 40). Good symbols and icons are difficult to design. There are extensive collections of "standardized" forms that have been developed and have come into common use. A number of these are depicted in this book, and there are many more in publications such as the ISO 7000 and IEC 60417.

The general public already understands some symbols and icons. Others are understood only by specific segments of the population (e.g., healthcare professionals). In all cases, however, it is a good idea to explain symbols and icons in the device labeling even when the writer believes them to be "generally recognized."

The use of tables and graphs should be minimized, particularly in documentation intended for the layperson. If a table or graph is necessary, label it clearly and include instructions on its use (Backinger p. 43).

COLOR

Color can be an excellent tool for creating attractive material that will hold the reader's attention. It can be effective for highlighting important material such as warning and caution statements.

TABLE 28.2
Safety Colors and Contrast Colors

Meaning or Objective	Safety Color	Contrast Color	Graphical Symbol Color	Examples of Use
Prohibition	Red	White	Black	Stop signs Emergency Stops Prohibition signs
Mandatory action	Blue	White	White	Obligation to wear personal protective equipment such as radiation-shielding garments
Warning	Yellow	Black	Black	Danger: hot surface Danger: acid Danger: High voltage
Safe condition	Green	White	White	Means of escape Safety equipment
Fire safety	Red	White	White	Fire-alarm call point Fire-fighting equipment Fire extinguisher
Supplementary information	White or the color of the safety sign	Black or the contrast color of the relevant safety sign	Symbol color of the relevant safety sign	As appropriate to reflect the message given by the graphical symbol

Source: ISO 3864-1 Table 1

Although color printing is somewhat more expensive, color illustrations are generally preferred over black and white graphics because of their lifelike character and greater ability to attract attention (Callan p. 12). However, color, like other highlighting techniques, can be overused easily.

Black printing on a white background provides the best contrast and is the easiest to read. Other color combinations that provide good contrast are (a) black on yellow, (b) blue on white, and (c) green on white (ISO 3864-1 p. 4). In using color, it is important to remember that different people perceive colors differently. Blue may be difficult for the elderly to see. Eighty percent of males cannot distinguish pastel colors, particularly blue/green (Backinger p. 43).

The ISO has established a color system to draw attention to objects and situations affecting safety and health. The general meaning of these safety colors and the contrasting color used with the safety colors are listed in Table 28.2. It is important to note that, in most countries, safety colors in certain fields (particularly labeling for the transport of dangerous goods) are covered by statutory regulations that may differ in some respect from those listed in Table 28.2.

THINGS TO REMEMBER

The design of labeling should complement and enhance the subject matter. It is universally true that good design cannot improve poorly written material, but a weak design can seriously compromise the best writing.

The designer of medical device labeling must consider the conditions under which the device will be stored and used. Careful consideration of these factors will lead the designer to the appropriate choices of materials, dimensions, print size, ink, finish, and mounting or binding methods so that the labeling will be durable enough to tolerate the expected usage and last as long as necessary to achieve the intended purpose of the device.

The designer must also determine the layout of the labeling and select the type fonts and sizes. The most commonly available and widely used typefaces (i.e. Baskerville, Bembo, Garamond, Janson, Palatino, and Times Roman) are about equally legible, although Times Roman is the most widely available.

Serif type is easier to read than sans serif type. A heavy sans serif type (also called "Gothic" type) is frequently used for titles, headlines, hazard signal words (e.g., DANGER, WARNING, CAUTION, POISON), and other required wording in hazard warnings to draw the reader's attention.

The type size selected for medical device labeling must be large enough to be easily read by the intended audience. Twelve-point type is recommended for the general audience. Fourteen-point type is recommended for the visually impaired and for the elderly reader. Frequent changes in font style or type size should be avoided.

Highlighting is a technique for emphasizing important words, thoughts, or phrases. Commonly used highlighting techniques include using boldfaced, underlined, or italicized type and using capital letters. To be effective, highlighting must be used sparingly. Overuse creates visual clutter and lessens the impact of the message. Judicious use of white space can be an important technique for improving the appearance of a document, making it easier to read, and emphasizing major divisions.

Using illustrations can greatly simplify the instructions for operating the device by augmenting text descriptions. Illustrations can also greatly help the semiliterate reader understand the operation of the device. Good illustrations should be clean, simple, uncluttered, and accurate. Each should convey one idea.

Color can be an excellent tool for creating attractive material that will hold the reader's attention. It can be effective for highlighting important material such as warning and caution statements. However, color, like other highlighting techniques, can be overused easily.

Black printing on a white background provides the best contrast and is the easiest to read. In using color, it is important to remember that different people perceive colors differently. Blue may be difficult for the elderly to see. Eighty percent of males cannot distinguish pastel colors.

Appendix A:
U.S. Department of Commerce Medical Device Product Categories/Classification

The Commerce Department defines the industrial sectors of the US economy using the North American Industry Classification System (NAICS). Beginning January 2002, the NAICS replaced the Standard Industrial Classification (SIC) codes for all US government statistical data. The NAICS codes are used to track key economic statistics, such as imports, exports, and total production in each sector. For the healthcare technology industry, the most important NAICS codes are (Trade p.1):

- **339111 Laboratory Apparatus and Furniture Manufacturing**—includes laboratory apparatus and laboratory and hospital furniture (except dental)
- **339112 Surgical and Medical Instrument Manufacturing**—includes medical, surgical, ophthalmic, and veterinary instruments and apparatus (except electrotherapeutic, electro-medical and irradiation apparatus)
- **339113 Surgical Appliance and Supplies Manufacturing**—includes orthopedic devices, prosthetic appliances, surgical dressings, crutches, surgical sutures, and personal industrial-safety devices (except protective eyewear)
- **339114 Dental Equipment and Supplies Manufacturing**—includes dental equipment and supplies used by dental laboratories and dentist's offices and includes dental chairs, dental-instrument delivery systems, dental hand instruments, and dental-impression material
- **339115 Ophthalmic Goods Manufacturing**—includes prescription eyeglasses (except those manufactured in a retail setting), contact lenses, sunglasses, eyeglass frames, and reading glasses made to standard powers
- **334510 Electromedical and Electrotherapeutic Apparatus**—includes Magnetic Resonance Imaging (MRI) equipment, medical ultrasound equipment, pacemakers, hearing aids, electrocardiographs (ECGs), and electromedical endoscopic equipment
- **334517 Irradiation Apparatus Manufacturing**—includes irradiation apparatus and tubes for applications, such as medical diagnostic, medical therapeutic, industrial, research, and scientific evaluation (Irradiation can take the form of beta-rays, gamma-rays, X-rays, or other ionizing radiation.)
- **325413 *In Vitro* Diagnostic Substance**—includes *in vitro* (i.e., not taken internally) diagnostic substances, such as chemical, biological, or radioactive substances (The substances are used for diagnostic tests that are performed in test tubes, petri dishes, machines, and other diagnostic test-type devices.)

Table A.1 lists the product groups that are included under the most important NAICS codes for the healthcare technology industry. Table A.1 also includes NAICS Code 325413, *In Vitro* Diagnostic Substances. The Department of Commerce designates product groups in this category as part of the drug industry. The category is included in Table A.1 because certain IVD products are regulated as medical devices. *In vitro* diagnostic products include immunology kits, allergy-measuring products, and drug-abuse tests that use chemical or biological reagents to test specimens (blood, urine, etc.) drawn from patients.

The trade data reported in Chapter 1 includes products that are injected into, or ingested by, the patient. These *in vivo* products are regulated as pharmaceuticals. Inclusion of these products in the trade figures, however, does not significantly alter the healthcare technology industry statistics.

TABLE A.1
U.S. Department of Commerce NAICS Classifications

NAICS Code	Products
339111 Laboratory Apparatus and Furniture Manufacturing	Bunsen burners; Hospital beds; Hospital furniture (e.g., operating room furniture); Laboratory-type autoclaves (except dental); Laboratory-type balances and scales; Laboratory-type benches; Laboratory-type blood-testing apparatus; Laboratory-type centrifuges; Laboratory-type distilling apparatus; Laboratory-type dryers; Laboratory-type evaporation apparatus; Laboratory-type freezers; Laboratory-type furnaces (except dental); Laboratory-type furniture (e.g., benches, cabinets, stools, tables); Laboratory-type incubators; Laboratory-type ovens; Laboratory-type sample preparation apparatus; Laboratory-type sterilizers (except dental); Laboratory-type vacuum pumps; Operating-room tables
339112 Surgical and Medical Instrument Manufacturing	Anesthesia apparatus; Blood-pressure apparatus; Blood-transfusion equipment; Bone drills; Bone plates and screws; Bone rongeurs; Bronchoscopes (except electromedical); Cannulae; Catheters; Corneal microscopes; Cystoscopes (except electromedical); Eye examining instruments and apparatus; Gastroscopes (except electromedical); Hypodermic and suture needles; Hypodermic syringes; Inhalation-therapy equipment; IV apparatus; Mechanical microsurgical instruments; Medical retractors; Medical thermometers; Medical thermometers; Medical tonometers; Ophthalmic instruments and apparatus (except laser surgical); Ophthalmometers and ophthalmoscopes; Optometers; Oxygen tents; Pelvimeters; Physiotherapy equipment (except electrotherapeutic); Retinoscopes (except electromedical); Skin grafting equipment; Speculums; Sphygmomanometers; Stethoscopes; Suction therapy apparatus; Surgical and medical inhalators; Surgical clamps; Surgical forceps; Surgical knife blades and handles; Surgical knives; Surgical probes; Surgical saws; Surgical stapling devices; Trocars; Veterinarians' instruments and apparatus
339113 Surgical Appliance and Supplies Manufacturing	Absorbent cotton and cotton balls; Artificial limbs; Biohazard protective clothing and accessories; Bulletproof vests; Cervical collars; Clean room suits and accessories; Colostomy appliances; Corn remover and bunion pad; Cotton tipped applicators; Crutches and walkers; Disposable surgical drapes; Elastic orthopedic hosiery; Firefighting suits and accessories; Gas masks; Gynecological supplies and appliances; Hard hats; Hospital and surgical sterilizers; Hydrotherapy equipment; Infant incubators; Intraocular lenses; Intrauterine devices; Iron lungs; Life preservers; Medical adhesive tape; Medical stretchers; Metal fabric and mesh safety gloves; Nose and ear plugs; Orthopedic arch supports; Orthopedic canes; Orthopedic devices; Orthopedic elastic hosiery; Orthopedic extension shoes; Orthopedic foot appliances; Orthopedic support hosiery; Orthopedic supports (e.g., abdominal, ankle, arch, kneecap); Patient restraints; Personal noise protectors; Personal safety devices, not specified elsewhere; Prosthetic appliances and supplies; Radiation shielding aprons, gloves, and sheeting; Respiratory protection mask; Rubber gloves (e.g., electrician's, examination, household-type, surgeon's); Safety helmets (e.g., motorized vehicle crash helmets, space helmets but not athletic helmets); Snake bite or burn first aid kits; Space suits; Splints; Surgical and orthopedic bandages and dressings; Surgical corsets; Surgical dressings; Surgical gauze made from purchased fabric; Surgical implants; Surgical supplies (except medical instruments); Surgical sutures; Tongue depressors; Traction apparatus; Ultrasonic medical cleaning equipment; Welder's hoods; Wheelchairs; Whirlpool baths (i.e., hydrotherapy equipment)
339114 Dental Equipment and Supplies Manufacturing	Dental abrasive points, wheels, and disks; Dental alloys for amalgams; Dental amalgams; Dental autoclaves; Dental chairs; Dental cutting instruments; Dental drills; Dental enamels; Dental equipment and instruments; Dental glues and cements; Dental hand instruments (e.g., forceps); Dental laboratory equipment; Dental laboratory furnaces; Dental sterilizer; Dental wax; Dental-impression materials; Dental-instrument delivery systems; Dentist's tools; Denture materials; Orthodontic appliances; Teeth (except customized); Ultrasonic dental equipment
339115 Ophthalmic Goods Manufacturing	Contact lenses; Corrective vision-type magnifiers; Glass and plastic eyes; Goggles (e.g., industrial, safety, sun, underwater); Ophthalmic eyeglass frames (i.e., fronts and temples); Ophthalmic lens grinding (except in retail stores); Ophthalmic lens mounts; Ophthalmic lens polishing; Ophthalmic lenses; Ophthalmic lens coating; Ophthalmic temples and fronts (i.e., eyeglass frames); Sunglasses and goggles

TABLE A.1
U.S. Department of Commerce NAICS Classifications

NAICS Code	Products
334510 Electromedical and Electrotherapeutic Apparatus	Automated blood and body-fluid analyzers (except laboratory); Cardiac sentinel; Cardiodynameter; Cardiographs; Electric cardiophone; Cardioscope; Cardiotachometer; Defibrilators; Diathermy units; Electric medical probes; Electrocardiographs; Electroencephalographs; Electrogastrograph; Electromedical audiological equipment; Electromedical bronchoscopes; Electromedical colonscopes; Electromedical cystoscopes; Electromedical diagnostic equipment; Electromedical dialysis equipment; Electromedical diathermy apparatus; Electromedical equipment; Electromedical gastroscopes; Electromedical laser equipment; Electromedical otoscopes; Electromedical respiratory analysis equipment; Electromedical retinoscopes; Electromedical therapy equipment; Electromyographs; Electronic hearing aids; Electrotherapeutic apparatus; Electrotherapeutic arc lamp units (except infrared, ultraviolet); Electrotherapeutic Carbon arc lamp units (except infrared and ultraviolet); Electrotherapy units; Heart-lung machine; Lithotripters; Medical laser systems and equipment; Medical ultrasonic scanning devices; Medical ultrasound equipment; MRI diagnostic equipment, MRI; Pacemakers; Patient monitoring equipment (e.g., intensive care, coronary care unit); PET scanners; Phonocardiographs; Surgical support systems (e.g., heart-lung machines) (except iron lungs); Transcutaneous electrical nerve stimulators (TENS); Ultrasonic medical cleaning equipment; Ultrasonic medical equipment
334517 Irradiation Apparatus Manufacturing	Beta-ray irradiation equipment; Computerized axial tomography (CT/CAT) scanners; Fluoroscopes; Fluoroscopic X-ray apparatus and tubes; Gamma ray irradiation equipment; Irradiation apparatus and tubes (e.g., industrial, medical diagnostic, medical therapeutic); Irradiation equipment; Medical radiation therapy equipment; Nuclear irradiation equipment; Radium equipment; Therapeutic X-ray apparatus and tubes (e.g., medical, industrial, research); X-ray apparatus and tubes (e.g., control, industrial, medical, research); X-ray generators; X-ray irradiation equipment; X-ray lamps
325413 *In Vitro* Diagnostic Substance	Blood-derivative IVD substances; Blood-glucose test kits; Coagulation IVD substances; Cytology and histology IVD substances; Enzyme and isoenzyme IVD substances; Hematology IVD substances; HIV test kits; Hormone IVD substances; IVD substances; Microbiology, virology, and serology IVD substances; Pregnancy test kits; Viral IVD test substances

Source: NAICS

Appendix B:
Australian Department of Health and Aging Classification of Devices

The Therapeutic Goods Act 1989 gives the Secretary of the Department of Health and Aging (DOHA) the power to classify therapeutic devices in various ways to facilitate the discharge of the objective of the act. The objective of the Therapeutic Goods Act 1989 was the establishment of a national system of controls relating to the quality, safety, efficacy, and timely availability of therapeutic goods in Australia. The various classification schemes are published in the Therapeutic Goods Regulations and other official publications of the Therapeutic Goods Administration (TGA). Since these regulations and orders can be amended at any time, the tables provided in this appendix should be taken only as a guide.

Under Section 7(1) of the Therapeutic Goods Act 1989, the Secretary of the DOHA may declare that certain goods are not therapeutic goods. The Secretary may also declare that goods, when used, advertised, or presented for supply in a particular way, are or are not therapeutic goods, for the purposes of this act. Such an order, when published in the *Commonwealth of Australia Gazette*, exempts the goods from regulation under the *Therapeutic Goods Act 1989*. The goods exempted from regulations under this provision at the time of publication are listed in Table B.1.

Certain therapeutic goods, which would be therapeutic devices under the definition in Section 2 of the Therapeutic Goods Act 1989, have been declared by the Secretary of the DOHA to not be therapeutic devices. This approach is adopted for devices where considerations of safety require them to be evaluated in the same manner as drugs. The expertise for evaluating these goods resides in the Drug Safety and Evaluation Branch of the TGA. The goods declared to not be therapeutic devices at the time of publication are listed in Table B.2.

Section 18 of the Therapeutic Goods Act 1989 allows the DOHA Secretary to exempt goods from the requirement that the therapeutic goods be registered or listed on the Australian Register of Therapeutic Goods (ARTG). The goods exempted under this provision of the Therapeutic Goods Act 1989 are listed in Schedule 5 of the Therapeutic Goods Regulations. Schedule 5 is reproduced in Table B.3.

Devices intended for experimental uses in humans are exempt goods under Section 19 of the Therapeutic Goods Act 1989 or the regulations to the act. Devices that would require registration under Part 3 of the Therapeutic Goods Act 1989 are exempt from registration if they are used in a clinical trial approved under the Clinical Trial Exemption (CTE) scheme. Goods subject to the Clinical Trial Notification (CTN) scheme are identified as exempt goods in Schedule 5A of Subregulation 12. The portion of Schedule 5A applicable to therapeutic devices is reproduced in Table B.4.

Under the Australian system, therapeutic devices that are perceived to present a significant risk of death or serious injury for a patient or user are required to be registered with the TGA. Registration means that these devices are subject to premarket evaluation of their quality, safety, and effectiveness. The devices that are required to be registered with the TGA are listed in Schedule 3 of the Therapeutic Goods Regulations. Schedule 3 is reproduced in Table 4.2. All other devices, unless they are exempt (Table B.3), must be listed on the ARTG. The listable devices are contained in Schedule 4 of the Therapeutic Goods Regulations. TGA has formulated policies for regulating specific listable devices. The list of devices for which TGA has specific policies at the time of publication is reproduced in Table B.5.

The Australian manufacturer of a registered or listed therapeutic device must be appropriately licensed to carry out the manufacture of the goods or class of goods within which the device is included. This provision of the Therapeutic Goods Act 1989 includes any subcontractor or testing facility in Australia that is contracted, or otherwise engaged, to carry out all or part of the manufacture of the registered or listed device. The licensing provision of the Therapeutic Goods Act 1989 applies unless the device is exempted by the regulations. The list of devices that can be manufactured without a license is given in Schedule 7 of the Therapeutic Goods Regulations and is reproduced in Table B.6. Certain persons are also exempted from the requirement to have a manufacturing license in order to carry on their profession. A list of the persons exempted from the requirement to be a licensed manufacturer is given in Table B.7.

TABLE B.1
Goods Declared to Not be Therapeutic Goods

Accupoint stimulators (noninvasive)

Aids to physical comfort

Apparel nonsterile, health/protective/safety for home, recreation, or occupation

Aromatherapy devices

Audio-therapy devices

Breathalyzers and blood-alcohol equipment

Color-therapy devices

Cosmetics without therapeutic claims

Dental chewing gum (unmedicated)

Dental whiteners and bleaches

Denture adhesives and aids

Depilatory preparations (dermal)

Deodorants (dermal or with therapeutic devices)

Electric blankets

Fitness equipment, excluding physiotherapy

Furniture, utensils, or personal aids for the disabled

Hair dyes, colorants, perms, bleaches

Hair-growth stimulators

Heat therapy devices

Household and personal aids for the disabled

Incontinence pads

Jewelry with reputed remedial powers but no therapeutic claims

Lipstick and facial makeup (not moisturizers), tinted with sunscreen as a secondary component

Magnetic-therapy devices

Massagers

Menstrual pads, but not tampons

Moisturizers, emollients, cleansers, and barrier products, not for dispensing

Muscle stimulators (electric) for cosmetic purposes

Nail hardeners, nail-biting deterrent products

Nitrogen as a power source

Oral-hygiene products unscheduled with restricted claims

Ostomy aids (i.e., adhesive removers and nonmedicated cleansers)

Sanitation, environmental-control, and detoxification equipment (not negative-ion generator or humidifier)

Slimming devices

Solaria and sun lamps

Spa baths and saunas

Soaps and detergents, unmedicated

Spa and mineral waters, with no therapeutic claims

Sterilant gases

Sunglasses, nonprescription spectacles

Sunscreens (as a secondary purpose with no claimed SPF)

Tissues, human, for direct donor-to-host transplant

Vibrators

Water purification/treatment/fluoridation equipment

Wigs and other nonimplantable aids

Source: DR4 p. 4

TABLE B.2
Goods Declared to be Drugs

Antiseptics
Transdermal patches
Contraceptives, intrauterine, containing hormones
Blood components, substitutes, and expanders
Diagnostics, *in vivo*, including imaging agents
Demulcents and absorbents, ingested
Dialysis solutions
Enemas, douches, laxatives, and irrigation fluids, unmedicated
Emollient/moisturizing products for dispensing
Homoeopathic preparations
Medical gases and chemical oxygen generators
Saline, water for irrigation or injection (excluding that for contact lenses and for device inflation)
Spermicidal and viricidal sponges and membranes
Sunscreen preparations

Source: DR4 p. 5

TABLE B.3
Therapeutic Goods Exempt From Registration or Listing on the Australian Registry of Therapeutic Goods

Clothing, nonsterile, except gloves, radiation shields, and shielding apparel
Goods for approved or notified clinical trials
Goods given special access for individual patients
Communications equipment (not monitoring)
Components (other than separate accessories or consumables, artificial-limb components, programmers for electronic implants)
Containers (except syringes, solution bags, and blood-collection tubes)
Dental devices fabricated outside the mouth
Dental-impression materials
Diagnostic tools and instruments, nonpowered, nonsterile
Disinfectants, household/commercial grade (with no specific claims)
Exports, noncommercial (not for clinical trials)
Extemporaneous products (one-off or custom-made devices)
Hot and cold packs
Imported devices (prior to February 15. 1991, and still used for treatment)
Imports pending approval for supply or remaining in customs control (e.g., Bond Store prior to reexport)
In vitro diagnostics (IVD) (except Pharmaceutical Benefits Scheme [PBS] goods, goods of human origin, intended for home use, or HIV or HCV diagnosis)
Linen and bedding, nonsterile
Manufacturing, laboratory, and dispensary equipment not for recycling human blood or tissue
Medicine dispensers, manual use
Furniture amd equipment, nonpowered for general patient care
Medical and dental instruments, nonpowered, nonsterile (except cannulae, endoscopes, etc.)
Orthoses and splints (simple)
Personal-use imports, most (if in small quantities or for visiting sports teams)
Samples not for human use
Tissue, human, for transplantation (stored and not further altered)

Source: DR4 p. 5

TABLE B.4

Therapeutic Goods Exempt from the Operation of Part 3 of the Act Subject to Conditions

Item No.	Therapeutic Goods	Conditions

1.	Medical device that is imported into Australia and is held under the direct control of the sponsor, until the device is:
	(a) the subject of a notification under item 3; or
	(b) approved for importation into Australia under Section 41HB of the act; or
	(c) authorised for supply under Section 41HC of the act; or
	(d) used for a Category A patient,[a)] within the meaning of Section 7.2 of the Therapeutic Goods Regulations 2002.
		(a) The supply of the device must be in accordance with the relevant notification, approval, authorisation or medical practitioner's direction.
		(b) The device must be kept in a warehouse or properly secured area under the control of the sponsor.
		(c) If the device is not used within twelve months of importation, the device must be destroyed or returned to the consignor of the device within one month after the end of that period.
		(d) The sponsor must:
		  (i) keep records relating to the source and supply of the device; and
		  (ii) if the device is destroyed under paragraph (c), keep records relating to the destruction; and
		  (iii) if requested by the Secretary of the Department of Health and Aging (DOHA), give the records to the Secretary.

2.	Medical device affected by Section 41FH of the act that is imported into Australia and is held under the direct control of the sponsor until a decision is made under section 41FI of the act in relation to the device	(a) The sponsor must:
		  (i) keep records relating to the source of the device; and
		  (ii) if requested by the Secretary of DOHA, give the records to the Secretary; and
		  (iii) before importing the device, have lodged an application under Section 41FC of the act for the device to be included in the Australian Register of Therapeutic Goods (ARTG).
		(b) If the application is not successful, the goods must be destroyed or returned to the consignor of the device within one month of the decision not to include the device in the ARTG.

3.	Medical device to be used solely for experimental purposes in humans	(a) before starting to use the device, the sponsor must notify the Secretary of DOHA:
		  (i) in a form approved by the Secretary, and
		  (ii) in accordance with any requirements determined by the Secretary for the form of notification, that the sponsor intends to sponsor a clinical trial using the device.
		(b) The notification must be accompanied by the notification fee specified in Item 1.8 of Schedule 5 of the TGR 2002.
		(c) The approval of the device for this purpose must be given by the sponsor (if the sponsor is conducting the trial), or by the body or organization conducting the trial for the sponsor, having regard to the advice of the ethics committee that has, or will assume, responsibility for monitoring the conduct of the trial.
		(d) The terms of the approval by the sponsor, body, or organization mentioned in paragraph (c) must be no less restrictive than the terms advised by the responsible ethics committee.
		(e) The trial must not be the subject of a direction by the Secretary that the trial not be conducted, or that it be stopped, because the Secretary has become aware that to conduct or continue the trial would be contrary to the public interest.
		(f) the sponsor (if the sponsor is conducting the trial), or the body or organization conducting the trial for the sponsor, must not receive, or have received, advice from the ethics committee that is inconsistent with the continuation of the trial.
		(g) The conditions stated in regulation 7.5 of the TGR 2002 must be complied with, as if that regulation applied to a person using a medical device under this item.

TABLE B.4 (CONTINUED)
Therapeutic Goods Exempt from the Operation of Part 3 of the Act Subject to Conditions

Item No.	Therapeutic Goods	Conditions
4.	Medical device that is imported into Australia by a member of a group of persons who are visiting Australia to participate in a national or international sporting event	(a) The device must be for use in the treatment of a member or members of the visiting group. (b) The importation of the device must not be prohibited under the Customs (Prohibited Imports) Regulations 1956. (c) The device must not be supplied to, or used in the treatment of, a person who is not a member of the visiting group. (d) The device must be destroyed or removed from Australia at the end of the visit. (e) A member of the group must be responsible for the control and custody of the device while the group is in Australia. (f) The person mentioned in paragraph (e) must: (i) carry a list, in English, of the quantity and nature of the device imported; and (ii) keep a record of the use of the device while the group is in Australia; and (iii) produce the list or the record for inspection at the request of a customs officer or a person who is an authorized officer for the purposes of a provision of Section 41FN of the act.
5	Medical device that is imported into Australia by a member of a group of persons, being members of the military forces of another country who are visiting Australia for military training	(a) The device must be for use in the treatment of a member or members of the visiting group. (b) The device must not be supplied to, or used in the treatment of, a person other than a member of: (i) the visiting group; or (ii) the Australian Defence Force. (c) The device must be destroyed or removed from Australia at the end of the visit. (d) A member of the group to whom the device has been issued must be responsible for the control and custody of the device while the group is in Australia. (e) The person mentioned in paragraph (d) must: (i) carry a list, in English, of the quantity and nature of the device imported; and (ii) keep a record of the use of the device while the group is in Australia; and (iii) produce the list or record for inspection at the request of a customs officer or a person who is an authorised person for the purposes of section 41FN of the Act.

[a] A Category A patient is a person who is seriously ill with a condition that is reasonably likely to lead to the person's death within one year or, without early treatment, to the person's premature death.

Source: TGR 2002, Schedule 2, Part 2, pp. 118–123

All legislation herein is reproduced by permission but does not purport to be the official or authorised version. It is subject to Commonwealth of Australia copyright.

TABLE B.5
Specific Listable Device Policies

Animal derivatives in listable devices	Endoscopes and accessories
Bandages, dressings, allied products, etc.	First-aid kits (see Kits)
Barium lime	Gloves examination/surgical
Blood bags	Hearing and speech aids
Catheters (urethral)	*In vitro* diagnostics (IVDs)
Condoms	Needleless injectors
Contact lenses and contact-lens care products	Insulin syringes
Contrast media injectors (powered)	Intraocular lenses (IOLs)
Dental products	Kits (therapeutic devices kits)
Dental-restorative materials	Oral-hygiene products
Devices for people with disabilities	Penile implants (inflatable)
Diaphragms (contraceptive)	Podiatry
Disinfectants:	Soda lime
without claims	Sutures and ligatures
hospital grade	Tampons (menstrual)
Drug-infusion injectors	
Ducted and wired systems	

Source: DR4 p. 184

TABLE B.6
Goods Exempt From Licensing to Manufacturer

Components
Containers except
  blood bags
  blood-collection tubes
  drainage bags
  nonglass containers for blood or injection
  parenteral-infusion bags
Dentrifices, with no active ingredient or low fluoride
Nonsterile devices except
  bandages, dressings, adhesive tapes, etc.
  contraceptive devices
  condoms
  contact lenses (soft)
  dental-restorative materials
  devices used for the prevention of disease transmission
  diaphragms
  disinfectants/sterilants (until December 31, 1997)
  gloves (examination and surgical)
  implants
IVDs for home use, PBS listing, HIV or HCV diagnosis, or lubricants of human origin for internal use

Source: DR4 p. 38

TABLE B.7
Persons Exempt from Licensing as Manufacturers

Biomedical engineers, pharmacists, and radiochemists in public hospitals
Registered healthcare workers dispensing for patients under their care
Herbalists/homoeopaths/nutritionists for clients
Supplementary labeling (address/ARTG No. only)
Pharmacists in their place of practice for supply other than by wholesale from their premises

Source: DR4 p. 38

Appendix C:
International Symbols for Medical Device Marking

Symbols are frequently preferred over words for use on medical devices. They are intended to obviate language differences and permit easier comprehension of a marking or indication, particularly where space is limited.

The symbols illustrated in in this appendix are frequently used on electromedical equipment, particularly for safety related aspects of the devices. There are, however, a wide variety of recognized international symbols available for use by device manufacturers. The recognized sources for internationally standardized symbols include IEC 60417-1, *Graphical symbols for use on equipment — Part 1: Overview and application*; ISO 7000, *Graphical symbols for use on equipment — Index and synopsis*; and ISO 7010, *Graphical symbols — Safety signs in workplaces and public areas*.

IEC Subcommittee 62A has developed a compendium of symbols used on medical electrical equipment. This compendium is published as IEC 60878, *Graphical symbols for electrical equipment in medical practice*.

TABLE C.1
International Symbols for Medical Device Marking

No.	Symbol[a]	Reference	Description
1.		IEC 60417-5031	Direct current
2.		IEC 60417-5032	Alternating current
3.		IEC 60417-5033	Both direct and alternating current
4.	$3\sim$	IEC 60417-5032-1	Three-phase alternating current
5.		IEC 60417-5017	Earth (ground)
6.		IEC 60417-5019	Protective earth (ground)
7.		IEC 60417-5020	Frame or chassis terminal
8.		IEC 60417-5021	Equipotentiality
9.		IEC 60417-5007	"On" (Supply)
10.		IEC 60417-5008	"Off" (Supply)
11.		IEC 60417-5172	Equipment protected throughout by double or reinforced insulation; Class II
12.		IEC 61010-1 Table 1 No. 12	Caution, risk of electrical shock
13.		IEC 60417-5041	Caution, hot surface
14.		ISO 7000-0434	Caution; Risk of danger; [b] Attention, consult accompanying documents [c]
15.		IEC 60417-5268	"In" position of a bi-stable push control
16.		IEC 60417-5269	"Out" position of a bi-stable push control

TABLE C.1 (CONTINUED)
International Symbols for Medical Device Marking

No.	Symbol [a)]	Reference	Description
17.		IEC 60417-5840	Type B equipment
18.		IEC 60417-5333	Type BF equipment
19.		IEC 60417-5335	Type CF equipment
20.		IEC 60417-5840	Defibrillation-proof type B applied part
21.		IEC 60417-5334	Defibrillation-proof type BF applied part
22.		IEC 60417-5336	Defibrillation-proof type CF applied part
23.		IEC 60417-5331	Category AP equipment (medical)
24.		IEC 60417-5332	Category APG equipment (medical)
25.		IEC 60417-5036	Dangerous voltage
26.		IEC 60417-5264	"On" for a part of equipment
27.		IEC 60417-5265	"Off" for a part of equipment
28.		IEC 60417-5010	"On"/"Off" (push-push)
29.		IEC 60417-5009	Stand-by
30.		IEC 60417-5011	"On"/"Off" (push button)
31.		IEC 60417-5266	Standby or preparatory state for a part of equipment
32.		ISO 7000-1140	Ready

TABLE C.1 (CONTINUED)
International Symbols for Medical Device Marking

No.	Symbol [a]	Reference	Description
33.		IEC 60417-5104	Start (of action)
34.		IEC 60417-5177	Fast start
35.		IEC 60417-5659	Start, test run
36.		IEC 60417-5335	Stop (of action)
37.		IEC 60417-5178	Fast stop
38.		IEC 60417-5638	Emergency stop
39.		IEC 60417-5111	Pause; interruption
40.		IEC 60417-5114	Foot switch
41.		ISO 7000-1853	Foot-operated
42.		ISO 7000-0096	Manual control
43.		IEC 60417-5140	Non-ionizing electromagnetic radiation
44.		IEC 60417-5134	Electrostatic sensitive devices
45.		IEC 60417-5001	Battery, general
46.		IEC 60417-5002	Positioning of cell
47.		IEC 60417-5546	Battery check
48.		IEC 60417-5639	Rechargeable battery

TABLE C.1 (CONTINUED)
International Symbols for Medical Device Marking

No.	Symbol [a]	Reference	Description
49.		IEC 60417-5016	Fuse
50.		IEC 60417-5090	Telephone, telephone adapter
51.		ISO 7000-1641	Operating instructions
52.		ISO 7000-0717	Call for maintenance
53.		IEC 60417-5152	Radiation of laser apparatus
54.		ISO 7000-0659	Biological risks
55.		IEC 60417-5307	Alarm, general
56.		IEC 60417-5308	Urgent alarm
57.		IEC 60417-5319	Alarm inhibit
58.		IEC 60417-5309	Alarm system clear
59.		IEC 60417-5013	Bell
60.		IEC 60417-5576	Bell cancel
61.		ISO 7000-0632	Temperature limitation [d]
62.		ISO 7000-0535	Upper limit of temperature [d]
63.		ISO 7000-0534	Lower limit of temperature [d]
64.		ISO 7000-0027	Cooling; air condition system

a) The symbols in this table are for convenience of reference only and have no official sanction. The reader is referred to ISO 7000 and IEC 60417.
b) See 5.4.1 of IEC 61010-1, which requires manufacturers to state that the documentation must be consulted in all cases where this symbol is marked.
c) See Appendix D of IEC 60601-1.
d) For additional symbols for use in labeling, see Table 14.3.

Source: Compiled from IEC 60417-1; IEC 60878; IEC 61010-1; ISO 7000; and ISO 7010

Appendix D:
Korean Medical Device Classification

The South Korean medical device regulations are intended to regulate the sale of products manufactured in or imported into the Republic of Korea. To accomplish that objective, the Korean Food and Drug Administration (KFDA) has classified all medical devices into one of three classes (see Table D.1) based on the degree of risk to the patient.

TABLE D.1
KFDA Categories of Medical Devices

Category	Device [1] Class I,[a] [2] Class II, [3] Class III
	(A) Medical Instruments
A01000	A01010[1, 2] Treatment table (manual-Class I, electronic-ClassII)
Operating and treatment tables	A01020 [1, 2] Operating table (manual-Class I, electronic-Class II)
	A01030 [2] Infant radiant warmer
	A01040 [1] Patient-position holder
	A01050 [2] Radiologic-therapy table
	A01060 [1] Operating and treatment chair
A02000	A02010[1, 2] Hospital bed (manual-Class I, electronic-Class II)
Beds for medical use	A02020 [2] Air-fluidized bed
	A02030 [2] Powered flotation-therapy bed
	A02040 [1, 2] Patient-rotation bed (manual-Class I, electronic-Class II)
	A02050 [2] Alternating-pressure air flotation mattress
A03000	A03010 [1] Shadowless operating light
Medical lights and lamps	A03020 [1, 2] Light source (general use-Class I, endoscopic use-Class II)
	A03030 [1] Operating headlamp
	A03040 [1] Examination light
	A03050 [1] Light-beam patient position indicator
A04000	A04010[2] Pressure steam sterilizer
Medical sterilizing apparatus	A04020 [1] Ultraviolet sterilizer
	A04030 [1] Boiling and steam sterilizer
	A04040 [2] Dry-heat sterilizer
	A04050 [2] Ethylene oxide (EtO) gas sterilizer
	A04060 [2] Formalin gas sterilizer
	A04070 [2] Chemical sterilizer
	A04080 [2] Ozone sterilizer
	A04090 [2] Low-temperature plasma sterilizer
A05000	A05010 [2] Medical water sterilizer
Medical water sterilizers	
A06000	A06010[2] Gas machine for anesthesia or analgesia
Anesthesia apparatus	A06020 [3] Electroanesthesia apparatus

(continued)

497

TABLE D.1 (CONTINUED)
KFDA Categories of Medical Devices

Category	Device [1] Class I,[a] [2] Class II, [3] Class III
A07000 Respiratory apparatus	A07010 [2] Ventilator
	A07020 [2] Manual emergency ventilator
	A07030 [2] Oxygen-supplying unit
	A07040 [2] Oxygen generator
	A07050 [2] Oxygen tent (manual - Class I, electronic - Class II)
	A07060 [2] Heat and moisture condenser (artificial nose)
	A07070 [2] External negative-pressure ventilator
	A07080 [2] Rocking bed
	A07090 [2] Medical gas-supply unit
	A07100 [2] Medical oxygen regulating supplier
A08000 Medical chambers	A08010 [2] Hyperbaric chamber
	A08020 [2] Topical oxygen chamber for extremities
	A08030 [3] Abdominal decompression chamber
	A08040 [2] Patient-care reverse-isolation chamber
A09000 Artificial internal organ apparatus	A09010 [2] Cardiopulmonary bypass heart-lung machine console
	A09020 [2] Extracorporeal blood circuit for cardiopulmonary bypass
	A09030 [2] Cardiopulmonary bypass blood reservoir
	A09040 [3] Cardiopulmonary bypass oxygenator
	A09050 [2] Cardiopulmonary bypass heat exchanger
	A09060 [2] Cardiopulmonary bypass temperature controller
	A09070 [3] Cardiopulmonary bypass blood filter
	A09080 [2] Cardiopulmonary prebypass filter
	A09090 [2] Cardiopulmonary bypass on-line blood gas monitor
	A09100 [2] Cardiopulmonary bypass level-sensing monitor and/or control
	A09110 [2] Cardiopulmonary bypass blood pump
	A09120 [2] Cardiopulmonary bypass gas-control unit
	A09130 [2] Cardiopulmonary bypass defoamer
	A09140 [2] Cardiopulmonary bypass bubble detector
	A09150 [3] Assisted-circulation apparatus
	A09160 [3] Membrane lung for long-term pulmonary support
	A09170 [2] Thermal-regulation system
	A09180 [2] Peritoneal-dialysis system
	A09190 [3] Ascites-treatment system by filtration, concentration and infusion
	A09200 [2] Hemodialysis system
	A09210 [3] High-permeability hemodialysis system
	A09220 [2] Extracorporeal blood circuit for hemodialysis
	A09230 [2] Water-purification system for hemodialysis
	A09240 [2] Dialyzers for hemodialysis
	A09250 [3] Hemoperfusion system
	A09260 [3] External pacemaker pulse generator
	A09270 [3] Implantable pacemaker pulse generator
	A09280 [2, 3] Pacemaker electrode disposable for external device-Class II, implantable electrode-Class III)
	A09290 [3] Pacemaker lead adaptor
	A09300 [3] Pacemaker polymeric mesh bag
	A09310 [3] Pacemaker repair or replacement material
	A09320 [3] Cochlear implant system
A10000 Neonatal incubators	A10010 [2] Neonatal incubator
	A10020 [2] Neonatal transport incubator
A11000 Medical X-ray systems	A11010[2] Computed tomography (CT) X-ray system
	A11020 [2] Radiographic X-ray system
	A11030 [2] General radiograpic and fluroscopic X-ray system
	A11040 [2] Fluoroscopic X-ray system

TABLE D.1 (CONTINUED)
KFDA Categories of Medical Devices

Category	Device [1] Class I,[a] [2] Class II, [3] Class III
	A11050 [2] Tomographic X-ray system
	A11060 [2] Mammograpic X-ray system
	A11070 [2] Photofluorographic X-ray system
	A11080 [2] Mobile X-ray system
	A11090 [2] Angiographic X-ray system
	A11100 [2] X-ray bone densitometer
	A11110 [2] Digital X-ray imaging system
	A11120 [2] Dental X-ray system
	A11130 [2] X-ray radiation-therapy system
	A11140 [2] Radiation-therapy simulation system
	A11145 [1] Photofluorographic X-ray camera
	A11150 [1] Radiographic grid
	A11160 [1] Radiographic film cassette
A12000 Nonionization diagnostic devices	A12010 [2] Magnetic resonance diagnostic device
A13000 Radiological devices	A13010 [2] Positron-emission CT system
	A13020 [2] Medical charged-particle radiation-therapy system
	A13030 [2] Medical neutron radiation-therapy system
	A13040 [2] Fluorescent scanner
	A13050 [2] Nuclear tomography system
	A13060 [2] Radionuclide rebreathing system
	A13070 [2] Radiation-therapy beam-shaping block
	A13080 [1] Manual radionuclide applicator system
	A13090 [2] Remote-controlled radionuclide applicator system
	A13100 [2] Radionuclide radiation-therapy system
	A13110 [2] Gamma camera
	A13120 [2] Radioisotope (RI) dynamic-function testing apparatus
	A13130 [1] Beta counter for clinical use
	A13140 [1] Gamma counter for clinical use
	A13150 [1] Beta or gamma counter for clinical use
	A13160 [2] Bone densitometer
	A13170 [2] Single-photon-emission CT system
A14000 Film developer for medical use	A14010 [1] Film developer for medical use
	A14020 [1] Medical imager
A15000 Radiation-protective devices	A15010 [1] X-ray-protective apron
	A15020 [1] X-ray-protective glove
	A15030 [1] X-ray-protective goggle
	A15040 [1] X-ray-protective partition
A16000 Physical devices for medical use	A16010 [2, 3] Electric neuromuscular stimulator (device for pain release and skin stimulation-Class II, other uses-Class III)
	A16020 [2] Ultraviolet radiator
	A16030 [2] Pediatric phototherapy unit
	A16040 [2, 3] Static-electricity stimulator for medical use (device delivering direct or alternating electric current for improving blood flow-Class II, other uses-Class III)
	A16050 [2, 3] Infrared radiator (device delivering infrared irradiation with wavelength of 700-50,000 nanometers for pain relief and skin stimulation-Class II, other uses-Class III)
	A16060 [3] Phototherapy unit
	A16070 [2, 3] Microwave diathermy (device delivering electromagnetic energy with frequency of 915~2450 MHz for pain release and skin stimulation-Class II, other uses-Class III)
	A16080 [2, 3] Shortwave diathermy (device delivering electromagnetic energy with frequency of 13~27.12 MHz for pain release and skin stimulation-Class II, other uses-Class III)

(continued)

TABLE D.1 (CONTINUED)
KFDA Categories of Medical Devices

Category	Device [1] Class I,[a] [2] Class II, [3] Class III
	A16085 [2, 3] High-frequency electromagnetic wave stimulator for medical use (device delivering high-frequency electromagnetic energy for pain release and skin stimulation-Class II, other uses-Class III)
	A16090 [2, 3] Ultrasonic diathermy (device delivering ultrasonic energy with frequency over 20 kHz for pain release and skin stimulation-Class II, other uses-Class III)
	A16100 [2] Paraffin bath
	A16110 [2, 3] Iontophoresis device (device using direct current to induce sweating for delivering saline solution or drug into skin. Device for fluoro-iontophoresis-Class II, other uses-Class III)
	A16120 [2] Moist-steam cabinet
	A16130 [2] Hydrotherapeutic equipment
	A16140 [1] Warm-bath equipment for medical use
	A16150 [2, 3] Heat for medical use (device delivering thermal energy for pain relief and skin stimulation-Class II, other uses-Class III) (NOTE: Devices using infrared ray or optic lamp are excluded from this classification.)
	A16160 [2, 3] Cold therapy apparatus (device maintaining low temperature for pain relief, skin stimulation and edema reduction-Class II, other uses and devices using ultralow temperature-Class III)
	A16170 [3] Electroconvulsive device for medical use in treating psychiatric disease
	A16180 [2, 3] Electric stimulator for medical use (devices for pain release and skin stimulation-Class II, other uses-Class III)
	A16190 [3] Inflator for cavernous artery
	A16250 [3] Hyperthermia apparatus
	A16260 [3] Resonance-therapy unit
	A16270 [2, 3] Combination stimulator for medical use (NOTE: A combination stimulator is one using two or more combined kinds of energies listed in the 16000 criteria. Classification is appointed to the upper Class.)
A17000 Cardiovascular devices	A17010 [2, 3] Defibrillator (device output under 360 J at 50 Ω testing load-Class II, device out put over 360 J at 50 Ω testing load-Class III)
	A17020 [3] External cardiac compressor
	A17030 [3] Carotid-sinus nerve stimulator
	A17040 [3] Automated blood-cell separator
	A17050 [3] Blood concentrator system
	A17060 [2] Autotransfusion apparatus
	A17070 [2] Automatic rotating tourniquet
	A17080 [1] Nonpneumatic tourniquet
	A17090 [2] Pneumatic tourniquet
	A17100 [2] Compressible limb sleeve
	A17110 [2] Steerable catheter-control system
	A17120 [2] Biopotential amplifier and signal conditioner
	A17130 [2] Blood-volume measuring device
	A17140 [2] Cutaneous blood-gas monitor
	A17150 [3] Indwelling blood hydrogen ion concentration (pH) analyzer
	A17160 [3] Indwelling blood oxyhemoglobin concentration analyzer
	A17170 [3] Indwelling blood-gas-pressure analyzer
	A17180 [2] Ultrasonic air-embolism monitor
	A17190 [2] Oximeter
	A17200 [2] Blood-flow meter
	A17210 [2] Plethysmograph
	A17220 [3] Rheoencephalograph
	A17230 [2] Cardiac output flowmeter
	A17250 [3] Blood-vessel embolization device
	A17260 [3] Cardiovascular intravascular filter
	A17270 [3] Ocular plethysmograph

TABLE D.1 (CONTINUED)
KFDA Categories of Medical Devices

Category	Device [1] Class I,[a] [2] Class II, [3] Class III
	A17280 [3] Implantable cardioverter defibrillator
	A17290 [3] Implantable defibrillator electrode
	A17300 [3] Abrasion catheter system
A18000	A18010 [3] Extracorporeal shockwave lithotripor
Urology devices	A18020 [3] Electrohydraulic lithotriptor
	A18030 [2] Water-jet renal-stone dislodger system
	A18040 [3] Implanted electrical urinary-continence device
	A18050 [3] Nonimplanted electrical continence device
	A18060 [3] Implantable spinal-cord stimulator for bladder evacuation
	A18070 [2] Urodynamics measurement system
	A18080 [1] Urine-flow or volume-measuring system
	A18090 [3] Electric stimulator for ejaculation
	A18100 [3] Urinary-continence device
A19000	A19010 [1, 2] Wheelchair (manual-Class I, electronic-Class II)
Patient transport devices	A19020 [1, 2] Patient transfer device (manual-Class I, electronic-Class II)
	A19030 [1, 2] Patient lift [manual-Class I, electronic-Class II]
A20000	A20010 [1] Stethoscope
Stethoscopes	A20020 [1] Fetal stethoscope
	A20030 [1,2] Esophageal stethoscope (manual-Class II, electronic-Class II)
A21000	A21010 [1, 2] Thermometers (device with mercury or liquid crystal display-Class I, electronic-Class II)
Clinical thermometric systems	A21015 [1] Color-change thermometer
	A21020 [3] Liquid crystal thermographic system
	A21030 [2] Telethermographic system
A22000	A22010 [1] Flow-type clinical chemistry automated analyzer
Hematological testing apparatus	A22011 [1] Discrete-type clinical chemistry automated analyzer
	A22012 [1] Centrifuge-type clinical chemistry automated analyzer
	A22013 [1] Dry-type clinical chemistry automated analyzer
	A22014 [1] Packed-type clinical chemistry automated analyzer
	A22020 [1] Blood-gases test system
	A22030 [1] Glucose test system
	A22040 [1] Lactic acid test system
	A22050 [1] Bilirubin test system
	A22060 [1] Manual-type blood-cell counter
	A22070 [1] Automated sedimentation-rate device
	A22080 [1] Multipurpose system for *in vitro* coagulation studies
	A22090 [2] Automated heparin analyzer
	A22100 [1] Automated hemoglobin system
	A22110 [1] Colorimeter for clinical use
	A22120 [1] Photometer for clinical use
	A22130 [1] Spectrophotometer for clinical use
	A22140 [1] Flame-emission photometer for clinical use
	A22150 [1] Flame photometer electrolyte analyzer
	A22151 [1] Ion selective electrode electrolyte analyzer
	A22152 [1] Coulometric electrolyte analyzer
	A22153 [1] Fluorometric electrolyte analyzer
	A22160 [1] Electrophoresis apparatus for clinical use
	A22170 [1] Immunofluorometer equipment
	A22180 [1] Immunonephelometer equipment
	A22190 [1] Photometric immunological analyzers
	A22200 [1] Enzyme analyzer for clinical use
	A22210 [1] Fluorometer for clinical use
	A22220 [1] Chromatography system for clinical use

(continued)

TABLE D.1 (CONTINUED)
KFDA Categories of Medical Devices

Category	Device [1] Class I,[a] [2] Class II, [3] Class III
	A22230 [1] Osmometer for clinical use
	A22240 [1] Plasma viscometer for clinical use
	A22250 [1] Blood-cell differential automatic analyzer
	A22260 [1] Automated blood-grouping and antibody test system
	A22270 [1] Immunological testing apparatus
	A22280 [1] Blood-sample processing device
	A22290 [1] Automated platelet-aggregation system
	A22300 [1] Automated Coombs test system
	A22310 [1] Endothelial-cell counter
A23000	A23010 [1, 2] Sphygmomanometer (device using aneroid or mercury-Class I, other devices-Class II)
Sphygmomanometers and sphygmographs	A23020 [2] Electronic direct blood-pressure meter
	A23030 [2] Ophthalmodynamometer
	A23035 [2] Blood-pressure monitor
	A23040 [2] Pulse-wave meter
	A23050 [3] Pulse-wave analyzer
A24000	A24010 [1] Urine analysis system
Urine or excrement analyzers	A24020 [1] Bacterial-identification apparatus
	A24030 [1] Bacterial-sensitivity test apparatus
	A24040 [1] Excrement analysis system
A25000	A25010 [1] Body-fluid testing apparatus
Body fluid testing apparatus	A25020 [1] Gastric-acidity test system
	A25030 [1] Viscometer
A26000	A26010 [2] Electrocardiograph (ECG)
Visceral function testing instruments	A26020 [2] ECG monitor
	A26030 [2] Vectorcardiograph (VCG)
	A26040 [2] Ambulatory ECG recording and analyzing system (Holter system)
	A26050 [2] Stress-test instrument for ECG
	A26060 [2] Apex cardiograph (vibrocardiograph)
	A26070 [2] Ballistocardiograph
	A26080 [2] Heart-rate meter
	A26090 [2] Patient monitor
	A26100 [2] Central monitor
	A26110 [2] Breathing-frequency monitor
	A26120 [2] Blood-pressure alarm meter
	A26130 [2] Airway-pressure monitor
	A26140 [3] Radio capsule system
	A26150 [2] Electroencephalograph
	A26160 [1] Electroencephalogram (EEG) signal spectrum analyzer
	A26170 [2] Alpha monitor
	A26180 [2] Phonocardiograph (PCG)
	A26190 [2] Fetal cardiac monitor
	A26200 [3] Fetal electroencephalographic monitor
	A26210 [2] Fetal PCG monitor
	A26220 [2] Fetal ultrasonic monitor
	A26230 [2] Perinatal monitoring system
	A26240 [2] Intrauterine-pressure monitor
	A26250 [2] External uterine-contraction monitor
	A26260 [2] Gastrointestinal-mobility monitoring system
	A26270 [2] Multipurpose polygraph
	A26280 [2] Evoked-potential monitoring system
	A26290 [2] Stimulator for evoked response
	A26300 [2] Aversive-conditioning device
	A26310 [2] Intracranial-pressure monitoring device

TABLE D.1 (CONTINUED)
KFDA Categories of Medical Devices

Category	Device [1] Class I,[a] [2] Class II, [3] Class III
	A26320 [2] Intrathoracic-pressure meter
	A26330 [2] Spinal-fluid pressure monitor
	A26340 [2] Magnetocardiograph
	A26350 [2] Electrocochleograph
	A26360 [2] Intracavitary phonocatheter system
	A26370 [2] Ultrasonic bone densitometer
	A26380 [2] Ultrasonic imaging system
	A26390 [2] Echocardiograph
	A26400 [2] Echoencephalograph
	A26410 [2] Ultrasonic pulsed Doppler diagnostic system
	A26420 [2] Ultrasonic pulsed echo diagnostic system
	A26430 [2] Medical image analyzer
A27000	A27010 [2] Diagnostic spirometer
Respiratory-function testing apparatus	A27020 [2] Monitoring spirometer
	A27030 [2] Peak-flow meter for spirometry
	A27040 [2] Rhinoanemometer
	A27050 [2] Diagnostic pulmonary-function interpretation calculator
	A27060 [1] Rebreathing device
	A27070 [2] Inspiratory airway-pressure meter
	A27080 [2] Oxygen-uptake computer
	A27090 [2] Pressure plethysmograph
	A27100 [2] Respiratory-resistance meter
	A27110 [2] Expiration-gas analyzer
	A27120 [3] Lung-water monitor
A28000	A28010 [2] Slit-lamp biomicroscope
Eye-testing instruments	A28020 [1] Keratoscope
	A28030 [2] Ophthalmic refractometer
	A28040 [1] Manual reflector
	A28050 [1] Stereoscope
	A28060 [2] Keratometer
	A28070 [1] Retinoscope
	A28080 [2] Corneal camera
	A28090 [1, 2] Anterior camera (device photographing anterior pupil using reflective beam-Class I, device including image display-Class II)
	A28100 [2] Fundus camera
	A28110 [2] Eye-movement monitor
	A28120 [1] Ophthalmoscope
	A28130 [1] Perimeter
	A28140 [2] Visual-field laser instrument
	A28150 [2] Photosimulator
	A28160 [1] Stereopsis-measuring instrument
	A28170 [1] Ophthalmic lens
	A28175 [1] Ophthalmic prism
	A28180 [1] Skiascopic rack
	A28190 [1] Ophthalmic trial lens set
	A28200 [1] Ophthalmic trial lens frame
	A28210 [2] Pupillometer
	A28220 [1] Light- and color-sensation test apparatus
	A28230 [1, 2] Electronic metal locator (manual-Class I, using electricity-Class II)
	A28240 [1] Schirmer strip
	A28250 [2] Electroretinograph (ERG)
	A28260 [2] Nystagmograph

(continued)

TABLE D.1 (CONTINUED)
KFDA Categories of Medical Devices

Category	Device [1] Class I,[a] [2] Class II, [3] Class III
	A28270 [2] Tonometer
	A28280 [2] Ocular-pressure applicator
	A28290 [2, 3] Iris camera (device photographing iris-Class II, device including analyzer-Class III)
	A28300 [2] Image visual-acuity tester
A29000	A29010 [2] Audiometer
Hearing-testing instruments	A29020 [2] Auditory impedance tester
	A29030 [1] Electronic noise generator for audiometric testing
	A29040 [1] Tuning fork
A30000	A30010 [2] Diagnostic electromyograph
Perception and organs diagnostic	A30020 [2] Diagnostic muscle stimulator
devices	A30030 [2] Nerve-conduction-velocity measurement device
	A30040 [2] Skin-potential measurement device
	A30050 [2] Galvanic-skin-response measurement device
	A30060 [2] Surgical nerve stimulator/locator
	A30070 [2] Tubal patency diagnosis equipment
	A30080 [2] Biofeedback device
	A30090 [2] Perinometer
	A30100 [1] Vibratory-sensation measuring apparatus
	A30110 [2] Apparatus for gait analysis
	A30120 [2] Sexual-function analyzer
	A30130 [2] Isokinetic testing and evaluation system
	A30140 [2] Rigidity analyzer
	A30150 [1] Ataxiagraph
	A30160 [2] Chronaximeter
	A30170 [1, 2] Algesimeter (maual-1, electronic-2)
	A30180 [1] Dynamometer
	A30190 [1] Bite-pressure meter
	A30200 [1] Percussor
	A30210 [1] Powered reflex hammer
	A30220 [1] Pinwheel
	A30230 [2] Pressure-measurement system
	A30240 [1, 2] Esthesiometer (maual-1, electronic-2)
	A30250 [1] Gustometer
	A30260 [2] Pulp tester
	A30270 [2] Caries-detection device
	A30280 [2] Root-canal meters
	A30290 [1] Dental-occlusal sound scope
	A30300 [2, 3] Skin-resistance measuring device (device measuring resistance-Class II, device including display-Class III)
	A30310 [2] Electroglottograph
	A30320 [1, 2] Body-fat measuring device (device using water bath-Class I, other devices-Class II)
	A30330 [2] Icterus measuring device
	A30340 [2] Noise generator
A31000	A31010 [1, 2] Endoscope for medical use; laryngoscope (without image monitor-Class I, with
Speculums for medical use	monitor-Class II), Bronchoscope (Class II), Esophagoscope (Class II), Mediastinoscope (Class II), Arthroscope (Class II), Choledochoscope (Class II), Anoscope (without image monitor-Class I, with monitor-Class II), Colonoscope (Class II), Cystoscope (Class II), Enteroscope (Class II), Esophagoscope (Class II), Cystourethroscope (Class II), Proctoscope (Class II), Resectoscope (Class II), Ureteroscope (Class II), Nephroscope (Class II), Sigmoidoscope (Class II), Culdoscope (Class II), Urethroscope (Class II), Hysteroscope (Class II), Gynecologiclaparoscope (Class II), fetoscope (Class II), Transabdominalamnioscope (Class II), Nasopharyngoscope (Class II), Esophgogast-roduodenoscopes (Class II), Biliary pancreatoscop (Class II), Transcervical endoscope (amnioscope) (Class II), Neurologicalendoscope (Class II)

TABLE D.1 (CONTINUED)
KFDA Categories of Medical Devices

Category	Device [1] Class I,[a] [2] Class II, [3] Class III
	A31020 [1] Instrument for endoscope
	A31025 [1, 2] Camera for medical use (device photographing exterior or open cavity of body- Class I, device photographing interior of body, device including light source or analyzing function- Class II)
	A31030 [1] Operating microscope
	A31040 [1] Colposcope
	A31050 [1] Biological microscope
	A31060 [1] Dental mirror
	A31070 [1] Headband mirror
A32000	A32010 [1] Medical centrifuge
Centrifuges for medical use	A32020 [2] Cooled centrifuge
	A32030 [1] Hematocrit centrifuge
	A32040 [2] Automated cell-washing centrifuge for immuno-hematology
A33000	A33010 [1] Dermatome
Tissue-processing devices	A33020 [1] Graft stretcher
	A33030 [1] Tissue-processing equipment
A34000	A34010 [1] Anaerobic chamber
Thermostats for medical use	A34020 [1] Bacterial-culture apparatus
	A34030 [1] Cell- and tissue-culture supplies and equipment
	A34040 [2, 3] Blood- and plasma-warming device (device using microwave-Class III, other devices- Class II)
	A34050 [2] Blood-storage refrigerator
	A34060 [2] Blood-storage freezer
	A34070 [2] Processing system for frozen blood
	A34080 [2] Environmental chamber for storage of platelet concentrate
	A34090 [2] Blood- and plasma-freezing device
	A34100 [2] Organ-storage cabinet for transportation
	A34110 [2] Isolated kidney-perfusion and transport system and accessories
A35000	A35010 [2] Electrosugical unit
Electrosugical devices	A35020 [2] Radiofrequency (RF) electrosurgical cautery apparatus
	A35025 [2] Electrode for electrosurgical unit
	A35030 [2] Thermal cautery unit
	A35040 [2] RF lesion generator
	A35050 [2] Lesion-temperature monitor
	A35060 [2] Vitreous aspiration and cutting instrument
	A35070 [1] Ophthalmic electrolysis unit
	A35080 [2] Phacofragmentation system
	A35090 [2] Photocoagulator
	A35100 [2] Ultrasonic surgical unit
	A35110 [3] Microwave surgical unit
	A35120 [2] Needle-type epilator
	A35130 [2] Tweezer-type epilator
A36000	A36010 [2] Cryogenic surgical device
Cryosurgery devices	
A37000	A37010 [2, 3] Laser surgical apparatus (devices using CO_2, Argon, Nd:YAG, Krypton, Ruby, Dye,
Laser apparatus for medical use	Copper Vapor, Alexandrite, Ho:YAG gas and diode-Class II, other devices-Class III)
	A37020 [2, 3] Laser apparatus for medical use (device for pain relief or skin stimulation-Class II, other devices-Class III)
	A37030 [2, 3] Ophthalmic laser apparatus (devices using Argon, Nd:YAG, Krypton, Diode-Class II, other devices-Class III)

(continued)

TABLE D.1 (CONTINUED)
KFDA Categories of Medical Devices

Category	Device [1] Class I,[a] [2] Class II, [3] Class III
A38000 Instruments for ligature and suture	A38010 [1, 3] Suture instrument (general-Class I, permanently implantable or biodegradable-Class III) A38020 [1] Ligature instrument A38030 [2] Hemorrhoidal ligator A38040 [1] Suture retention device A38070 [1] Ligature carrier and passer A38080 [1] Ligature guide A38090 [1, 2, 3] Clip for medical use (temporary for skin and nose-Class I, temporary for blood vessels and internal tissue-Class II, permanently implantable or biodegradable-Class III) A38170 [1, 2, 3] Staple for medical use (temporary for skin and nose-Class I, temporary for blood vessels and internal tissue-Class II, permanently implantable or biodegradable-Class III) A38190 [1] Clip instrument A38200 [1] Staple instrument A38210 [1] Umbilical occlusion device A38220 [2] Fetal scalp electrode A38230 [3] Phimosis binder
A39000 Aspirators for medical use	A39010 [1, 2, 3] Aspirator for medical use (manual-Class I, motorized or vacuum- Class II, devices for tissue expansion-Class III) A39020 [2] Fetal vacuum extractor A39030 [2] Drainage and continuous aspirator A39050 [1, 2] Breast pump (manual Class I, motorized-Class II) A39060 [1] Cement monomer vapor evacuator
A40000 Pneumothorax and pneumoperitoneum apparatus	A40010 [2] Pneumothorax apparatus A40020 [2] Pneumoperitoneum apparatus A40030 [2] Hysteroscopic insufflator
A41000 Knives for medical use	A41010 [1, 2] Knife for medical use (manual-Class I, motorized-Class II)
A42000 Scissors for medical use	A42010 [1, 2] Scissors for medical use (manual-Class I, motorized-Class II)
A43000 Curettes for medical use	A43010 [1, 2] Curettes for medical use (manual-Class I, motorized-Class II)
A44000 Clamps for medical use	A44010 [1, 2] Clamp for medical use (manual-Class I, motorized-Class II)
A45000 Forceps for medical use	A45010 [1, 2] Dressing forceps for medical use (manual-Class I, motorized-Class II) A45020 [1, 2] Forceps for medical use (manual-Class I, motorized-Class II) (NOTE: Including tongeur, cutting forceps (nippers), holder, plier)
A46000 Saws for medical use	A46010 [1, 2] Saw for medical use (manual-Class I, motorized-Class II) A46020 [1, 2] Casting cutter (manual-Class I, motorized-Class II) A46030 [1] Plaster cutter for dental use
A47000 Chisels for medical use	A47010 [1, 2] Chisel for medical use (manual-Class I, motorized-Class II) A47020 [1, 2] Gouge for medical use (manual-Class I, motorized-Class II)
A48000 Raspatories for medical use	A48010 [1, 2] Elevator for medical use (manual-Class I, motorized-Class II) (NOTE: Including elevator, periosteum elevator, lifter) A48020 [1, 2] Raspatory (manual-Class I, motorized-Class II) A48030 [1, 2] Stripper for medical use (manual-Class I, motorized--Class II) A48040 [1, 2] Urethrotome (manual-Class I, motorized--Class II)
A49000 Mallets for medical use	A49010 [1] Mallet for medical use
A50000 Files for medical use	A50010 [1, 2] File for medical use (manual-Class I, motorized--Class II)
A51000 Levers for medical use	A51010 [1, 2] Lever for medical use (manual-Class II, motorized-Class II)
A52000 Snare for medical use	A52010 [1, 2] Snare for medical use (manual-Class II, motorized-Class II)

TABLE D.1 (CONTINUED)
KFDA Categories of Medical Devices

Category	Device [1] Class I,[a] [2] Class II, [3] Class III
A53000 Needle for syringe and puncture	A53010 [1, 2] Needle for syringe (nonsterilized-Class I, sterilized-Class II)
	A53020 [1, 2] Suture needle (nonsterilized-Class I, sterilized-Class II)
	A53030 [1, 2] Needle for transfusion (nonsterilized-Class I, sterilized-Class II)
	A53040 [1, 2] Puncture needle (nonsterilized-Class I, sterilized-Class II)
	A53050 [1, 2] Trocar (nonsterilized-Class I, sterilized-Class II)
	A53060 [1, 2] Biopsy needle (nonsterilized-Class I, sterilized-Class II)
	A53070 [1, 2] Lancet (nonsterilized-Class I, sterilized-Class II)
A54000 Syringes	A54010 [2] Syringe for general use
	A54020 [2] Jet injector
	A54030 [2] Cartridge syringe
	A54040 [1] Irrigating syringe
	A54050 [1] Syringe for dental use
A55000 Puncturing, abrading, perforating instruments for medical use	A55010 [1, 2] Puncture instrument (manual or dental use-Class I, motorized or sterilized-Class II)
	A55020 [1, 2] Trephine for medical use (manual-Class I, motorized or sterilized-Class II) (NOTE: including excavator)
	A55030 [1, 2] Handpiece (manual or dental use-Class I, others-Class II)
	A55040 [1, 2] Scaler (manual-Class I, motorized-Class II)
	A55050 [1] Abrasive device
	A55060 [2] Powered corneal burr
	A55070 [2] Airbrush
	A55080 [2] Dermabrasion device
A56000 Wound retractors and speculums	A56010 [1, 2] Retractor (manual-Class I, motorized-Class II) NOTE: Including hooks, retractors, separators, retainers)
	A56020 [1] Speculum
A57000 Tube and catheter for medical use	A57010 [1] Continent-ileostomy tube and catheter
	A57020 [2] Gastrointestinal tube and catheter
	A57030 [2] Tracheal tube and catheter
	A57040 [2] Biliary tube and catheter
	A57050 [2] Suprapubic urological tube and catheter
	A57060 [2] Urological tube and catheter
	A57070 [1] Oxygen tube and catheter
	A57080 [1] Disposable sucker
	A57090 [1] Suction tube and catheter
	A57110 [2] Cardiopulmonary bypass vascular catheter, cannula, or tubing
	A57120 [2] Tube and catheter for open heart surgery
	A57130 [2] Intravascular catheter
	A57140 [2] Catheter cannula
	A57150 [3] Intravascular occluding catheter
	A57160 [2] Septostomy catheter
	A57170 [2] Flow-directed catheter
	A57180 [2] Steerable catheter
	A57190 [2] Electrode recording catheter or electrode recording probe
	A57200 [2] Continuous-flush catheter
	A57210 [2] Cerebrospinal catheter devices
	A57220 [1] Introduction-drainage tube and catheter
	A57230 [2] Thermodilution probe
	A57240 [3] Embolectomy catheter
	A57250 [2, 3] Blood-access device for dialysis (nonimplantable-Class II, implantable-Class III) (NOTE: Parts are Class II [adapter, stopcock, connector, fixing devices, manifold, etc.])
	A57260 [2] Percutaneous catheter
	A57265 [2] Catheter for general use
	A57270 [2] Fiberoptic catheter

(continued)

TABLE D.1 (CONTINUED)
KFDA Categories of Medical Devices

Category	Device [1] Class I,[a] [2] Class II, [3] Class III
	A57275 [2] Balloon catheter
	A57280 [3] Catheter-balloon repair kit
	A57290 [1] Epistaxis balloon
	A57300 [2] Anesthesia-conduction catheter
	A57310 [1] Nasopharyngeal catheter
	A57320 [3] Peritoneo-venous shunt
	A57330 [3] Pleuro-peritoneal shunt
	A57340 [2] Central-nervous-system fluid shunt and components
	A57350 [2] Cervical drain
	A57360 [2] Endolymphatic shunt
	A57370 [3] Endolymphatic shunt tube with valve
	A57380 [2] Tympanostomy tube
	A57390 [3] Tympanostomy tube with semipermeable membrane
A58000	A58010 [1, 2] Sound for medical use (manual-Class I, motorized-Class II)
Probes and sounds for medical use	A58020 [1] Probe for medical use
	A58030 [1] Packer for medical use
	A58040 [1] Cleaning brush for medical use
	A58050 [1] Pusher for medical use
	A58060 [1, 2] Electrode for medical use (external body-Class I, internal body or implantable-Class II) (NOTE: Except when other criteria are applicable)
A59000	A59010 [1, 2] Vessel dilator (manual-Class I, motorized-Class II)
Dilators and expanders for medical use	A59020 [1, 2] Esophageal dilator (manual-Class I, motorized-Class II)
	A59030 [1, 2] Rectal dilator (manual-Class I, motorized-Class II)
	A59040 [1, 2] Urethral dilator (manual-Class I, motorized-Class II)
	A59060 [1] Expandable cervical dilator
	A59070 [2] Hygroscopic-*Laminaria* cervical dilator
	A59080 [3] Vibratory cervical dilator
	A59090 [3] Tissue expander
A60000	A60010 [1] Applicator for medical use
Applicators for medical use	
A61000	A61010[1] Spreader
Dispensers and mixing instruments	A61020 [1] Mixing spatula
	A61030 [1] Cement-mixing slab
	A61040 [1] Dental amalgam capsule
	A61050 [1] Amalgamator
	A61060 [1] Mixer for impression materials
	A61070 [1] Mercury and alloy dispenser
	A61080 [1] Cement mixer
	A61090 [1] Cement dispenser
A62000	A62010 [1] Filling instrument
Filling instruments for medical use	A62020 [1] Drop carrier
	A62030 [1] Amalgam carrier
	A62040 [1] Amalgam plugger
A63000	A63010 [1] Tongue depressor
Depressors for medical use	A63020 [1] Depressor
	A63030 [1] Spatula for depressor
A64000	A64010 [1, 2] Goniometer (manual-Class I, motorized-Class II)
Measuring and introducing instruments	A64020 [1] Gauge for medical use
	A64030 [1] Caliper for medical use
	A64040 [1, 2] Marker for medical use (marker external-Class I, marker internal-Class II)
	A64050 [1] Guide for medical use
	A64060 [1] Guard for medical use
	A64070 [1] Sizer for medical use

TABLE D.1 (CONTINUED)
KFDA Categories of Medical Devices

Category	Device [1] Class I,[a] [2] Class II, [3] Class III
	A64080 [1] Pelvicmeter
	A64090 [1] Template for clinical use
	A64100 [1] Cephalometer
	A64110 [1, 2] Stereotaxic instrument (without frame-Class I, with frame-Class II)
	A64120 [1] Distometer
	A64130 [1] Exophthalmometer
	A64140 [1] Nearpoint ruler
	A64150 [1] Interpupillary distance meter
	A64160 [2] Guide wire
	A64170 [2] Introducer
	A64180 [2] Stylet
	A64190 [1] Tip occluder
	A64200 [1] Catheter guide wire
A65000	A65010 [1, 2] Irrigator for medical use (manual-Class I, motorized-Class II)
Douche instruments for medical use	A65020 [2] Colonic irrigation system
	A65030 [1, 2] Ostomy irrigator (manual-Class I, motorized-Class II)
	A65040 [2] Jet lavage
	A65050 [2] Endometrial washer
	A65060 [2] Therapeutic vaginal douche apparatus
	A65070 [2] Urethral washer for medical use
A66000	A66010 [2] Empty container for the collection of blood and blood components
Blood donor or transfusion and biopsy set	A66020 [2] Blood-collection set
	A66025 [1, 2] Blood-collection instrument (collector without needle-Class I, collector with needle-Class II)
	A66030 [2] Blood-donor set
	A66040 [2] Blood-collection and -donor set
	A66050 [1] Capillary blood-collection tube
	A66060 [2] Vacuum-assisted blood-collection system
	A66070 [1] Vacuum-assisted blood-collection tube
	A66080 [2] Blood-specimen collection device
	A66090 [2] Fetal blood sampler
	A66100 [2] Arterial blood-sampling kit
	A66110 [2] Biopsy instrument kit
A67000	A67010 [1, 2] Traction equipment (manual-Class I, motorized-Class II)
Medical devices for orthopedics and restoration	A67020 [1, 2] Powered exercise equipment (manual-Class I, motorized-Class II)
	A67025 [2, 3] Powered orthopedic device (device for pain relief-Class II, Other uses-Class III)
	A67030 [2] Orthopedics appliance
	A67040 [1, 2] Breast percussor (manual-Class Class I, motorized-Class II)
	A67050 [1] Orthopedics and bone operating instruments
	A67060 [2] Hair implant system
A68000	A68010 [2] Dental unit and chair
Dental units and chairs	A68020 [2] Dental unit
	A68030 [2] Dental chair
A69000	A69010 [1] Dental engine
Dental engines	
A70000	A70010 [1] Broach for dental use
Broaches for dental use	
A71000	A71010 [1] Explorer for dental use
Explorers for dental use	
A72000	A72010 [1] Air syringe
Moisture-excluding instruments for dental use	A72020 [1] Rubber dam

(continued)

TABLE D.1 (CONTINUED)
KFDA Categories of Medical Devices

Category	Device [1] Class I,[a] [2] Class II, [3] Class III
A73000 Impression-taking and articulating instruments for dental use	A73010 [1] Articulator
	A73020 [1] Facebow
	A73030 [1] Articulation paper
	A73040 [1] Pantograph
	A73050 [1] Impression tray
A74000 Vulcanizers and curing units for dental use	A74010 [1] Ultraviolet-light-curing unit for polymerization
	A74020 [1] Visible-light-curing unit for polymerization
A75000 Casting machines for dental use	A75010 [1] Casting machine for dental use
A76000 Sight-corrective spectacles	A76010 [1] Sight-corrective spectacles
A77000 Sight-corrective ophthalmic lenses	A77010 [1] Sight-corrective ophthalmic lens
	A77020 [3] Rigid gas-permeable contact lens
	A77030 [3] Soft (hydrophilic) contact lens
	A77040 [2] Therapeutic contact lens
	A77050 [2] Diagnostic contact lens
A78000 Hearing aids	A78010 [2] Hearing aid
A79000 Infusion instruments	A79010 [2] Infusion pump
	A79020 [2] IV container
	A79030 [2] Intravascular-administration set
	A79040 [1] Pressure infusor for an IV bag
	A79050 [3] Insulin injector
	A79060 [2] Angiographic injector and syringe
	A79070 [2] Indicating injector
	A79080 [2] Syringe actuator for an injector
	A79090 [1] Powder blower
	A79100 [2] Automatic fluid-delivery system
	A79110 [2] Electronic monitor for gravity-flow infusion system
	A79120 [2] Anesthesia-conduction kit
	A79130 [1] Enema kit
	A79140 [2] Infusion filter
	A79150 [3] Implantable infusion device
	A79160 [2] Infusion device
A80000 Hernia supporters	A80010 [1] Hernia supporter
	A80020 [1] Therapeutic scrotal supporter
A81000 Inhalators for medical use	A81010 [2] Nebulizer
	A81020 [2] Medicinal nonventilatory nebulizer (atomizer)
A82000 Vibrators	A82010 [2, 3] Vibrators for medical use (devices for pain relief, physical therapy and skin stimulation-Class II, other devices including home use under physician's prescription-Class III)
A83000 Electric stimulator for medical use by personal [Separated from A16000 criteria]	A83010 [2, 3] Low-frequency electric stimulator for personal medical use (device for pain relief and skin stimulation including home use under physician's prescription-Class II, other applications, including home use under physician's prescription-Class III)
	A83020 [2, 3] Statics electricity generator for personal electric stimulation (device for improving blood circulation, including home use under physician's prescription-Class II, other applications, including home use under physician's prescription-Class III)
	A83030 [2, 3] Infrared radiator for personal medical use (device for pain relief and skin stimulation including home use under physician's prescription-Class II, other applications including home use under physician's prescription-Class III)
	A83040 [2] Ultraviolet radiator for personal medical use

TABLE D.1 (CONTINUED)
KFDA Categories of Medical Devices

Category	Device [1] Class I,[a] [2] Class II, [3] Class III
	A83050 [3] Visible-light irradiator for personal medical use
	A83060 [2] Heat for personal medical use
	A83070 [2, 3] Shortwave diathermy for personal use (device for pain relief, including home use under physician's prescription-Class II, other applications, including home use under physician's prescription-Class III)
	A83080 [2, 3] Combinational stimulator for medical use by personal [device for pain relief including home use under physician's prescription-Class II, other applications including home use under physician's prescription-Class III) (NOTE: A combination stimulator is adevice using a combination of two or more kinds of energies listed in category A83000. Classification is appointed to the upper Class.)
	A83090 [2, 3] Electric stimulator for personal medical use (device for pain relief and skin stimulation, including home use under physician's prescription-Class II, other applications, including home use under physician's prescription-Class III)
	A83100 [2, 3] Ultrasonic diathermy (device for pain relief and skin stimulation, including home use under physician's prescription-Class II, other applications, including home use under physician's prescription-Class III)
A84000	A84010 [1, 2] Needles for acupuncturist (nonsterilized-Class I, sterilized-Class II)
Acupuncture and moxibustion apparatus	A84020 [1] Cupping apparatus for medical use
	A84030 [1, 2, 3] Electric moxibustion apparatus (ignition type-Class I, electric type for pain relief, improving blood circulation, supository bathing-Class II, other electric-type devices-Class III)
A85000	A85010 [3] Magnetic-induction apparatus with permanent magnet
Magnetic-induction apparatus for medical use	A85020 [3] Electromagnetic-induction apparatus
A86000	A86010 [2] Electrolyte-water-producing apparatus
Medicinal-substance-producing equipment	
A87000	A87010 [1] Film-viewing device for medical use
Film-viewing devices for medical use	
A88000	A88010 [2] Treatment table and chair for ear, nose and throat
Treatment tables for ear, nose and throat	A88020 [2] Treatment table for ear, nose and throat
	A88030 [1, 2] Chair for ear, nose and throat (manual-Class I, motorized-Class II)
A89000	A89010 [2] Ophthalmic-instrument table and chair
Ophthalmic-instrument tables and chairs	A89020 [2] Ophthalmic-instrument table
	A89030 [1, 2] Ophthalmic-instrument chair (manual-Class I, motorized-Class II)
(B) Medical Supplies	
B01000	B01010 [1] Medical X-ray film
Radiographic supplies	B01020 [1] Intensifying screen
	B01030 [1] Fluorescent screen
B02000	B02010 [2] Silk suture
Sutures and ligatures	B02020 [2] Plastic suture
	B02030 [2] Stainless-steel suture
	B02040 [3] Absorbable surgical gut suture
	B02050 [2, 3] Ligature (temporary-Class II, permanently implantable or biodegradable-Class III)
B03000	B03010 [2, 3] Ankle-joint prosthesis (device with approved safety and structure-Class II, device made of ceramics or biodegradable materials and surface coated for improving biocompatibility-Class III)
Orthopedic materials	B03020 [2, 3] Elbow-joint prosthesis (device with approved safety and structure-Class II, device made of ceramics or biodegradable materials and surface coated for improving biocompatibility-Class III)
	B03030 [2, 3] Finger-joint prosthesis (device with approved safety and structure-Class II, device made of ceramics or biodegradable materials and surface coated for improving biocompatibility-Class III)

(continued)

TABLE D.1 (CONTINUED)
KFDA Categories of Medical Devices

Category	Device [1] Class I,[a] [2] Class II, [3] Class III
	B03040 [2, 3] Hip-joint metal-constrained cemented or uncemented prosthesis (device with approved safety and structure-Class II, device made of ceramics or biodegradable materials and surface coated for improving biocompatibility-Class III)
	B03050 [2, 3] Knee-joint prosthesis [device with approved safety and structure-Class II, device made of ceramics, biodegradable materials and surface coated for improving biocompatibility-Class III]
	B03060 [2, 3] Shoulder joint prosthesis (device with approved safety and structure-Class II, device made of ceramics or biodegradable materials and surface coated for improving biocompatibility-Class III)
	B03070 [2, 3] Toe-joint prosthesis (device with approved safety and structure-Class II, device made of ceramics or biodegradable materials and surface coated for improving biocompatibility-Class III) (NOTE: Including artificial joint for replacing metatarsophalangeal joint or proximal phalanx)
	B03080 [2, 3] Wrist-joint prosthesis (device with approved safety and structure-Class II, device made of ceramics or biodegradable materials and surface coated for improving biocompatibility-Class III) (NOTE: Including artificial joint for replacing lunate, scaphoid, trapezium, and ulnar)
	B03090 [2, 3] Bone plate (device with approved safety and structure-Class II, device made of ceramics or biodegradable materials and surface coated for improving biocompatibility-Class III)
	B03100 [2, 3] Bone screw (device with approved safety and structure-Class II, device made of ceramics or biodegradable materials and surface coated for improving biocompatibility-Class III)
	B03110 [2] Intramedullary-fixation nail
	B03120 [2] Bone-fixation cerclage
	B03130 [2] Bone cap
	B03140 [2] Single/multiple-component metallic bone-fixation appliance
	B03160 [2, 3] Spinal interlaminal-fixation orthosis (device with approved safety and structure-Class II, device made of ceramics or biodegradable materials and surface coated for improving biocompatibility-Class III)
	B03180 [2] External bone fixator
	B03190 [3] Bone cement
	B03200 [2] Cranioplasty plate
	B03210 [2] Cranioplasty-plate fastener
	B03220 [2] Preformed craniosynostosis strip
	B03230 [2, 3] Mandibular-implant facial prosthesis (device with approved safety and structure-Class II, device made of ceramics or biodegradable materials and surface coated for improving biocompatibility-Class III)
	B03240 [2] Ophthalmic conformer
	B03260 [2, 3] Eye-sphere implant (nondegradable device with approved safety and structure-Class II, biodegradable or device made of materials without approved safety-Class III)
	B03280 [2] Scleral shell
	B03290 [3] Eye-valve implant
	B03300 [2, 3] Stent for medical use (devices uses in vagina, trachea, renal duct, esophagus, urethra-Class II, device contacting blood-Class III)
	B03330 [2] Surgical mesh
	B03340 [2] Methyl methacrylate cuff for aneurysmorrhaphy
	B03350 [2] Nerve cuff
	B03360 [3] Adhesives
	B03370 [2] Sacculotomy tack (Cody tack)
	B03380 [2, 3] Synthetic polymer implant material (nondegradable-Class II, biodegradable-Class III)
	B03390 [2, 3] Silicon material (solid type-Class II, gel type-Class III)
	B03400 [2] Artificial eye
B04000 Human tissue and organ substitutes	B04010[3] Vascular-graft prosthesis
	B04020 [3] Patch prosthesis
	B04030 [3] Replacement heart valve
	B04040 [3] Annuloplasty ring
	B04050 [3] Silicone breast prosthesis

TABLE D.1 (CONTINUED)
KFDA Categories of Medical Devices

Category	Device [1] Class I,[a] [2] Class II, [3] Class III
	B04060 [3] Esophageal prosthesis
	B04070 [3] Tracheal prosthesis
	B04080 [2] Ossicular-replacement prosthesis
	B04090 [3] Artificial tympanic membrane
	B04100 [3] Fallopian-tube prosthesis
	B04110 [3] Artificial vocal chord
	B04120 [3] Testicular prosthesis
	B04130 [3] Keratoprosthesis
	B04140 [3] Intraocular lens (IOL)
	B04150 [2] Laryngeal prosthesis (Taub design)
	B04160 [3] Ligament prosthesis
	B04170 [2] Tendon prosthesis
	B04180 [3] Dura substitute
	B04190 [3] Penile inflatable implant
	B04200 [3] Penile rigidity implant
	B04210 [3] Implanted mechanical/hydraulic urinary-continence device
	B04220 [3] Bone heterograft
	B04230 [3] Biomaterial graft and prosthesis
	B04235 [3] Graft and prosthesis
	B04240 [2] Middle-ear mold
	B04250 [2] Punctum plug
	B04260 [3] Collagen implant
B05000	B05010 [1] Splint
Splints	B05020 [1] Inflatable extremity splint
	B05030 [1] Denis Browne splint
	B05040 [1] External nasal splint
	B05050 [1] Congenital hip-dislocation abduction splint
B06000	B06010 [1] Visual-acuity chart
Test charts for visual acuity and color blindness	B06020 [1] Color-vision tester
B07000	B07010 [1] Surgical gloves
Surgical supplies	B07020 [1] Finger cots
	B07030 [1] Ostomy pouch
	B07040 [1] Organ bag
	B07050 [1. 2] Surgical drape (goods do not contact wound or internal body-Class I, goods contacting wound, internal body or body cavity-Class II)
	B07060 [1] Drape adhesive
	B07070 [2, 3] Wound dressing (wound coverage-Class II, wound dressing made of biodegradable, antibacterial, or biological materials-Class III)
	B07080 [1] Skin-pressure protectors
	B07090 [1] Elastic bandage
B08000	B08010 [2] Condom
Condoms	B08020 [2] Condom for female
B09000	B09010 [2] Vaginal pessary
Contraceptive devices	B09020 [2] Contraceptive diaphragm and accessories
	B09030 [3] Contraceptive intrauterine device (IUCD) and introducer
	B09040 [3] Contraceptive tubal occlusion device (TOD) and introducer
(C) Dental Materials	
C01000	C01010 [1] Dental pure metal
Dental metals	C01020 [2] Dental precious-metal alloy
	C01030 [2] Base-metal alloy
	C01040 [2] Amalgam alloy

(continued)

TABLE D.1 (CONTINUED)
KFDA Categories of Medical Devices

Category	Device [1] Class I,[a] [2] Class II, [3] Class III
	C01050 [1] Dental mercury
	C01060 [1] Dental fusible alloy
C02000	C02010 [2] Porcelain powder
Teeth-crown materials	C02020 [2] Synthetic resin for dental crown
	C02030 [1] Preformed gold denture tooth
	C02040 [2] Porcelain tooth
	C02050 [2] Preformed plastic denture tooth
	C02060 [1] Preformed crown
	C02070 [1] Backing and facing for an artificial tooth
C03000	C03010 [2] Denture-base resin
Denture-base materials	C03020 [1] Tray resin
	C03030 [1, 2] Denture pad (cushion) (wax-based, temporary use-Class I, nonwax material-Class II)
	C03040 [2] Denture-base glazing and hardening agent
	C03050 [2] Denture-base repair resin
	C03060 [2] Denture relining resin
C04000	C04010 [1] Gutta percha
Dental root-canal filling and treatment	C04020 [1] Endodontic silver point
materials	C04030 [2] Other dental root-canal filling material
	C04040 [2] Dental root-canal sealer
	C04050 [1] Endodontic paper point
	C04060 [1] Dental cleanser
	C04070 [1] Dental materials for exclusion of moisture
C05000	C05010 [2] Dental cement
Dental bonding and filling materials	C05020 [2] Dental-filling materials
	C05030 [1, 2] Base and liner (varnish-Class I, base and liner-Class II)
	C05040 [2] Temporary-filling materials
	C05050 [2] Dental adhesive
	C05060 [2] Coating material for resin fillings
	C05070 [2] Tooth-shade resin material
C06000	C06010 [2] Impression material
Dental-impression materials	C06020 [1, 2] Treatment material for dental impression (preimpression use-Class I, postimpression use-Class II)
	C06030 [1] Dental-impression material for duplicating model
C07000	C07010 [1] Dental wax
Dental waxes	C07020 [1] Base-plate shellac
C08000	C08010 [1] Dental gypsum product
Dental gypsum products and investments	C08020 [1] Dental investment
C09000	C09010 [1] Dental abrasive polishing agent
Dental abrasive and polishing materials	
C10000	C10010 [1] Orthodontic bracket and band
Orthodontic materials	C10020 [1] Orthodontic wire
	C10030 [2] Orthodontic cement
	C10040 [2] Orthodontic resin for appliance
	C10050 [2] Extraoral orthodontic
	C10060 [1] Other orthodontic appliance
C11000	C11010 [2] Precision attachment
Prosthodontic materials	C11020 [1] Preformed clasp
	C11030 [1] Root-canal post
	C11040 [1] Retentive and splinting pin
	C11050 [2] Fit checker
	C11060 [2] Bite-registration material

TABLE D.1 (CONTINUED)
KFDA Categories of Medical Devices

Category	Device [1] Class I,[a] [2] Class II, [3] Class III
C12000 Dental implant	C12020 [2] Intraosseous-fixation plate
	C12030 [2] Intraosseous-fixation screw
	C12040 [2] Intraosseous-fixation wire
	C12050 [2, 3] Dental implant (external parts: screw, cylinder, abutment, etc.-Class II, implanting parts-Class III)
	C12070 [3] Bone-graft materials
	C12080 [3] Periodontium regeneration induced materials
	C12090 [1] Instrument and material for dental implanting
C13000 Preventive materials	C13010 [2] Pit and fissure sealant

[a] Class I devices are subject to premarket notification. While the KFDA will review the product information including the labeling, no product testing is required (ITA, Korea p. 2).

Source: KFDA, Regulation

Appendix E: Japanese Medical Device Classification

Medical devices are defined in the Pharmaceutical Affairs Law (PAL) of Japan as "equipment or instruments intended for use in the diagnosis, cure, or prevention of disease in humans or animals, or intended to affect the structure or functions of the body of humans or animals, and which are designated by cabinet order" (Guide p. 2). Not all instruments and apparatus intended for these purposes are regulated by the PAL. Only those devices listed in a Cabinet order are subject to regulation.

In Japan, medical devices that have been listed in a Cabinet order are classified based on their purpose. Devices are divided into five types: (a) medical instruments and apparatus, (b) medical products, (c) dental materials, (d) sanitary products, and (e) medical devices exclusively for animals. Under authority of the PAL, the Ministry of Health, Labor, and Welfare (MHLW) issued MHLW Notification No. 1008 on November 1, 1995. MHLW Notification No. 1008 establishes the Generic Names of Medical Devices and Their Classification. The categories and generic names of the human-use devices are listed in Table E.1. Devices that are intended for animal use only fall under the jurisdiction of the Ministry of Agriculture, Forestry, and Fisheries (Guide p. 4) and are not covered in this appendix.

Not all of the devices listed in Table E.1 require a product-by-product approval. There are 76 categories, listed in Table E.2, that are exempt from approval on a product-by-product basis because of their exclusive use by specialists, established effectiveness and safety, and reliable operating technique. Medical devices that conform to certain Japanese Industrial Standards (JIS) as noted in Table E.3 are also exempt from a product-by-product approval (Guide, pp. 41–42).

The medical device classifications to which a partial license (*kubun-kyoka*) can be applied are listed in Attached Table 1-3 of the Enforcement Regulations of the Pharmaceutical Affairs Law (ERPALs). That table is reproduced in Table E.4

The 1994 revision to the PAL added requirements for manufacturing control and quality control at manufacturing plants. Referred to as "medical device good manufacturing practices," or simply as "medical device GMPs," this ordinance establishes the requirement for a quality assurance (QA) system for manufacturers of certain medical devices. Some medical devices are exempted from the requirements of medical device GMPs. These are listed in Table F.5.

Some medical devices are approved by the prefectural government without the intervention of the MHWL. These products are listed in Table E.6. For these products, the prefectural governor is completely responsible for the examination of both the *kyoka* licenses and *shonin* product-approval applications. If the application is acceptable, the prefectural governor issues the letter of license and/or letter of product approval.

For some medical devices, the ERPALs allows a degree of flexibility in providing the information required by PAL Article 63. Those devices are listed in Table 4 attached to the ERPALs, which is reproduced in Table E.8.

TABLE E.1
Japanese MHWL Categories of Medical Devices

Category	Classification	Products
		Medical Instruments and Apparatus
IA-1	Operating and treatment tables	Treatment tables; operating tables for surgical use; operating tables for orthopedic use; operating tables for obstetric and gynecologic use; operating tables for ophthalmic use; operating tables for ear, nose, and throat use; treatment tables; radiation-therapy couches
IA-2	Medical lights and lamps	Operating lights, examination lights, head mirrors
IA-3	Disinfecting devices for medical use	Dry-heat sterilizers, automatic pressure sterilizers, nonautomatic pressure sterilizers, ultraviolet sterilizers, automatic ethylene oxide (EtO) gas sterilizers, nonautomatic EtO gas sterilizers, formalin gas sterilizers, boiling and steam sterilizers
IA-4	Medical water sterilizers	Water sterilizers
IA-5	Anesthesia apparatus, rebreathing bags, and gas-absorption apparatus	Disposable anesthesia instruments, rebreathing bags, carbon dioxide absorbers, closed-circuit anesthesia apparatus, to-and-fro anesthesia apparatus, mixing-anesthesia apparatus
IA-6	Respiratory-therapy apparatus	Disposable respiratory tubes and catheters, resuscitators, oxygen inhalators, oxygen tents, single hyperbaric oxygen chambers, multiple hyperbaric oxygen chambers, oxygen generators and oxygen condensers, oxygen-supplying units
IA-7	Artificial internal organ apparatus	Mechanical valve prostheses; tissue valve prostheses; vascular grafts; vascular prostheses; single-patient dialysate-control apparatus; artificial heart-lung machines (pump oxygenators); bubble oxygenators; membrane oxygenators; implantable cardiac pacemakers; external cardiac pacemakers; parts and accessories of cardiac pacemakers; extracorporeal blood circuits for hemodialysis; extracorporeal blood circuits for artificial heart-lung machines; hemofilters; hemofiltration apparatus; hemoperfusion adsorption columns; hemoperfusion apparatus; membrane plasma separators; membrane plasma fractionators; blood-separation apparatus; plasma-separation apparatus; peritoneal-dialysis systems; hemodialysis monitors; multi-patient dialysate-delivery equipment; hollow-fiber dialyzers; coil dialyzers; plate dialyzers; artificial endocrine pancreases; ascites-treatment systems by filtration; concentration; and infusion; assisted-circulation apparatus
IA-8	Infant incubators	Closed-circulation incubators, open incubators, simple transportable incubators, thermostat transportable incubators, accessories for incubators
IA-9	Medical X-ray apparatus and X-ray tubes	General radiographic X-ray apparatus, fluoroscopic X-ray apparatus, tomographic X-ray apparatus, mobile X-ray apparatus, portable X-ray apparatus, skull radiographic apparatus, pediatric X-ray apparatus, urological and gynecological X-ray apparatus, mammographic X-ray apparatus, cardiovascular X-ray apparatus, fixed X-ray apparatus for operating-theater use, mobile surgical X-ray apparatus, mass-survey X-ray apparatus, head and neck X-ray CT scanners, whole-body X-ray CT scanners, general dental X-ray apparatus, special-purpose dental X-ray apparatus, medical X-ray tube assemblies, medical X-ray-tank units, medical X-ray high-voltage generators, X-ray-tube/image-intensifier supporting structures, conventional fluoroscopic tables, remote-controlled fluoroscopic tables, photofluorographic tables, direct radiographic units, indirect radiographic units, tomographic stands, skull radiographic stands, urological and gynecological X-ray tables, special-purpose radiographic stands, X-ray image-intensifier assemblies, X-ray television systems, digital radiographic apparatus, medical linear accelerators, medical betatrons, medical microtrons, medical X-ray beam-limiting devices, X-ray exposure reduction devices, X-ray automatic exposure controllers, film/cassette changers, patient restrainers, X-ray film cassettes, X-ray grids, radiation-therapy planning apparatus

TABLE E.1 (CONTINUED)
Japanese MHWL Categories of Medical Devices

Category	Classification	Products
IA-10	RI therapy apparatus	Single-photon-emission CT equipment, positron-emission CT equipment, scintillation cameras, tomographic scintigram apparatus, nuclear medicine data processors, 60 Co teletherapy apparatus, 137 Cs teletherapy apparatus, automatic well counters, scintillation counters, liquid scintillation counters, radioisotope (RI) dynamic-function testing apparatus, thyroid-uptake measuring apparatus, RI renogram testing apparatus, RI regional cerebral blood-flow measuring apparatus, RI cardiac-function testing apparatus, bone-density measuring apparatus, remote controlled after-loading therapy apparatus, whole-body counters, RI blood-volume computers, sealed sources for 60 Co intracavitary irradiation-therapy apparatus, sealed sources for 137 Cs intracavitary irradiation-therapy apparatus, sealed sources for 192 Ir intracavitary irradiation-therapy apparatus, 60 Co small sealed sources for RI interstitial irradiation therapy, 137 Cs small sealed sources for RI interstitial irradiation therapy, 192 Ir small sealed sources for RI interstitial irradiation therapy, 226 Ra small sealed sources for RI interstitial irradiation therapy, 90 Sr small sealed sources for RI interstitial irradiation therapy, radioimmunoassay instruments
IA-11	Radioactivity-protective equipment	X-ray-protective aprons, X-ray-protective gloves, X-ray-protective goggles, patient-protective devices, X-ray-protective partitions, X-ray-protective curtains, photofluorographic-protective boxes, dental X-ray-protective booths
IA-12	Physical-therapy instruments	Pulse-method ultrasonic diagnostic apparatus, Doppler-method ultrasonic diagnostic apparatus, laser-ray therapy equipment, infrared radiators, ultraviolet radiators, defibrillators, massagers, VHF-therapy equipment, microwave therapy equipment, low-frequency therapy equipment, ultrasonic therapy instruments, static-electricity therapy apparatus, hydrotherapeutic equipment, warm-bath therapy equipment, hyperthermia equipment, tubal patency diagnosis equipment, heart massagers, ultrasonic surgical apparatus, dental ionizers, cardiovascular ultrasonic diagnostic imaging apparatus, abdominal ultrasonic diagnostic imaging apparatus, mammary and thyroid ultrasonic diagnostic imaging apparatus, universal ultrasonic diagnostic imaging apparatus, medical thermographic apparatus
IA-13	Stethoscopes	Stethoscopes, Electronic stethoscopes
IA-14	Percussion hammers	
IA-15	Tongue depressors	
IA-16	Clinical thermometers	Clinical thermometers, clinical mercury thermometers, electronic thermometers, deep-body thermometers
IA-17	Hematological testing apparatus	Oxymeters, automatic analyzers (flow) for clinical chemistry, automatic analyzers (discrete) for clinical chemistry, automatic analyzers (centrifuge) for clinical chemistry, automatic analyzers (dry chemistry) for clinical chemistry, clinical spectrophotometers, flame photometers, electrolyte analyzers with ion selective electrode, coulometric electrolyte analyzers, electrophoresis apparatus, immunological analyzers, immunological analyzers (photometric), immunological analyzers (turbidimetric, nephrometric), immunological analyzers (fluorometric), special-purpose analyzers for clinical chemistry, blood-cell counters, blood-cell-differential automatic analyzers, blood-clot analyzers, blood-type analyzers, immunological testing apparatus, parts and accessories of serological testing apparatus
IA-18	Blood-pressure and pulse-rate measurement instruments	Aneroid sphygmomanometers, mercurial sphygmomanometers, electronic invasive blood-pressure meters, electronic noninvasive blood-pressure meters, electronic sphygmomanometers, ultrasonic blood-pressure meters, ophthalmodynamometers, pulse-wave meters (plethysmographs)
IA-19	Urine or excrement analyzers	Urine analyzers, bacterial-identification apparatus, bacterial-sensitivity test apparatus, urine gravimeters
IA-20	Body-fluid testing instruments	

(continued)

TABLE E.1 (CONTINUED)
Japanese MHWL Categories of Medical Devices

Category	Classification	Products
IA-21	Visceral-function testing instruments	Oscilloscopes for medical use, multibasic health testing systems, electrocardiogram analyzing systems, phono- and electrocardiogram analyzing systems, single-channel electrocardiographs (ECGs), multichannel ECGs, vectorcardiographs (VCGs), stress-test instruments for ECG, ambulatory ECG recording and analyzing systems (Holter systems), telephonic ECG-transmission systems, electronystagmographs (ENGs), electroretinographs (ERGs), photostimulators, acoustic stimulators, electroencephalographs (EEG), EEG (frequency) analyzers, phono- and electrocardiographs, multichannel phonocardiographs (PCGs), fetal phono- and electrocardiographs, heart-rate meters, intracranial-pressure meters, airway pressure meters, intrathoracic-pressure meters, cystometers, intrauterine-pressure meters, electromagnetic blood-flow meters, ultrasonic Doppler blood-flow meters, tonometers, thermal dilution cardiac output flow meters, dye-dilution cardiac output flow meters, electrocochleographs, magnetocardiographs, multipurpose polygraphs, central monitors, bedside monitors, surgical monitors, tocomonitors, radiocapsule systems, electrocardiogram telemetering systems (ECG telemeters), multipurpose telemetering systems, expiration-gas analyzers, electronic spirometers, percutaneous blood-gas analyzers, blood-gas analyzers, respiratory-resistance meters, cell classification and sorting apparatus, nuclear magnetic resonance CT equipment, body plethysmographs, basal metabolism meters
IA-22	Ophthalmologic testing instruments	Trial lenses, fundus cameras, corneal cameras, anterior cameras, slit-lamp microscopes, testing instruments and apparatus for oculomotor function, testing instruments and apparatus for light sensation and color sensation, testing instruments and apparatus for fundus, objective refractometers, subjective refractometers, perimeters
IA-23	Hearing testing instruments	Audiometers, auditory feedback recorders
IA-24	Testing instruments and apparatus for perception organs or locomotion organs	Gait-analysis systems, electromyographs (EMGs), electric stimulators, back-dynamometers, goniometers, aethesiometers, hand dynamometer, equilibrium-function test equipment, vibratory-sensation measuring equipment, dental pulp testers, root-canal meters, occlusal sound checkers, electric periodontal pocket probes, mandibular-movement measuring apparatus, bite-pressure meters, occlusal analyzers, aethesiometers
IA-25	Scopes and speculums for medical use	Video image endoscope, rigid endoscopes, nonactive treatment devices for endoscopes, fiberscopes, gastrocameras, biological microscopes, colposcopes, operating microscopes, dental mirrors and mirror holders
IA-26	Centrifuges for medical use	Medical centrifuges, hematocrit centrifuges, cooled centrifuges
IA-27	Microtomes for medical use	Microtomes, strops for medical microtomes
IA-28	Thermostats for medical use	Citrated whole-blood cabinets, organ-storage cabinet for transportation, bacterial-culture apparatus
IA-29	Electrosurgical units	Electrosurgical units, microwave surgical units
IA-30	Apparatus for ligature and suture	Suture needles, needle holders, ligature carriers, suture instruments, automatic suture instruments, nonabsorbable wound clips, aneurysm needles
IA-31	Medical cautery instruments	Cautery instruments, carbon dioxide gas laser surgical apparatus and coagulators, YAG laser surgical apparatus and coagulators, argon laser surgical apparatus and coagulators, cryosurgery apparatus
IA-32	Aspirators for medical use	Aspirators for medical use, continuous-suction units, electric drainage and continuous aspirators, vacuum extractors, accessories for aspirators, dental suction apparatus
IA-33	Pneumothorax apparatus and pneumoperitoneum apparatus	
IA-34	Knives for medical use	
IA-35	Scissors for medical use	
IA-36	Dressing forceps for medical use	
IA-37	Curettes for medical use	
IA-38	Hooks for medical use	
IA-39	Forceps for medical use	
IA-40	Saws for medical use	
IA-41	Chisels for medical use	
IA-42	Raspatories for medical use	Spatula for medical use, periosteal elevators and raspatories

TABLE E.1 (CONTINUED)
Japanese MHWL Categories of Medical Devices

Category	Classification	Products
IA-43	Mallets for medical use	
IA-44	Files for medical use	
IA-45	Levers for medical use	Levers, dental levers and elevators
IA-46	Snares for medical use	Snares and excision instruments
IA-47	Needles for syringe and puncture needles	Injection needles for general use, disposable injection needles for general use, disposable anesthesia needles, disposable hemodialysis needles, reusable puncture needles, injection needles for dental use, disposable injection needles for dental use, sterile plastic indwelling cannulas with needle
IA-48	Syringes	Sterile syringes for general use, sterile microsyringes, sterile insulin syringes, syringes for general use, microsyringes, insulin syringes, syringes for dental use, disposable syringes for dental use
IA-49	Puncturing, cutting, and perforating instruments	Perforators, disposable lumbar-puncture instruments, disposable thorax-puncture instruments, reusable puncture instruments, medical puncture instruments, dental burs, dental drills, mandrels, reamers, root-canal files
IA-50	Wound retractors and speculums	Entropion forceps, eye speculums and fixation forceps, mouth speculums, ear speculums, nasal speculums, vaginal speculums, anoscopes, wound retractors
IA-51	Rostral tubes and body-fluid guidance tubes	Sterile therapeutic tubes, catheters, introducers and guide wires for blood vessels, sterile angiographic and diagnostic tubes and catheters for blood vessels, sterile feeding tubes and catheters, sterile tubes and catheters for open heart surgery, sterile angiographic and diagnostic tube and catheter for blood vessels, sterile tubes and catheters for ureterolithotomy, sterile bile-duct tubes and catheter, sterile fistula tubes and catheters, sterile balloon tubes and catheters (Foley catheters), sterile urethral tubes and catheters (Nelation catheters), sterile tubes and catheters for esophagus and stomach, sterile tracheotomy tubes and catheters, sterile tracheal tubes and catheters, sterile intestine tubes and catheters, sterile tubes and catheters for intravenous hyperalimentation, sterile indwelling cannulas with needle and catheter for hemodialysis, sterile suction tubes and catheters, sterile cerebrospinal drains, manual inflation devices for PTCA balloons, sterile implantable tubes and catheters, sterile drain tubes and catheters, sterile indwelling cannulas for blood access
IA-52	Dilators	
IA-53	Directors, probes and sounds for medical use	
IA-54	Applicators for medical use	
IA-55	Douche instruments for medical use	Vaginal douches for home use, nasal douches, vaginal douches, irrigators, ultrasonic washers, water syringes and minimum syringes for dental use, electromotive douches
IA-56	Blood-donor sets and blood-transfusion sets	Blood-collecting instruments, blood-collecting instruments for test, blood-donor and transfusion sets, parts and related devices for blood donation and blood transfusion
IA-57	Vaccination instruments	
IA-58	Orthopedic equipment	Orthodontic face bows, ceramic appliances for orthodontic use, elastic appliances for orthodontic use, other traction equipment, measuring devices for orthodontic use, extension and traction equipment, appliances for cephalofacial photo, automatic intermittent traction equipment, pneumatic bone-operating instruments, polymeric appliances for orthodontic use, orthodontic chin caps, bone-setting instruments, electric bone-operating instruments, orthodontic head gear, powered exercise equipment, metallic appliances for orthodontic use, self-exercising nonpowered equipment with measuring device, ligament-tunneling devices, bone-setting and bone-operating implements, instruments for ligature tying and band pushing
IA-59	Dental units	Periedodontic units, orthodontic units, dental units for general use
IA-60	Dental engines	Dental laboratory trimmers, CAD/CAM for dental laboratory, dental laboratory electric lathes, air rotary cutting dental apparatus, electric rotary cutting dental apparatus, dental laboratory handpiece, tools for electric dental engine, dental laboratory vacuum spatulators, dental laboratory high-speed lathes, belt arms of dental engines, wrist and slip-joint K_4s for dental engines, dental engine belts
IA-61	Dental handpieces	Handpieces for engine, handpieces for light

(continued)

TABLE E.1 (CONTINUED)
Japanese MHWL Categories of Medical Devices

Category	Classification	Products
IA-62	Dental cutting instruments	Chisels, excavators, scalers, and curettes
IA-63	Broaches for dental use	Broaches, cleansers
IA-64	Explorers for dental use	
IA-65	Filling instruments for dental use	Plastic filling instruments, carvers, amalgam pluggers, amalgam carvers, mallets, amalgam carriers, composite resin carriers
IA-66	Mixing apparatus for dental use	Mixing spatulas, cement-mixing tablets, amalgamators, mixers for impression materials
IA-67	Moisture-excluding instruments for dental use	Hot-air syringes, rubber-dam clamps, rubber-dam punches, clamp forceps, rubber-dam frames
IA-68	Impression and articulation instruments	Impression trays for dentulous instruments, impression trays for edentulous cases, partial-impression trays, partial-impression trays (adjustable type), articulators, face bows
IA-69	Vulcanizers and curing units for dental use	Ultraviolet-curing units, visible-light-curing units, hydraulic presses, curing units
IA-70	Casting machines for dental use	
IA-71	Sight-corrective spectacles	Spectacles for nearsightedness, spectacles for farsightedness, spectacles for astigmatism, spectacles for presbyopia, special spectacles
IA-72	Sight-corrective ophthalmic lenses	Monofocal ophthalmic lenses of optical glass, monofocal ophthalmic lenses of plastic, multifocal ophthalmic lenses of optical glass, multifocal ophthalmic lenses of plastic, hard contact lenses, soft contact lenses, therapeutic contact lenses, diagnostic contact lenses, intraocular lenses (IOLs)
IA-73	Hearing aids	Pocket-type hearing aids, behind-the-ear-type hearing aids, spectacle-type hearing aids, in-the-ear-type hearing aids
IA-74	Infusion instruments	Infusion pumps, automatic fluid-delivery systems, powder blowers for medical use, transfusion solution sets, winged needles, gingival applicators, cartridge-type injection devices for dental use, contrast medium injectors, sterile infusion set
IA-75	Hernia supporters	
IA-76	Aspirators for medical use	Ultrasonic vaporizer for home use, electric vaporizer for home use, ultrasonic nebulizers, spray nebulizers
IA-77	Vibrators	Electric massagers for home use, suction massagers for home use, air massagers for home use, ultrasonic air-bubble baths for home use, air-bubble baths for home use, whirlpool baths for home use
IA-78	Electric-therapy apparatus for home use	Low-frequency-therapy apparatus for home use, static-electricity-therapy apparatus for home use, VHF-therapy apparatus for home use, compound-electric-therapy apparatus for home use, infrared lamps for home use, ultraviolet lamps for home use, electric hotpacks for home use
IA-79	Digital compression apparatus	Digital compressors with heating for home use, digital compressors with rollers for home use
IA-80	Acupuncture and moxibustion apparatus	Vital-point heat-therapy apparatus, sticking needles, acupuncture-point detectors, acupuncture needles, electro-acupuncture equipment, sticking contact needles
IA-81	Magnetic-induction-therapy apparatus	Electromagnetic-induction-therapy apparatus for home use, magnetic-induction-therapy apparatus with permanent magnet for home use
IA-82	Vision-corrective devices	
IA-83	Medicinal-substance-producing equipment	Electrolyte-water-producing apparatus with tank, continuous electrolyte-water-producing apparatus
IA-84	Accessories to the above products designated by the Ministerial Ordinance of Health, Labor, and Welfare	Fluorescent screens for photofluorograph, fluorescent screens for fluoroscope, photostimulable phosphor plates, medical X-ray-intensifying screen. film viewing devices, related devices for anesthesia apparatus

Medical Products

Category	Classification	Products
MP-1	X-ray films	Medical X-ray films, dental X-ray films
MP-2	Sutures	Absorbable sutures, silk sutures, plastic sutures, stainless-steel sutures
MP-3	Surgical gloves and finger sacks	Surgical gloves, disposable surgical gloves, finger sacks, dental gloves

TABLE E.1 (CONTINUED)
Japanese MHWL Categories of Medical Devices

Category	Classification	Products
MP-4	Orthopedic materials	Metallic artificial bones, polymeric artificial bones, ceramic artificial bones, metallic artificial joints, polymeric artificial joints, ceramic artificial joints, bone-setting plates, bone nails, bone-setting screws, bone cements, polymeric materials, waxes, adhesives, mammary prostheses, artificial tympanic membranes, gypsums for medical use, gypsum bandages, dental implants, first-aid bandages
MP-5	Splints	
MP-6	Test charts for visual acuity and color blindness	Test charts for visual acuity and astigmatism, test charts for color blindness

Dental Material

Category	Classification	Products
DM-1	Dental metals	Dental gold ingots, dental high-gold alloys, dental low-gold alloys, gold alloys for porcelain bonding, dental casting gold-silver-palladium alloys, dental wrought-gold-silver-palladium alloys, dental gold-silver-palladium alloy solders, dental silver ingots, dental casting silver alloys (type I), dental casting silver alloys (type II), dental casting silver alloys (type III), dental silver solders, plus metals for dental casting 14K-gold alloys, dental casting nickel-chromium alloys (type I), dental casting nickel-chromium alloys (type II), dental wrought-nickel-chromium alloys, dental wrought-cobalt-chromium alloys, dental casting cobalt-chromium alloys, stainless-steel wires for dental use, silver alloys for dental amalgam, dental mercury, dental copper alloys, dental tin-antimony alloys
DM-2	Teeth-crown material	Vacuum-fired porcelain teeth, dental low-fusing porcelains, dental high-fusing porcelains, dental metal-bonding porcelains, synthetic resins for crown, acrylic-resin teeth
DM-3	Denture-base materials	Acrylic denture-base resins, acrylic resins for denture, resins for impression trays, resins for dental patterns
DM-4	Dental root-canal filling materials	Dental root-canal filling points, dental root-canal filling sealing materials, dental root-canal solid materials
DM-5	Dental bonding and filling materials	Dental zinc phosphate cements, dental silicophosphate cements, dental polycarboxylate cements, dental resin cements, dental zinc oxide-eugenol cements, dental glass ionomer cements for bonding use, dental composite resins for filling, dental composite resins for bonding, dental glass ionomer cements for filling, dental silicate cements, dental temporary stoppings, denture adhesives
DM-6	Dental-impression materials	Alginate impression materials, elastomeric precision impression materials, agar impression materials, impression compounds
DM-7	Dental waxes	Dental inlay casting waxes, dental base-plate waxes, casting waxes, sticky waxes, dental base waxes, utility waxes, shellac base waxes
DM-8	Dental gypsums and gypsum products	Dental plasters, dental stones, dental-impression plasters, dental gypsum-bonded investments, dental phosphate-bonded and silicate-bonded investments
DM-9	Dental abrasive and polishing materials	Diamond abrasive materials, rubber polishing materials, carborundum abrasive materials

Sanitary Products

Category	Classification	Products
SM-1	Sanitary tampons	
SM-2	Condoms	
SM-3	Contraceptives	Pessaries, intrauterine contraceptive devices
SM-4	Sex devices	

Source: Guide pp. 4–29.

TABLE E.2
Japanese MHLW Medical Devices Exempted From Approval on a Product-by-Product Basis

Category	Classification	Products
		Medical Instruments and Apparatus
IA-1	Operating tables and treatment tables	Excluding electromotive or radiotherapy couches
IA-2	Medical lamps and lights	
IA-3	Medical disinfection apparatus	Boiling or steam disinfection apparatus, ultraviolet sterilizers
IA-7	Artificial organs	Sizers
IA-8	Incubators	Simple transportable incubators
IA-9	Medical X-ray equipment and X-ray tube	Film changers, bucky devices, automatic processors for medical or dental film, medical imagers, patient restrainer, X-ray-film cassettes, X-ray grids
IA-10	RI therapy apparatus	Radioimmunoassay instruments, Liquid scintillation counters
IA-12	Physical-therapy instruments	Medical thermographic equipment (excluding contact-type medical thermographic)
IA-13	Stethoscopes	Stethoscopes, electronic stethoscopes
IA-14	Percussion hammers	
IA-15	Tongue depressors	
IA-16	Clinical thermometers	Clinical mercury thermometers
IA-17	Hematological testing apparatus	Hematological analyzer (excluding oxymeter as well as exclusive analyzer with reaction system)
IA-18	Blood-pressure and pulse-rate measurement instruments	Aneroid sphygmomanometers, mercurial sphygmomanometers
IA-19	Urine or excrement analyzers	Excluding exclusive analyzer with reaction system as primer
IA-21	Visceral-function testing instruments	Lead wire for electrocardiograph (ECG) electrode, blood-gas analyzers, histological analysis instruments
IA-22	Ophthalmologic testing instruments	Trial lenses, Slit-lamp microscope
IA-24	Testing instruments and apparatus for perception organs or locomotion organs	Back dynamometers, goniometers, aethesiometers, hand-dynamometers
IA-25	Scopes and speculums for medical use	Nonactive treatment devices for endoscopes, biological microscopes, colposcopes, operating microscopes, dental mirrors and mirror holders
IA-26	Centrifuges for medical use	
IA-27	Microtomes for medical use	Microtomes, strops for medical microtomes
IA-28	Thermostats for medical use	
IA-30	Apparatus for ligature and suture	Suture needles, needle holders, ligature carriers, suture instruments, automatic suture instruments, nonabsorbable wound clips, aneurysm needles
IA-32	Aspirators for medical use	Aspirators for manual use only
IA-34	Knives for medical use	Excluding electromotive ones
IA-35	Scissors for medical use	Excluding electromotive ones
IA-36	Dressing forceps for medical use	Excluding electromotive ones
IA-37	Curettes for medical use	Excluding electromotive ones
IA-38	Hooks for medical use	Excluding electromotive ones
IA-39	Forceps for medical use	Excluding electromotive ones
IA-40	Saws for medical use	Excluding electromotive ones
IA-41	Chisels for medical use	Excluding electromotive ones
IA-42	Raspatories for medical use	Excluding electromotive ones
IA-43	Mallets for medical use	Excluding electromotive ones
IA-44	Files for medical use	Excluding electromotive ones
IA-45	Levers for medical use	Excluding electromotive ones
IA-46	Snares for medical use	Excluding electromotive ones
IA-47	Needles for syringe and puncture needles	Excluding sterile ones
IA-48	Syringes	Excluding sterile ones

TABLE E.2 (CONTINUED)
Japanese MHLW Medical Devices Exempted From Approval on a Product-by-Product Basis

Category	Classification	Products
IA-49	Puncturing, cutting, and perforating instruments	Medical puncture instruments, dental burs, dental drills, dental reamers, dental files, dental root-canal filling spreaders and pluggers, dental mandrels, dental laboratory burs, dental laboratory mandrels, medical perforators (excluding sterile ones)
IA-50	Wound retractors and speculums	Excluding electromotive ones
IA-51	Rostral tubes and body-fluid guidance tubes	Metallic tubes and catheters (excluding electromotive ones), manual inflation devices for PTCA balloons
IA-52	Dilators	Excluding electromotive ones
IA-53	Directors, probes, and sounds for medical use	Excluding electromotive ones
IA-54	Applicators for medical use	
IA-55	Douche instruments for medical use	Manual-type douches (excluding sterile ones), washers
IA-56	Blood-donor sets and blood-transfusion sets	Blood-collecting tubes
IA-57	Vaccination instruments	
IA-58	Orthopedic equipment	Self-exercising nonpowered equipment with measuring device, ligament-tunneling devices, bone-setting and bone-operating implements, instruments for ligature tying and band pushing, appliances for cephalofacial photo, measuring devices for orthodontic use, orthodontic headgear, orthodontic chin straps
IA-60	Dental engines	Dental engines (excluding air and electric rotary cutting dental apparatus), dental laboratory handpieces, belt arms of dental engines, wrist and slip-joint K_4s for dental engines, dental engine belts
IA-61	Dental handpieces	
IA-62	Dental cutting instruments	Excluding automated scaler
IA-63	Broaches for dental use	
IA-64	Explorers for dental use	
IA-65	Filling instruments for dental use	
IA-66	Mixing apparatus for dental use	
IA-67	Moisture-excluding instruments for dental use	Excluding dental rubber-dam sheets
IA-68	Impression and articulation instruments	
IA-69	Vulcanizers and curing units for dental use	
IA-70	Casting machines for dental use	
IA-71	Sight-corrective spectacles	
IA-74	Infusion instruments	Manual injectors, manual powder blowers for medical use, dental gingival applicators
IA-75	Hernia supporters	
IA-80	Acupuncture and moxibustion apparatus	Acupuncture needles
IA-84	Accessories to the above products designated by the Ministerial Ordinance of Health, Labor, and Welfare	Film-viewing devices, medical X-ray fluorescent screens, medical X-ray-intensifying screens

Medical Products

Category	Classification	Products
MP-1	X-ray films	
MP-3	Surgical gloves and finger sacks	Finger sacks
MP-4	Orthopedic materials	Sterile nonwoven surgical dressings, gypsum bandages
MP-5	Splints	
MP-6	Test charts for visual acuity and color blindness	

(continued)

TABLE E.2 (CONTINUED)
Japanese MHLW Medical Devices Exempted From Approval on a Product-by-Product Basis

Category	Classification	Products
		Dental Material
DM-1	Dental metals	Dental fusible alloys
DM-3	Denture-base materials	Resins for impression trays, resins for dental patterns
DM-6	Dental-impression materials	Elastomeric precision impression materials, agar impression materials
DM-7	Dental waxes	
DM-8	Dental gypsums and gypsum products	
DM-9	Dental abrasive and polishing materials	

Source: Guide pp. 38-41.

TABLE E.3
Japanese MHLW Japanese Industrial Standards (JIS) for Medical Devices

Standard	Product
	Medical Electric Equipment
C 6310	Low frequency therapy equipment [a]
T 1001	General requirements for safety of medical electric equipment (Technically equivalent to IEC 60601-1)
T 1002	General rules of testing methods for safety of medical electric equipment (Technically equivalent to IEC 60601-1)
T 1005	Style manual for instructions manuals
T 1006	Graphical symbols for medical electrical equipment
T 1011	Glossary of terms used in medical electrical equipment (General terms)
T 1031	General rules for alarms of medical equipment
T 1107	Clinical electronic spirometers
T 1113	Phonocardiographs (PCG) [a]
T 1114	Vectorcardiographs (VCG) [a]
T 1115	Noninvasive electronic sphygmomanometers [a]
T 1116	Direct blood-pressure meters for clinical use [a]
T 1117	Ambulatory ECG recording systems (Holter systems) [a]
T 1140	Clinical electronic thermometers [a]
T 1141	Medical infrared thermographs [a]
T 1150	Electromyographs [a]
T 1160	Electronystagmographs [a]
T 1161	Electroretinographs [a]
T 1170	Clinical electronic spirometers [a]
T 1171	Rhinomanometers [a]
T 1190	Stabilometers [a]
T 1201-1	Audiometers – Part 1: Pure-time audiometers
T 1201-2	Audiometers – Part 2: Equipment for speech audiometry
T 1202	Electrocardiographs [a] (Technically equivalent to IEC 60601-2-25)
T 1203	Electroencephalographs [a] (Technically equivalent to IEC 60601-2-26)
T 1204	Laser photocoagulators [a]
T 1205	Ultrasonic ophthalmic biometers [a]
T 1206	Computer-aided perimeters [a]
T 1213	Photic stimulators for electroencephalographs [a]
T 1214	Medical treadmills [a]
T 1301	General requirements for patient monitors
T 1303	Fetal monitors [a]
T 1304	Electrocardiographic monitoring equipment [a] (Technically equivalent to IEC 60601-2-27)
T 1305	Direct blood-pressure monitors [a]
T 1306	Continuous measuring clinical electrical thermometers [a]
T 1307	Wireless ECG monitors [a]
T 1308	Neonatal monitors [a]
T 1309	Polygraphs for clinical use [a]
T 1353	Microwave therapy equipment [a]
T 1355	Defibrillators [a] (Technically equivalent to IEC 60601-2-4)
T 1356	External cardiac pacemakers [a]
T 1453	Electrosurgical units [a] (Technically Equivalent to IEC 60601-2-2)
T 1501	General methods of measuring the performances of ultrasonic pulse-echo diagnostic equipment
T 1503	A-mode ultrasonic diagnostic equipment [a]
T 1504	Manual scanning B-mode ultrasonic diagnostic equipment [a]
T 1505	M-mode ultrasonic diagnostic equipment [a]
T 1506	Ultrasonic Doppler fetal diagnostic equipment [a]
T 1507	Electronic linear scanning ultrasonic diagnostic equipment [a]
T 1553	Endoscopic apparatus for medical use [a]
T 1603	Electric motor-driven blood pump for cardiopulmonary bypass [a]

(continued)

TABLE E.3 (CONTINUED)
Japanese MHLW Japanese Industrial Standards (JIS) for Medical Devices

Standard	Product
T 1653	Infusion pumps [a]
T 1701	Centrifuge for medical use
T 1702	Incubator

General Medical Instruments and Apparatus (knives, clubs, forceps)

T 0201	Cut and durability test method for surgical knives
T 0202	Cut and durability test method for dressing scissors
T 2106	Bone chisels
T 2603	Medical hooks
T 2604	Medical sharp curettes
T 2605	Probes
T 2606	Periosteal raspatories
T 2607	Periosteal elevators
T 2608	Uterine dilators
T 2609	Uterine curettes, sharp
T 2610	Uterine curettes
T 2615	Ophthalmic retractors
T 2616	Eye knives
T 2617	Eye scissors
T 2619	Lacrimal probe
T 2802	Tonsilotomes

General Medical Instruments and Apparatus (needles, tubes, containers)

T 3101	Injection needles
T 3102	Surgical needle, sutures
T 3201	Glass syringes
T 3202	Urethral bougies
T 3203	Urethral catheters [a]
T 3204	Cannula for maxillary antrum and frontal sinus [a]
T 3205	Urethral injection syringe
T 3206	Urethral glass syringe
T 3208	Skin trephines [a]
T 3501	Alcohol-cotton case
T 3502	Cryocauter [a]
T 3505	Dressing trays

General Medical Instruments and Apparatus (sutures, meters, optical devices)

T 4101	Surgical silk sutures [a]
T 4102	Surgical suture catgut [a]
T 4201	Petechiometers [a]
T 4202	Standard scale for medical bougie and catheter
T 4203	Sphygmomanometer
T 4204	Hemocytometer
T 4205	Hemometers
T 4206	Clinical thermometers, mercury-in-glass, with maximum device
T 4402	Test lens set
T 4403	Anoscopes
T 4404	Abdominal speculum
T 4406	Reflectors with head band
T 4408	Gastroscopes
T 4901	Tuning forks (for medical use) [a]
T 4902	Surgical saws
T 4903	Bone files

TABLE E.3 (CONTINUED)
Japanese MHLW Japanese Industrial Standards (JIS) for Medical Devices

Standard	Product

Dental Instruments and Apparatus

Standard	Product
T 5106	Belts for electric dental engines
T 5109	Electric dental engines
T 5201	Dental burs
T 5202	Pulp canal reamers (Superseded by JIS T 5216)
T 5204	Dental mandrels
T 5205	Dental barbed broaches (Technically equivalent to ISO 3630-1)
T 5206	Dental cotton broaches (Technically equivalent to ISO 3630-1)
T 5207	Pulp canal reamers for angle hand-piece (Superseded by JIS T 5216)
T 5208	Peeso's root reamers for dental use
T 5209	Dental carborundum wheels
T 5210	Dental carborundum points (Technically equivalent to ISO 7786)
T 5211	Dental root canal K files
T 5212	Dental root canal H files
T 5213	Dental root canal paste carriers
T 5214	Dental root canal rasps
T5215	Dental root canal enlargers – Type G
T 5216	Dental root-canal reamers
T 5301	Dental rubber dam clamps
T 5302	Dental impression trays
T 5401	Dental pincettes
T 5402	Dental explorers (Technically equivalent to ISO 7492)
T 5404	Dental spoon excavators
T 5406	Dental scalers (Technically equivalent to ISO 13397-1 and 13397-3)
T 5407	Dental elevators
T 5408	Dental bone files
T 5409	Dental broach holders
T 5410	Dental forceps
T 5413	Dental curettes
T 5415	Gingival applicators
T 5416	Dental root canal filling spreaders
T 5417	Dental root canal filling pluggers
T 5418	Periodontal pocket probes
T 5419	Dental root canal finger pluggers
T 5420	Periodontal curettes – GR Type (Technically equivalent to ISO 13397-1 and –2)
T 5501	Dental rotary instruments – Number coding system
T 5901	Dimension for dental hand-piece
T 5902	Dental spittoons [a]
T 5903	Dental mirrors and dental mirror holders
T 5904	Dental hand-pieces – Coupling dimensions
T 5905	Dental hand-pieces – Hose connectors (Technically equivalent to ISO 9169)
T 5906	Dental hand-piece – Part 1: High-speed air-turbine hand-pieces
T 5907	Dental hand-piece – Part 2: Straight and geared angle hand-pieces (Technically equivalent to ISO 7785-1)

Dental Materials

Standard	Product
T 6101	Nickel-chromium alloy wires for dental use [a]
T 6102	Nickel-chromium alloy plates for dental use [a]
T 6103	Stainless steel wires for dental use [a]
T 6104	Cobalt-chromium alloy wires for dental use [a]
T 6105	Dental wrought gold-silver-palladium alloy [a]
T 6106	Dental casting gold-silver-palladium alloy [a]
T 6107	Dental gold-silver-palladium alloy solders for dental use [a]

(continued)

TABLE E.3 (CONTINUED)
Japanese MHLW Japanese Industrial Standards (JIS) for Medical Devices

Standard	Product
T 6108	Dental casting silver alloy [a]
T 6109	Dental amalgam alloy [a]
T 6110	Dental fusible alloy
T 6111	Dental silver solders [a]
T 6112	Dental mercury [a] (Technically equivalent to ISO 1560)
T 6113	Dental casting 14K gold alloy [a]
T 6114	Plus metals for dental casting 14K gold alloys [a]
T 6115	Dental casting cobalt-chromium alloy [a] (Technically equivalent to ISO 6871-1)
T 6116	Dental casting gold alloy [a] (Technically equivalent to ISO 1562)
T 6117	Dental gold alloy solders [a]
T 6118	Dental casting precious alloys for metal ceramics (Technically equivalent to ISO 9693)
T 6119	Dental brazing materials – Test method (Technically equivalent to ISO 9333)
T 6501	Acrylic denture base resin [a]
T 6502	Dental base plate wax
T 6503	Dental inlay casting wax (Technically equivalent to ISO 1561)
T 6504	Dental impression compound [a]
T 6505	Dental alginate impression material [a] (Technically equivalent to ISO 1563)
T 6506	Synthetic resin teeth [a] (Technically equivalent to ISO 3336)
T 6507	Dental temporary stopping [a]
T 6508	Dental heat curing acrylic resin for crown [a]
T 6509	Dental self-curing acrylic resin for crown [a]
T 6510	Dental base plates
T 6511	Porcelain teeth for plate denture [a] (Technically equivalent to ISO 4824)
T 6512	Dental agar impression material [a]
T 6513	Dental elastomeric impression material [a] (Technically equivalent to ISO 4823)
T 6514	Dental composite resin for filling [a] (Technically equivalent to ISO 4049)
T 6515	Dental root canal obturating points [a]
T 6516	Dental ceramic fused to metal to metal restorative materials (Technically equivalent to ISO 9693)
T 6517	Dental synthetic resins for crown and bridge (Technically equivalent to ISO 10477)
T 6518	Dental acrylic resins for crown and bridge
T 6519	Short-term resilient lining materials for removable dentures
T6520	Long term soft lining materials for removable dentures
T 6601	Gypsum-bonded dental investments for casting (Technically equivalent to ISO 7490)
T 6602	Dental zinc phosphate cement [a]
T 6603	Dental silicate cement [a] (Technically equivalent to ISO 9917)
T 6604	Dental plaster (Technically equivalent to ISO 6873)
T 6605	Dental stone (Technically equivalent to ISO 6873)
T 6606	Dental zinc polycarboxylate cements [a]
T 6607	Dental glass polyalkenoate cement [a]

Medical Apparatus, Utensils, etc.

Standard	Product
T 7201	Anesthetic equipment [a]
T 7202	Oxygen tents [a]
T 7203	Oxygen analyzers for medical use [a] (Technically equivalent to ISO 5356-1, ISO 7767, and IEC 60601-1)
T 7204	Lung ventilators for medical use [a] (Technically equivalent to ISO 5369 and ISO 8185)
T 7205	Medical (operator-powered) resuscitators [a]
T 7206	Gas-powered resuscitators [a]
T 7302	Schimmelbusch's bandage sterilizers
T 7303	Baby incubators [a]
T 7304	Gastrointestinal sewing tools
T 7305	Holding stretchers
T 7306	Ophthalmoscopes [a]
T 7307	Major amblyoscopes [a]
T 7308	Refractor heads [a]

TABLE E.3 (CONTINUED)
Japanese MHLW Japanese Industrial Standards (JIS) for Medical Devices

Standard	Product
T 7309	Visual acuity testing equipments [a]
T 7310	Chart projectors [a]
T 7311	Retinoscopes [a]
T 7312	Tonometers [a]
T 7313	Single-vision corrective lenses [a] (Technically equivalent to ISO 8980-1)
T 7314	Multifocal corrective lenses [a]
T 7315	Progressive power corrective lenses [a]
T 7316	Slit-lamp microscopes
T 7317	Operation microscopes
T 7318	Ophthalmometers [a]
T 7319	Refractometers [a]
T 7320	Fundus cameras [a]
T 7321	Hyperbaric oxygen chambers [a]
T 7322	High-pressure steam sterilizers for medical use [a]
T 7323	Ethylene oxide gas sterilizers for medical use [a]
T 7324	High-pressure steam sterilizers for medical use (small size) [a]
T 7325	Ethylene oxide gas sterilizers for medical use (small size) [a]
T 7326	Hermatocrit centrifuges (Cancelled)
T 7327	Electric suction equipment for medical use [a]

Miscellaneous Medical Devices and Sanitary Goods

Standard	Product
T 9101	Rubber sheets for medical purposes
T 9102	Rubber hot water bags
T 9103	Rubber ice water bottles
T 9104	Rubber ice water bags
T 9105	Rubber patient cushions
T 9106	Rubber nipples
T 9107	Surgical rubber gloves (Cancelled) See T 9113 through T 9116
T 9108	Finger sack for medical use
T 9111	Rubber condoms (Cancelled) See T 9111 Parts 1 to 10
T 9111-1	Rubber condoms – Part 1: Requirements (Technically equivalent to ISO 4074-1)
T 9111-2	Rubber condoms – Part 2: Determination of length (Technically equivalent to ISO 4074-2)
T 9111-3	Rubber condoms – Part 3: Determination of width (Technically equivalent to ISO 4074-3)
T 9111-5	Rubber condoms – Part 5: Testing for holes – Water leak test
T 9111-6	Rubber condoms – Part 6: Determination of bursting volume and pressure
T 9111-7	Rubber condoms – Part 7: Oven conditioning (Identical to ISO 4074-7)
T 9111-9	Rubber condoms – Part 9: Determination of tensile properties (Identical to ISO 4074-9)
T 9111-10	Rubber condoms – Part 10: Packaging and labelling (Technically equivalent to ISO 4074-10)
T 9112	Feeding bottles
T 9113	Single-use rubber glovers for dentistry (Technically equivalent to ISO 11193)
T 9114	Single-use polyvinyl chloride gloves for dentistry
T 9115	Single-use rubber examination gloves (Technically equivalent to ISO 11193)
T 9116	Single-use polyvinyl chloride examination gloves
T 9201	Manually-propelled wheelchairs (Technically equivalent to ISO 6440, ISO 7193, ISO 7176-1, -3, -8, -11, and -13)
T 9203	Motorized wheelchairs (Technically equivalent to ISO 6440 and ISO 7193)
T 9204	Wooden axilla crutches
T 9205	Hospital beds
T 9206	Wheelchairs – Part 21: Requirements and test methods for electromagnetic compatibility of electrically powered wheelchairs and motorized scooters
T 9231	Urine collection systems
T 9232	Glossary of terms used in ostomy aids (Technically equivalent to ISO 8670-1 and ISO 8670-2)
T 9233	Testing methods for ostomy aids (Technically equivalent to ISO 8670-2)

(continued)

TABLE E.3 (CONTINUED)
Japanese MHLW Japanese Industrial Standards (JIS) for Medical Devices

Standard	Product

Orthopedic Equipment

Standard	Product
T 0101	Glossary of terms used in prosthetics and orthoptics (Technically equivalent to ISO 8549-1, -2 and –3)
T 0111-1	Prosthetic – Structural testing of lower-limb prostheses – Part 1: Test configurations (Identical to ISO 10328-1)
T 0111-2	Prosthetic – Structural testing of lower-limb prostheses – Part 2: Test samples (Identical to ISO 10328-2)
T 0111-3	Prosthetic – Structural testing of lower-limb prostheses – Part 3: Principal structural tests (Identical to ISO 10328-3)
T 0111-4	Prosthetic – Structural testing of lower-limb prostheses – Part 4: Parameters for principal structural tests (Identical to ISO 10328-4)
T 0111-5	Prosthetic – Structural testing of lower-limb prostheses – Part 5: Supplemental structural tests (Identical to ISO 10328-5)
T 0111-6	Prosthetic – Structural testing of lower-limb prostheses – Part 6: Loading parameters for supplemental structural tests (Identical to ISO 10328-6)
T 0111-7	Prosthetic – Structural testing of lower-limb prostheses – Part 7: Test submission document (Identical to ISO 10328-7)
T 0111-8	Prosthetic – Structural testing of lower-limb prostheses – Part 8: Test report (Identical to ISO 10328-8)
T 9212	Artificial feet and ankle joints
T 9213	Artificial knees
T 9214	Metallic ankle joints for lower extremity orthoses
T 9215	Stirrups for lower extremity orthoses
T 9216	Metallic knee joints for lower extremity orthoses
T 9217	Voluntary opening hooks
T 9218	Voluntary opening hands
T 9219	Functional elbow units
T 9220	Outside locking elbow hinge units
T 9221	Control cable system units
T 9222	Wrist units
T 9223	Cosmetic gloves for artificial hands
T 9224	Cosmetic hands for arm prostheses

Radiation-related Apparatus

Standard	Product
C 3407	High tension cables for X-ray apparatus
Z 4004	Graphical symbols for medical radiological equipment
Z 4005	Medical radiology – Terminology
Z 4102	General rules on medical X-ray tubes [a]
Z 4301	X-ray film badges
Z 4302	R-ray and hard X-ray film badges
Z 4320	Thermoluminescence dosimeter system for X-ray and Gamma-ray personal monitoring
Z 4323	Film badge cases for X-rays, Gamma-rays and thermal neutrons
Z 4324	Area monitors for X- and Gamma-rays (Technically equivalent to IEC 60532)
Z 4331	Calibration phantom for X- and Gamma-ray personal dosimeters
Z 4332	General requirements for personal dosimeters X and Gamma radiation
Z 4333	Portable photon ambient dose equivalent ratemeters for radiation protection
Z 4606	Industrial X-ray apparatus for radiographic testing
Z 4607	Industrial X-ray apparatus for radiography (Non-portable type)
Z 4608	Rotational type cobalt 60 (60 Co) teletherapy apparatus [a]
Z 4613	Capacitor discharge X-ray apparatus for chest indirect radiography [a]
Z 4701	General rules for medical X-ray equipment (Technically equivalent to IEC 60601-1-3)
Z 4702	General rules of high potential generators of medical x-ray apparatus (Technically equivalent to IEC 60601-2-7 and IEC 60601-2-15)
Z 4703	General rules of mechanical units for medical X-ray apparatus (Technically equivalent to IEC 60601-2-32)
Z 4704	General rules on medical X-ray tube assemblies [a]
Z 4705	Medical electron accelerators
Z 4711	Mono-tank X-ray units for diagnostic use [a] (Technically equivalent to IEC 60601-1-3 and IEC 60601-2-7)
Z 4712	Variable X-ray beam limiting devices for diagnostic use [a]
Z 4713	Shielding devices for stand-by radiation [a]
Z 4721	Medical X-ray image intensifiers (Technically equivalent to IEC 61262-1 through –7)
Z 4731	High voltage cable plug and socket for medical X-ray equipment (Technically equivalent to IEC 60526)
Z 4732	High voltage cables with plugs for medical X-ray equipment (Technically equivalent to IEC 60526)

TABLE E.3 (CONTINUED)
Japanese MHLW Japanese Industrial Standards (JIS) for Medical Devices

Standard	Product
Z 4806	Medical X-ray protective screens [a]
Z 4808	Glove box for handling of radioactive substances
Z 4814	Protective boxes for mass chest indirect radiography [a]
Z 4816	Back scattering X-ray reducing paints
Z 4817	Radiation shielding lead bricks
Z 4818	Reducing materials for back scattering X-rays
Z 4821	Classification and testing of sealed radioactive sources
Z 4821-1	Sealed radioactive sources – General requirements and classification
Z 4821-2	Sealed radioactive sources – Leakage test method
Z 4831	Protective devices against diagnostic medical X-radiation
Z 4901	Mirror cameras for chest indirect radiography [a]
Z 4905	Radiographic cassettes (Technically equivalent to IEC 60406)
Z 4907	Film marker for radiography
Z 4908	Disc type number markers for photo fluorography
Z 4910	Anti-scatter grids (Technically equivalent to IEC 60627)
Z 4912	Dimensions for X-ray intensifying screens
Z 4915	X-ray water phantom for chest and abdomen
Z 4916	Resolution test charts for X-ray apparatus
Z 4917	Test charts for modulation transfer function of X-ray equipment
Z 4918	Viewing cabinets for medical X-ray films
Z 4920	Name printers for radiographic films
Z 4921	Measuring devices for X-ray tube voltage
Z 4922	Phantoms for single photon emission computed tomograph
Z 4923	Phantoms for X-ray computed tomography
Z 4924	Phantoms for magnetic resonance equipment for medical diagnosis
Z 4950	Graphical symbols and signs for magnetic resonance equipment for medical diagnosis
Z 4951	Particular requirements for the safety of magnetic resonance equipment for medical diagnosis (Identical to IEC 60601-2-33)

Miscellaneous	
C 6802	Radiation safety standards for laser products (Technically equivalent to IEC 60825-1)
K 0609	Methods for quantitative analysis of immobilized trypsin
K 0610	Tentative method for quantitative analysis by immunoturbidimetry of human serum albumin
K 3604	Medium for tissue culture (Minimum essential medium)
K 3605	General rules for autoclaving
T 0102	Glossary of terms used in rehabilitation equipment for physically disabled or handicapped persons
Z 2801	Antimicrobial products – Test for antimicrobial activity and efficacy
Z 8500	Ergonomics – Basic human body measurement for technological design

[a] A device complying with this Japanese Industrial Standard (JIS) does not require approval under Article 14, Paragraph 1 of the PAL providing the structure, usage, efficacy, effect, performance, etc. is not clearly different from an already approved device.

Sources: Guide pp. 41–43 and ILI Standards InfoBase, Updated 9 January 2003.

TABLE E.4
Japanese MHLW Medical Devices Subject to the Partial License
(*Kubun-kyoka*) **System**

Class A: Medical Devices Specified in No. 1 to 32 of Attached Table 1-3

Anesthesia apparatus and breathing bag and gas-absorption can for anesthesia apparatus

Respiratory-assistance apparatus

Artificial organs

Tongue depressor

Apparatus for ligature and suture

Knife for medical use

Scissor for medical use

Dressing forceps for medical use

Curette for medical use

Hook for medical use

Forceps for medical use

Saw for medical use

Chisel for medical use

Raspatory for medical use

Mallet for medical use

File for medical use

Lever for medical use

Snare for medical use

Injection or puncture needle

Syringe

Medical puncture, cutting, and perforating instrument

Retractor or speculum for medical use

Medical metallic tube and catheter

Medical dilator

Director, probe, and sound for medical use

Blood-collecting and -transfusion device

Orthopedic equipment

Infusion apparatus

Acupuncture and moxibustion apparatus

Silk surgical suture

Orthopedic equipment (excluding first-aid adhesive tape and nonwoven gauze)

Class B: Medical Devices Specified in No. 33 to 35 of Attached Table 1-3

Contraceptive

Surgical glove and finger sack for medical use

Orthopedic equipment (limited to first-aid adhesive tape and nonwoven gauze)

Sanitary tampon

Class B: Medical Devices Specified in No. 36 of Attached Table 1-3

Sight-corrective spectacles

Sources: Guide p 252.

TABLE E.5
Japanese MHLW Medical Devices Exempted from the Requirements of Medical Device Good Manufacturing Practice (GMP)

Category	Classification	Products
		Medical Instruments and apparatus
IA-1	Operating tables and treatment tables	Excluding electromotive or radiotherapy couches
IA-2	Medical lamps and lights	
IA-3	Medical disinfection apparatus	Boiling or steam disinfection apparatus, ultraviolet sterilizers
IA-7	Artificial organs	Sizers
IA-8	Incubators	Simple transportable incubators
IA-9	Medical X-ray equipment and X-ray tubes	Film changers, bucky devices, automatic processors for medical or dental film, medical imagers, patient restrainers, X-ray film cassettes, X-ray grids
IA-11	Radioactivity-protective equipment	X-ray-protective aprons, X-ray-protective gloves, X-ray-protective partitions, photofluorographic-protective boxes
IA-13	Stethoscopes	
IA-14	Percussion hammers	
IA-15	Tongue depressors	
IA-16	Clinical thermometers	Clinical mercury thermometers
IA-18	Blood-pressure and pulse-rate measurement instruments	Aneroid sphygmomanometers, mercurial sphygmomanometers
IA-21	Visceral-function testing instruments	Lead wires for ECG electrodes
IA-22	Ophthalmologic testing instruments	Trial lenses
IA-23	Audiometry equipment	Tuning forks for medical use
IA-24	Testing instruments and apparatus for perception organs or locomotion organs	Back dynamometers, goniometers, aethesiometers, hand dynamometers
IA-25	Scopes and speculums for medical use	Nonactive treatment devices for endoscopes, biological microscopes, colposcopes, operating microscopes, dental mirrors and mirror holders
IA-26	Centrifuges for medical use	
IA-27	Microtomes for medical use	Microtomes, strops for medical microtomes
IA-28	Thermostats for medical use	
IA-30	Apparatus for ligature and suture	Suture needles, needle holders, ligature carriers, suture instruments, automatic suture instruments, nonabsorbable wound clips, aneurysm needles
IA-32	Aspirators for medical use	Aspirators for manual use only
IA-34	Knives for medical use	Excluding electromotive ones
IA-35	Scissors for medical use	Excluding electromotive ones
IA-36	Dressing forceps for medical use	Excluding electromotive ones
IA-37	Curettes for medical use	Excluding electromotive ones
IA-38	Hooks for medical use	Excluding electromotive ones
IA-39	Forceps for medical use	Excluding electromotive ones
IA-40	Saws for medical use	Excluding electromotive ones
IA-41	Chisels for medical use	Excluding electromotive ones
IA-42	Raspatories for medical use	Excluding electromotive ones
IA-43	Mallets for medical use	Excluding electromotive ones
IA-44	Files for medical use	Excluding electromotive ones
IA-45	Levers for medical use	Excluding electromotive ones
IA-46	Snares for medical use	Excluding electromotive ones
IA-47	Needles for syringe and puncture needles	Excluding sterile ones
IA-48	Syringes	Excluding sterile ones

(continued)

TABLE E.5 (CONTINUED)
Japanese MHLW Medical Devices Exempted from the Requirements of Medical Device Good Manufacturing Practice (GMP)

Category	Classification	Products
IA-49	Puncturing, cutting, and perforating instruments	Medical puncture instruments, dental burs, dental drills, dental reamers, dental files, dental root-canal filling spreaders and pluggers, dental mandrels, dental laboratory burs, dental laboratory mandrels, medical perforators (excluding sterile ones)
IA-50	Wound retractors and speculums	Excluding electromotive ones
IA-51	Rostral tubes and body-fluid guidance tubes	Metallic tubes and catheters (excluding electromotive ones), manual inflation devices for PTCA balloons
IA-52	Dilators	Excluding electromotive ones
IA-53	Directors, probes, and sounds for medical use	Excluding electromotive ones
IA-54	Applicators for medical use	
IA-55	Douche instruments for medical use	Manual-type douches (excluding sterile ones), washers
IA-57	Vaccination instruments	
IA-58	Orthopedic equipment	Ligament tunneling device, bone setting and bone operating implement, instrument for ligature tying and band pushing, appliances for cephalofacial photo, measuring device for orthodontic use, orthodontic headgear, orthodontic chin strap
IA-60	Dental engines	Dental engines (excluding air and electric rotary cutting dental apparatus), belt arms of dental engines, wrist and slip-joint K_4s for dental engines, dental engine belts
IA-61	Dental handpieces	For dental laboratory use only
IA-62	Dental cutting instruments	Excluding automated scalers
IA-63	Broaches for dental use	
IA-64	Explorers for dental use	
IA-65	Filling instruments for dental use	
IA-66	Mixing apparatus for dental use	
IA-67	Moisture-excluding instruments for dental use	Excluding dental rubber-dam sheets
IA-68	Impression and articulation instruments	
IA-69	Vulcanizers and curing units for dental use	Excluding visible-light activators and ultraviolet activators
IA-70	Casting machines for dental use	
IA-71	Sight-corrective spectacles	
IA-72	Sight-corrective ophthalmic lenses	Monofocal vision-corrective lenses (limited to those conforming to JIS), multifocal vision-corrective lenses (limited to those conforming to JIS), progressive-power corrective lenses (limited to those conforming to JIS)
IA-74	Infusion instruments	Manual injectors, manual powder blowers for medical use, dental gingival applicators
IA-75	Hernia supporters	
IA-80	Acupuncture and moxibustion apparatus	Acupuncture needles
IA-84	Accessories to the above products designated by the Ministerial Ordinance of Health, Labor, and Welfare	Film-viewing devices, medical X-ray fluorescent screens, medical X-ray-intensifying screens

Medical Products

Category	Classification	Products
MP-1	X-ray films	
MP-2	Sutures	Silk sutures (excluding sterile ones)
MP-3	Surgical gloves and finger sacks	Finger sacks
MP-4	Orthopedic materials	Sterile nonwoven surgical dressings, gypsum bandages
MP-5	Splints	
MP-6	Test charts for visual acuity and color blindness	

TABLE E.5 (CONTINUED)
Japanese MHLW Medical Devices Exempted from the Requirements of Medical Device Good Manufacturing Practice (GMP)

Category	Classification	Products
		Dental Material
DM-1	Dental metals	Dental fusible alloys
DM-3	Denture-base materials	Resins for impression trays, resins for dental patterns
DM-6	Dental-impression materials	Elastomeric precision impression materials, agar impression materials
DM-7	Dental waxes	
DM-8	Dental gypsums and gypsum products	
DM-9	Dental abrasive and polishing materials	

Source: Guide pp. 271–273.

TABLE E.6
Japanese MHLW Medical Devices to be Approved by Prefectural Governor

Category	Classification	Products
		Medical Instruments
IA-47	Needles for syringe and puncture needles	Injection needles other than disposable injection needles, with needles made of cold strip stainless-steel sheets under SUS 304, SUS 304L, or SUS 321 of Japanese Industrial Standard G 4305
IA-48	Glass syringes	Syringes made of glass
IA-62	Dental cutting instruments	Scalers for removal of dental calculus and bacterial plaque, with blade made of stainless-steel bar under SUS 420J2, SUS 420F, SUS 440A, SUS 440B, SUS 440C, or SUS 440F of Japanese Industrial Standard G4303 and with a handle made of stainless-steel bar under SUS 304, SUS 304L, SUS 304LN, SUS 305, SUS 309S, SUS 310S, SUS 316, SUS 316L, SUS 316LN, SUS 316 J1L, SUS 317, SUS 317L, SUS 317J1, SUS 321, SUS 347, SUS 420J2, SUS 420F, SUS 440A, SUS 440B, SUS 440C, SUS 440F, SUSXM7, or SUSXM 15J1 of Japanese Industrial Standard G4303
IA-65	Dental filling instruments	Composite resin carriers for injection of composite resin into the hole in dental repair, made of aluminum, brass, synthetic resin, or stainless steel
IA-68	Impressions or articulation instruments	Impression trays for the teeth and jaw, made of aluminum, brass, stainless steel, or britannia metal
IA-72	Vision-corrective lenses	Single-vision corrective lenses made of plastic with refractive index not less than 1.40 and not more than 1.89 with: spherical refractive power from -60.0 D to +60.0 D; cylindrical refractive power from -60.0 D to +60.0 D; prismatic refractive power from 0.00 D to +60.0 D; measurement of lenses: round lenses from 30.0 mm to 100.0 mm in diameter; other lenses: from 10.0 mm to 150.0 mm across in any direction; thickness at the center from 0.4 mm to 30.0 mm

Source: Guide p. 259.

TABLE E.7

Japanese MHLW Devices for Which Labeling Exceptions are Permitted

Category	Classification	Products
Medical Instruments		
IA-14	Percussion hammers	
IA-15	Tongue depressors	
IA-25	Scopes and speculums for medical use	Dental mirrors
IA-30	Apparatus for ligature and suture	Suture needles, needle holders, ligature carriers, suture instruments
IA-34	Knives for medical use	
IA-35	Scissors for medical use	
IA-36	Dressing forceps for medical use	
IA-37	Curettes for medical use	
IA-38	Hooks for medical use	
IA-39	Forceps for medical use	
IA-40	Saws for medical use	
IA-41	Chisels for medical use	
IA-42	Raspatories for medical use	Periosteal elevators and raspatories
IA-43	Mallets for medical use	
IA-44	Files for medical use	
IA-45	Levers for medical use	Dental elevators
IA-46	Snares for medical use	Snares and excision instruments
IA-49	Puncturing, cutting, and perforating instruments	Perforators, disposable lumbar-puncture instruments, disposable thorax-puncture instruments, reusable puncture instruments, dental burs, mandrels, reamers, root-canal files
IA-50	Wound retractors and speculums	Entropion forceps, eye speculums and fixation forceps, mouth speculums, ear speculums, nasal speculums, vaginal speculums, anoscopes, wound retractors
IA-52	Dilators	
IA-53	Directors, probes, and sounds for medical use	
IA-54	Applicators for medical use	
IA-62	Dental cutting instruments	Chisels, excavators, scalers, ultrasonic scalers
IA-63	Broaches for dental use	Broaches, cleansers
IA-64	Explorers for dental use	
IA-65	Filling instruments for dental use	Plastic filling instruments, amalgam pluggers, amalgam carvers, mallets, amalgam carriers, composite resin carriers
IA-66	Mixing apparatus for dental use	Mixing spatulas, cement-mixing tablets, amalgamators, mixers for impression materials
IA-67	Moisture-excluding instruments for dental use	Hot-air syringes, rubber-dam clamps, rubber-dam punches, clamp forceps, rubber-dam frames
IA-68	Impression and articulation instruments	Impression trays for dentulous instruments, impression trays for edentulous cases, partial impression trays, partial-impression trays (adjustable type), articulators, face bows
IA-71	Sight-corrective spectacles	Spectacles for nearsightedness, spectacles for farsightedness, spectacles for astigmatism, spectacles for presbyopia, special spectacles
IA-72	Sight-corrective ophthalmic lenses	Monofocal ophthalmic lenses of optical glass, monofocal ophthalmic lenses of plastic, multifocal ophthalmic lenses of optical glass, multifocal ophthalmic lenses of plastic, hard contact lenses, soft contact lenses, therapeutic contact lenses, diagnostic contact lenses, intraocular lenses (IOLs)
Medical Supplies		
MP-4	Orthopedic materials	Metallic artificial bones, polymeric artificial bones, ceramic artificial bones, metallic artificial joints, polymeric artificial joints, ceramic artificial joints, bone-setting plates, bone nails, bone-setting screws, bone cements, polymeric materials, waxes, adhesives, mammary prostheses, artificial tympanic membranes, gypsums for medical use, gypsum bandages, dental implants, first-aid bandages
MP-5	Splints	

Source: ERPALs, Article 61, Attached Table 4.

Appendix F:
U.S. Department of Health and Human Services Medical Device Classification

Under authority of Section 513 of the Federal Food, Drug, and Cosmetic Act (FD&CA), as amended, the Secretary of the Department of Health and Human Services (DHHS) shall classify medical devices intended for human use into one of three categories based on the required level of regulatory controls necessary to protect public health and safety. For this purpose, the safety and effectiveness of a device are determined:

- With respect to the persons for whose use the device is represented or intended;
- With respect to the conditions of use prescribed, recommended, or suggested in the labeling of the device; and
- Weighing any probable benefit to health from use of the device against any probable risk of injury or illness from such use.

The effectiveness of a device is determined on the basis of well-controlled investigations, including clinical investigations where appropriate, by experts qualified by training and experience to evaluate the effectiveness of the device. To be considered effective, qualified experts must be able to conclude fairly and responsibly that the device will have the effect it purports, or is represented to have, under the conditions of use prescribed, recommended, or suggested in the labeling.

The Secretary may determine that there exists valid scientific evidence (other than evidence derived from well-controlled investigations) that is sufficient to determine the effectiveness of a device. With this information, a qualified expert can conclude that the device will have the effect it purports, or is represented to have, under the conditions of use prescribed, recommended, or suggested in the labeling. The Secretary may authorize the effectiveness of the device to be determined on the basis of such evidence.

To assist in the process of classifying devices, the Secretary may establish panels of experts who are qualified by training and experience to evaluate the safety and effectiveness of the devices to be referred to the panel. The panels are organized according to the various fields of clinical medicine and fundamental sciences in which devices intended for human use are used.

When reviewing a device, a panel, to the maximum extent practicable, shall provide an opportunity for interested persons to submit data and views on the classification of the device. Upon completion of its review, the panel submits its recommendation to the Secretary. The Secretary, upon receipt of the panel's recommendations, publishes in the *Federal Register* the panel's recommendation and a proposed regulation classifying the device. After appropriate review of public comments, the Secretary publishes a final regulation.

The classification process is carried out by the Commissioner of Food and Drugs, or his/her designee, under authority from the Secretary of DHHS. The procedures for implementing the classification process are found in 21 CFR Part 860.

The regulations classifying medical devices are published in 21 CFR Parts 862 through 892. Currently, there are sixteen categories, each of which is divided into several subparts. These sixteen categories together contain classifications for over 1,600 medical devices. New device classifications are being added regularly. In the Safe Medical Devices Act (SMDA), Congress encouraged the FDA to reclassify Class III devices as Class II devices. Therefore, the reader is encouraged to check the latest version of 21 CFR for information regarding specific device classification. For reference, the sixteen categories are summarized in the Table F.1.

TABLE F.1
U.S. DHHS Medical Device Classifications

Part	Description
862	Clinical Chemistry and Clinical Toxicology Devices; Subparts

B—Clinical Chemistry Test Systems: This subpart covers systems intended to detect or measure the level or concentration of specific chemicals or compounds produced by various natural and disease processes. These are predominantly Class I and Class II devices. The human chorionic gonadotropin (hCG) test system intended for any use other than the early detection of pregnancy is Class III (§862.1155).

C—Clinical Laboratory Instruments: This subpart covers general-purpose laboratory systems that are labeled or promoted for specific medical use. These are Class I devices.

D—Clinical Toxicology Test Systems: This subpart covers systems intended to measure or monitor the levels of toxic substances in human specimens. These are predominantly Class II devices.

| 864 | Hematology and Pathology Devices; Subparts |

B—Biological Stains: This subpart covers dyes and chemical-solution stains for medical purposes. These are predominately Class I devices. Immunohistochemistry reagents and kits intended for uses other than those described in the regulation are Class III devices (§864.1860(b)(3)).

C—Cell and Tissue Culture Products: This subpart covers cells and tissue media, supplies, and equipment used to culture animal and human cells and tissues for medical purposes. These devices are Class I.

D—Pathology Instrumentation and Accessories: This subpart covers equipment that is used to prepare, store, transport, and process tissue specimens. These are Class I devices.

E—Specimen-Preparation Reagents: This subpart covers reagents used to collect, prepare, and examine specimens from the human body for diagnostic histopathology, cytology, and hematology. These are predominately Class I devices. Analyte-specific reagents intended as a component in a test intended for use in the diagnosis of a contagious condition that is highly likely to result in a fatal outcome and for which prompt, accurate diagnosis offers the opportunity to mitigate the public-health impact of the condition are Class III devices (§864.4020(b)(3)).

F—Automated and Semiautomated Hematology Devices: This subpart covers automated and semiautomated devices used to analyze blood. These are predominantly Class II devices. The automated differential cell counter, when intended to flag or identify and classify one or more of the formed elements of blood, is a Class II device. The intended use of the automated differential cell counter for any other purpose, including to count or classify abnormal blood cells, renders it a Class III device (§864.5220(b)(2)).

G—Manual Hematology Devices: This subpart covers manually operated devices used to analyze blood. These are Class I and Class II devices.

H—Hematology Kits and Packages: This subpart covers devices intended to detect and measure specific compounds or properties in blood and blood serum. These are predominantly Class II devices.

I—Hematology Reagents: This subpart covers reagents used in the analysis of blood and plasma. These are Class I and Class II devices.

J—Products Used In Establishments That Manufacture Blood and Blood Products: This subpart covers devices and supplies used in the collection and processing of blood and blood products. These are predominantly Class II devices. The electromagnetic blood and plasma warning device (§864.9205(b)(2)) and the automated blood-cell separator (§864.9245) are Class III devices.

TABLE F.1 (CONTINUED)
U.S. DHHS Medical Device Classifications

Part	Description

866 Immunology and Microbiology Devices; Subparts

B—Diagnostic Devices: This subpart covers devices intended to determine the susceptibility of bacterial pathogens to therapeutic agents. These are Class II devices.

C—Microbiology Devices: This subpart covers devices and culture media intended to cultivate and identify microorganisms. These are predominantly Class I devices. The oxidase screening test for gonorrhea (§866.2420) is a Class III device.

D—Serological Reagents: This subpart covers reagents that are used to identify specific bacteria. These are predominantly Class I and Class II devices. The gonococcal antibody test (GAT) (§866.3290) and the herpes simplex virus serological reagents (§866.3305) are Class III devices.

E—Immunology Laboratory Equipment and Reagents: This subpart covers devices and reagents used in the analysis of proteins and antigen-antibody reactions. These are Class I devices.

F—Immunological Test Systems: This subpart covers devices and reagents used to measure specific substances in serum or other body fluids by immunochemical techniques. These are Class I and Class II devices.

G—Tumor-Associated Antigen Immunological Test Systems: This subpart covers devices intended to aid in the detection and management of tumors. These are Class II devices. As of April 1, 2002, the only device in this category is the tumor-associated antigen immunological test system (§866.6010).

868 Anesthesiology Devices; Subparts

B—Diagnostic Devices: This subpart covers devices used to measure the concentration of various gases and to analyze various patient parameters such as tidal volume, etc. These are predominantly Class II devices. The indwelling blood oxyhemoglobin concentration analyzer (§868.1120), indwelling blood carbon dioxide partial pressure (PCO_2) analyzer (§868.1150), indwelling blood hydrogen ion concentration (pH) analyzer (§868.1170), and the indwelling blood oxygen partial pressure (PO_2) analyzer (§868.1200) are Class III devices.

C—Monitoring Devices: This subpart covers devices used to control anesthetic gases and monitor the effects of anesthetic drugs and gases. These are predominantly Class II devices. The lung-water monitor that monitors the fluid volume changes in a patient's lung by measuring changes in thoracic electrical impedance by means of electrodes placed on the patient's chest is a Class III device (§868.2450).

F—Therapeutic Devices: This subpart covers devices used to administer anesthesia gases, control the breathing circuit, or provide support for the patient's breathing. These are predominantly Class II devices. Electroanesthesia apparatus (§868.5400), and the membrane lung for long-term pulmonary support (§868.5610) are Class III devices.

G—Miscellaneous: This subpart covers devices used to support anesthesiology procedures. These are predominantly Class I devices.

870 Cardiovascular Diagnostic Devices; Subparts

B—Cardiovascular Diagnostic Devices: This subpart covers devices used in the detection or diagnosis of cardiovascular disease. These are predominantly Class II devices. The arrhythmia detector and alarm (§870.1025), the catheter balloon repair kit (§870.1350), and the trace microsphere (§870.1360) are Class III devices.

C—Cardiovascular Monitoring Devices: This subpart covers devices used to monitor the cardiovascular system. These are Class II devices.

D—Cardiovascular Prosthetic Devices: This subpart covers devices that assist the functioning of the cardiovascular system either permanently or on an emergency basis. These are predominantly Class III devices.

E—Cardiovascular Surgical Devices: This subpart covers devices and instruments used during cardiovascular surgical procedures. These are predominantly Class II and Class III devices.

F—Cardiovascular Therapeutic Devices: This subpart covers devices that are used primarily to treat cardiovascular problems externally, e.g., external defibrillators. These are Class II and Class III devices.

872 Dental Devices; Subparts

B—Diagnostic Devices: This subpart covers devices used in the examination and diagnosis of diseases of the teeth, jaw, and oral structures. These are predominantly Class II devices.

D—Prosthetic Devices: This subpart covers materials and devices used in the repair or replacement of teeth. These are predominantly Class II and Class III devices.

E—Surgical Devices: This subpart covers devices, instruments, and accessories used in oral surgery. This subpart also includes powered devices used during dental cleaning. These are predominantly Class II devices.

F—Therapeutic Devices: This subpart covers devices primarily used during orthodontic treatment. These are predominantly Class I devices.

G—Miscellaneous Devices: This subpart covers materials and devices used in dental procedures or for the care of the teeth and gums, e.g., dental floss. These are predominantly Class I devices. The endodontic dry-heat sterilizer (§872.6730) is a Class III device.

(continued)

TABLE F.1 (CONTINUED)
U.S. DHHS Medical Device Classifications

Part	Description

874   Ear, Nose, and Throat Devices; Subparts

       B—Diagnostic Devices: This subpart covers devices used in auditory testing. These are Class I and Class II devices.

       D—Prosthetic Devices: This subpart covers materials and devices used to compensate for impaired hearing or to supplement or replace structures in the ear, nose, and throat. These are predominantly Class II devices. The endolymphatic shunt tube with valve (§874.3850) is a Class III device.

       E—Surgical Devices: This subpart covers devices and instruments used during eye, ear, and throat surgery. These are Class I and Class II devices.

       F—Therapeutic Devices: This subpart covers devices to treat conditions of the ear, nose, and throat including antichoke devices. These are predominately Class I devices. The suction antichoke device (§874.5350) and the tongs antichoke device (§874.5370) are Class III devices.

876   Gastroenterology and Urology Devices; Subparts

       B—Diagnostic Devices: This subpart covers devices used for the diagnosis of conditions of the gastrointestinal and urology systems. These are Class I and Class II devices.

       C—Monitoring Devices: This subpart covers devices used to monitor functioning of the gastrointestinal and urology systems. These are Class II devices.

       D—Prosthetic Devices: This subpart covers implanted prosthetics such as the penile implant. Of the three devices in this part, two are Class III: the penile inflatable implant (§876.3350) and the testicular prosthesis (§876.3750).

       E—Surgical Devices: This subpart covers devices and instruments used during gastrointestinal and urology surgery. These are predominantly Class II devices.

       F—Therapeutic Devices: This subpart covers devices used to facilitate drainage, dilate gastrointestinal and urology tracts, or control urinary continence. These are predominantly Class II and Class III devices.

878   General and Plastic-Surgery Devices; Subparts

       B—Diagnostic Devices: This subpart covers devices that are inserted into a body cavity to aid observation. These are Class I devices.

       D—Prosthetic Devices: This subpart covers materials and devices used in reconstructive surgery. These are predominantly Class I and Class II devices. The silicone inflatable breast prosthesis (§878.3530) and the silicone gel-filled breast prosthesis (§878.3540) are Class III devices.

       E—Surgical Devices: This subpart covers devices, materials, and instruments used during surgical procedures. These are Class I and Class II devices. Absorbable powder for lubricating a surgeon's glove (§878.4480), absorbable hemostatic agents and dressings (§878.4490), and polytetrafluoroethylene injectables (§878.4520) are Class III devices.

       F—Therapeutic Devices: This subpart covers devices used in and around the operating room, for removal of hair, and for treatment of chronic skin ulcers or bed sores. These are predominantly Class II devices. The topical oxygen chamber for extremities (§878.5650) is a Class III device.

880   General Hospital and Personal-Use Devices; Subparts

       C—General Hospital and Personal-Use Monitoring Devices: This subpart covers devices used for monitoring patients in a general hospital environment, e.g., thermometers, scales, etc. These are predominantly Class II devices.

       F—General Hospital and Personal-Use Therapeutic Devices: This subpart covers devices used to care for patients in a general hospital environment, e.g., hospital beds, bandages, IV containers, etc. These are predominately Class I and Class II devices. The chemical cold pack snakebite kit (§880.5760) is a Class III device.

       G—General Hospital and Personal-Use Miscellaneous Devices: This subpart covers a variety of devices used for medical purposes. These are Class I and Class II devices.

882   Neurological Devices; Subparts

       B—Neurological Diagnostic Devices: This subpart covers devices and instruments used for diagnosing neurological disorders. These are predominantly Class II devices. The ocular plethysmograph (§882.1790) and the rheoencephalograph (§882.1825) are Class III devices.

       E—Neurological Surgical Devices: This subpart covers devices and instruments used during neurosurgical procedures. These are Class I and Class II devices.

       F—Neurological Therapeutic Devices: This subpart covers devices used for treating disorders of the brain, spinal cord, or other parts of the nervous system, or to repair skull defects. These are Class II and Class III devices.

TABLE F.1 (CONTINUED)
U.S. DHHS Medical Device Classifications

Part	Description

884 Obstetrical and Gynecological Devices; Subparts

B—Obstetrical and Gynecological Diagnostic Devices: This subpart covers devices used for gynecological or fetal examinations. These are predominately Class II devices. The transabdominal amnioscope (fetoscope) and its accessories (§884.1600) are Class III devices.

C—Obstetrical and Gynecological Monitoring Devices: This subpart covers devices used primarily for fetal monitoring. These are predominantly Class II devices. The obstetric data analyzer (fetal status data analyzer) (§884.2050), the fetal electroencephalographic monitor (§884.2620), the fetal scalp clip electrode and applicator (§884.2685), the elethermographic system intended for use alone in diagnostic screening for detection of breast cancer or other uses (§884.2980(b)(2), and the liquid crystal thermographic system intended for use alone in diagnostic screening for detection of breast cancer or other uses (§884.2982(b)(2)) are Class III devices.

D—Obstetrical and Gynecological Prosthetic Devices: This subpart covers devices intended to facilitate draining or support of pelvic organs following surgery. These are Class II devices.

E—Obstetrical and Gynecological Surgical Devices: This subpart covers devices and instruments used in obstetric-gynecologic surgery. These are predominantly Class II devices. The expandable cervical dilator (§884.4250) and the vibratory cervical dilator (§884.4270) are Class III devices.

F—Obstetrical and Gynecological Therapeutic Devices: This subpart covers various devices used in obstetric-gynecologic procedures including prevention of pregnancy, e.g., condoms, and intrauterine contraceptive devices (IUCDs). These are Class II and Class III devices.

886 Ophthalmic Devices; Subparts

B—Diagnostic Devices: This subpart covers devices used to measure and evaluate conditions of the eye. These are Class I and Class II devices.

D—Prosthetic Devices: This subpart covers devices used temporarily or permanently to treat conditions such as glaucoma—as permanent replacements, or to facilitate recovery following surgery. These are predominately Class II devices. The intraocular lens (IOL) (§886.3600) is a Class III device.

E—Surgical Devices: This subpart covers devices and instruments used for ophthalmic surgery. These are predominantly Class II devices. Intraocular gas (§886.4270), intraocular fluid (§886.4275), and the intraocular-pressure measuring device (§886.4280) are Class III devices.

F—Therapeutic Devices: This subpart covers devices and material used for various conditions, e.g., prescription spectacle lenses, contact lenses, etc. These are predominantly Class I and Class II devices. Rigid gas-permeable contact lenses intended for extended wear (§886.5916(b)(2)), and soft (hydrophilic) contact lenses intended for extended wear (§886.5925(b)(2)) are Class III devices.

888 Orthopedic Devices; Subparts

B—Diagnostic Devices: This subpart covers devices used to evaluate joints and neuromuscular functions. These are Class I and Class II devices.

D—Prosthetic Devices: This subpart covers devices used for the repair or replacement of bones and joints. These are Class II and Class III devices.

E—Surgical Devices: This subpart covers devices and instruments used during orthopedic surgery. These are predominantly Class I devices.

890 Physical Medicine Devices; Subparts

B—Physical Medicine Diagnostic Devices: This subpart covers various diagnostic devices used for medical purposes. These are predominantly Class II devices.

D—Physical Medicine Prosthetic Devices: This subpart covers various devices used to support, protect, or aid the patient. These are predominantly Class II devices. The rigid pneumatic structure orthosis (§890.3610) and the stair-climbing wheelchair (§890.3890) are Class III devices.

F—Physical Medicine Therapeutic Devices: This subpart covers devices that provide a variety of therapies for medical conditions. These are predominantly Class II and Class III devices. Microwave diathermy for uses other than applying therapeutic deep heat for selected medical conditions (§890.5275(b)(2)), shortwave diathermy for uses other than applying therapeutic deep heat for selected medical conditions (§890.5290(b)(2)), ultrasonic diathermy for uses other than applying therapeutic deep heat for selected medical conditions (§890.5300(b)(2)), the iontophoresis device used for purposes other than those specified in the identification regulations (§890.5525(b)(2)), and ultrasound and muscle stimulator for uses other than applying therapeutic deep heat for selected medical conditions (§890.5860(b)(2)) are Class III devices.

(continued)

TABLE F.1 (CONTINUED)
U.S. DHHS Medical Device Classifications

Part	Description

892   Radiology Devices; Subparts

B—Diagnostic Devices: This subpart covers devices intended to present images of body structures from analysis of reflected energy or energy emitted by radionuclides. These are predominantly Class II devices. The transilluminator for breast evaluation device (§892.1990) is a Class III device.

F—Therapeutic Devices: This subpart covers devices that provide various kinds of radiation therapy. These are predominantly Class II devices.

G—Miscellaneous Devices: This subpart covers personal protection equipment that protects the patient, operator, or other personnel from unnecessary exposure to radiation. These are Class I devices.

Source: 21 CFR, *Parts 800 to 1299* pp. 177–506

Appendix G:
Useful Web Sites

ANMAT
http://www.anmat.gov.ar/

Australian Therapeutic Goods Administration (TGA)
http://www.health.gov.au/tga/

Brazil
http://portalweb02.saude.gov.br/saude/

Canadian Therapeutic Products Directorate
http://www.hc-sc.gc.ca/hpb-dgps/therapeut/htmleng/index.html

European Active Implant Harmonized Standards
http://europa.eu.int/comm/enterprise/newapproach/standardization/harmstds/reflist/implmedd.html

European Committee for Electrotechnical Standardization (CENELEC)
http://www.cenelec.org/

European Committee for Standardization (CEN)
http://www.cenorm.be/

European Electromagnetic Compatibility Harmonized Standards
http://europa.eu.int/comm/enterprise/newapproach/standardization/harmstds/reflist/emc.html

European *In Vitro* Diagnostic Medical Device Harmonized Standards
http://europa.eu.int/comm/enterprise/newapproach/standardization/harmstds/reflist/invimedd.html

European Medical Device Harmonized Standards
http://europa.eu.int/comm/enterprise/newapproach/standardization/harmstds/reflist/meddevic.html

European Packaging and Packaging Waste Harmonized Standards
http://europa.eu.int/comm/enterprise/newapproach/standardization/harmstds/reflist/packagin.html

European Personal Protective Equipment Harmonized Standards
http://europa.eu.int/comm/enterprise/newapproach/standardization/harmstds/reflist/ppe.html

European Radio Equipment and Telecommunications Terminal Equipment Harmonized Standards
http://europa.eu.int/comm/enterprise/newapproach/standardization/harmstds/reflist/radiotte.html

Global Harmonization Task Force
http://www.ghtf.org/

International Electrotechnical Commission (IEC)
http://www.iec.ch/

International Standards Organization (ISO)
http://www.iso.ch/iso/en/ISOOnline.openerpage

Interarntional Standard Paper Sizes
http://www.cl.cam.ac.uk/~mgk25/iso-paper.html

Japanese Ministry of Health, Labor and Welfare (MHLW)
http://www.mhlw.go.jp/english/index.html

Korea Food and Drug Administration
http://www.kfda.go.kr/

Mexico
http://www.ssa.gob.mx/

Thai Food and Drug Administration (TFDA)
http://www.fda.moph.go.th/enginfo.htm

US Food and Drug Administration (FDA)
http://www.fda.gov/default.htm

US FDA, Center for Devices and Radiological Health (CDRH)
http://www.fda.gov/cdrh/index.html

US Trade Statistics
http://www.ita.doc.gov/td/mdequip/384stats.html

Glossary of Acronyms

a.c. Alternating current

AIDS Acquired Immune Deficiency Syndrome

AIMD Active Implantable Medical Device

AIMDD Active Implantable Medical Device Directive (European Union)

ALADI *Asociation Lationamericana de Integracion*

ANMAT National Administration of Drugs, Foods, and Medical Technology (Argentina)

ANVISA *Agência Nacional de Vigilância Sanitária* (Brazilian National Health Vigilance Agency)

AQIS Australian Quarantine and Inspection Service

AQSIQ State General Administration for Quality Supervision and Inspection and Quarantine (China)

ARTG Australian Register of Therapeutic Goods

AUA Authorized User Approval (Australia)

BCE Bureau of Compliance and Enforcement (Canada)

BCP Brazilian Customer Protection

BGTD Biologics and Genetic Therapies Directorate (Canada)

CAB Conformity Assessment Branch (Australia)

CAT Computerized Axial Tomography

CC&CR Canadian Consumer Chemicals and Containers Regulations

CCC China Compulsory Certification

CDRH Center for Devices and Radiological Health (United States)

CEC Commission of the European Communities

CEN *Comité Européene de Normalisation* (European Committee for Standardization)

CENELEC *Comité Européene de Normalisation Electrotrotechnique* (European Committee for Electrotechnical Standardization)

CFC Chlorofluorocarbon

CFG Certificate for Foreign Government

CFR Code of Federal Regulations (United States)

cGMP Current Good Manufacturing Practice

CMD Certification Commission for Medical Devices (China)

CMDR Canadian "Medical Device Regulations"

CME Continuing Medical Education

CNCA Certification and Accreditation Administration (China)

COE Certificate of Exportability (United States)

CPCS Compulsory Product Certification System (China)

CPR Controlled Products Regulations (Canada)

CPSC Consumer Product Safety Commission (United States)

CRT Cathode Ray Tube

CSA Canadian Standards Association

CSE Sterility Safety Coefficient

CT Computed Tomography

CTDI Computed Tomography Dose Index

CTE Clinical Trial Exemption (Australia)

CTN Clinical Trial Notification (Australia)

d.c. Direct current

DCB Designated Certification Bodies (China)

DHHS Department of Health and Human Services (United States)

DLCB Device Labeling Compliance Branch (United States)

DNA Deoxyribonucleic Acid

DOHA Department of Health and Aging (Australia)

DPIE Department of Primary Industries and Energy (Australia)

DSD Classification, Packaging and Labelling of Dangerous Substances Directive (European Union)

EAS Electronic Article Surveillance

EC European Communities (The EC was succeeded by the European Union on November 1, 1993)

ECG Electrocardiograph

EEA European Economic Area

EEC European Economic Community

EEG Electroencephalogram

EFTA European Free Trade Association

EMA Mexican Accreditation Entity

EMC Electromagnetic Compatibility

EMCD Electromagnetic Compatibility Directive (European Union)

EMEA European Agency for the Evaluation of Medicinal Products

EMI Electromagnetic Interference

EN European Norm

ENG Electronstagmograph

EOPAL Enforcement Ordinance of the Pharmaceutical Affairs Law (Japan)

EPA Environmental Protection Agency (United States)

ER Essential Requirement (from Annex I of an EU medical device directive)

ER&EA Export Reform and Enhancement Act of 1996 (United States)

ERG Electroretinograph

ERPALs Enforcement Regulations of the Pharmaceutical Affairs Law (Japan)

EtO Ethylene oxide

EU European Union

F&DA Food and Drugs Act, as amended (Canada)

FCC Federal Communications Commission (United States)

FCPA Federal Caustic Poison Act (United States)

FD&CA Federal Food, Drug, and Cosmetic Act, as amended (United States)

FDA Food and Drug Administration (United States)

FDAMA Food and Drug Administration Modernization Act of 1997 (United States)

FHSA Federal Hazardous Substances Act (United States)

FPLA Fair Packaging and Labeling Act (United States)

FTCA Federal Trade Commission Act (United States)

GCP Good Clinical Practice

GHTF Global Harmonization Task Force

GMP Good Manufacturing Practice

HAM Amateur Radio

HBV Hepatitis B Virus

HCFC Hydrochlorofluorocarbon

HCI Hydrochloric Acid

HCV Hepatitis C Virus

HDs Harmonization Documents (European Union)

HDE Humanitarian Device Exemption (United States)

HGC Human Chorionic Gonadotropin

HIV Human Immunodeficiency Virus

HPB Health Protection Branch (Canada)

HPFB Health Products and Food Branch (Canada)

HTSUS Harmonized Tariff Schedule of the United States

HTLV Human T-lymphotropic Virus

HUD Humanitarian Use Device (United States)

Hz Hertz

IDE Investigational Device Exemption (United States)

IEC International Electrotechnical Commission

IFU Instructions For Use

IMSS *Institito Mexicano del Seguro Social* (Mexico)

INPI National Industrial Property Institute (Brazil)

IOL Intraocular Lens

IPU Individual Patient Use (Australia)

IRB Institutional Review Board

ISM Industrial, Scientific, and Medical

ISO International Standards Organization

ISSSTE Mexican Institute for Healthcare for Government Employees

IUCD Intrauterine Contraceptive Device

IVD *In Vitro* Diagnostic

IVDD *In Vitro* Diagnostic Medical Devices Directive (European Union)

JAAME Japan Association for the Advancement of Medical Equipment

JIS Japanese Industrial Standards

JSA Japanese Standards Association

KFDA Korean Food and Drug Administration

KMIIC Korean Medical Instrument and Industry Corporation

MDB Medical Devices Bureau (Canada)

MDD Medical Device Directive (European Union)

MDED Medical Device Evaluation Department (South Korea)

MDR Regulation on Supervision and Administration of Medical Devices (China)

MDRs Medical Device Reports (United States)

MDRC Canadian Medical Devices Review Committee

MDUFMA Medical Device User Fee and Modernization Act of 2002 (United States)

Mercosur Southern Cone Common Market (South America)

MHLW Ministry of Health, Labor, and Welfare (Japan)

MOHW Ministry of Health and Welfare (South Korea)

MPD Medicinal Products Directive (European Union)

MRA Mutual Recognition Agreement

MRI Magnetic Resonance Imaging

NAFTA North American Free Trade Agreement

NAICS North American Industry Classification System (United States)

NMX *Normas Mexicanas* (Mexico)

NOM *Normas Oficiales Mexicanas* (Mexico)

NRL National Serological Reference Laboratory (Australia)

OB/GYN Obstetric/Gynecologic

OCC *Organismo de Controle e Certificação* (Control and Certification Laboratory, Brazil)

ODE Office of Device Evaluation (United States)

OTC Over-the-Counter

PAB Pharmaceutical Affairs Bureau (Japan)

PAL Pharmaceutical Affairs Law (Japan)

PBS Pharmaceutical Benefits Scheme (Australia)

PET Positron Emission Tomography

PGL poly (glycolide/1-lactide)

PHSA Public Health Service Act (United States)

PLDD Percutaneous Laser Disc Decompression

PMA Premarket Approval

PMEDC Pharmaceutical and Medical Device Examination Center (Japan)

PMSB Pharmaceutical and Medical Safety Bureau (Japan)

ppm Parts per Million

PRC People's Republic of China

PVC Polyvinyl chloride

QA Quality Assurance

QSR Quality System Regulation (United States)

REDA Radiation Emitting Devices Act (Canada)

REDR Radiation Emitting Devices Regulations (Canada)

RPB Radiation Protection Bureau (Canada)

R&TTED Radio and Telecommunications Terminal Equipment Directive (European Union)

RCHSA Radiation Control for Health and Safety Act of 1968 (United States)

RF Radio Frequency

RI Radioisotope

SI International System of Units (Metric system)

SAIQ State Administration for Entry/Exit Inspection and Quarantine (China)

SAL Sterility Assurance Level

SBTS State Bureau of Quality and Technical Supervisions (China)

SDA State Drug Administration (China)

SIC Standard Industrial Classification (United States)

SINLAP National Accreditation System for Testing Laboratories (Mexico)

SMDA Safe Medical Devices Act of 1990 (United States)

SPAC State Pharmaceutical Administration of China

SQLO Safety Quality Licensing Office of Boiler and Pressure Vessels (China)

SSA *Secretaria de Salud* (Mexico Ministry of Health)

STDs Sexually Transmitted Diseases

SVS National Secretary of Sanitary Vigilance (Brazil)

TCAM Telecommunication Conformity Assessment and Market Surveillance Committee (European Union)

TFDA Thai Food and Drug Administration

TGA Therapeutic Goods Administration (Australia)

TGAC Therapeutic Goods Advertising Code (Australia)

TGO Therapeutic Goods Order (Australia)
TODs Tubal Occlusion Devices
TPD Therapeutic Products Directorate (Canada)
TPP Therapeutic Products Programme (Canada)
TSS Toxic Shock Syndrome
UV Ultraviolet
V Volt
VHF Very High Frequency
WLF Washington Legal Foundation
WTO World Trade Organization

References

16 CFR Code of Federal Regulations, *Title 16 – Commercial practices*, Chapter II – Consumer Product Safety Commission, Part 1500 – Hazardous substances and articles, administration and enforcement Regulations. January 1, 2001. Washington, DC: Government Printing Office.

19 CFR Code of Federal Regulations, *Title 19 – Customs duties*, Chapter I, United States Customs Service, Department of the Treasury, Part 134 – Country of origin marking. April 1, 2001. Washington DC: Government Printing Office.

21 CFR Code of Federal Regulations, *Title 21 – Food and Drugs*. April 1, 2002. Washington DC: Government Printing Office.

40 CFR Code of Federal Regulations, *Title 40 – Protection of Environment – Chapter I – Environmental Protection Agency*, Part 82 – Protection of Stratospheric Ozone. July 1, 2001. Washington, DC: Government Printing Office.

47 CFR Code of Federal Regulations, *Title 47 –Telecommunications*, Chapter I – Federal Communications Commission. October 2, 2002. Washington DC: Government Printing Office.

510(k) US Department of Health and Human Services, Food and Drug Administration. 1994. 510(k) sign-off procedures. Blue Book Memorandum No. K94-2. (June 1). URL: http://www.fda.gov/cdrh/k942.html [updated May 7, 1996, cited January 19, 2002].

Active Active implantable devices will get 19-nation marker. 1993. *Europe Drug & Device Report* (March 8): 5.

Alpert Alpert, S. 1994. Memorandum from Acting Director, Office of Device Evaluation to ODE division directors. April 19.

Alpert, Wheelchair Alpert, S. 1994. Letter to powered wheelchair/scooter or accessory/component manufacturer. May 26. URL: http://www.fda.gov/cdrh/ode/883.pdf [updated May 26, 1994, cited July 20, 2002].

ANMAT National Administration of Drugs, Food, and Medical Technology. 1995. *Boletin Oficial*. Buenos Aries. Resolution 255/95.

Archer Archer, J. D. 1984. The FDA does not approve uses of drugs, *Journal of the American Medical Association* 252 (8) 1054–1055.

Availability US Department of Health and Human Services, Food and Drug Administration. 1999. *Preparing notices of availability of investigational medical devices and for recruiting study subjects*. (March 19). URL: http://www.fda.gov/cdrh/comp/2229.html [issued March 19, 1999, cited January 20, 2002].

Backinger Backinger, C. L., and P. A. Kingsley. *Write it write: Recommendations for developing user instruction manuals for medical devices used in home healthcare*. Washington, DC: US Department of Health and Human Services.

Bracey Bracey, A. 1997. *In vitro* diagnostic devices: Guidance for the preparation of 510(k) submissions. Department of Health and Human Services Publication FDA 97-4224. URL: http://www.fda.gov/cdrh/manual/ivdmanul.html [updated January 1997, cited January 2, 2002].

Brazil US Department of Commerce, Office of Microelectronics, Medical Equipment and Instrumentation. Medical device regulatory requirements for Brazil. 2002. URL: http://www.ita.doc.gov/td/mdequip/brazilregs.pdf [updated March 21, 2002, cited May 19, 2002].

Briones Briones, M. N. 1992. 1993 reference guide for the healthcare technology industry. Alexandria, VA: Healthcare Technology Institute.

Britannica, 2002 *Encyclopædia Britannica 2002 book of the year.* 2002. s.v. Britannica world data "Thailand". Chicago, Illinois: Encyclopædia Britannica, Inc.

Britannica, Chinese *Encyclopædia Britannica.* 1969. s.v. "Chinese language," Vol. 5. Chicago, Illinois: Encyclopædia Britannica, Inc. pp. 630–634.

Callan Callan, J. R., and J. W. Gwynne. 1993. *Human factors principles for medical device labeling.* FDA Contract No. 223-89-6022. Pacific Science & Engineering Group. San Diego:

Cardamone Cardamone, T. 1989 *Labeling—Regulatory requirements for medical devices.* US Department of Health and Human Services, Food and Drug Administration. Washington, DC: Government Printing Office.

CC&CR Consumer chemicals and containers regulations, 2001. 2001. *Canada Gazette,* Part II 135, no. 17 (August 15): 1552–16271.

CCC Mark Certification and Accreditation Administration of the People's Republic of China. 2001. *Regulation for compulsory product certification mark,* URL: www.cqc.com.cn/ccc/markeng.pdf [created December 3, 2001, cited July 9, 2002].

CEC, Amended Proposal Commission of the European Communities. 1993. *Amended proposal for a council directive on packaging and packaging waste.* COM(93) 416 final–SYN 436 (September 9) Brussels: Commission of the European Communities.

CEC, Classification Commission of the European Communities. 1998. *Guidelines for the classification of medical devices (Working draft).* MEDDEV 2.4/1 (November) Brussels: Commission of the European Communities.

CEC, Definitions Commission of the European Communities. 1994 *Guidelines relating to the application lf the Council Directive 90/385/EEC on active implantable medical devices; the Council Directive 93/42/EEC on medical devices.* Definition of "medical devices"; Definition of "accessory"; Definition of "manufacturer." MEDDEV 2.1/1 (April) Brussels: Commission of the European Communities.

CEC, Demarcation Commission of the European Communities. 1994. *Guidelines relating to the application of the Council Directive 90/385/EEC on active implantable medical devices; the Council Directive 93/42/EEC on medical devices. Demarcation with other directives: Directive 89/336/EEC relating to electromagnetic compatibility; Directive 89/686/EEC relating to personal protective equipment. MEDDEV 1.14 (March) Brussels:* Commission of the European Communities.

CEC, Field Commission of the European Communities. 1994. *Guidelines relating to the application of the Council Directive 90/385/EEC on active implantable medical devices; the Council Directive 93/42/EEC on medical devices.* Field of application of Directive 90/385/EEC. MEDDEV 2.1/2 Rev. 2 (April) Brussels: Commission of the European Communities.

CEC, Guide Commission of the European Communities. 1999. *Guide to the implementation of directives based on the new approach and the global approach.* Directorate-General for Enterprise. (September) Brussels: Commission of the European Communities.

CEC, Guidelines 65/65/EEC Commission of the European Communities. 1998. *Guidelines relating to the demarcation between Directive 90/385/EEC on active implantable medical devices, Directive 93/42/EEC on medical devices and Directive 65/65/EEC relating to medicinal products and related directives.* MEDDEV. 2.1/3 Rev. 5.1 (March) Brussels: Commission of the European Communities.

CEC, R&TTED FAQ Commission of the European Communities. 2001. *Radio & telecommunication terminal equipment – Frequently asked questions.* (June 26) URL: http://europa.eu.int/comm/enterprise/rtte/faq.htm. [cited: May 12, 2002]

CEC, Vigilance Commission of the European Communities. 1998, *Guidelines on a medical devices vigilance system,* MEDDEV 2.12/1 Rev. 3 (March) Brussels: Commission of the European Communities.

cGMP Medical devices; current good manufacturing practice (cGMP) final rule; quality system regulation. 1996. *Federal Register* 61, no. 195 (October 7): 52601-52662.

Chicago *The Chicago manual of style'* 1993. 14th ed. Chicago: University of Chicago Press.

China Classification *Regulation for medical device classification.* 2000. State Drug Administration Decree No. 15. (April 10). Beijing: State Drug Administration.

China MDR *Regulation on supervision and administration of medical devices.* 2000. State Council of the People's Republic of China Order No. 276. (January 4). Beijing: State Council of the People's Republic of China.

China Registration *Management provisions or registration of medical devices.* 2000. State Drug Administration Decree No. 16. (April 10). Beijing: State Drug Administration.

China Standards *Measures for the management of medical appliance standards.* 2002. State Drug Administration Decree No. 31. (January 1) Beijing: State Drug Administration.

Chinese *Oxfords Chinese Minidictionary.* 2001. "Chinese characters." Eds. Boping Yuan, Sally Church. New York: Oxford University Press.

Chlamydiae US Department of Health and Human Services, Food and Drug Administration. 1992. *Review criteria for assessment of in vitro diagnostic devices for direct detection of chlamydiae in clinical specimens.* URL: http://www.fda.gov/cdrh/ode/778.pdf. [updated January 1992, cited January 2, 2002].

CMDR Medical Devices Regulations. 1998. *Canada Gazette* Part II 132, no. 11 (May 27): 1645–1698.

CME Final guidance on industry-supported scientific and educational activities. 1997. *Federal Register,* 62, no. 237 (December 3): 64074–64100.

CNCA Certification and Accreditation Administration of the People's Republic of China. Circular relevant to the implementation of the compulsory product certification system. URL: www.cqc.com.cn/ccc/changeeng.pdf [created December 3, 2001, cited July 9, 2002].

Commercialization of IVDs US Department of Health and Human Services, Food and Drug Administration. 1998. FDA Compliance Policy Guide, Chapter 3 – Devices, Subchapter 300 - General/Processes, Commercialization of In Vitro Diagnostic Devices (IVDs) Labeled for Research Use Only or Investigational Use Only (Draft). (January 5). URL: http://www.fda.gov/cdrh/comp/ivddrfg.pdf [released January 5, 1998, cited January 20, 2002].

Contraceptive Labeling US Department of Health and Human Services, Food and Drug Administration. 1998. Uniform contraceptive labeling. (July 23). URL: http://www.fda.gov/cdrh/ode/contrlab.pdf [updated July 27, 1998, cited February 19, 2002].

Contract Manufacturers Contract manufacturers and sterilizers getting FDA attention. *The GMP Letter,* no. 168 (January 1994): 6–7.

CPCS State General Administration for Quality Supervision and Inspection and Quarantine of the People's Republic of China, 2001. *First catalogue of products subject to compulsory certification.* (December 3). URL: www.cqc.com.cn/ccc/catalogureeng.pdf [created December 3, 2001, cited July 9, 2002].

CPS Reg State General Administration for Quality Supervision and Inspection and Quarantine of the People's Republic of China. 2001. *Regulations for compulsory product certification*, URL: www.cqc.com.cn/ccc/regulationseng.pdf (Created 3 December 2001, cited 9 July 2002).

C-Reactive US Department of Health and Human Services, Food and Drug Administration. 1998. *In Vitro* Diagnostic C-Reactive Protein Immunological Test System. (July 20). URL: http://www.fda.gov/cdrh/ode/78.html [updated September 21, 1998, cited January 2, 2002].

CSA Canadian Standards Association. 1990. *Medical electrical equipment – Part 1: General requirements for safety.* C22.2.601.1-M90. Etobicoke, Ontario (November). Revised 2001.

CSA Collateral Canadian Standards Association. 2002. *Medical electrical equipment – Part 1-1: General requirements for safety – Collateral standard: Safety requirements for medical electrical systems.* C22.2.601.1.1. Etobicoke, Ontario (April 1).

Deciding When to Submit US Department of Health and Human Services, Food and Drug Administration. 1997. *Deciding when to submit a 510(k) for a change to an existing device.* (January 10). URL: http://www.fda.gov/cdrh/ode/510kmod.pdf. [issued January 10, 1997, cited January 19, 2002].

Decision 93/465/EEC Council Decision of 22 July 1993 concerning the modules for the various phases of the conformity assessment procedures and the rules for the affixing and use of the CE conformity marking, which are intended to be used in the technical harmonization directives. 1993. *Official Journal of the European Communities*, 36, no. L 220 (August 30).

Decision Decision in *Washington Legal Foundation v. Henney.* 2000. *Federal Register* 65, no. 52 (March 16): 14286–14288.

Densitometer Makers Densitometer makers promote unapproved uses: FDA. 1993. *Devices & Diagnostics Letter* 20, no. 31 (August 6): 3.

Device Labeling US Department of Health and Human Services, Food and Drug Administration, Center for Devices and Radiological health, Office of Device Evaluation. 1991. *Device labeling guidance*, General Program Memorandum #G91–1 (March 6). URL: http://www.fda.gov/cdrh/g91-1.html [updated May 13, 1996, cited January 19, 2002].

Dialysis FDA meets with dialysis community on disinfectants. 1992. *Medical Device Report* 21, no. 41 (October 15): 1.

Directive 90/385/EEC Council Directive 90/385/EEC of 20 June 1990 on the approximation of the laws of the member states relating to active implantable medical devices. 1990. *Official Journal of the European Communities*, 33, no. L 189 (July 20).

Directive 92/32/EEC Council Directive 92/32/EEC of 30 April 1992 amending for the seventh time Directive 67/548/EEC on the approximation of the laws, regulations and administrative provisions relating to the classification, packaging and labelling of dangerous substances. 1992. *Official Journal of the European Communities*, 35, no. L 154 (May 5).

Directive 93/42/EEC Council Directive 93/42/EEC of 14 June 1993 concerning medical devices. 1993. *Official Journal of the European Communities*, 36, no. L 169 (July 12).

Directive 94/62/EC European Parliament and Council Directive 94/62/EC of 20 December 1994 on packaging and packaging waste. 1994. *Official Journal of the European Communities*, 37, no. L 365 (December 31).

Directive 97/69/EC Commission Directive 97/69/EC of 5 December 1997 adapting to technical progress for the 23rd time Council Directive 67/548/EEC on the approximation of the laws, regulations and administrative provisions relating to the classification, packaging and labelling of dangerous substances (Text with EEA relevance). 1997. *Official Journal of the European Communities*, 40, no. L 343 (December 13).

Directive 98/79/EC Directive 98/79/EC of the European Parliament and of the Council of 27 October 1998 on *in vitro* diagnostic medical devices. 1998. *Official Journal of the European Communities*, 41, no. L 331 (July 12).

Directive 1999/5/EC Directive 1999/5/EC of the European Parliament and of the Council of 9 March 1999 on radio equipment and telecommunications terminal equipment and the mutual recognition of their conformity. 1999. *Official Journal of the European Communities*, 42, no. L 91 (April 7).

Directive 2000/70/EC Directive 2000/70/EC of the European Parliament and of the Council of 16 November 2000 amending Council Directive 93/42/EEC as regards medical devices incorporating stable derivates of human blood or human plasma. 2000. *Official Journal of the European Communities*, 43, no. L 313 (December 13).

Directive 2001/59/EC Commission Directive 2001/59/EC of 6 August 2001 adapting to technical progress for the 28th time Council Directive 67/548/EC on the approximation of the laws, regulations and administrative provisions relating to the classification, packaging and labelling of dangerous substances (Text with EEA relevance). 2001. *Official Journal of the European Communities*, 44 no. L 225 (August 21): 225.

DoC, 2001 U.S. Department of Commerce. 2000. *Mexico country commercial guide FY2001*. URL: http://www.usatrade.gov/Website/CCG.nsf/CCGurl/CCG-MEXICO2001-CH [created October 2, 2000, cited April 6, 2001].

DR4 Australian Department of Health and Aging, Therapeutic Goods Administration, Conformity Assessment Branch. 1998. *Australian Medical Device Requirements Version 4*. (May). URL: http://www.health.gov.au/tga/docs/html/dr4.htm [created August 18, 1999, cited 30 March 2002].

EN 375 European Committee for Standardization. 2001. *Information supplied by the manufacturer with* in vitro *diagnostic reagents for professional use*. EN 375. (January). Brussels: Comite Européen de Normalisation.

EN 376 European Committee for Standardization. 2002. In vitro *diagnostic systems – Requirements for labelling of* in vitro *diagnostic reagents for self-testing*. EN 376. (April). Brussels: Comite Européen de Normalisation.

EN 556-1 European Committee for Standardization. 2001. *Sterilization of medical devices – Requirements for terminally sterilized devices to be labeled 'Sterile' – Requirements for terminally sterilized medical devices*. EN 556. (October). Brussels: Comite Européen de Normalisation.

EN 591 European Committee for Standardization. 2001. *Instructions for use for* in vitro *diagnostic instruments for professional use*. EN 591. (March). Brussels: Comite Européen de Normalisation.

EN 980 European Committee for Standardization. 1996. *Graphical symbols for use in the labeling of medical devices*. EN 980. (May). including Amendment 1 (August 1999) and Amendment 2 (October 2001). Brussels: Comite Européen de Normalisation.

EN 1041 European Committee for Standardization. 1998. *Information supplied by the manufacturer with medical devices*. EN 1041. (February). Brussels: Comite Européen de Normalisation .

EN 1658 European Committee for Standardization. 1996. *Requirements for marking of* in vitro *diagnostic instruments*. EN 1568. (December). Brussels: Comite Européen de Normalisation.

EN 45502-1 European Committee for Electrotechnical Standardization. 1997. *Active implantable medical devices – Part 1: General requirements for safety, marking and information to be provided by the manufacturer.* EN 45502-1. (August) Brussels: Comite Européen de Normalisation Electrotechnique.

EN 46001 European Committee for Standardization/European Committee for Electrotechnical Standardization. 1996. *Quality systems – Medical devices – Particular requirements for the application of ISO 9001.* EN 46001. (August). Brussels: Comite Européen de Normalisation.

EN 60601-1 European Committee for Electrotechnical Standardization. 1990. *Medical electrical equipment—Part 1: General requirements for safety.* EN 60601-1. Subclause 6.1. including Amendment 1 (1990) and Amendment 2 (1995). Brussels: Comite Européen de Normalisation Electrotechnique.

EN 60601-1-1 European Committee for Electrotechnical Standardization. 2001. *Medical electrical equipment—1: Collaateral standard: Safety requirements for medical electrical systems.* EN 60601-1-1. Subclause 6.1.201. Brussels: Comite Européen de Normalisation Electrotechnique.

EN 61010-1 European Committee for Electrotechnical Standardization. 2001. *Safety requirements for electrical equipment for measurement, control, and laboratory use –Part 1: General requirements (IEC 61010-1).* EN 61010-1. (March). Brussels: Comite Européen de Normalisation Electrotechnique.

ERPALs Japanese Ministry of Health, Labor, and Welfare. 1996. *Enforcement regulations of the pharmaceutical affairs law.* Ordinance No. 1. (February 1) including amendments up to Ordinance No. 1 (March 30, 2000). Tokyo: Japanese Ministry of Health, Labor, and Welfare.

Estrogen US Department of Health and Human Services, Food and Drug Administration. 1997. *Premarketing approval review criteria for premarket approval of estrogen (ER) or progesterone (PGR) receptors in vitro diagnostic devices using steroid hormone binding (SBA) with dextran-coated charcoal (DCC) separation, histochemical receptor binding assays, or solid phase enzyme immunoassay (EIA) methodologies* (Draft). URL: http://www.fda.gov/cdrh/ode/odecl603.html [updated September 5, 1997, cited January 3, 2002].

EUCOMED EUCOMED Conformity Assessment Technical Committee, Labelling Subcommittee. 1994. *Guidance on interpretations of requirements relating to labelling and instructions for use.* (March 16). Brussels: European Medical Technology Industry Association.

EU Expands EU to expand into 25-nation market for devices May 1, 2004. *Devices & Diagnostics Letter*, 29 no. 50 (December 23): 5.

Exporting Medical Devices US Department of Health and Human Services, Food and Drug Administration, Center for Devices and Radiological Health. 1998. *Exporting medical devices.* URL: http://www.fda.gov/cdrh/devadvice/39.html [updated May 23, 2001, cited December 29, 2001].

Exports and Imports US Department of Health and Human Services, Food and Drug Administration, Center for Devices and Radiological Health, 1998. Exports and imports under the FDA export reform and enhancement act of 1996. Guidance for industry. URL: http://www.fda.gov/opacom/fedregister/frexport.html [updated February 1998, cited December 29, 2001].

F&DA *Food and Drugs Act. Revised Statutes*, c. F-27, §§1 *et seq.*(1985).

FDA Enforcement Report US Department of Health and Human Services, Food and Drug Administration. 1993. *FDA enforcement report*, no. 93-26 (June 30): 7.

FDAMA Guidance US Department of Health and Human Services, Food and Drug Administration, Center for Devices and Radiological Health. 1998. *FDA Modernization Act of 1997,*

Guidance for the device industry on implementation of highest priority provisions. URL: http://www.fda.gov/cdrh/modact/dayone.pdf [created February 6, 1998, cited December 8, 2001).

FDA Reply Dotzel to Messers. Popeo and Samp, Washington D.C., 28 January 2002. US Department of Health and Human Services, Food and Drug Administration. Docket No. 01P-0250. URL: http://www.fda.gov/ohrms/dockets/dailys/02/Jan02/013002/01p-0250_pdn0001_01_vol2.pdf [issued January 30, 2002, cited June 2, 2002].

FD&CA *Federal Food, Drug, and Cosmetic Act*, as amended. US Code, vol. 21, §§321 *et seq.* (1938).

FHSA *Federal Hazardous Substances Act.* US Code, vol. 15, §§1261 *et seq.* 1960. as amended by *The Child Protection Act of 1966* (US Public Law 89-756), *The Child Protection and Toy Safety Act of 1969* (US Public Law 91-113), and *The Poison Prevention Packaging Act of 1970* (Public Law 91-601).

Final Draft US Department of Health and Human Services, Food and Drug Administration. 1991. *Review of final draft medical device labeling.* Blue Book Memorandum No. P91-4. (August 29, URL: http://www.fda.gov/cdrh/p91-4.html [updated November 11, 2000, cited January 20, 2002].

FPLA *Fair Packaging and Labeling Act.* US Code, vol. 15 §§1451 *et seq.* (1996).

FTCA *Federal Trade Commission Act*, as amended. US Code, vol. 15, §§52 *et seq.* (1914).

Gibbs, Medical Device Promotion Gibbs, J. N. 1992. Medical device promotion activities and private litigation. *Food and Drug Law Journal* 47 (3): 295–307.

Gibbs, New Medical Devices Gibbs, J. N. 1990. New medical device legislation enacted. *Regulatory Affairs* 2: 505–536.

Glycohemoglobin US Department of Health and Human Services, Food and Drug Administration. 1991. *Review Criteria for Assessment of Glycohemoglobin (Glycated or Glycosylated) Hemoglobin* In Vitro *Diagnostic Devices.* URL: http://www.fda.gov/cdrh/ode/odecl658.html [updated September 5, 1997, cited January 2, 2002].

GMP Medical devices: Good manufacturing practice (GMP) regulations. 1987. *Federal Register* 43, no. 141 (July 21): 31508–31532.

GMP Letter GMP letter. 1993. *Editorial* 160 (May): 1.

Guide *Guide to medical device registration in Japan.* 1997. 6th ed. Tokyo: Yakuji Nippo, Ltd.

Graphics *Graphics master 4.* 1988. 4th ed. Los Angeles: Dean Lem Associates, Inc.

Hasegawa Hasegawa, Hirokazu. 2000. Regulation of medical devices in Japan. *The Regulatory Affairs Journal (Devices)* 8(2): 194–197.

HDE US Department of Health and Human Services, Food and Drug Administration. 2001. *Humanitarian device exemptions (HDE) regulation: questions and answers.* Final Guidance for Industry. URL: http://www.fda.gov/cdrh/ode/guidance/1381.html [released July 12, 2001, cited January 20, 2002].

Hearing Aid FDA warns hearing aid manufacturers to stop making misleading advertising claims. 1993. *CDRH Medical Devices Bulletin* 11, no. 5 (May): 1.

Higson Higson, G. R. 1993. *The Medical device directives—A manufacturer's handbook.* Egham, Surrey: Medical Technology Consultants Europe, Ltd.

Holstein Holstein, H. M. 1991. The Safe Medical Devices Act of 1990. *Regulatory Affairs* 3: 91–104.

Home-Use IVDs U.S. Department of Health and Human Services, Food and Drug Administration. 1988. *Assessing the safety and effectiveness of home-use in vitro diagnostic devices*

(IVDs): draft points to consider regarding labeling and premarket submissions. URL: http://www.fda.gov/cdrh/ode/odec1272.html [updated September 5, 1997, cited January 3, 2002].

House, Government Operations US House. 1991. Committee on Government Operations. *Promotion of Drugs and Medical Devices for Unapproved Uses: Hearing before the Human Resources and Intergovernmental Relations Subcommittee.* 102nd Cong., 1st sess. June 11.

House, Interstate and Foreign Commerce US House. 1976. Committee on Interstate and Foreign Commerce. *Medical Device Amendment of 1976.* 84th Cong. 2nd sess. H. Rept. 883.

HTSUS *Harmonized Tariff Schedule of the United States.* US Code, vol. 19, §1202.

IEC 60417-1 International Electrotechnical Commission. 1998. Graphical symbols for use on equipment—Part 1: Overview and application. IEC 60417-1. Geneva: International Electrotechnical Commission.

IEC 60601-1 International Electrotechnical Commission. 1988. *Medical electrical equipment – Part 1: General requirements for safety.* 2nd ed. as amended. Geneva: International Electrotechnical Commission.

IEC 60601-1-1 International Electrotechnical Commission. 2000. *Medical electrical equipment – Part 1-1: General requirements for safety – Collateral standard: Safety requirements for medical electrical systems.* IEC 60601-1-1. 2nd ed. (December). Geneva: International Electrotechnical Commission.

IEC 60878 International Electrotechnical Commission. 1988. *Graphical symbols for electrical equipment in medical practice.* IEC 60878. Geneva: International Electrotechnical Commission.

IEC 61010-1 International Electrotechnical Commission. 2001. *Safety requirements for electrical equipment for measurement, control and laboratory use – Part 1: General requirements.* IEC 61010-1. Geneva: International Electrotechnical Commission.

IEC, A2 International Electrotechnical Commission. Amendment 2 to IEC 60601-1, 2nd ed. 1995. *Medical electrical equipment—Part 1: General requirements for safety* (March 7). Geneva: International Electrotechnical Commission.

ISO 3864-1 International Standards Organization. 2002. *Graphical symbols -- Safety colours and safety signs -- Part 1: Design principles for safety signs in workplaces and public areas.* ISO 3864-1. Geneva: International Standards Organization.

ISO 7000 International Standards Organization. 2000. *Graphical symbols for use on equipment – Index and synopsis.* ISO 7000. Geneva: International Standards Organization.

ISO 7010 International Standards Organization. 2002. *Graphical symbols – Safety signs in workplaces and public areas.* ISO/FDIS 7010. Geneva: International Standards Organization.

ITA, China US Department of Commerce, Office of Microelectronics, Medical Equipment and Instrumentation. 2001. *Medical device regulatory requirements for China.* International Trade Administration Web site. URL: www.ita.doc.gov/td/mdequip/chinaregs.html [updated March 16, 2001, cited March 3, 2002].

ITA, Korea US Department of Commerce, Office of Microelectronics, Medical Equipment and Instrumentation. 2001. *Medical device regulatory requirements for korea (South Korea).* International Trade Administration Web site. URL: http://www.ita.doc.gov/td/mdequip/korearegs.htm [updated November 15, 2001, cited March 7, 2002].

JIS T 1001 Japanese Standards Association. 1992. *Medical electrical equipment – Part 1: General requirements for safety.* Tokyo: Japanese Standards Association.

JIS T 1005 Japanese Standards Association. 1983. *Style Manual for Instruction Manuals.* Tokyo: Japanese Standards Association.

Justice US Department of Justice. 1994. Heart valve maker will pay $10.75 million plus medical costs to settle U.S. claim it made false statements to market device. Press release. (June 30). Washington, DC.

Kuhn Kuhn, M. 1996. *International Standard Paper Sizes.* URL: http://www.cl.cam.ac.uk/~mgk25/iso-paper.html [updated May 6, 2002, cited: July 27, 2002].

KFDA, Regulation *Regulations regarding designation of medical devices.* 2000. Korean Food and Drug Administration. Notification 2000-37. (August 1). Soul: Korean Food and Drug Administration.

Labels Labels, Complaint Management Key to Limit Liability *MDR Watch*, no. 83 (September): 3–4.

Larkin Larkin, L. 1995. South Korea—Medical device regulations—Impact on U.S. companies. Paper presented at symposium, Doing Business in Japan and Asia: A Medical Technology Update. Health Industry Manufacturers Association, December 7–8, Washington, DC.

Lowery Lowery, A., J. Strojny, and J. Puleo. 1996. *Medical device quality system manual: A small entity compliance guide.* Department of Health and Human Services Publication FDA 97-4179. URL: http://www.fda.gov/cdrh/dsma/gmp_man.html [published December 1996, cited January 26, 2002].

Lowery, Puleo Lowery, A., and J. V. Puleo. 1991. *Medical devices—Good manufacturing practices manual.* 5th ed. US Department of Health and Human Services, Food and Drug Administration. Washington, DC: Government Printing Office.

MDR Japan *Medical device regulations in Japan: Questions and answers.* 1994. 1st ed. Tokyo: Yakuji Nippo, Ltd.

MDRC Canadian Ministry of National Health and Welfare, Medical Devices Review Committee. 1992. (May) Ottawa, Ontario: Health Canada.

Medical Devices, Exemptions Medical devices: Exemptions from premarket notification for certain classified devices. 1994. *Federal Register*, 59, no. 234 (December 7): 63005–63015.

Medical Devices, Substantial Equivalence Medical devices: Substantial equivalence; 510(k) summaries and 510(k) statements; class III summaries; confidentiality of information. *Federal Register*, 59, no. 239 (December 14): 64287–64294.

Medical Laser U.S. Department of Health and Human Services, Food and Drug Administration. 1995. *Guidance on the content and organization of a premarket notification for a medical laser* (Draft). URL: http://www.fda.gov/cdrh/ode/386.pdf [published June 1995, cited January 13, 2002].

Metric Labeling Metric labeling: Quantity of contents labeling requirement for food, human and animal drugs, animal foods, cosmetics, and medical devices. 1993. *Federal Register*, 58, no. 243 (December 21): 67444–67464.

MDUFMA Summary US Department of Health and Human Services, Food and Drug Administration. 2002. *Summary of the medical device user fee and modernization act of 2002*, URL: http://www.fda.gov/cdrh/mdufma/mdufmasummary.pdf [updated: November 7, 2002, cited: December 4, 2002].

Ministry *Reglamentase actividad – Art. 10 Ley no. 16,463.* 1985. Ministerio de Salud y Acción Social Decree No. 2,505/85. *Boletín Oficial* No. 25849 (January 17, 1986): 1

Munsey Munsey and Samuel. 1984. *Medical Device Regulation in Transition*, The Seventy-Fifth Anniversary Commemorative Volume of Food and Drug Law. Washington, DC: Food and Drug Law Institute, Inc.

NAICS US Department of Commerce, US Census Bureau. 1998. *1997 NAICS and 1987 SIC Correspondence Tables.* URL: http://www.census.gov/epcd/www/naicstab.htm [updated September 8, 1998, cited: May 19, 2002].

NOM-137 Secretaria de Salud. 1998. Información regulatoria – Especificaciones generales de etiquetado que deberán los dispositivos médicos, tanto de manufactura nacional de procedencia extranjera. NORMA Oficial Mexicana NOM-137-SSA-1-1995. *Diario Oficial,* (November 18): 47–52.

O'Keefe O'Keefe, D. F., and R. A. Spiegel. 1976. *An analytical legislative history of the Medical Device Amendments of 1976.* Washington, DC: The Food and Drug Law Institute, Inc.

Orthopedic Device US Department of Health and Human Services, Food and Drug Administration. 1987. *Orthopedic device approval and labeling.* Washington, DC: Department of Health and Human Services.

Overseas GMP Australian Department of Health and Aging, Therapeutic Goods Administration. 2000. *Guidelines on standard of overseas manufacturers,* 12th ed. (September). Canberra: Therapeutic Goods Administration.

PAB No. 127 Japanese Ministry of Health and Welfare, Pharmaceutical Affairs Bureau. 1989. *The guidelines for exhibition for not yet approved medical devices.* PAB Notification No. 127. (February 13). Tokyo: Printing Bureau, Ministry of Finance.

PAB No. 333 Japanese Ministry of Health and Welfare, Pharmaceutical Affairs Bureau. 1994. *Enforcement of the law for partial amendment to the Pharmaceutical Affairs Law and the Adverse Drug Reaction Suffering Relief and Research Promotion Fund Law.* PAB Notification No. 333. (March 31). Tokyo: Printing Bureau, Ministry of Finance.

PAB No. 615 Japanese Ministry of Health and Welfare, Pharmaceutical Affairs Bureau. 1992. *Clinical practice for trials on medical devices.* PAB Notification No. 615. (July 1). Tokyo: Printing Bureau, Ministry of Finance.

PAB No. 662 Japanese Ministry of Health and Welfare, Pharmaceutical Affairs Bureau, 1985 *On in vitro test drugs.* PAB Notification No. 662. (June 29). Tokyo: Printing Bureau, Ministry of Finance.

PAB No. 752 Japanese Ministry of Health and Welfare, Pharmaceutical Affairs Bureau. 1993. *Generic names of medical devices and their classification,* PAB Notification No. 752. (September 28). Tokyo: Printing Bureau, Ministry of Finance.

PAL *Pharmaceutical Affairs Law of Japan.* Law No. 145 of August 10, 1960, as amended (1999).

Palmer Palmer, J. 1993. *European Affairs,* 1993 Britannica Book of the Year, Chicago: Encyclopaedia Britannica, Inc. 419–420, 422–423.

Park, Melvin Park, J., M. Melvin, and A. Barcome. 1996. *Investigational device exemptions manual,* Department of Health and Human Services Publication FDA 96-4159. URL: http://www.fda.gov/cdrh/manual/idemanul.pdf [published June 1996, cited January 20, 2002].

Park, Poneleit Park, J., K. Poneleit, and R. Parr. 1997. *Premarket approval (PMA) manual.* Department of Health and Human Services Publication FDA 97-4214. URL: http://www.fda.gov/cdrh/dsma/pmaman/front.html [published July 1997, cited January 20, 2002].

Parr, Barcome, Import of Medical Devices Parr, R., and A. Barcome. 1988. *Import of medical devices—A workshop manual.* US Department of Health and Human Services, Food and Drug Administration. Washington DC: Government Printing Office.

Parr, Barcome, Regulatory Requirements Parr, R., and A. Barcome. 1989. *Regulatory requirements for medical devices—A workshop manual.* US Department of Health and Human Services, Food and Drug Administration. Washington, DC: Government Printing Office.

Peck Peck, J. C., and K. H. Rabin. 1989. *Regulating change: The regulation of foods, drugs, medical devices and cosmetics in the 1990s.* Washington, DC: The Food and Drug Law Institute, Inc.

Pharmaceutical *Pharmaceutical Administration in Japan.* 2001. 10th ed. Tokyo: Yakuji Nippo, Ltd.

Public Law 78-410 *Public Health Service Act, Biological; Products.* US Public Law 78-410. US Code, vol. 42, §§201 *et seq.*

Public Law 94–295 *The Medical Device Amendments of 1976.* US Public Law 94–295. May 28, 1976.

Public Law 101-629 *Safe Medical Device Act of 1990.* US Public Law 101-629. November 28, 1990.

QSR US Department of Health and Human Services, Food and Drug Administration, Center for Devices and Radiological Health. 1996. *The quality system regulation,* URL: http://www.fda.gov/cdrh/dsma/gmp_man.html [updated April 14, 1999, cited January 20, 2002].

RCHSA *Radiation Control for Health and Safety Act.* US Code 42 §§263b *et seq.* (1968).

Reusable U.S. Department of Health and Human Services, Food and Drug Administration, Center for Devices and radiological Health, Office of Device Evaluation. 1996. *Labeling reusable medical devices for reprocessing in healthcare facilities: FDA reviewer guide.* URL: http://www.fda.gov/cdrh/ode/198.pdf [issued April 1996, cited July 20, 2002].

Revocation US Department of Health and Human Services, Food and Drug Administration. 1997. *Dear sponsor letter concerning the revocation of 21 CFR Part 813 - IOL IDE regulations.* URL: http://www.fda.gov/cdrh/iolmay97.pdf [issued May 20, 1997, cited January 20, 2002].

Rice Rice, L., and A. Lowery. 1995. Premarket notification 510(k): Regulatory requirements for medical devices. Department of Health and Human Services Publication FDA 95-4158. URL: http://www.fda.gov/cdrh/manual/510kprt1.html [published August 1995, cited January 19, 2002].

Schöpflin Schöpflin, G. 1994. *European affairs.* 1994 Britannica Book of the Year. Chicago: Encyclopaedia Britannica, Inc. 419–421.

Scurfield Scurfield, E. 1992. *Chinese.* Lincolnwood, Illinois: NTC Publishing Group.

Spinal System US Department of Health and Human Services, Food and Drug Administration. 2000. *Guidance for spinal system 510(k)s.* URL: http://www.fda.gov/cdrh/ode/guidance/636.pdf [issued Spetember 27, 2000, cited January 20, 2002].

Standards *Standards & certification systems concerning drugs in Japan.* 1992. 3rd ed. Tokyo: Yakugyo Jiho Co., Ltd.

Dr. Hwal SUH SUH, Hwal. 2001. Letter to author (November 11).

Swiss MRA EU/Swiss MRA belatedly comes into force. 2002. *Clinica,* no. 1010 (June 5): 2.

TGA, General Information Australian Department of Health and Aging, Therapeutic Goods Administration. 2002. *Medical devices – A new approach to regulation: General information,* TGA Web site. URL: http://www.health.gov.au/tga/docs/pdf/devinfo2.pdf [issued May 24, 2002, cited June 23, 2002].

TGA, New Legislation Australian Department of Health and Aging, Therapeutic Goods Administration. 2002. *Medical devices – A new approach to regulation: Proposed new legislation,* TGA Web site. URL: http://www.health.gov.au/tga/docs/pdf/devinfo1.pdf [issued May 24, 2002, cited June 23, 2002].

TGA, Restricted Australian Department of Health and Aging, Therapeutic Goods Administration. 2002. *Restricted representations.* URL: http://www.health.gov.au/tga/docs/pdf/advrestr.pdf [updated January 16, 2002, cited June 24, 2002].

TGAC Therapeutic Goods Advertising Code Council. 2001. *Therapeutic goods advertising code.* URL: http://www.tgacc.com.au/codeList.cfm [issued July 25, 2001, cited June 23, 2002].

TGO 37 Australian Department of Health and Aging, Therapeutic Goods Administration, 1991. *General requirements for labels of therapeutic devices.* Therapeutic Goods Order No. 37. (February 7). Canberra: Australian Government Publishing Service.

Thai FDA About the Thai Food and Drug Administration. 2002. Thai FDA Web site. URL: www.fda.moph.go.th/aboutthaifda43.htm [cited February 25, 2002].

Thai HSA *Hazardous Substance Act*, B.E. 2535. 1992. Thai FDA Web site. URL: http://www2.fda.moph.go.th/law/Law_Book_1.asp?productcd=6&lawid=600003_1&lawname=HAZARDOUS%20SUBSTANCE%20ACT%20B.E.2535&language=e&Contents=1&v_call=lawlink&historylink=/law&arg_language=e [cited February 25, 2002].

Thai MDA *Medical Device Act*, B.E. 2531. 1988. Thai FDA Web site. URL: http://www2.fda.moph.go.th/law/Law_Book_1.asp?productcd=5&lawid=500010&lawname=MEDICAL%20DEVICE%20ACT&language=e&Contents=1&v_call=lawlink&historylink=/law&arg_language=e [cited February 25, 2002].

Thai Medical Device Medical device control in Thailand. 2001. Thai FDA Web site. URL: www.fda.moph.go.th/fda-net/html/product/other/kbs3/homepage2.htm [updated May 2001, cited February 25, 2002[.

Therapeutic Goods Act *Therapeutic Goods Act 1989.* Act No. 21 of 1990 including amendments up to Act No. 81 of 2001. Commonwealth of Australia. (September 22, 2001) Canberra: Australian Government Printing Service.

TGR 2002 Therapeutic Goods (Medical Devcies) Regulations 2002. 2002. Australian Attorney's General Department. URL: http://scaleplus.law.gov.au/html/numrul/19/9530/pdf/2002No236.pdf [updated October 3, 2002, cited November 24, 2002].

Tichy Tichy, H.J. 1988. *Effective writing for engineers, managers, scientists.* 2nd ed. New York: John Wiley & Sons, Inc.

TPD IVD Guidance Therapeutic Products Programme. 1998. *Guidance for the labelling of in vitro diagnostic devices* (Draft). Document GD012/RevDR-MDB. (March 24). URL: http://www.hc-sc.gc.ca/hpb-dgps/therapeut/zfiles/english/guides/meddev/labl_ivd_e.pdf [issued june 24, 1998, cited July 3, 2001]

TPD Labelling Guidance Therapeutic Products Programme. 1999. *Guidance for the labelling of medical devices, Sections 21 to 23 of the medical devices regulations, appendices for labelling: Prolonged wear contact lenses, menstrual tampons, contraceptive devices, and medical gloves* (Draft). Document GD011/RevDR-MDB. (January 11). URL: http://www.hc-sc.gc.ca/hpb-dgps/therapeut/zfiles/english/guides/meddev/labl_dv9_e.html [created January 11, 1999, cited July 3, 2001].

TPD Standards Policy Health Canada, Therapeutic Products Directorate. 2002. *Policy on recognition and use of standards under the medical devices regulations.* (April 11). URL: http://www.hc-sc.gc.ca/hpb-dgps/therapeut/zfiles/english/policy/issued/standards_e.pdf [issued April 11, 2002, cited June 5, 2002].

TPP Draft Standards Policy Health Canada, Therapeutic Products Programme. 1999. *Draft policy on recognition and use of standards under the medical devices regulations.* (June 10) Ottawa, Ontario: Therapeutic Products Directorate.

Trade US Department of Commerce, International Trade Administration, Office of Microelectronics, Medical Equipment and Instrumentation. 2001 *Industry & Trade Statistics—Medical Equipment.* URL: http://www.ita.doc.gov/td/medequip/384stats.html [updated December 31, 2001, cited May 5, 2002].

Treverton Treverton, G. 1993. *Europe's single market.* 1993 Britannica Book of the Year. Chicago: Encyclopaedia Britannica, Inc. 421–422.

Unsafe Food and Drug Administration Press Office. 2002. FDA announces the destruction of unsafe medical devices. URL: http://www.fda.gov/bbs/topics/NEWS/2002/MEW00823.html [updated 16 July 16, 2002, cited July 18, 2002].

USP United States Pharmacopeia Convention, Inc. 2002. *The united states pharmacopeia—The national formulary.* USP 25-NF 20. (January 1) Rockville, MD: United States Pharmacopeia Convention, Inc.

Warning Letters Warning letters tied to promotion likely to rise. 1993. *Devices & Diagnostics Letter,* 20, no. 25 (September 10): 1–2.

Warning, September 1999 US Department of Health and Human Services, Food and Drug Administration, Center for Devices and Radiological Health, Office of Compliance. 1999. *Warning letter, Axon Instruments, Inc., September 1, 1999.* URL: http://www.fda.gov/foi/warning_letters/m2910n.pdf [issued September 1, 1999, cited June 1, 2002].

Warning, March 2000 US Department of Health and Human Services, Food and Drug Administration, Center for Devices and Radiological Health, Office of Compliance. 2000. *Warning letter, Medtronic, Inc., March 23, 2000.* URL: http://www.fda.gov/foi/warning_letters/m3573n.pdf [issued March 23, 2000, cited June 1, 2002].

Warning, May 2000 US Department of Health and Human Services, Food and Drug Administration, Center for Devices and Radiological Health, Office of Compliance. 2000. *Warning letter, Focus Imaging Group, Inc., May 11, 2000.* URL: http://www.fda.gov/foi/warning_letters/m3754n.pdf [issued May 11, 2000, cited June 1, 2002].

Warning, March 2001 US Department of Health and Human Services, Food and Drug Administration, Center for Devices and Radiological Health, Office of Compliance. 2001. *Warning letter, Microcurrent Research, Incorporated, March 29, 2001.* URL: http://www.fda.gov/foi/warning_letters/g1311d.pdf [issued March 29, 2001, cited June 1, 2002].

WLF Washington Legal Foundation. 2001. *Citizens petition regarding manufacturer dissemination of non-misleading information concerning off-label uses of FDA-approved products.* (May 23). URL: http://www.fda.gov/ohrms/dockets/dailys/01/May01/053001/cp00001.pdf [updated May 25, 2001, cited June 2, 2002].

World Almanac *The world almanac and book of facts.* 1996. s.v. "Lunar Calendar." Mahwak, New Jersey: World Almanac Books.

Zinsser Zinsser, W. 1990. *On writing well.* 4th ed. New York: Harper Collins Publishers.

Index

A

Accompanying documentation, *see also* Instructions for use
label design, 471–472
United States, 380
Accuracy, degree of, 65
Acknowledgment symbol, 169, *see also* Conformity markings
Acronyms, glossary, 547–551
Active Implantable Medical Device Directive (AIMDD)
basics, 5, 213, 237–238, 247
CE conformity marking, 246
combination of devices, 241, 243
conformity marking, 207
control panels, 239–240
dates manufacture, 241
device identification, 240, 242
display panels, 239–240
energy sources, 247
European Union, 191, 195, 201, 205, 207–208
expiration dates, 241, 243
exposure to environmental conditions, 247
handling conditions, 243
implantable devices, 246
instructions for use, 244–246
labeling requirements, 238–243
manufacturers, 240, 242
Medical Device Directive, 228
medicinal products, 246, 247
misbranding, 238
noninvasive identification, 239, 246
operation of device, 245
patient information, 245, 246–247
performance expected, 244, 247
product identification, 239
radioactive substances, 243, 246
reciprocal interference, 245, 246, 247
reusable devices, 246
selection of device, 245
shelf life, 247
special purpose devices, 241, 243
sterile conditions and products, 240–241, 243, 245
storage conditions, 243
year of authorization, 246
Active implantable medical devices (AIMDs), *see also* Implantable devices; specific devices
Australia, 39, 53–54, 61
Canada, 97
Adulteration
Argentina, 12–13
Australia, 28
Canada, 74
Japan, 279

Thailand, 182–183
United States, 324–325
Adverse effects, *see* Contraindications; Safety information; Warning statements
Advertising
Australia, 28, 29–31
Canada, 74, 75–76
China, 168
Japan, 279–281
Korea, 175, 179
Thailand, 186
United States, 324, 326–336, 421–423
AIMDD, *see* Active Implantable Medical Device Directive (AIMDD)
AIMDs, *see* Active implantable medical devices (AIMDs)
Air purifiers, misbranding, 74
Amateur radio (HAM) transmitters, 456
Amendment of 2002, *see* Therapeutic Goods Amendment of 2002
AM/FM radios, 456
AMNAT web site, 545
Ampules, *see* Packaging containers and materials
Analytical X-ray equipment, 136, *see also* X-ray equipment
Animal origin products, *see also* Human origin products; Origin of products
Australia, 41, 53, 54
Canada, 83
quarantine, 41
ANMAT, *see* National Administration of Drugs, Foods, and Medical Technology (ANMAT)
Antennas, 455
ANVISA, *see* Brazilian National Health Vigilance Agency (ANVISA)
Approval of devices
Japan, 283–284, 286, 287, 293
United States, 417–419, 434
Argentina
adulteration, 12–13
arrangement of label, 14
background, 9
basics, 9, 15–16
bringing devices to market, 10–12
contents of label, 14
importing products, 11–12
intent of law, 9
labeling requirements, 13–14
language requirements, 14, 16
materials for label, 13
misbranding, 12–13
packaging materials, 13
regulation scope, 9
reprocessing, 15
reuse, 15
sterile packing, 14

technical director, 12, 14
ARTG, *see* Australian Register of Therapeutic Goods (ARTG)
Articles, *see* Literature references
Assembler information, 387
AUA program, *see* Authorized User Approval (AUA) program
Audience consideration, 461–462, *see also* User information
Australia, *see also* Therapeutic Goods Act (TGA); Therapeutic Goods Amendment of 2002
active implantable medical devices, 53–54
animal origin devices, 54
Australian Register of Therapeutic Goods, 61
barrier contraceptive devices, 56
basics, 45, 56–57
batch number, 48
breast prostheses, 54, 56
classification of devices, 60–61, 483–490
components and kits, 52–53
contraceptive devices, 55–56
directions for use, 45
drug-infusion system, 54
European Union, 194
heart-valve prostheses, 53
hepatitis C diagnosis, devices for, 52, 56
HIV diagnosis, devices for, 52, 56
home use devices, 52
human origin products, 52, 54–55
implantable devices, 48
individually wrapped goods, 50
industry regulation, 5
intraocular lenses, 55
intrauterine contraceptive devices, 55
in vitro diagnostic goods, 51–52
labeling requirements, 46–51
manufacturer's name and location, 47–48
misbranding, 45
name of device, 47
nonsterile devices, 48
outer packages, 50
packages, 49–51
presentation requirements, 46–47
quantity in package, 48
registration, 52, 53–56
serial number, 48
sterile devices, 48
Australian Register of Therapeutic Goods (ARTG), *see also* Australia
Australia, 61
bringing devices to market, 32–33
exporting, 40
grandfathering products, 26
importing, 37–38, 40–41
Authorizations, *see* Licenses
Authorized User Approval (AUA) program, 43

B

Babies, *see* Children
Baby products, 60, 195, *see also* Children
Baggage inspection devices, 136
Bandages, *see also* Sutures
    Australia, 30, 60
    Canada, 95
    common knowledge, 351
Banned devices, 363–364
Barrier contraceptive devices, *see*
    Contraceptives
Batch numbers, *see* Control numbers
Batteries, *see also* Active Implantable Medical
    Device Directive (AIMDD)
    Canada, 117
    In Vitro Diagnostic Medical Device
        Directive, 259
    Japan, 302
    Medical Device Directive, 231, 233
BCP, *see* Brazilian Customer Protection (BCP)
    code
Bilayer neoprene-latex gloves, 112
Bleaching products for teeth, 195
Blister packs, *see* Packaging containers and
    materials
Blood and blood products, *see also* Body
    fluids, human
    Australia, 51
    Canada, 72
    European Union, 192
    Korea, 174
    Medical Device Directive, 224
    United States, 322, 323, 430
Blood glucose test kits, 304–305
Blood purification equipment, 156, *see also*
    Dialysis equipment
Body fluids, human, 41, *see also* Blood and
    blood products
Book of Procedures, 15
Bottles, *see* Packaging containers and
    materials
Branched DNA amplification, 52
Brazil
    background, 17
    basics, 17, 20–21
    bringing devices to market, 18–19
    distributors, 19
    importing, 18, 19
    intent of law, 17
    labeling requirements, 20
    language requirements, 20
    manufacturers, 18, 19
    misbranding, 20
    product registration, 18
    regulations scope, 17
    web site, 545
Brazilian Customer Protection (BCP) code, 20
Brazilian National Health Vigilance Agency
    (ANVISA), 17–20
Breast prostheses, 47, 53, 54, 56
Bringing devices to market
    Argentina, 10–12
    Australia, 32–40
    Brazil, 18–19
    Canada, 76–85

    China, 158–168
    European Union, 198–203
    Japan, 282–287
    Korea, 175–177
    Thailand, 183–185
Bringing devices to market, United States, *see*
    also United States
    advertising, 421–423
    approval, 417–419
    basics, 413, 426–427
    changes to existing devices, 415–416
    custom devices, 425
    exemptions, 419–426
    humanitarian exemptions, 425–426
    intended use, 416–417
    intraocular lenses, 424–425
    investigational uses, 419–425
    licensing, 418–419
    manufacturing practices, 423
    notification review program, 414–415
    premarket requirements, 414–417
    prohibited practices, 420–421
    study subjects, recruiting, 422–423
Buffers, 99

C

Cabinet X-ray equipment, 136, *see also* X-ray
    equipment
Calendar, China, 171, *see also* Date
Calibration, *see also* Instructions for use
    Australia, 65
    European Union, 191
    In Vitro Diagnostic Medical Device
        Directive, 266–268
Canada, *see also* Food and Drugs Act
    (F&DA); Radiation-emitting
    devices, Canada
    basics, 93, 134
    classification of devices, 81–85
    class III and IV devices, 97, 103, 104
    clinical trials, 106
    condoms, 110
    contact lenses, soft, 107–108
    contraceptives, 109–110
    corrosive products, 127–129
    custom made devices, 106–107
    device identification, 94–95, 98, 102,
        103–104
    display panel, 118–119
    electrical equipment, 112–118
    explosive materials, 104
    exporting, 107
    first aid statements, 122
    flammable products, 124–127
    gloves, medical, 110–112
    hazard symbols and statements, 121, 122
    home use devices, 122
    immediate container label, 102–103
    implant-registration card, 105–106
    industry regulation, 5
    information format, 119–121
    instructions for use, 95–96, 98–101,
        102–103, 104
    investigational devices, 106

    in vitro diagnostic devices, 98–104
    labeling requirements, 93–97
    language requirements, 97
    menstrual tampons, 108–109
    misbranding, 93
    package inserts, 98–103
    pressurized containers and contents, 122,
        123–124
    prominence of information, 97
    prophylactics, 110
    reagent label, 10–104
    self-service display labeling, 97
    signal words, 121
    small packages, 122
    special access devices, 106–107
    sterile devices, 96–97, 101–102, 103, 104
    toxic products, 129–133
Canadian Therapeutic Products Directorate,
    545
Capacitors, 231
Cardiovascular implants, *see* Heart valves;
    Pacemakers
Cassette players, 456
Catalogue numbers, *see* Control numbers
Catalytic ingredients, 99, 104
Catheters, 417
CAT scanners, 168
Caution statements, *see also* Hazard symbols
    and statements; Signal words;
    Warning statements
    In Vitro Diagnostic Medical Device
        Directive, 252, 254, 263
    label problems, reducing, 468–469
CB radio, *see* Citizens Band (CB) radio
CD players, 456
CE, *see* Conformity markings
Cell lines, *see* Human origin products
Cellular telephones, 454, 455, 456, 457
CEN, *see* Committee for Standardization
    (CEN)
CENELEC, *see* Committee for
    Electrotechnical Standardization
    (CENELEC)
Center for Devices and Radiological Health
    (CDRH), US FDA, 546
Certificate of Confirmation, 176
Certificate of Free Sale
    China, 167–168
    Korea, 177
    Mexico, 146, 148
    Thailand, 184
Certificate of Listing, 38
Certificate of Product for Export, 177
Certification, 384
Cervical cap, 56, 453
Changes
    Australia, 35–39
    Canada, 84
    China, 163–164
    documents, 434–435
    In Vitro Diagnostic Medical Device
        Directive, 266–268
    Japan, 285–286
    United States, 415–416
Chemistry products, 372–373

Children
  advertising to, 30
  child-proof closures, 123
  child-resistant containers, 123
  diapers, 195
  hearing loss, 445
  incubators, 60
  warning statements, 324
China
  acknowledgment symbol, 169
  advertising, 168
  background, 155–156
  basics, 155, 171–172
  bringing devices to market, 158–168
  calendar, 171
  changes, 163–164
  classification of devices, 157, 158–160
  compulsory product certification system,
    164–166
  device registration, 158
  drug/device combinations, 160
  exporting, 167–168
  importing, 160–162, 179
  intent of law, 155–156
  investigational use, 168
  labeling requirements, 168–169
  language requirements, 169–171
  licensing, 162–163
  manufacturers, 158–160
  misbranding, 157
  postmarket surveillance, 167
  product manual, 168–169
  promotion, 168
  quality-system approval, 162–163
  registration, 158
  registration certificate alteration, 163–164
  regulation scope, 156–157
  standards, 166–167
China Certification Commission for Medical
    Devices (CMD), 169
China Compulsory Certification (CCC) mark,
    164–166
Circuit breakers, 259
Citizens Band (CB) radio, 454, 455
Claiming equivalence, 35–37, 38–39
Claims about products
  Australia, 29, 30
  Canada, 73, 74
Classification of devices
  Australia, 60–61, 483–490
  Canada, 76–77, 81–85
  China, 157, 158–160
  Japan, 275, 517–538
  Japanese, 517–538
  Korea, 497–515
  Korean, 173, 497–515
  Medical Device Directive, 217–218
  Thailand, 186–187
  United States, 318–319, 372–373,
    539–544
Cleaning, reducing labeling problems, 463
Clinical products, 372–373
Clinical trials, see also Exemptions;
    Investigational use
  Australia, 42–44, 46, 66
  Canada, 88–89, 106

China, 159
  Clinical Trial Exemption (CTE) scheme,
    42, 43
  Clinical Trial Notification (CTN) scheme,
    42–43
  conformity marking, 203, 210, 213–214
  European Union, 202, 210
  Japan, 299
  Korea, 174, 175, 176
  United States, 419, 423
Clutter words, 465
CMD, see China Certification Commission for
    Medical Devices (CMD)
Cold-cathode gas discharge tubes, 386
Color additives, 325, 419
Color on labels, 475–476, see also
    Presentation requirements
Combination of devices, see Compatibility of
    devices and drugs
Combustible products, 125, see also
    Flammable products
Commencing the supply, 40
Committee for Electrotechnical
    Standardization (CENELEC),
    193, 198, 228, 545
Committee for Standardization (CEN), 193,
    198, 545
Commonly known directions, 351
Comparison of devices, 33
Compatibility of devices and drugs
  Active Implantable Medical Device
    Directive, 241, 243
  Australia, 65
  Canada, 80, 99, 100
  China, 160
  Medical Device Directive, 225
Complaint handling, 438, see also Problem
    reporting
Compliance with standards, see Standards
Component manufacturers, 431–432
Components in kits, see kits
Composition of healthcare industry, 4–5
Composition of products
  Australia, 32
  Brazil, 20
  Canada, 73
Compulsory product certification system
    (CPCS), 164–166
Computed tomography equipment
  China, 168
  United States, 390–393
Conception, see Contraceptives
Conditions for use, label design, 471–472
Condoms, see also Contraceptives
  Australia, 30, 56, 60
  Canada, 109–110
  United States, 450
Conformance to standards, see Standards
Conformity markings, see also Markings
  Active Implantable Medical Device
    Directive, 238, 246
  China, 169
  European Union, 200, 202, 203, 206–210,
    213–214
  In Vitro Diagnostic Medical Device
    Directive, 208

Medical Device Directive, 207
Consumption, medical devices, 4, 6
Contact lenses and products
  Australia, 60
  Brazil, 19
  Canada, 107–108
  European Union, 195
  Japan, 305
  United States, 317, 327, 329, 441
Containers, see Packaging containers and
    materials
Content development, 434
Contents by volume, 252, 254
Continuing education, 333–334
Contraceptives, see also Intrauterine
    contraceptive devices (IUCDs)
  Australia, 53, 55–56
  Canada, 71, 109–110
  European Union, 191
  United States, 451–453
Contract manufacturers, 431
Contraindications, see also Harmful
    statements; Instructions for use;
    Safety information
  Argentina, 14
  Australia, 45
  Canada, 96, 99
  Japan, 305–306, 310
  reducing labeling problems, 468
  United States, 349, 350, 401
Control numbers
  Argentina, 12–13, 14
  Australia, 48, 49, 50, 52, 53, 54, 56, 67
  Canada, 102, 103, 104, 105, 109, 111
  China, 164, 167
  In Vitro Diagnostic Medical Device
    Directive, 252, 254, 256
  Japan, 299, 307
  Korea, 178
  Mexico, 150
  Thailand, 185
  United States, 437, 438
Control panels
  Active Implantable Medical Device
    Directive, 239–240
  Medical Device Directive, 220
Controls for labeling, 433–438
Cooper Committee, 317
Copolymer gloves, 112
Cordless phones, 456
Corrosive products, 127–129
Cosmetics
  Brazil, 19
  Canada, 75
  Korea, 174
  United States, 340
Count, see Quantitative information
Country of origin, see Origin of products
Covers, see Packaging containers and
    materials
Creutzfeldt-Jakob disease, 53
CTE, see Clinical trials
CTN, see Clinical trials
Custom made devices
  Australia, 43, 64, 66
  Canada, 89–90, 106–107

European Union, 202–203, 210–211, 213
Medical Device Directive, 224
United States, 413, 425, 431

D

Dangerous substances, 269–271, *see also*
Hazardous substances
Dangerous Substances Directive, 211
Danger statements, *see* Hazard symbols and
statements; Signal words
Date
of authorization, 246
format of, 97, 171, 186, 187
of implantation, 105
issue for instructions for use, 264, 268
of procedures, 53
revisions, 464
Date of manufacture, *see also* Expiration dates
Active Implantable Medical Device
Directive, 241, 244
Japan, 299
Korea, 178
Medical Device Directive, 223
Mexico, 150
United States, 437
Death, *see* Problem reporting
Deceptive labels, *see* False information;
Misbranding; Omission of facts
and information
Declaration of Conformity, 87–88
Decontamination, 55
Defective products, 406–407
Defibrillators, 356, 457–458
Demonstrations
European Union, 198, 203
Japan, 280–281
Demonstration-type gas discharge devices,
136
Densitometers, 328
Dental Equipment and Supplies
Manufacturing codes, 479, 480
Dental products and equipment
Australia, 46
Canada, 136–138
European Union, 195
Japan, 304, 309, 311
United States, 441–442
Department of Health and Aging (DOHA),
25–44
Designated devices, 287, *see also* Custom
made devices; Special access
devices
Design specifications, *see also* Manufacturing
processes; Sizes
Australia, 32
Canada, 73
Japan, 295
Device classification, *see* Classification of
devices
Device identification, *see also* Control
numbers
Active Implantable Medical Device
Directive, 240, 242
Australia, 47, 49, 50, 51

Canada, 82, 94–95, 98, 102, 103–104, 105
In Vitro Diagnostic Medical Device
Directive, 254
Korea, 178
Medical Device Directive, 222
Thailand, 185
United States, 342
Device interference
Active Implantable Medical Device
Directive, 245, 246, 247
Japan, 295
Medical Device Directive, 227
United States, 355–356, 453–458
Device label design, *see* Label design
Device marking, international symbols,
491–495
Device tracking, *see* Tracking of devices
Diagnostic products, 3, *see also* specific
products
Diagnostic X-ray equipment, 139–141, *see
also* X-ray equipment
Dialysis equipment
Australia, 60
China, 156
Japan, 304–305
United States, 355, 416
Diapers, *see* Children
Diaphragms, 453
Directions for use, *see* Instructions for use
Directives, overlapping, 209–210, *see also*
specific directives
Disease, sexually transmitted, *see* Sexually
transmitted diseases (STDs)
Disinfectants, 37, 53
Display panels
Active Implantable Medical Device
Directive, 239–240
Canada, 118–119
In Vitro Diagnostic Medical Device
Directive, 251
Medical Device Directive (MDD), 220
United States, 341–342, 378–379
Disposal, *see also* Instructions for use
Canada, 101
Japan, 280, 281, 304–305, 305
Medical Device Directive, 227
Mexico, 151
Distributors and distribution, *see also*
Manufacturers
Argentina, 16
Australia, 47, 49
Brazil, 19
Canada, 78, 94, 109
China, 160
good manufacturing practices, 437–438
improper labeling, 6
Korea, 175, 178
Thailand, 182
United States, 437–438
Doctor indentifiers, *see* Healthcare provider
identifiers
Documentation, *see also* Instructions for use;
Presentation requirements
changes, 434–435
label design, 471–472
United States, 380

Document control, 434–435
DOHA, *see* Department of Health and Aging
(DOHA)
Dosage amounts, *see also* Instructions for use
Australia, 45
Japan, 294, 299
Korea, 178
Dressings, *see* Bandages; Sutures
Drug and device combinations, *see*
Compatibility of devices and
drugs
Drug identification, 307–308
Drug-infusion systems
Australia, 53, 54
faults, 26
Duration of treatment, 45, *see also* Instructions
for use

E

Effectiveness of product, *see also* Safety
information
Canada, 79–80
China, 158
Korea, 175–176
United States, 319, 333
Efficacy, *see* Safety information
Electrical devices
Canada, 112–118
Japan, 300–302, 306
Electric shavers, 456
Electrocardiographs, 156
Electromagnetic interference
Active Implantable Medical Device
Directive, 245, 246, 247
Japan, 295
Medical Device Directive, 227
United States, 355–356, 453–458
Electromedical and Electrotherapeutic
Apparatus codes, 479, 481
Electronic labeling, 356
Electronic products, 385
Enforcement Ordinance of PAL (EOPAL), 278
Enforcement Regulations of PAL (ERPAL),
278
Environmental conditions
Active Implantable Medical Device
Directive, 247
Japan, 295
label design, 471
Medical Device Directive, 227
reducing labeling problems, 463
United States, 355–356
EOPAL, *see* Enforcement Ordinance of PAL
(EOPAL)
Equivalence
Australia, 35–37, 38–39
Japan, 284–582
Establishment license, 77–78
European Active Implant Harmonized
Standards, 545
European Electromagnetic Compatibility
Harmonized Standards, 545
European industry regulation, 5

European *In Vitro* Diagnostic Medical Device Harmonized Standards, 545
European Medical Device Harmonized Standards, 545
European Packaging and Packaging Waste Harmonized Standards, 545
European Personal Protective Equipment Harmonized Standards, 545
European Union
    Active Implantable Medical Device Directive, 195, 207–208
    background, 192–193
    basics, 191–192, 213–214
    bringing devices to market, 198–203
    clinical evaluations, 202, 210
    competent authorities, 201
    conformity, 200, 206–210
    custom made devices, 202–203, 210–211
    Dangerous Substances Directive, 211
    demonstrations, 203
    exhibitions, 203
    intent of law, 192–193
    investigational marking, 210
    In Vitro Diagnostic Device Directive, 195, 208
    labeling requirements, 196–197
    language requirements, 197
    local requirements, 213
    Low-Voltage Directive, 211
    Machinery Directive, 211
    Medical Device Directive, 195, 205, 207
    mutual recognition agreements, 193–194
    notified bodies, 196, 200–201
    overlapping directives, 209–210
    packaging and packaging waste, 212
    performance evaluation, 202, 210
    postmarket requirements, 203–206
    Proprietary Medicinal Products Directive, 211
    Radio Equipment Telecommunications Terminal Equipment Directive, 195–196, 208–209, 545
    reevaluation, 202
    regulation scope, 194–196
    safeguard clause, 205–206
    standards, 197–198
    trade fairs, 198, 203
    vigilance, 203, 204–205
Evaluation of labels, 469
Examination gloves, *see* Gloves
Exemptions, *see also* Clinical trials
    Australia, 26, 32, 42–44, 486–490
    Canada, 88–89
    In Vitro Diagnostic Medical Device Directive, 271
    United States, 347–352, 371, 419–426, 423, 430–432
Exhibitions
    European Union, 198, 203
    Japan, 280–281
Expected test values, *see* Performance expected
Experienced gained reviews, 62
Experimental usage, *see* Clinical trials

Expiration dates, *see also* Date of manufacture
    Active Implantable Medical Device Directive, 241, 243, 244
    Australia, 56, 64
    Canada, 96, 109, 110
    In Vitro Diagnostic Medical Device Directive, 252, 254, 256–257
    Korea, 178
    Medical Device Directive, 223
    Mexico, 148–149, 150
    Thailand, 182, 185, 186
    United States, 352, 381, 437
Explosive materials, 104
Exporting
    Argentina, 16
    Australia, 37–38, 40, 46, 47–48, 62
    Canada, 82, 83, 90, 107
    China, 167–168
    Japan, 297–298
    United States, 319–320, 322, 357–360, 385
Exposure to environmental conditions
    Active Implantable Medical Device Directive (AIMDD), 247
    Japan, 295
    Medical Device Directive (MDD), 227
    reducing labeling problems, 463
    United States, 355–356
Extracorporeal-therapy systems, 53
Extreme danger statements, *see* Hazard symbols and statements; Signal words

F

Fabrication Book, 13
Failure to inform, *see* Omission of facts and information
Fairs, trade
    European Union, 198, 203
    Japan, 280–281
False information, *see also* Misbranding; Omission of facts and information
    Australia, 29, 35
    Canada, 74–75, 93
    Japan, 279
    United States, 324, 325–326
FDA, *see* U.S. Food and Drug Administration (FDA)
Feces, *see* Human origin products
Field-wired terminal boxes, 259
Fire transceivers, 455
First aid statements, 119, 122
Flammable products, 124–127
Flowcharts, label problems, 467
Fluids, human body, 41
Fluoroscopic equipment, 390
Follow-up action, 263, 266
Fonts, 474, *see also* Instructions for use; Label design
Food, Drug, and Cosmetic Act (FD&CA), *see also* United States
    adulteration, 324–325
    advertising, 324, 326–336

article reprints, 330–333
background, 316–321
basics, 315–316, 336
continuing education, 333–334
false information, 325–326
intent of law, 316–321
Internet, 334–336
labeling requirements, 323
misbranding, 324–325
misleading information, 325–326
off-label uses, 329–330
promotion, 326–336
regulations and scope, 322–323
related laws, 321–322
United States, 322, 323, 351
Food and Drug Administration, *see* specific country
Food and Drugs Act (F&DA), *see also* Canada
    adulteration, 74
    advertising, 74, 75–76
    background, 71–73
    basics, 71, 90–91
    bringing devices to market, 76–85
    classification of devices, 76–77, 81–85
    clinical trials, 88–89
    complaint handling, 78
    custom devices, 89–90
    distribution records, 78
    effectiveness requirements, 79–80
    establishment license, 77–78
    exemptions, 88–89
    exporting, 90
    false information, 74–75
    foreign manufacturers, 85
    implant registration, 79
    intent of law, 71–73
    investigational exemptions, 88–89
    labeling, 73–74
    licenses, 77–78, 81–85
    mandatory problem reporting, 78–79
    misbranding, 74
    misleading information, 74–75
    problem reporting, 78–79
    promotion, 75–76
    regulations, 73
    regulation scope, 73
    safety measures, 79–80
    special access devices, 89–90
    standards, 85–88
Foreign manufacturers, *see* Manufacturers
Format of label, *see* Label arrangement and format
Formulation of products, *see* Composition of products
Fractionation Book, 13
Frequency of treatment, 45, *see also* Instructions for use

G

Generally known directions, 351
Global Harmonization Task Force, 546
Global market healthcare industry, 4
Glossary of Acronyms, 547–551
Gloves, 60, 110–112

Good clinical practice (GCP), 176
Good manufacturing practices (GMPs), *see also* Qualitative information
   Australia, 39
   Brazil, 3
   European Union, 193
   Japan, 282–283
   Korea, 173
   Mexico, 146, 148
Good manufacturing practices (GMPs), United States, *see also* Qualitative information; United States
   applications, 430–433
   approval, 434
   basics, 429–430, 439–440
   changes to document, 434–435
   clinical trials, 419, 423
   complaint handling, 438
   component manufacturers, 431–432
   content development, 434
   contract manufacturers, 431
   controls for labeling, 433–438
   custom made devices, 431
   distribution, 437–438
   document control, 434–435
   exemptions, 419, 423, 430–432
   handling conditions, 436
   inspection of components, 435–436
   integrity of label, 435
   laboratories, contract, 431
   manufacturers, 431–432
   nonconforming products, 437
   overlabeling, 439
   purchasing, 435
   quality, 433–438
   receipt of components, 435–436
   relabelers, 432
   remanufacturers, 432
   repackers, 432
   specification developer, 433
   storage conditions, 436
   testing laboratories, 431
   traceability, 438
   transport conditions, 439
Government, Mexico, 147–149
Grandfathering, 26, 177, 179
Graphics, 475, 491–495
Guarantee
   Brazil, 20
   China, 159, 160, 161

H

HAM radios, 456
Handheld portable transceivers, 455
Handling conditions
   Active Implantable Medical Device Directive, 243
   Medical Device Directive, 223
   United States, 436
Handling requirements, *see* Instructions for use

Harmful statements, *see also* Hazard symbols and statements; Safety information; Signal words; Warning statements
   home-use devices, 123
   toxic products, Canada, 132–133
Harmonized standards, 197–198, 239
Hazardous Substance Act (Thailand), 186–187, 188
Hazardous substances
   In Vitro Diagnostic Medical Device Directive, 269–270
   United States, 375–381
Hazard symbols and statements, 100, 104, 119–121, 122, *see also* Signal words; Warning statements
HCV, *see* Hepatitis C Virus (HCV)
Healthcare facilities, 78, 89, 106
Healthcare industry
   basics, 3
   composition, 4–5
   global market, 4
   regulation, 5–6
   technology, 3–4
Healthcare provider identifiers
   Australia, 53, 54, 66
   Canada, 105
Hearing aids, 327, 443–447
Heart-rate monitors, 60
Heart valves
   Australia, 53, 60
   faults, 26
   Mexico, 146, 148
Hematology products, 373
Hepatitis C Virus (HCV), 51, 52, 53, 56
High-intensity mercury vapor lamps, 136, 405
Highlighting, label design, 474
Home-use devices, *see* Household-use devices
Hospital identifiers, 53, 54
Household-use devices
   Australia, 52
   Canada, 122
   Japan, 306, 310
   United States, 374–375
Human Immunodeficiency Virus (HIV)
   Australia, 43, 51, 52, 53, 56
   United States, 452–453
Humanitarian exemptions, 413, 425–426
Human origin products, *see also* Animal origin products; Origin of products
   Australia, 41, 46, 52, 53, 54–55
   Canada, 83
Hydrocarbon polymer gloves, 112

I

Illustrations, label design, 475
Immediate containers, *see* Packaging containers and materials
Immunology products, 373
Implantable devices, *see also* Active implantable medical devices (AIMDs)
   Active Implantable Medical Device Directive, 246

Australia, 46, 47, 48, 49, 50, 65
Korea, 177
Medical Device Directive, 226
United States, 457
Implant registration, 78, 79, 105–106
Importing, *see also* Manufacturers
   Argentina, 11–12, 16
   Australia, 37–38, 40–41, 47–48, 62
   Brazil, 18, 19
   Canada, 78–79, 87, 89, 94
   China, 158, 160–162, 168
   European Union, 198
   improper labeling, 6
   Japan, 282–283, 286, 287, 292, 293, 298
   Korea, 176–177, 178, 179
   Mexico, 146, 149, 151
   Thailand, 181, 182
   United States, 320, 322, 360–363
Improper labeling, *see* False information; Misbranding; Misleading information; Omission of facts and information
Inactivation procedures, 55
Inch-pound system, 344–345
Incidental radiators, 409
Incidents, *see* Problem reporting
Incompatibilities, *see* Contraindications
Incubators, baby, *see* Children
Index, reducing labeling problems, 464
Indications for use, *see* Instructions for use
Individually wrapped goods, 50, *see also* Packaging containers and materials
Individual Patient Use (IPU) scheme, 43
Individuals recruited for studies, 422–423, *see also* Patient identifiers
Industrial equipment, 409–410
Industry composition, 4–5
Infectious agents, 103, 104
Information format, *see* Presentation requirements
Injury, *see* Problem reporting
Inspections, 62, 435–436
Installation of devices
   Australia, 65
   In Vitro Diagnostic Medical Device Directive, 266–268
   Medical Device Directive, 225
Instructions for use, *see also* Contraindications; Harmful statements; Safety information
   Active Implantable Medical Device Directive, 244–246, 245
   Argentina, 14
   Australia, 45, 64–65
   Canada, 95–96
   contraceptives, 55, 56, 110
   dental X-ray equipment, 138
   designing labels, 471–477
   diagnostic X-ray equipment, 140–141
   electrical equipment, 227
   electromedical equipment, 116–117
   European Union, 196–197
   immediate container labels, 102–103
   in vitro diagnostic devices, 98–101

In Vitro Diagnostic Medical Device
Directive, 258–290, 260–267
Japan, 294–296
Korea, 178
label designing, 471–477
label problems, reducing, 465–467
label requirements, 63–64, 185
Medical Device Directive, 220–221,
224–226, 227
medical device license application, 83, 84
packages, 49, 50
reagent labels, 104
reducing labeling problems, 463
registration of devices, 35
sterile devices, 353
Thailand, 185
United States, 340, 348–349, 353, 401,
441–451
Instruments and equipment, *see* specific
device
Insulation, 259
Insulin syringes, *see* Syringes
Integrity of label, 435
Intellectual property, protecting, 18
Intended use, *see also* Instructions for use
In Vitro Diagnostic Medical Device
Directive, 250, 252, 254, 257,
261, 264
Medical Device Directive, 220–221
reducing labeling problems, 463
United States, 339–340, 366, 416–417
Intentional radiators, 408–409
Intent of law
Argentina, 9
Australia, 26
Brazil, 17
Canada, 71–73, 135
China, 155–156
European Union, 192–193
Korea, 173–174
Thailand, 181
United States, 316–321
Interactions, *see* Compatibility of devices and
drugs
Interference from electrical fields
Active Implantable Medical Device
Directive, 245, 246, 247
Japan, 295
Medical Device Directive, 227
United States, 355–356, 453–458
International Electrotechnical Commission
(IEC), 546
International Standard Paper Sizes, 546
International Standards Organization (ISO),
174, 546
International symbols, medical device
marking, 491–495
Internet, 328, 334–336, *see also* Web sites
Intraocular lenses
Australia, 53, 55
faults, 26
Japan, 305–306
United States, 424–425
Intraocular visco-elastic fluids, 53

Intrauterine contraceptive devices (IUCDs),
see also Contraceptives
Australia, 51, 53, 55
faults, 26
United States, 317, 452
Intravascular catheters, 417
Invasive devices, *see* specific device
Investigational use, *see also* Clinical trials
Australia, 42
Canada, 83, 84, 87, 88–89, 106
China, 168
European Union, 210
In Vitro Diagnostic Medical Device
Directive, 253, 255, 260
Medical Device Directive, 224
United States, 419–425
Investigators, 421–422, *see also* Clinical trials
In Vitro Diagnostic Medical Device Directive
(IVDD)
basics, 5, 213, 249, 271–272
calibration, 266–268
caution statements, 252, 254, 263
changes in performance, 266–268
conformity marking, 208
contents, 252, 254
dangerous substances, 269–271
date of issue for IFU, 264, 268
device identification, 254
display panels, 251
European Union, 191–192, 194, 195, 202,
205, 208
exemptions, 271
expiration dates, 252, 254, 256–257
follow-up action, 263, 266
hazardous substances, 269–270
immediate containers, 251–253
implementation of labeling requirements,
270–271
improper labeling, 249
installation, 266–268
instructions for use, 258–290, 260–267
instruments and equipment, 264–268
intended purpose, 250, 252, 254, 257, 261,
264
investigational use, 253, 255, 260
in vitro diagnostic devices, 255–260
labeling requirements, 250, 270–271
literature references, 263, 266
lot numbers, 252, 254, 256
manufacturers, 252, 253–254, 256
markings, 258–260
measuring function, 250–251
monitoring, 251
omission of information, 263–264, 268
operating instructions, 255
outer containers, 253–255
performance expected, 262, 265
product identification, 250, 252, 253, 256
quality-control, 266
radiation-emitting devices, 263
reagents, 251–255, 261
results, explanation, 262–263, 266
self-testing, 250
shelf life, 262
specimens, 262, 265

sterile devices and conditions, 253, 255,
263
storage information, 253, 254–255, 262
technical specifications, 268
testing, 262, 265–266
warnings, 257–258
In vitro diagnostic products
Australia, 43, 46, 51–52
basics, 3
Canada, 98–104
instrumentation and equipment, 255–260
Japan, 282, 306–309
NAICS codes, 479–481
Thailand, 186–187
In vitro diagnostic products, United States
accompanying documents, 380
basics, 365–370, 381–382
chemistry products, 372–373
classifications, 372–373
clinical products, 372–373
display panels, 378–379
exemptions, 371
expiration dates, 381
Food, Drug, and Cosmetic Act, 322, 323,
351
hazardous substances, 375–381
hematology products, 373
household-use devices, 374–375
immediate containers, 366–367
immunology products, 373
labeling requirements, 366, 376, 435
microbiology products, 373
outer containers, 380
package inserts, 367–370
pathology products, 373
poisons, 380–381
presentation requirements, 379–380
prominence of labeling, 376–377
reagents, 371
special cases, 370–372
storage information, 381
toxicology products, 372–373
wrappings, 380
In Vitro Diagnostic Substance codes, 479, 481
In vivo diagnostic products, 480–481
Ionizing radiation, 385–394, *see also*
Radiation-emitting devices
IPU scheme, *see* Individual Patient Use (IPU)
scheme
Irradiation Apparatus Manufacturing codes,
479, 481
Irritants, 129, *see also* Corrosive products
IUCDs, *see* Intrauterine contraceptive devices
(IUCDs)
IVDD, *see* In Vitro Diagnostic Medical Device
Directive (IVDD)

J

Japan, *see also* Pharmaceutical Affairs Law
(PAL)
approval standards, 293
basics, 291, 310–311
classification of devices, 517–538
clinical trials, 299

contact lenses, 305
dental materials, 304, 309, 311
disposable products, 304–305
electrical devices, 300–302, 306
exporting, 297–298
household-use devices, 306, 310
immediate containers, 292–293
importing, 292, 293, 298
industry regulation, 5
intraocular lenses, 305–306
in vitro diagnostic devices, 306–309
labeling requirements, 291–297
language requirements, 296–297, 308, 309
laser devices, 303–304
manufacturers, 292, 293, 294
misbranding, 291
package inserts, 294–296, 308–309
packaging materials, 292–293, 307–309
performance standards, 292
prominence of information, 296
radiation-related devices, 303
reagents, 307–309
standards, 292–293
steel devices, 305
testing, 298–299
trade names, 297
ultrasonic equipment, 303
warning statements, 294–296, 299–306
Japanese Ministry of Health, Labor, and
Welfare (MHLW), 546

K

Kits
Australia, 52–53, 56, 67
Canada, 81, 95, 99, 102, 103
European Union, 192
Japan, 304–305
Korea
advertising, 175, 179
background, 173–174
basics, 173, 178–179
bringing devices to market, 175–177
classification of devices, 497–515
clinical trials, 174, 175, 176
Food and Drug Administration, 174, 546
grandfathering, 177, 179
importing, 176–177, 178, 179
instructions for use, 178
intent of law, 173–174
labeling, 178
language requirements, 178, 179
premarket requirements, 175–176
presentation requirements, 178
regulations and scope, 5, 174–175

L

Label arrangement and format, *see also*
Presentation requirements;
Prominence of information
Argentina, 14
Australia, 63
Canada, 119–122

Label design
basics, 471, 476–477
color, 475–476
conditions for use, 471–472
device labels, 471
documentation, 471–472
graphics, 475
highlighting, 474
illustrations, 475
layout, 472
physical attributes, 472–474
typefaces, 474
white space, 474–475
Labeling materials, 13, 472–474
Labeling problems, reducing
audience consideration, 461–462
basics, 461, 470
caution statements, 468–469
choosing words carefully, 465
clear writing, 464–465
clutter, avoiding, 465
flowchart, 467
instructions, 465–467
labeling evaluation, 469
lay users, 461–462
liability, limiting, 469
list, 467
organization of information, 462–464
professional users, 462
text, 466–467
warning statements, 468–469
Labeling requirements
Active Implantable Medical Device
Directive, 238–243
Argentina, 13–14
Australia, 27–28, 46–51, 62–67
basics, 6
Brazil, 20
Canada, 73–74, 93–97, 136–142
China, 168–169
European Union, 196–197
In Vitro Diagnostic Medical Device
Directive, 250, 270–271
Japan, 291–297
Korea, 177, 178
Medical Device Directive, 218–224,
228–235
Thailand, 185–186
United States, 323, 338–341, 366, 376,
383–385, 395–399
Laboratories, contract, 431
Laboratory Apparatus and Furniture
Manufacturing codes, 479, 480
Lancets, 304–305
Language requirements
Argentina, 14, 16
Australia, 47, 52, 56, 63
Brazil, 20
Canada, 79, 93, 97, 106, 121, 123, 124
China, 161, 167, 169–171
European Union, 197, 214
Japan, 296–297, 308, 309
Korea, 178, 179
Mexico, 145, 149, 150, 151
Thailand, 186

United States, 341
Lanham Act, 328
Laptop computers, 456
Lasers
Australia, 60
Canada, 136
Japan, 303–304
United States, 394–400
Latex products, 111–112, 450
Law, intent of
Argentina, 9
Australia, 26
Brazil, 17
Canada, 71–73, 135
China, 155–156
European Union, 192–193
Korea, 173–174
Thailand, 181
United States, 316–321
Law enforcement devices, 351–352
Layout of label, *see* Label arrangement and
format; Presentation
requirements; Prominence of
information
Lay users, *see* User information
Liability, limiting, 469
Licenses
Australia, 39
Canada, 77–78, 81–85, 87
China, 161, 162–163
Japan, 282–283
United States, 418–419
Life of device, *see* Shelf life
Life-support devices, 287
Light-emitting products, 394–405
Lists, reducing label problems, 467
Literature, *see* Documentation; Instructions
for use
Literature references, *see also* Report
references
In Vitro Diagnostic Medical Device
Directive, 263, 266
United States, 330–333, 370, 402
Long-range transmitters and transceivers, 455
Lot numbers, *see* Control numbers
Low-energy electron microscopes, 136
Low-Voltage Directive, 211
Lubricants, personal, 30

M

Machinery Directive, 211
Maintenance, *see also* Instructions for use
Australia, 65
In Vitro Diagnostic Medical Device
Directive, 267
Medical Device Directive, 225
United States, 400
Maintenance, reducing labeling problems, 464
Mandatory problem reporting, *see* Problem
reporting
Manual, *see* Product manuals

Manufacturers, *see also* Distributors and
distribution; Importing; Origin of
products
Active Implantable Medical Device
Directive, 240, 242
Argentina, 13, 16
Australia, 32, 37–38, 39, 47–48, 49, 50,
53, 56, 60, 62, 67
Brazil, 18, 19
Canada, 78–79, 85, 94, 102, 103, 109
China, 158–160, 159
European Union, 198–200, 214
improper labeling, 6
In Vitro Diagnostic Medical Device
Directive, 252, 253–254, 256
Japan, 282–283, 287, 292, 293, 294
Korea, 178
Medical Device Directive, 222, 225
Mexico, 146
Thailand, 185
United States, 318, 321, 338–339, 375,
431–432
Manufacturing processes, 83, 84, *see also*
Design specifications
Markings, *see also* Conformity markings
international symbols, 491–495
In Vitro Diagnostic Medical Device
Directive, 258–260
Japan, 300
Materials in device, 83, *see also* specific type
of device
Measuring functions and equipment, *see also*
Units of measure
Australia, 65
China, 157
In Vitro Diagnostic Medical Device
Directive, 250–251
Medical Device Directive, 221, 227
United States, 342, 344–347
Medical Device Directive (MDD)
Active Implantable Medical Device
Directive, 228
basics, 5, 213, 215–217, 235
blood derivatives, 224
classification of devices, 217–218
compatibility, 225
conformity marking, 207
controls and displays, 220
custom made devices, 224
date of manufacture, 223
device identification, 222
disposal, 227
European Union, 195, 205, 207
expiration date, 223
exposure to environmental conditions, 227
handling conditions, 223
implantable devices, 226
installation, 225
instructions for use, 220–221, 224–226,
227
intended purpose, 220–221
investigational uses, 224
labeling requirements, 218–224, 228–235
maintenance, 225
manufacturers, 222, 225
measuring function, 221, 227

medical electrical equipment, 228–235
medicinal information, 227
misbranding, 218
packaging, 226
patient information, 227
performance expected, 225
precautions, 224
preparation directions, 226
product identification, 223
radiation-emitting devices, 226
reciprocal interference, 226
reuse of devices, 226
single-use devices, 223
special operating instructions, 224
special purpose devices, 224
sterile conditions, 222, 226
storage conditions, 223
units of measure, 221
warnings, 224
Medical equipment
basics, 4, 6
electrical, 228–235
international marking symbols, 491–495
United States, 409–410
Medical gloves, 60, 110–112
Medium-range mobile transceivers, 455
Menstrual tampons
Canada, 108–109
European Union, 195
labeling failure, 29
United States, 447
Mercury vapor lamps
Canada, 136
United States, 405
Metric units, *see* Units of measure
Mexico
background, 144
basics, 143–144, 151
bringing devices to market, 145–149
intent of law, 144
labeling, 149–151
minimum labeling requirements, 150–151
misbranding, 149
packaging types, 149–150
product registration, 145–147
registry of medical devices, 5
regulations, 144–145
selling to government, 147–149
web site, 546
Microbiological products, 41, 373
Microscopes, 136
Microwaves, 136, 394
Minors, *see* Children
Misbranding, *see also* False information;
Omission of facts and
information
Active Implantable Medical Device
Directive, 238
Argentina, 12–13
Australia, 28, 45
Brazil, 20
Canada, 74, 93
China, 157
Japan, 279, 291
Medical Device Directive, 218
Thailand, 182–183

United States, 324–325, 338
Misdirection, *see* False information;
Misbranding; Misleading
information
Misleading information
Australia, 29, 35
Canada, 74–75
Japan, 279
United States, 325–326, 340
Model of devices, 53, 54
Monitoring, 251
Motorized scooters, 456
Multiple devices, package of, *see* Quantitative
information
Mutual recognition agreements (MRA),
193–194

N

Name of devices, *see* Device identification
Names of patients, *see* Patient identifiers
National Administration of Drugs, Foods, and
Medical Technology (ANMAT),
9–16
Natural rubber products, 111–112, 450–451
Neoprene gloves, 112
Net contents statement, 346, 372, *see also*
Quantitative information
Nonconforming products, 437
Noninvasive devices, 76, 239, 246
Nonreactive ingredients, 99
Nonsterile products, 48, 49, *see also* Sterile
conditions and products
North American Industry Classification
System (NAICS), 4, 479–481
Notarization
China, 161
Korea, 177
Mexico, 146
Notification review program, 414–415
Notified Bodies, 196, 200–201

O

Off-label uses, 329–330, 335
Omission of facts and information, *see also*
False information; Misbranding
Australia, 29
Canada, 75
European Union, 205
In Vitro Diagnostic Medical Device
Directive, 263–264, 268
United States, 324, 326
Operating instructions, *see* Instructions for use
Ophthalmic Goods Manufacturing codes, 479,
480
Organization, reducing label problems,
462–464
Organs, 41
Origin of products, *see also* Animal origin
products; Human origin
products; Manufacturers
Argentina, 14
Australia, 32, 54

Brazil, 20
Korea, 178
Thailand, 185
United States, 322, 323, 361–362
Outer containers, *see* Packaging containers
        and materials
Overlabeling, 439
Overlapping directives, 209–210, *see also*
        specific directives
Over-the-counter devices, 341–347
Ozone, 443, 448–450

P

Pacemakers
    Active Implantable Medical Device
        Directive, 247
    Australia, 47, 60
    China, 156
    United States, 356, 457–458
Package inserts
    Canada, 73, 98–103
    Japan, 294–296, 308–309
    Korea, 177
    United States, 367–370
Packaging containers and materials, *see also*
        Presentation requirements
    Argentina, 13
    Australia, 27–28, 50
    Canada, 83, 102–103
    European Union, 212
    In Vitro Diagnostic Medical Device
        Directive, 251–255
    Japan, 292–293, 307–308, 307–309, 309
    United States, 366–367, 380
Paper, glossy, 474, *see also* Product manuals
Pathology products, 373
Patient identifiers
    Australia, 53, 54, 66
    Canada, 79, 106
Patient information, 245, 246–247
People's Republic of China, *see* China
Performance changes, 266–268, *see also*
        Product changes
Performance expected
    Active Implantable Medical Device
        Directive, 244, 247
    Canada, 101
    European Union, 202, 210
    In Vitro Diagnostic Medical Device
        Directive, 262, 265
    Japan, 292
    Medical Device Directive, 225
    United States, 370
Personal lubricants, 30
Pharmaceutical Affairs Law (PAL), *see also*
        Japan
    adulteration, 279
    advertising, 279–281
    approval of device, 283–284, 286, 287
    basics, 275, 288–289
    bringing devices to market, 282–287
    changes in products, 285–286
    designated devices, 287
    Enforcement Ordinance of PAL, 278

Enforcement Regulations of PAL, 278
    equivalence, 284–582
    exhibitions, 280–281
    false information, 279
    importing, 282–283, 286, 287
    in vitro diagnostic products, 282
    licenses, 282–283
    life-support devices, 287
    manufacturers, 282–283, 287
    misbranding, 279
    misleading information, 279
    prefecture government role, 286
    problem reporting, 287–288
    promotion, 279–281
    reexamination, 284–582
    regulations, 276–278
    revisions, 275–276
    tracking of devices, 288
    vigilance, 287–288
Pharmacist degree, 12
Photofluorographic X-ray equipment,
        138–139
Physical attributes of labels, 472–474
Physician identifiers, *see* Healthcare provider
        identifiers
Plant origin material, 41
Plasma, *see* Human origin products
Poisons, 380–381, *see also* Hazardous
        substances; Toxic products
Police transceivers, 455
Polymerase chain reaction, 52
Polyolefin gloves, 112
Polyvinyl chloride, 112
Postmarket requirements
    Australia, 62
    Canada, 86
    China, 160, 167
    European Union, 203–206
    Korea, 173
    Thailand, 184–185
Powdered natural rubber latex gloves, 111
Powderless natural rubber latex gloves, 111
Precautions, *see* Harmful statements; Safety
        information; Warning statements
Prefecture government role, 286
Pregnancy, 71, *see also* Contraceptives
Premarket requirements
    Korea, 175–176
    United States, 318, 414–419
Preparation directions, *see also* Instructions
        for use
    Canada, 99
    Japan, 295
    Medical Device Directive, 226
    reducing labeling problems, 463
Prescription devices, 348
Presentation requirements, *see also* Label
        arrangement and format;
        Packaging containers and
        materials; Prominence of
        information
    Australia, 32
    Canada, 119–121, 120
    In Vitro Diagnostic Medical Device
        Directive, 270–271
    Japan, 294, 308

Korea, 178
    label design, 471–477
    Mexico, 145–146, 151
    United States, 46–47, 332–333, 343,
            376–377, 379–380, 395–399
Preservatives, 99
Pressurized containers and contents
    Canada, 122, 123–124
    China, 156
Preventive maintenance, *see* Instructions for
        use; Maintenance
Price, 20, 176
Problem reporting
    Australia, 62
    Canada, 78–79
    European Union, 203, 204–205
    United States, 319
Procedure dates, 53, *see also* Date
Procedure packs, *see* Kits
Product changes, 35–39, *see also* Performance
        changes
Product identification
    Active Implantable Medical Device
        Directive, 239
    In Vitro Diagnostic Medical Device
        Directive, 250, 252, 253, 256
    Medical Device Directive, 223
    United States, 384
Production, *see* Manufacturers
Product manuals
    China, 159, 160, 161, 168–169
    design, 472–474
    European Union, 197
    In Vitro Diagnostic Medical Device
        Directive, 266
Product registration, *see* Registration
Prohibited practices, 420–421
Prominence of information, *see also* Label
        arrangement and format;
        Presentation requirements
    Canada, 93, 97
    Japan, 296
    United States, 340–341, 376–377
Promotion, *see also* Advertising
    Australia, 29–31
    Canada, 75–76
    China, 168
    Japan, 279–281
    United States, 326–336
Prophylactics, 110, condoms, *see also*
        Contraceptives
Proprietary Medicinal Products Directive, 211
Public-health threats, 62
Publicly known directions, 351
Purchasing, 400, 435
Purpose of device, *see* Instructions for use;
        Intended use

Q

Qualitative information, *see also* Good
        manufacturing practices (GMPs)
    Argentina, 32, 33
    Australia, 39
    Brazil, 20

Canada, 73, 83, 84, 99
China, 162–163
In Vitro Diagnostic Medical Device
    Directive, 266
Japan, 283
Mexico, 151
United States, 372, 433–438
Quality Control Book, 13
Quantitative information, 48
Australia, 48, 49, 50, 51
Brazil, 20
Canada, 99
Japan, 299, 307, 309
Korea, 176
Mexico, 149, 151
United States, 342–344, 346–347, 367,
    372
Quarantine of products, 41, 51

R

Radiation-emitting devices
Australia, 65
In Vitro Diagnostic Medical Device
    Directive, 263
Korea, 174
Medical Device Directive, 226
radio-frequency radiation, 394, 407–410
United States, 321–322
Radiation-emitting devices, Canada, *see also*
    Canada
background, 135
basics, 135, 142
dental X-ray equipment, 136–138
diagnostic X-ray equipment, 139–141
intent of law, 135
labeling requirements, 136–142
photofluorographic X-ray equipment,
    138–139
regulations, 135–136
ultrasound-therapy devices, 141–142
Radiation-emitting devices, United States, *see*
    also United States
assembler information, 387
basics, 383, 410–411
certification, 384
cold-cathode gas discharge tubes, 386
computed tomography equipment,
    390–393
defective products, 406–407
electronic products, 385
exporting, 385
fluoroscopic equipment, 390
high-intensity mercury vapor lamps, 405
industrial equipment, 409–410
ionizing radiation, 385–394
labeling requirements, 383–385, 395–399
lasers, 394–400
light-emitting products, 394–405
medical equipment, 409–410
microwave radiation, 394
product identification, 384
purchasing information, 400
radio-frequency radiation, 394, 407–410
radiographic equipment, 389–390

repair of equipment, 389
scientific equipment, 409–410
service information, 400
sunlamp products, 403–405
television receivers, 385–386
ultrasonic products, 405–406
ultraviolet lamps, 403–405
user information, 387–388, 399–400,
    404–405
variances, 384–385
warning statements, 389
X-ray systems, 393–394
x-ray systems, 386–393
Radiation-related devices, 303
Radioactive substances, 243, 246
Radio Equipment and Telecommunications
    Terminal Equipment Directive
    (R&TTED), 195–196, 208–209
Radio Equipment and Telecommunications
    Terminal Equipment
    Harmonized Standards, 545
Radio-frequency radiation, 394, 407–410
Radiographic equipment, 389–390, *see also*
    X-ray equipment
Radiopharmaceuticals, 72
Reactive ingredients, *see* Catalytic ingredients
Reagents
Australia, 56
Canada, 99, 103–104
China, 157
European Union, 191
In Vitro Diagnostic Medical Device
    Directive, 251–255, 261
Japan, 307–309
United States, 367, 368, 371
Recalls, *see* Misbranding
Receipt of components, 435–436
Recipients, *see* Patient identifiers
Record Book of Intake of Imported Products,
    13
Recruiting study subjects, 422–423
Reevaluation, 202
Reexamination, 284–582
References, 553–565
Registration, *see also* Control numbers
Australia, 33–35, 44, 53–56, 61
Brazil, 18
China, 158
Thailand, 183
Registration certificate alteration, 163–164
Registration numbers, *see* Control numbers
Regulations and scope
Argentina, 9
Australia, 27, 60–62
Brazil, 17
Canada, 73, 135–136
China, 156–157
European Union, 194–196
healthcare industry, 5–6
Japan, 276–278
Korea, 5, 174–175
Thailand, 181–182
United States, 322–323
Regulatory history, Australia, 32
Relabelers, 432
Remanufacturers, 432

Repackers, 351, 352–353, 432
Reportable incidents, *see* Problem reporting
Report references, *see also* Literature
    references
Canada, 83, 84, 102
China, 159
Reprocessing of devices, *see also* Reuse of
    devices
Argentina, 15
United States, 321, 355
Republic of Korea, *see* Korea
Research devices, 351–352
Resterilization, 15, 16, *see also* Sterile
    conditions and products
Results of tests, *see also* Tests
Canada, 101
In Vitro Diagnostic Medical Device
    Directive, 262–263, 266
Retail theft prevention systems, 458
Reuse of devices, *see also* Instructions for use;
    Reprocessing of devices
Active Implantable Medical Device
    Directive, 246
Argentina, 15, 16
Medical Device Directive, 226
United States, 355
Revisions, *see* Changes
R&TTED, *see* Radio Equipment and
    Telecommunications Terminal
    Equipment Directive (R&TTED)
Rubber latex gloves, 111–112

S

Sachets, *see* Packaging containers and
    materials
Safeguard Clause, 205–206, 238
Safety information, *see also*
    Contraindications; Harmful
    statements; Instructions for use;
    Signal words
Australia, 32, 33, 65, 67
Brazil, 20
Canada, 73, 79–80
China, 158
Korea, 175–176
purpose of labeling, 6
United States, 319, 333
Sans serif typeface, 474, *see also* Label design
Scalpels, 95
Scientific equipment, 409–410
Security transceivers, 455
Selection of device, 245, *see also* User
    information
Self-service display labeling, 97
Self-testing devices, 250, 261–266, 268
Semen, *see* Human origin products
Serial numbers, *see* Control numbers
Serif typefaces, 474, *see also* Label design
Serum, *see* Human origin products
Service manuals, 35, 400, *see also* Instructions
    for use
Sexually transmitted diseases (STDs), 110,
    451–453

Shelf life
  Active Implantable Medical Device
      Directive, 247
  Brazil, 20
  Canada, 80, 97
  In Vitro Diagnostic Medical Device
      Directive, 262
Shoplifting prevention systems, 458
Signal words, *see also* Hazard symbols and
      statements; Safety information;
      specific products
  Canada, 100, 119–120, 121
  reducing labeling problems, 468
SI labeling, 345
Single-use devices, 223, 355, *see also*
      Instructions for use
Sizes, 49, 50
Small packages, 50, 122
Software in devices
  Canada, 80, 100
  China, 157
  European Union, 195
Special access devices
  Canada, 89–90, 106–107
  United States, 416
Special labeling requirements, United States,
      see also United States
  basics, 441, 458
  cellular telephones, 457
  condoms, 450
  contact lens solutions and tablets, 441
  contraceptives, 451–453
  defibrillators, 457–458
  dentures, 441–442
  electromagnetic interference, 453–458
  hearing aids, 443–447
  menstrual tampons, 447
  natural rubber, 450–451
  ozone, 443, 448–450
  pacemakers, 457–458
  sexually transmitted diseases, 451–453
  theft prevention systems, 458
  use-related statements, 441–451
  wheelchairs, electric, 453–457
Special operating instructions, 224
Special purpose devices, *see also* Custom
      made devices
  Active Implantable Medical Device
      Directive, 241, 243
  Australia, 66
  Canada, 89–90, 106–107
  Medical Device Directive, 224
Specification developer, 433
Specific devices, 356–357, *see also* Special
      labeling requirements, United
      States
Specimen collection and handling
  Canada, 100
  In Vitro Diagnostic Medical Device
      Directive, 262, 265
  United States, 369
Stabilizers, 99
Standards
  Australia, 32, 40, 41, 61
  Canada, 85–88
  China, 166–167

European Union, 197–198
  Japan, 292–293
  United States, 320–321
Steel devices, 305
Sterile conditions and products, *see also*
      Nonsterile products
  Active Implantable Medical Device
      Directive, 240–241, 243, 245
  Argentina, 14
  Australia, 37, 46, 47, 48, 49, 51, 53, 65, 67
  Canada, 80, 96–97, 101–102, 103, 104
  immediate containers, 103
  in vitro diagnostic devices, 101–102
  In Vitro Diagnostic Medical Device
      Directive, 253, 255, 263
  Medical Device Directive, 222, 226, 229
  Mexico, 146, 150
  reagent label, 104
  resterilization, 15, 16
  United States, 353–355, 438, 439
Storage conditions, *see also* Instructions for
      use
  Active Implantable Medical Device
      Directive, 243
  Argentina, 14
  Australia, 64
  Canada, 80, 96, 99, 101, 103
  In Vitro Diagnostic Medical Device
      Directive, 253, 254–255, 262
  Japan, 298, 299
  label design, 471
  Medical Device Directive, 223
  reducing labeling problems, 464
  Thailand, 181, 185
  United States, 381, 436
Strip packs, *see* Packaging containers and
      materials
Studies affecting devices, 83, 84, 176
Study subjects, recruiting, 422–423
Summary, reducing labeling problems, 464
Sunlamps, 136, 403–405
Surgeon identifiers, *see* Healthcare provider
      identifiers
Surgical and Medical Instrument
      Manufacturing codes, 479, 480
Surgical Appliance and Supplies
      Manufacturing codes, 479, 480
Surgical lasers, *see* Lasers
Sutures, 47, *see also* Bandages
Switches, 259
Switzerland, 194
Symbols, *see also* Hazard symbols and
      statements
  Canada, 100
  international, 491–495
  label design, 475
  reducing labeling problems, 467
Synthetic hydrocarbon polymer gloves, 112
Syringes, 47, 60, *see also* Packaging
      containers and materials
System packs, 67, *see also* Kits

T

Table of contents, labeling design, 463

Tampons, *see* Menstrual tampons
Tattooing instruments, 195
Teaching devices, 351–352
Technical director, 12, 14
Technical information, 268, 464, 465
Technology, 3–4, 99
Teeth products, *see* Dental products and
      equipment
Television receivers, 136, 385–386
Terminals, 259
Terms of Technical Responsibility, 19
Testing laboratories, 431
Tests
  Canada, 100–101
  China, 159, 160, 161
  In Vitro Diagnostic Medical Device
      Directive, 262–263, 265–266,
      266
  Japan, 298–299
  Korea, 173, 174, 175, 176
  Mexico, 145
  reducing labeling problems, 469
  United States, 368
Text, reducing label problems, 466–467
TGA, *see* Therapeutic Goods Act (TGA);
      Therapeutic Goods
      Administration (TGA)
TGAC, *see* Therapeutic Goods Advertising
      Code (TGAC)
Thailand
  adulteration, 182–183
  advertising, 186
  background, 181
  basics, 181, 187
  bringing devices to market, 183–185
  classification of devices, 186–187
  documents, 185–186
  expiration dates, 186
  Food and Drug Administration, 546
  Hazardous Substance Act, 186–187, 188
  intent of law, 181
  in vitro diagnostic devices, 186–187
  labeling requirements, 185–186
  language requirements, 186
  misbranding, 182–183
  notification, 184
  postmarket requirements, 184–185
  registration, 183
  regulations scope, 181–182
Theft prevention systems, 458
Therapeutic Goods Act (TGA), *see also*
      Australia; Therapeutic Goods
      Amendment of 2002
  adulteration, 28
  advertising, 28, 29–31
  background, 26
  basics, 25–26, 44
  bringing devices to market, 32–40
  claiming equivalence, 35–37, 38–39
  clinical trials, 42–43, 46
  commencing the supply, 40
  custom devices, 43
  directions for use, 45
  exemptions, 42–44
  experimental usage, 42–44
  exporting, 37–38, 40, 46

false or misleading labeling, 29
importing, 37–38, 40–41
intent of law, 26
investigation use, 42
labeling, 27–28
listing, 37–38
manufacturers, 32, 37–38, 39
misbranding, 28, 45
promotion, 29–31
registration, 33–35, 44
regulations scope, 27
standards, 41
Therapeutic Goods Administration (TGA), 46,
545, see also Australia;
Therapeutic Goods Act (TGA);
Therapeutic Goods Amendment
of 2002
Therapeutic Goods Advertising Code
(TGAC), 39–31
Therapeutic Goods Amendment of 2002, see
also Australia; Therapeutic
Goods Act (TGA)
approach, 60
basics, 26–27, 59, 67–68
clinical trials, 66
custom made devices, 66
device classification, 60–61
experimental purposes, 66
exporting, 62
format of label, 63
importing, 62
instructions for use requirements, 64–65
labeling principles, 62–67
language requirement, 63
postmarket requirements, 62
principles, 61
procedure packs, 67
regulation scope, 60–62
special purpose devices, 66
standards, 61
system packs, 67
Therapeutic serums, 322, 323
Thin film copolymer gloves, 112
Tissues, 41, 72, 192
Tongue depressors, 95
Toxicology products, 372–373
Toxic products, 123, 129–133, see also
Hazardous substances
Toxic shock syndrome (TSS)
Canada, 108
labeling failure, 29
United States, 447
Tracking of devices
Australia, 35
Japan, 288
United States, 320, 438
Trade fairs, 198, 203, 280–281
Trademarks, see Manufacturers
Trade names, 297, see also Manufacturers
Training information, see Instructions for use
Transparent packages, 50, 51, see also
Packaging containers and
materials
Transportation of devices
Canada, 80
Japan, 295

label design, 471
Thailand, 181
United States, 439
Troubleshooting, 267–268, 464
TSS, see Toxic shock syndrome (TSS)
Tubes, see Packaging containers and materials
TV sets, 456
Two-way radios, 455, 456
Typefaces, 474, see also Presentation
requirements

U

Ultrasonic devices
Canada, 136, 141–142
China, 156
Japan, 303
United States, 328, 405–406
Ultraviolet lamps, 403–405
Unintentional radiators, 409
United States, see also Bringing devices to
market, United States; Food,
Drug, and Cosmetic Act
(FD&CA); Radiation-emitting
devices, United States
adverse reactions, 350
banned devices, 363–364
basics, 337–338, 364
classification of devices, 318–319,
539–544
commonly known directions, 351
contraindications, 349
device identification, 342
display panels, 341–342
electromagnetic compatibility, 355–356
electronic labeling, 356
exemptions, 347–352
expiration dates, 352
exporting, 357–360
importing, 360–363
inch-pound system, 344–345
indications for use, 348–349
instructions for use, 340
intended uses, 339–340
in vitro diagnostic products, 351
labeling requirements, 338–341
language requirements, 341
law enforcement devices, 351–352
manufacturers, 338–339
measurement units, 342, 344–347
misbranding, 338
misleading statements, 340
net contents statement, 346
omission of information, 324, 326, 338
over-the-counter devices, 341–347
precautions, 350
prescription devices, 348
processing, devices for, 351, 352–353
prominence of statements, 340–341
quantity, 342–344, 346–347
repacking, devices for, 351, 352–353
reprocessed single-use devices, 355
research devices, 351–352
SI labeling, 345
single-use devices, 355

specific devices, 356–357
sterile devices, 353–355
teaching devices, 351–352
units of measure, 344–347
veterinary use, 364
warning statements, 349–350
Unit packages, 49, see also Packaging
containers and materials
Units of measure, see also Measuring
functions and equipment
Australia, 47
Brazil, 20
Medical Device Directive, 221
United States, 344–347
Urine, see Human origin products
U.S. Department of Commerce NAICS codes,
479–481
U.S. Department of Health and Human
Services, 539–544
U.S. Food and Drug Administration (FDA), 5,
85, 546
U.S. Trade Statistics, 546
Usage instructions, see Instructions for use
Useful life of device, see Shelf life
Use-related statements, 441–451, see also
Instructions for use
User information
common knowledge, 351
label design, 471
label problems, reducing, 461–462
United States, 387–388, 399–400,
404–405

V

Vaccines, 72, 174, 323
Values expected, test, see Tests
Vapor lamps, 136, 405
Variances, 384–385
Vessels, see Packaging containers and
materials
Veterinarians, products for
Australia, 46
United States, 325, 364
Vials, see Packaging containers and materials
Vibration, 471
Videocassettes, see Instructions for use
Vigilance, see also Problem reporting
European Union, 203, 204–205
Japan, 287–288
Vinyl gloves, 112
Virus, 322, 323, see also Decontamination;
Infectious agents
Visual-aid materials, see Instructions for use
Vitamins, 19

W

Walkie-talkies, 455, 456
Warning statements, see also
Contraindications; Harmful
statements; Instructions for use;
Safety information
Australia, 33, 35, 56, 64

Canada, 95, 96, 100
design of label, 472
In Vitro Diagnostic Medical Device
    Directive, 257–258
Japan, 294–296, 299–306, 310
Korea, 178
label problems, reducing, 468–469
Medical Device Directive, 224
Mexico, 151
Thailand, 185
United States, 349–350, 389, 401,
    444–445, 448–450
Web sites, 545–546, *see also* Internet
Weighing devices, *see* Measuring functions
    and equipment

Weight of product, 178
Weight-reduction products, 317
Wheelchairs, 453–457
White space, label design, 474–475
Words, effect of, 465
Wrappers, *see* Packaging containers and
    materials
Writing, label design, 464–465

X

X-ray equipment, *see also* Radiation-emitting
    devices, Canada

Active Implantable Medical Device
    Directive, 239
Australia, 60
China, 156
Japan, 303
United States, 386–393, 393–394

Y

Year of authorization, 246, *see also* Date

9 780367 395162

T - #0247 - 071024 - C0 - 279/216/33 - PB - 9780367395162 - Gloss Lamination